MINERAL TABLES

Hand-Specimen Properties of 1500 Minerals

RICHARD V. DIETRICH
Professor of Geological Sciences
Virginia Polytechnic Institute

McGRAW-HILL BOOK COMPANY
New York   St. Louis   San Francisco   Toronto   London   Sydney

MINERAL TABLES

Copyright © 1969 by McGraw-Hill, Inc. All rights reserved. Printed in the United States of America. No part of this publication may be reproduced, stored in a retrieval system, or transmitted, in any form or by any means, electronic, mechanical, photocopying, recording, or otherwise, without the prior written permission of the publisher.

16895
234567890WH721069

To Frances

PREFACE

In recent years, more and more people have joined the ranks of those who pursue avocations relating to minerals. Probably the most gratifying aspect of this increase has been a broadening of the scope of the activities, capabilities, and aspirations of a notable number of the interested individuals. One consequence has been the expressed desire by several devotees for more inclusive hand-specimen determinative tables. In fact, scattered requests of a decade ago have swelled to a common plea of today. The belief has apparently been that with such tables even those who do not have access to equipment like the polarizing microscope and x-ray diffractometer could more readily identify many more than the 250 or so minerals included on hand-specimen tables given in available mineralogy books. The determinative tables that constitute most of this book should aid the mineral collector, the student, the geologist, the mining engineer, and others interested in minerals to do exactly that.

Properties of more than 1500 minerals were dealt with in making the tabulations. The two main tables are based almost wholly on hand-specimen characteristics -- appearances and physical properties that can be determined by performing simple tests. Anyone with good observational powers and even a basic knowledge of the diverse hand-specimen properties of minerals should be able to use the tables with relative ease.

Nonetheless, it seems only prudent to note that the tables should not be expected to offer any panacea. Consider, for example, how the relatively large number of minerals in some sections of Table II might serve initially to confuse rather than to clarify. This, however, is as it should be -- as anyone delves more deeply into a scientific field, he should become increasingly aware of how much he needs to consult source and reference literature and also how necessary it is to check his results with those of others, especially the results of professionals who do have access to the relatively sophisticated equipment of a modern laboratory.

If the gaining of such acumen, alone, accrues as a result of the use of these tables, they will have served a well worthwhile purpose. If, in addition, the tables are used as a ready source of certain information about most known minerals, if they help those who use them to enlarge their mineral repertoires more easily, and if they enhance greater insight into mineral search and mineral research, they will have functioned as was hoped from the outset of their preparation.

Paul Desautels, Marion Godshaw, J. B. Jago, D. E. Jensen, and Drs. A. J. Boucot, Clifford Frondel, G. V. Gibbs, R. L. Heller, F. H. Pough, C. E. Sears, C. G. Tillman, and W. S. White kindly gave personal opinions and advice relating to certain data presented in this book. Compilation of the data for the computer sorts and several other of the especially tedious editorial tasks were greatly facilitated because of aid given by my wife, Frances S. Dietrich. Final preparation of the computer programs used for sorting the data was by P. N. Bergstresser. Computer time was made available by the Virginia Engineering Experiment Station (now Research Division) at the Virginia Polytechnic Institute. Most of the data upon which the tables were based were taken from *Dana's System of Mineralogy (Seventh Edition), Volumes I and II* by Palache, Berman, and Frondel; *Rock Forming Minerals: Volumes I - IV* by Deer, Howie, and Zussman; and *Elements of Optical Mineralogy (Fourth Edition) - Part II* by Winchell and Winchell. Permission to list the Dana numbers and the page references to the silicate descriptions in the Deer, Howie, and Zussman and the Winchell and Winchell books was granted by John Wiley & Sons, Inc. Each of these contributions is gratefully acknowledged.

Blacksburg, Virginia   R. V. Dietrich
July, 1968

CONTENTS

	Page
PREFACE	1
INTRODUCTION	5
TABLE I. MINERALS WITH METALLIC LUSTER	11

        Color: Red 12
               Orange 14
               Yellow 14
               Green 16
               Blue 17
               Purple 18
               Colorless 18
               White 19
               Gray 22
               Black 29
               Brown 34

TABLE II. MINERALS WITH NON-METALLIC LUSTER 37

        Color: Red 38
               Orange 51
               Yellow 54
               Green 72
               Blue 89
               Purple 97
               Colorless 100
               White 114
               Gray 129
               Black 140
               Brown 148

TABLE III. MINERALS ARRANGED ACCORDING TO
          CHEMICAL COMPOSITIONS 165

               Aluminum-bearing 166
               Antimony-bearing 171
               Arsenic-bearing 171
               Barium-bearing 173
               Beryllium-bearing 173
               Bismuth-bearing 174
               Boron-bearing 174
               Cadmium-bearing 175
               Calcium-bearing 175
               Cesium-bearing 181
               Chromium-bearing 181
               Cobalt-bearing 181
               Columbium-bearing 181
               Copper-bearing 182
               Gold-bearing 183
               Iron-bearing 184

Page

    Lead-bearing 188
    Lithium-bearing 190
    Magnesium-bearing 191
    Manganese-bearing 194
    Mercury-bearing 196
    Molybdenum-bearing 196
    Nickel-bearing 196
    Platinum-bearing 197
    Potassium-bearing 197
    Rare Earths-bearing 199
    Silver-bearing 199
    Sodium-bearing 200
    Strontium-bearing 203
    Thallium-bearing 204
    Thorium-bearing 204
    Tin-bearing 204
    Titanium-bearing 204
    Tungsten-bearing 205
    Uranium-bearing 205
    Vanadium-bearing 206
    Yttrium-bearing 207
    Zinc-bearing 207
    Zirconium-bearing 208
    Addendum 209

APPENDIX I, ABBREVIATIONS 211

APPENDIX II, GLOSSARY 213

INDEX 223

INTRODUCTION

The original tabulation from which these tables were generated included the numbered minerals in the first two volumes of *Dana's System of Mineralogy (Seventh Edition)*, most of the varieties of silica given in the third volume of *Dana's System of Mineralogy (Seventh Edition)*, and all but some of the dubious, very rare, and relatively insignificant varieties of the silicate minerals listed in Strunz' *Mineralogische Tabellen (3 Auflage)*. Three tables are given -- two determinative tables and one table based on chemical compositions.

The two determinative tables, Tables I and II, constitute a pair: Table I includes minerals with a metallic luster; Table II includes those with a nonmetallic luster. There is a sub-table for each color on both the metallic and the nonmetallic tables. On the color sub-tables, the minerals are arranged in order of least to greatest hardness. Within each hardness category, the minerals are arranged in order of least to greatest specific gravity. In addition, the crystal system, chemical class, remarks, and one or two references are given for each mineral.

To elaborate:

Minerals with METALLIC lusters are separated from those with NONMETALLIC lusters to give the two major divisions, Table I and Table II. Minerals with submetallic lusters appear on both tables and have the submetallic aspect given as one of the first two entries under "remarks." (It is because of these dually entered, submetallic minerals that such apparently anomalous groups as "Metallic Colorless" occur.)

COLORS are arranged on both the metallic and the nonmetallic tables in the following order: red, orange, yellow, green, blue, purple, colorless, white, gray, black, and brown. Each mineral is listed under all colors reported for it. Gold-colored and brass-colored minerals are reported under both orange and yellow; silver-colored minerals are under white and gray; bronze-colored minerals are under both yellow and brown; chartreuse minerals are under yellow and green; salmon-colored ones are under red and orange; *etc.*

HARDNESS is the ability of a mineral to resist abrasion or rupture. In Tables I and II, the minerals are arranged in order of least to greatest hardness under each color. The basis is the lower hardness for the minerals for which a range is reported so that the order is as follows -- 1, 1 - 1.5, 1 - 2, 1 - >2, 1.5, 1.5 - 2, ...etc. The cited hardness values are according to the Mohs Scale:

          1 - talc,   6 - orthoclase,
          2 - gypsum,   7 - quartz,
          3 - calcite,   8 - topaz,
          4 - fluorite,   9 - corundum, and
          5 - apatite,   10 - diamond.

Other objects (and their Mohs hardness) commonly used for determining Mohs hardness rank are the fingernail (2.5), copper coin (3+), knife blade (5 - 5.5), and metal file (6.5).

The Mohs scale is used with full cognizance of the fact that it lacks the precision of some of the alternatively employed, more modern, quantitative systems. The reason it is used in these tables is that it is by far the most convenient, extant scale. One must, however, keep its limitations in mind.

To determine Mohs hardness, one generally tries either to scratch fresh surfaces of materials having known hardnesses with a sharp corner of the unknown or to scratch a fresh surface of the unknown with sharp points of materials with known hardnesses. A mineral will scratch materials softer than itself and will be scratched by minerals harder than itself. Minerals of approximately equal hardnesses may scratch each other.

A few minerals exhibit notably different hardnesses in different directions. Kyanite, with a hardness of 4 - 5 parallel to the length of its blades and 6 - 7 at right angles thereto, is an especially good and oft-cited example. This phenomenon can be very helpful in identifying unknowns so hardness determinations should be made in order that the property is discovered if existant.

There are two additional precautions that must be taken when determining hardness: 1) do not mistake the powder of a softer mineral left on the surface of a harder material as a scratch - generally such a powder can be rubbed off rather easily; and 2) do not confuse a tearing apart of the grains of an aggregate or the breaking off of small cleavage fragments as a scratch - hand lens examination of the surface will usually suffice to distinguish true scratches from such disaggregation or cleavage.

SPECIFIC GRAVITY is the ratio of the weight of a substance to the weight of an equal volume of water (at 4°C, strictly speaking). On Tables I and II the specific gravity for each named mineral is given as either a single value or as a range. The minerals are arranged from light to heavy within individual hardness categories with the basis for the order being a single value or the higher figure of the recorded range. A few of the values are preceded on the tables by a letter: "C" indicates a calculated value; "P" the value for pure material; and "A" the value for artificial material.

The approximate Specific Gravity of most specimens can be determined rather simply. One commonly used method consists in essence of weighing the specimen in air and again in water and substituting the proper weights in the appropriate places of this equation:

$$\text{Specific Gravity of } x = \frac{\text{weight of } x \text{ in air}}{\text{weight of } x \text{ in air} - \text{weight of } x \text{ in water}}$$

The main point to recall, of course, is that the loss of weight in water is equal to the weight of an equal volume of water (Archimedes' principle). The most commonly encountered problems center around failures to recognize that a given mineral specimen is impure or porous.

Simply constructed apparatuses for determining Specific Gravity are described in a number of books (*e.g.*, Sinkankas, 1966, *Mineralogy: A First Course*, Van Nostrand, p. 182-190). Other methods for finding specific gravity values are outlined in some introductory mineralogy books and in most elementary physics books. Some people are able to develop the capability of hefting minerals and thereby estimating their Specific Gravity with remarkable accuracy.

Specific Gravity is an especially useful property for identifying specimens that one does not wish to submit tests that might harm them. The exceptions, of course, are minerals that are soluble in water or other fluids used for specific gravity determinations.

MINERAL NAMES are given as spelled in the source references except for those minerals the names of which have been changed as the result of actions of the Commission on New Minerals and Mineral Names of the International Mineralogical Association. Group names (such as amphibole) and queries (such as "mixture") are given in parentheses below the names.

CRYSTAL SYSTEMS are indicated by four-letter abbreviations, as follows:

   ISOM. - isometric,
   TETR. - tetragonal,
   HEXA. - hexagonal,
   ORTH. - orthorhombic,
   MNCL. - monoclinic, and
   TRCL. - triclinic.

AMOR. for amorphous and UNKN. for unknown are the other designations used in the crystal system column.

CHEMICAL CLASS is indicated by an appropriate three-letter designation:

   ELE - native element,
   SLD - sulfide (including arsenides, selenides, *etc.*),
   SST - sulfosalt,
   OXD - oxide or hydroxide,
   HAL - halide,
   CBT - carbonate,
   NIT - nitrate,
   IOD - iodate,
   BOR - borate,
   SUF - sulfate,
   SEL - selenate,
   TEL - tellurate,
   SLI - selenite,
   CHR - chromate,
   PHO - phosphate, *etc.*
   ARS - arsenate,
   ASI - arsenite,
   ANT - antimonate,
   ATI - antimonite,
   VAN - vanadinate,
   VAS - vanadium oxysalt,
   WOS - tungstate,
   MBS - molybdate,
   CCC - salt of an organic acid,
   SIL - silica mineral,
   NSI - nesosilicate,
   SSI - sorosilicate,
   CSI - cyclosilicate,
   ISI - inosilicate,
   PSI - phyllosilicate, and
   TSI - tectosilicate.

REMARKS include the more pertinent features in the following general order: Submetallic (see above); Diaphaneity (if opaque or subtranslucent); Color modifiers (*e.g.*, tin-white); Tarnished surface colors; Relationship within any series or group; Streak (if different from color); Habit, if

7

diagnostic (but also including the notation "diverse" for minerals with several relatively common habits); Luster; Breakage characteristics (cleavage and/or fracture); Tenacity (*e.g.*, brittle, malleable, sectile); Easily checked chemical properties (*e.g.*, solubility, with or without gelatinization or effervescence, in common solutions and, for some reactions, with a notation of the resulting product(s) or solution color); fluorescence, phosphorescence, and other readily determinable luminescence data; magnetism and electrical properties; miscellaneous properties, such as flame color and fusion values and products; and finally, typical, common, or single occurrence and/or association information.

Within the Remarks section, commas separate terms within categories (*e.g.*, Vitreous, admantine) whereas periods are used between categories (*e.g.*, Vitreous, Fluorescent). Dashes may substitute for different prepositions (*e.g.*, vitr-admn should be read vitreous *to* admantine and sol-HCl means soluble *in* hydrochloric acid). Plus signs may mean *and* as well as plus. The symbols ⊥ (perpendicular to), // (parallel to), ∼ (approximately), < (less than), > (greater than), ↓ (deposited - *i.e.*, "Gyp ↓" means gypsum is deposited), and ↑ (evolved - *i.e.*, "H2S ↑" means hydrogen sulfide gas is given off) are also used. It is in this "Remarks" section that most of the abbreviations listed in Appendix I are used.

REFERENCES consist of Dana numbers for the nonsilicate minerals, D-111 plus a page number in the third volume of *Dana's System of Mineralogy (Seventh Edition)* for the silica minerals, and the page numbers in Winchell and Winchell (preceded by a W, as W342) and in Deer, Howie, and Zussman (indicated as to volume number and page, as 4-401) for the silicates. Silicates not dealt with by the Winchells are given the notation "WN" and silicates not treated by Deer, Howie, and Zussman are designated by an "N" in the sixth position of the reference column. These references were chosen because of their rather complete coverage.

A ● preceding an entry indicates the mineral to be "common", *i.e.*, relatively easily obtained in the field or from some mineral dealer(s). The matter of whether or not certain minerals are common is quite subjective. It depends not only upon the definition selected but also upon contingencies relating to personal experiences of the person making the decisions and, of course, for some minerals upon the time the decision is made.

In the first edition of these tables, "decisions relating to... whether a mineral is or is not common were avoided by marking as common those minerals described in most elementary mineralogy books." In preparing this second edition of the tables, a different tact was taken. First, in consultation with Paul Desautels of the Smithsonian Institution, it was decided that a mineral might be considered "common" if it would be possible to go to a field occurrence or mineral supply house and get a sample with relative ease. Next, this "definition" and a copy of the index of the first edition of the tables were sent to Arthur J. Boucot, Paul Desautels, Marion Godshaw, John B. Jago, David E. Jensen, and Frederick H. Pough along with the request that each check the minerals he(she) would consider "common" according to the supplied definition. All responded and, independently, the writer also checked the list. On the basis of the seven opinions, each mineral checked by four or more persons is indicated as "common" on Tables I and II. [It is noteworthy that there was fairly good overall agreement: of those checked by four or more, 121 were marked by all seven, 68 were marked by six, 44 were marked by five, and 57 were marked by four. In addition, for the record, 67 were marked by three, 100 by two, and 191 by one only. This is a total of 648, of which 290 are marked on the tables. Strictly in accordance with the definition, each of the 648 minerals is probably "common", *i.e.*, they can be considered available if it is assumed that

8

one can go anywhere in the world to collect -- unfettered by property owners, political boundaries, or difficulty or remoteness of access (including seldom tapped caches of some dealers) -- and if one keeps in mind the fact that some minerals are rather widespread in occurrence even though they typically occur as only minute grains in trace amounts in certain rocks. Nonetheless, it appeared that the arbitrary division chosen as noted above could be justified, even if only on the basis of its being indicative of the probable future experiences of most of those seeking and identifying relatively available minerals.]

The suggested procedure for using the tables is to note whether the mineral is metallic or nonmetallic and its color, to determine its Mohs hardness, and then to observe or determine other properties as would appear necessary from entries in the "Remarks" column. An example may be given: If the mineral is nonmetallic and colorless and has a hardness of seven, it should be one of eight minerals on pages 112 and 113. If, in addition, it has a vitreous luster and no apparent cleavage, it would appear probably to be tridymite, quartz, or tourmaline but possibly danburite. Thence, some of the, $e.g.$, chemical tests and occurrence data might be found to be diagnostic. BUT, this also would be a good time to check additional features of these minerals as reported in one or more of the more complete, standard references.

(About 175 of the 1504 minerals originally entered into the memory of the computer do not appear in either Table I or Table II. This is the result of a lack of data relating to these minerals' metallic versus nonmetallic luster, their color, or their hardness.)

Table III consists of a chemical listing of all of the minerals except for the salts of organic acids ($e.g.$, whewellite) and a few other minerals like graphite (C), quartz (SiO_2), and salammoniac (NH_4Cl) neither the anionic nor cationic elements of which are included in the elements chosen for the tabulation headings. Each of the minerals containing the element of the heading in its formula, even if only as an alternative constituent, is listed. The minerals are listed alphabetically under each heading.

Most of the formulae for the non-silicate minerals are those given in the Dana's System volumes. Although most of those given for the silicates are from the other main references (Winchell and Winchell and Deer, Howie, and Zussman), a few were taken from other sources. The apparent inconsistencies in the types of formulae reflect their different sources. As noted on page 165, discrepancies between any formula given on this list and that given on the lists presented by Fleischer (1966, $American\ Mineralogist$, Vol. 51, p. 1247-1357) are noted. In addition, the alternative formulae given by Fleischer are listed as an addendum to Table II, on pages 209-210.

There are two appendices. Appendix I explains the abbreviations used in Tables I and II. Appendix II is a glossary of terms used in this book.

Table I.

MINERALS

with

METALLIC LUSTER

METALLIC RED

HARDNESS	NAME	XL. SYS.	SPECIFIC GRAVITY	CHEM. CLASS	REMARKS	REFERENCES
1 -1.5	KERMESITE	MNCL.	4.68	SLD	LATHS, HAIRLK TUFTS. ADMN. 2 CLVGS. SECT. ALTER PROD OF STIBNITE	2.8.3.
1 -1.5	MELONITE	HEXA.	7.35	SLD	OPQ. REDDISH WHT, TARN-BRWN. PLTS. BASAL CLVG. SOL-HNO3. LO-T VNS	2.9.8.
1.5-2	HUTCHINSONITE	ORTH.	4.6	SST	PRISM, ACICULAR. ADMN. IN DOLO WITH SULFIDES AT VALAIS,SWITZERLAND	3.8.1.
1.5-2	COVELLITE	HEXA.	4.6 -4.76	SLD	SOME IRID. GRAY STRK. 1 PERF CLVG-FLEX PLTS. ALTER OF CU-SULFIDES	2.6.8.1
1.5-2	MALDONITE	ISOM. C	15.70	ELE	OPQ. PINKISH AG-WHT. CU-RED TO BLK TARN. GRNLR. VEINS + GREISEN	1.1.1.2
2 -2.5	LORANDITE	MNCL.	5.53	SST	RED, EXPOS-DK PB-GRAY COVERED BY YEL POWDER. STRK-RED. PRISM-STRIATED. ADMN. 3 CLVG. FLEX-SEPARATING TO FIBERS. WITH REALGAR,ETC.	3.5.10
2 -2.5	AIKINITE	ORTH.	7.06-7.08	SST	OPQ. PB-GRAY, TARN-BRWN, CU-RED, COMMONLY WITH YELLOWISH GREEN COATING. PRISM-ACICULAR, STRIATED. IN GOLD-QUARTZ VEINS	3.4.1.3
2 -2.5	CINNABAR	HEXA.	8.09	SLD	SCARLET STRK. ADMN-DULL. NEAR VOLS + HOT SPRINGS-VNS + INCRUSTS	2.6.9.
2 -2.5	BISMUTH	HEXA.	9.7 -9.83	ELE	OPQ-REDDISH AG-WHT, IRIDESCENT TARN. SECT. SOL-HNO3. VEINS + PEGS	1.2.1.5
2 -3	HEMATOPHANITE	TETR.	7.70	OXD	SUBMET. YEL-RED STRK. TBLR, LAMELLAR AGGS. 1 CLVG. SOL-HCL, HNO3. RARE-WITH PLUMBOFERRITE, ETC. (VERMLAND, SWEDEN)	7.3.4.
2.5	CANFIELDITE	ISOM.	6.28	SST	OPQ. CHEM DISTINGUISHED FROM ARGYRODITE. BLUISH-PURPLISH BLK, RED-DISH STL-GRAY. OCTAH + DODEC,MSV RADIATING AGGS, LO-T VEINS	3.1.3.2
2.5	ARGYRODITE	ISOM.	6.26-6.29	SST	OPQ. CHEM DISTINGUISHED FROM CANFIELDITE. BLUISH-PURPLISH BLK, RED-DISH STL-GRAY. OCTAH - DODEC,MSV RADIATING AGGS, LO-T VEINS	3.1.3.1
2.5-3	COPPER	ISOM. C	8.94	ELE	OPQ-BRANCHING XLS. MAL, DUCT. SOL-HNO3. BASALTIC VOLS. VEINS	1.1.1.4
3	BORNITE	ISOM.	5.06-5.08	SLD	OPQ.PURPLISH IRID TARN. GRAY-BLK STRK. MSV. SOL-HNO3. CU-DEPS	2.4.3.
3	LIVEINGITE	MNCL.	5.3	SST	OPQ. LITTLE KNOWN-FROM BINNENTAL, VALAIS, SWITZERLAND	3.7.3.
3	UMANGITE	UNKN.	5.62	SLD	OPQ.RED-VIOLET TINT. BLK STRK. MSV. SOL-HNO3.WITH OTHER SELENIDES	2.4.2.
3 -3.5	LAUTITE	ORTH.	4.8 -5.0	SLD	SUBMET. OPQ. BLK-GRAYIRED TINGE).SOL-HNO3. VNS-WITH NATIVE AS,ETC.	2.9.5.5
3 -4	HAMMARITE	MNCL.		SST	OPQ. REDDISH STL-GRAY. BLK STRK. PRISM-CURVED FACES, ACICULAR. RARE - ON DRUSY QUARTZ (GLADHAMMAR, SWEDEN)	3.5.13.
3 -4	ALLEMONTITE	HEXA.	5.8 -6.2	ELE	OPQ.SN-WHT, REDDISH GRAY, DK TARN. 1 PERF CLVG. WITH AS, SB,VEINS	1.2.1.3
3 -4.5	TENNANTITE	ISOM.	4.62	SST	SOFTER + LIGHTER THAN TETRAHEDRITE. TETRA, MSV. DCMP-HNO3. VNS	3.2.4.2
3 -4.5	TETRAHEDRITE	ISOM.	4.99	SST	HARDER + HEAVIER THAN TENNANTITE. TETRA, MSV. DCMP-HNO3.WDSPRD-VNS	3.2.4.1
3.5	HEMATOLITE	HEXA.	3.49	ARS	SUBMET. TBLR,RHOMB,STRIATED. VITR. 1 CLVG. SOL-ACIDS. WITH BARITE, JACOBSITE, ETC. IN VNS IN LIMESTONE (NORDMARK, SWEDEN)	41.1.3.
3.5	BENJAMINITE	UNKN.	6.34	SST	OPQ. TARN-DULL, YEL, CU-RED, GRNLR. 1 CLVG. SOL-HCL(HOT), HNO3. RARE-QTZ,MUSCOVITE,FLUORITE,SULFIDE VN.NORTH OF MANHATTAN, NEVADA	3.5.12.
3.5	RICKARDITE	UNKN.	7.54	SLD	OPQ.FRESH-RESEMBLES TARN BORNITE. MSV. SOL-HNO3. WITH NATIVE TE	2.5.2.
3.5-4	WARWICKITE	ORTH.	3.34-3.36	BOR	SUBMET-PEARLY. BLUE-BLK STRK. SLENDER PRISMS, ROUNDED ENDS. DCMP-H2SO4. RARE - CONTACT METAMORPHOSED LIMESTONE	24.1.4.
3.5-4	SPHALERITE	ISOM.	3.9 -4.1	SLD	RESIN. CURVED XL FACES COMMON. FLUO-OR. TRIBOLUM. SOL-HCL. VEINS	2.6.2.1
3.5-4	FAMATINITE	UNKN.	4.47-4.57	SST	OPQ. GRAY-RED TINGE. CRUSTS, MSV. WITH ENARGITE,ETC.-SULFIDE DEPS	3.3.2.1

METALLIC RED

Hardness	Mineral	Crystal System	Sp.Gr.		Description	Ref.
3.5-4	CUPRITE	ISOM.	6.14	OXD	SUBMET(SOME). OCTAH, HAIRLK, MSV. COMMON-OXIDIZED ZONES OF CU-DEPS	4.1.1.
4	HAUERITE	ISOM.	3.46	SLD	REDDISH BRWN-BLK. SUBCONCH, CUBIC CLVG. SOL-HCL. WITH GYP, SULFUR	2.9.1.5
4	GERMANITE	ISOM.	4.46-4.59	SST	OPQ. REDDISH GRAY. DK GRAY STRK. SOL-HNO3.WITH OTHER SULFIDES-DEPS	3.3.1.2
4	HUEBNERITE	MNCL.	7.12	WOS	SUBMET. TARN-IRID. PRISM-STRIATED,//-RADIATING XL GRPS, 1 CLVG. DCMP-AQ-REG. H2SO4 OR HCL(SLOWLY), DIVERSE-VNS,CNTCT META-PLACERS	48.1.1.1
-5	STIBIOPALLADINITE	ISOM.	9.51	SLD	OPQ. AG-WHT, STL-GRAY. UNEVN FRACT. SOL-HOT AQUA REGIA. PT-DEPS	2.2.2.
4.5-5.5	LINNAEITE	ISOM.	4.5 -4.8	SLD	OPQ. MINERALS OF LINNAEITE SERIES RANGE IN COMPOSITION BETWEEN	2.7.1.1
4.5-5.5	SIEGENITE	ISOM.	4.5 -4.8	SLD	LINNAEITE AND POLYDYMITE. LIGHT GRAY TO STEEL GRAY (VIOLARITE-	2.7.1.2
4.5-5.5	CARROLLITE	ISOM.	4.5 -4.8	SLD	VIOLET GRAY), TARNISH TO COPPER-RED OR VIOLET GRAY. OCTA XLS,	2.7.1.3
4.5-5.5	VIOLARITE	ISOM.	4.5 -4.8	SLD	MSV, GRNLR. IMPERF CUBIC CLVG. UNEVEN-SUBCONCH FRACT. SOL-HNO3	2.7.1.4
4.5-5.5	POLYDYMITE	ISOM.	4.5 -4.8	SLD	YIELDING SULFUR. RARE, ASSOC WITH OTHER CU+NI+FE-SULFIDES, VNS	2.7.1.5
5	LEPIDOCROCITE	ORTH.	4.05-4.13	OXD	SUBMET. ORANGE STRK. SCALY, FIBR MSV. 3 CLVGS. WITH GOETHITE, ETC.	6.1.2.1
5	GLAUCODOT	ORTH.	6.28-6.52	SLD	OPQ. GRAYISH SN-WHT, REDDISH AG-WHT. BLK STRK. PRISM, STRIATED ON PRISM ZONE FACES, DCMP-HNO3(PINK SOLUTION). WITH COBALTITE	2.9.5.2
5	MAUCHERITE	TETR.	8.0	SLD	OPQ.REDDISH GRAY. UNEVN. SOL-HNO3,HCL,H2SO4 WITH NICCOLITE, VEINS	2.4.1.
5	BISMUTOTANTALITE	ORTH.	8.26	OXD	SUBMET. EXPOS-PINKISH YEL. PRISM. RARE-PEGMATITE(SW UGANDA)	8.1.9.
5 -5.5	GOETHITE	ORTH.	3.3 -4.29	OXD	SUBMET. YELLOWISH STRK. MSV, FIBR, ETC. SILKY-ADMN. SOL-HCL. WXING	7.1.2.2
-5.5	NICCOLITE	HEXA.	7.78	SLD	OPQ. PALE CU-RED,GRAY TARN, BRWNISH STRK. SOL-AQUA REGIA. AS,S-VNS	2.6.5.3
-6	PYROPHANITE	HEXA.	4.54	OXD	SUBMET. YEL STRK.SCALY.RHOMB CLVG. RARE-CAVITIES(VERMLAND, SWEDEN)	4.4.1.5
-6	HEMATITE	HEXA.	5.26	OXD	MET-EARTHY. STRK-RED. PLTY,GRNLR. SOL-HCL. SED FE-FMS, RARE IN VNS	4.4.1.2
5.5	COBALTITE	ISOM.	6.33	SLD	OPQ. TYPICALLY REDDISH AG-WHT, ALSO VIOLET STL-GRAY + DK GRAY. HAB- IT-LIKE PYRITE. CUBIC CLVG, DCMP-HNO3. HI-T DISSEM DEPS + VEINS	2.9.2.1
5.5	BREITHAUPTITE	HEXA.	8.23	SLD	OPQ. VIOLET CU-RED, RED-BRWN STRK. SOL-HNO3.AG-BEARING CALCITE VNS	2.6.5.4
5.5-6	ANATASE	TETR.	3.90	OXD	SUBMET. PALE YEL STRK. ACUTE PYRAMIDAL, ADMN. BASAL + PYRAMIDAL CLVG. CONVERTS TO RUTILE ON HEATING. VNS, ACCESS-IG RKS, DETRITAL	4.5.2.
5.5-6	RAMMELSBERGITE	ORTH.	7.0 -7.2	SLD	OPQ. REDDISH SN-WHT. GRAY STRK. MSV. WITH CO-NI MINS-MESO-T VNS	2.9.3.3
6	PSEUDOBROOKITE	ORTH.	4.33-4.39	OXD	SUBMET. TBLR-STRIATED. GRSY. 1 CLVG. VOLCANIC AREAS FUMAROLIC,ETC.	7.5.1.
6	COLUMBITE	ORTH.	5.15-5.25	OXD	SUBMET. IN SERIES WITH TANTALITE. RED-BLK STRK. TARN-IRID. TBLR, PRISM, 2 CLEAVAGES, BRITTLE. GRANITIC PEGMATITES	8.3.2.1
6 -6.5	RUTILE	TETR.	4.21-4.25	OXD	SUBMET. PRISM-STRIATED,ADMN.POOR CLVGS. VNS, ACCESS-META + IG RKS	4.5.1.1
6 -6.5	TANTALITE	ORTH.	7.90-8.00	OXD	SUBMET. IN SERIES WITH COLUMBITE, WHICH SEE. PEGMATITES	8.3.2.2
6 -7	CASSITERITE	TETR.	6.99	OXD	SUBMET. RADIAL CONCRETIONARY MASSES. ADMN-DULL. HI-T VNS, GREISENS	4.5.1.5

METALLIC ORANGE

HARDNESS	NAME	XL. SYS.	SPECIFIC GRAVITY	CHEM. CLASS	REMARKS	REFERENCES
1.5	DIMORPHITE	ORTH.	2.58	SLD	OR-YEL. GROUPS PARALLEL XLS. ADMN. SOL-WARM HNO3 + CS2. FUMAROLES	2.5.1.
2.5-3	GOLD	ISOM.	19.31	ELE	OPAQUE. BRANCHING XLS, NUGGETS. MAL, DUCT. VEINS + PLACERS	1.1.1.1

METALLIC YELLOW

HARDNESS	NAME	XL. SYS.	SPECIFIC GRAVITY	CHEM. CLASS	REMARKS	REFERENCES
1.5	DIMORPHITE	ORTH.	2.58	SLD	OR-YEL. GROUPS PARALLEL XLS. ADMN. SOL-WARM HNO3 + CS2. FUMAROLES	2.5.1.
1.5-2	ORPIMENT	MNCL.	3.49	SLD	FOL, FIBR, GRNLR. RESIN. 1 CLVG-INELAST, LO-T VNS, ALTER PROD.	2.8.1.
1.5-2	COVELLITE	HEXA.	4.6 -4.76	SLD	SOME IRID. GRAY STRK. 1 PERF CLVG-FLEX PLTS. ALTER OF CU-SULFIDES	2.6.8.1
1.5-2	SYLVANITE	MNCL.	8.16	SLD	OPQ. YELLOWISH STL-GRAY TO AG-WHT. PRISM(SOME SKELETAL). DCMP-HNO3 (RUSTY COLORED GOLD RESID), WITH CALAVERITE, LO-T VEINS, ETC.	2.9.7.3
2	BISMUTHINITE	ORTH.	6.75-6.81	SLD	OPQ. PB-GRAY SN-WHT, YEL-IRID TARN. FIBR, BLDD. SOL-HNO3. HI-T VNS	2.8.2.2
2 -2.5	LORANDITE	MNCL.	5.53	SST	RED, EXPOS-DK PB-GRAY COVERED BY YEL POWDER. STRK-RED. PRISM-STRI-ATED. ADMN. 3 CLVGS. FLEX-SEPARATING TO FIBERS. WITH REALGAR, ETC.	3.5.10.
2 -2.5	AIKINITE	ORTH.	7.06-7.08	SST	OPQ. PB-GRAY. TARN-BRWN, CJ-RED. PRISM-ACICULAR, STRIATED. COMMONLY WITH YELLOWISH GREEN COATING. IN GOLD-QUARTZ VEINS	3.4.1.3
2 -3	KRENNERITE	ORTH.	8.62	SLD	OPQ. AG-WHT, BRASS-YEL. STRIATED PRISMS.BASAL CLVG. TELLURIDE DEPS	2.9.7.1
2.5	FULOPPITE	MNCL.	5.23	SST	OPQ. PLAGIONITE GRP. S.G.-DEFINITIVE IN GRP. TARN-STL BLUE, BRONZY. STRK-REDDISH GRAY. STRIATED PRISM. RARE-WITH SPHAL(NAGYAG,ROUMANIA)	3.7.4.1
2.5	MUTHMANNITE	UNKN.	5.60	SLD	OPQ.BRASS YEL. FRESH SURF + STRK GRAY. TABULAR, 1 PERF CLVG. SOL-HNO3 YIELDING GOLD RESIDUE, WITH OTHER TELLURIDES	2.6.14.2
2.5	KLAPROTHITE (MIXTURE ?)	ORTH.	6.01	SST	OPQ. STL-GRAY. TARN-BRASSY, IRID. PRISM, MSV. SOL-HCL, HNO3. WITH OTHER BISMUTH MINERALS IN SEVERAL DEPOSITS	3.4.5.
2.5	OMYHEEITE	UNKN.	6.03	SST	OPQ. STL-GRAY. AG-WHT.TARN-YEL.STRK-GRAY(PAPER), REDDISH-BRWN(POR-CELAIN). ACICULAR, FIBR. SOL-HOT HCL. LO-T AG-BEARING VEINS	3.5.2.
2.5-3	BOULANGERITE	MNCL.	6.23	SST	OPQ. YEL-SPOTTED BLUISH PB-GRAY. STRK-BRWN. PRISM-ACICULAR(STRIA-TED), FIBR (FLEX), 1 GOOD CLVG. SOL-HOT HCL(H2S*). LO- TO MOD-T VNS	3.5.1.
2.5-3	CALAVERITE	MNCL.	9.22-9.26	SLD	OPQ. BRASS-YEL, AG-WHT, YEL- TO GREEN-GRAY STRK. STRIATED LATHS. LO-T VNS + LO-T PORTIONS OF HI- + MOD-T VEINS	2.9.7.2
2.5-3	GOLD	ISOM.	19.31	ELE	OPAQUE. BRANCHING XLS, NUGGETS. MAL, DUCT. VEINS + PLACERS	1.1.1.1
2.5-3.5	GALENOBISMUTITE	ORTH.	7.04	SST	OPQ. TARN-YEL, IRID. STRK-BLK. MSV, FIBR, LATHLK XLS(COMMONLY BENT AND STRIATED), 1 CLVG. SOL-HCL. WITH SULFIDES IN QTZ-SULFIDE VNS	3.8.3.
3	ASTROPHYLLITE	TRCL.	3.3 -3.4	SSI	SUBMET-PEARLY. BRONZY-GOLDN YEL. CLVG-BRTL FLKS. DCMP-HCL. ALK RKS	W480 N
3	ALTAITE	ISOM.	8.15	SLD	DPQ.YELLOWISH SN-WHT, BRONZE TARN. MSV. PERF CUBIC CLVG. AU,S-VNS	2.6.1.3
3 -3.5	ANDORITE	ORTH.	5.33-5.37	SST	OPQ. BLK STRK. TARN-YEL,IRID. STRIATED PRISM, MSV.SOL-HCL. S-VNS	3.7.1.1

METALLIC YELLOW

Hardness	Mineral	Crystal	SG	Class	Description	Reference
3 -3.5	MILLERITE	HEXA.	5.3 -5.7	SLD	OPQ. BRASS-BRONZE YEL, GRN-BLK STRK. HAIRLK XLS, LO-T, IN CAVITIES	2.6.5.5
3 -3.5	EMPRESSITE	UNKN.	7.51	SLD	OPQ.BRONZY, GRAY-BLK STRK. MSV. FRIABLE. SOL-HNO3.WITH GALENA + TE	2.6.14.1
3 -3.5	GOLD AMALGAM	ISOM.	15.47	ELE	OPQ. GRAINS + LUMPS. CONCHOIDAL FRACTURE. SOL-HNO3. WITH PT. RARE	1.1.4.
3 -4	COLUSITE	ISOM.	4.50	SST	OPQ. BRONZE. BLK STRK. MSV. WITH PYRITE, CU-MINS,ETC.-SULFIDE VNS	3.3.1.3
3.5	SULVANITE	ISOM.	3.86-4.00	SST	OPQ. BRONZE-YEL. BLK STRK. CUBIC.MSV. WITH MALACHITE,ETC.-ORE DEPS	3.3.1.1
3.5	CUBANITE	ORTH.	4.03-4.18	SLD	OPQ. BRASS-BRONZE YEL. CONCH. MAG. SOL-HNO3. WITH CHALCOPY + PYRRH	2.6.6.
3.5	BENJAMINITE	UNKN.	6.34	SST	OPQ. TARN-DULL, YEL, CU-RED. GRNL. 1 CLVG. SOL-HCL(HOT), HNO3. RARE-QTZ,MUSCOVITE,FLUORITE,SULFIDE VN,NORTH OF MANHATTAN,NEVADA	3.5.12.
3.5-4	SPHALERITE	ISOM.	3.9 -4.1	SLD	RESIN. CURVED XL FACES COMMON. FLUO-OR. TRIBOLUM. SOL-HCL. VEINS	2.6.2.1
3.5-4	CHALCOPYRITE	TETR.	4.1 -4.3	SLD	OPQ.BRASSY, IRID TARN, GRN-BLK STRK. SOL-HNO3. MESO- TO HI-T VNS	2.6.3.1
3.5-4	PENTLANDITE	ISOM.	4.6 -5.0	SLD	OPQ. BRONZE-YEL. BLK STRK. MSV. INTIMATELY ASSOC WITH PYRRHOTITE	2.6.5.6
3.5-4	DYSCRASITE	ORTH.	9.6 -9.81	SLD	OPQ. AG-WHT, TARN GRAY, YEL, BLK. MSV. SECT. DECOMP-HNO3. AG-DEPS	2.2.1.
3.5-4.5	PYRRHOTITE	HEXA.	4.58-4.79	SLD	OPQ. BRONZY. TARN. MSV. UNEVN. MAG. DCMP-HCL. BASALTS, S-VEINS	2.6.5.1
4	HUEBNERITE	MNCL.	7.12	WOS	SUBMET. TARN-IRID. PRISM-STRIATED, -RADIATING XL GRPS. 1 CLVG. DCMP-AQ-REG, H2SO4 OR HCL(SLOWLY). DIVERSE-VNS,CNTCT META,PLACERS	48.1.1.1
4 -5.5	BETAFITE	ISOM.	3.7 -5.	OXD	SUBMET. OCTAH. WAXY. CONCH. METAMICT. DCMP-ACIDS. GRANITIC PEGS	8.4.1.
5	BISMUTOTANTALITE	ORTH.	8.26	OXD	SUBMET. EXPOS-PINKISH YEL. PRISM. RARE-PEGMATITE(SW UGANDA)	8.1.9.
5 -5.5	GOETHITE	ORTH.	3.3 -4.29	OXD	SUBMET. YELLOWISH STRK. MSV, FIBR, ETC. SILKY-ADMN. SOL-HCL. WXING PRODUCT	7.1.2.2
5 -6	PRIORITE	ORTH.	4.85-5.05	OXD	SUBMET. IN SERIES WITH ESCHYNITE(SEE). RED-YEL STRK. GRANITIC PEGS	8.3.5.2
5 -6	ESCHYNITE	ORTH.	5.14-5.24	OXD	SUBMET. IN SERIES WITH PRIORITE. BLK-BRWN STRK. PRISM, MSV. WAXY.	8.3.5.1
5 -6	"SAMARSKITE"	UNKN.	5.67	OXD	SUBMET. DISTINGUISHED FROM SAMARSKITE ONLY BY X-RAY ANALYSIS	8.3.6.2
5 -6	SAMARSKITE	ORTH.	5.69	OXD	SUBMET. RED-BRWN TO BLK STRK. PRISM, MSV. VITR. RDACTV. PEGS	8.3.6.1
5.5	PEROVSKITE	MNCL	3.97-4.05	OXD	SUBMET. CUBIC, OCTAH, GRNLR. ADMN. 1 CLVG. DCMP-HF, H2SO4.BASIC IG	7.4.2.1
5.5	MAGNESIOCHROMITE	ISOM.	4.1 -4.3	OXD	CHROMITE SERIES OF SPINEL GROUP. MSV, GRNLR, OCTAH. BRWN STRK. MAG (FEEBLY), WITH OLIVINE-RICH(COMMONLY SERPENTINIZED) IGNEOUS ROCKS	7.2.1.11
5.5-6	ANATASE	TETR.	3.90	OXD	SUBMET. PALE YEL STRK. ACUTE PYRAMIDAL. ADMN. BASAL + PYRAMIDAL CLVG. CONVERTS TO RUTILE ON HEATING. VNS,ACCESS-IG RKS, DETRITAL	4.5.2.
5.5-6	COHENITE	ORTH.	7.20-7.65	ELE	OPQ.BRONZY-GLD ON EXPOS. BRTL. MAG. SOL-CU(NH3)6CL2. METEOR + TERRES FE	1.1.7.3
5.5-6.5	FERGUSONITE	TETR. C	5.38	OXD	SUBMET. IN SERIES WITH FORMANITE. PRISM. COMMONLY HEMIHEDRAL, IR-REGULAR MASSES. VITR. METAMICT. DCMP-HF. GRANITIC PEGS	8.1.3.1
5.5-6.5	FORMANITE	TETR. C	7.03	OXD	SUBMET.IN SERIES WITH FERGUSONITE,WHICH SEE. PLACERS(W. AUSTRALIA)	8.1.3.2
6 -6.5	RUTILE	TETR.	4.21-4.25	OXD	SUBMET. PRISM-STRIATED.ADMN.POOR CLVGS. VNS, ACCESS-META + IG RKS	4.5.1.1

METALLIC YELLOW

HARDNESS	NAME	XL. SYS.	SPECIFIC GRAVITY	CHEM. CLASS	REMARKS	REFERENCES
● 6 -6.5	MARCASITE	ORTH.	4.89	SLD	OPQ. SN-WHT. BRONZE-YEL. TBLR. COCKSCOMB-LK. LO-T, ACID DEPOSITION	2.9.4.
● 6 -6.5	PYRITE	ISOM.	4.82-5.02	SLD	OPQ.BRASSY. CUBES, DODEC, MSV. UNEVN. SOL-(POWDER)-HNO3.WIDESPREAD	2.9.1.1
● 6 -7	CASSITERITE	TETR.	6.99	OXD	SUBMET. RADIAL CONCRETIONARY MASSES. ADMN-DULL. HI-T VNS, GREISENS	4.5.1.5
6 -7	TANTALUM	ISOM. C	16.72	ELE	OPQ. GRAYISH YELLOW. MINUTE XLS + GRAINS. IN GOLD WASHINGS	1.1.8.
6 -7	PLATINIRIDIUM	ISOM. X	22.84	ELE	OPQ. YELLOWISH AG-WHT. FRACTURES GRAY. SOMEWHAT MAL. PLACERS.RARE	1.1.6.3
6.5-7	SCHREIBERSITE	TETR.	7.0 -7.3	ELE	OPQ.BRASS-YELLOW OR BROWN TARNISH. BRITTLE. MAG. DIFFICULTLY SOL-HCL + HNO3. METEORITES, COMBUSTION PRODUCT IN COAL MINES (FRANCE)	1.1.7.6

METALLIC GREEN

HARDNESS	NAME	XL. SYS.	SPECIFIC GRAVITY	CHEM. CLASS	REMARKS	REFERENCES
2 -2.5	AIKINITE	ORTH.	7.06-7.08	SST	OPQ. PB-GRAY. TARN-BRWN, CU-RED, COMMONLY WITH YELLOWISH GREEN COATING. PRISM-ACICULAR, STRIATED. IN GOLD-QUARTZ VEINS	3.4.1.3
3 -4	CHALCOSTIBITE	ORTH.	4.90-5.00	SST	OPQ. BLDS, MSV. 3 GOOD CLVGS. DCMP-HNO3. QTZ-PYRITE VNS-(COMMONLY ALTERED ON EXPOSED SURFACES TO MALACHITE AND AZURITE)	3.5.9.1
3.5-4	SPHALERITE	ISOM.	3.9 -4.1	SLD	RESIN. CURVED XL FACES COMMON. FLUO-OR. TRIBOLUM. SOL-HCL. VEINS	2.6.2.1
4 -5.5	BETAFITE	ISOM.	3.7 -5.	OXD	SUBMET. OCTAH. WAXY. CONCH. METAMICT. DCMP-ACIDS. GRANITIC PEGS	8.4.1.
● 5 -6	BRONZITE (ORTHOPYROXENE)	ORTH.	3.3 -3.45	ISI	SUBMET.(BRONZY). FS12-30. IN BASIC + ULTRABSIC IG RKS, SOME METAS	W4OS 2-9
5.5	MAGNESIOCHROMITE	ISOM.	4.1 -4.3	OXD	CHROMITE SERIES OF SPINEL GROUP. MSV, GRNLR, OCTAH. BRWN STRK. MAG (FEEBLY).WITH OLIVINE-RICH(COMMONLY SERPENTINIZED) IGNEOUS ROCKS	7.2.1.11
● 5.5-6	ANATASE	TETR.	3.90	OXD	SUBMET. PALE YEL STRK. ACUTE PYRAMIDAL. ADMN. BASAL + PYRAMIDAL CLVGS. CONVERTS TO RUTILE ON HEATING. VNS, ACCESS-IG RKS. DETRITAL	4.5.2.
5.5-6.5	EUXENITE	ORTH.	4.9 -5.9	OXD	SUBMET. IN SERIES WITH POLYCRASE. STRK YEL,GRAY,RED-BRWN. PRISM, MSV,GRSY. CONCH. DCMP-HOT HCL,HF,H2SO4. GRANITIC PEGMATITES	8.3.3.1
5.5-6.5	POLYCRASE	ORTH.	4.9 -5.9	OXD	SUBMET. IN SERIES WITH EUXENITE, WHICH SEE. PEGMATITES	8.3.3.2
● 6 -6.5	RUTILE	TETR.	4.21-4.25	OXD	SUBMET. PRISM-STRIATED.ADMN.POOR CLVGS. VNS, ACCESS-META + IG RKS	4.5.1.1
9.5	MOISSANITE	HEXA.	3.1 -3.21	ELE	SUBMET. TABULAR. CONCHOIDAL FRACT. METEOR. (ARTIF=CARBORUNDUM)	1.1.7.4

METALLIC BLUE

HARDNESS	NAME	XL. SYS.	SPECIFIC GRAVITY	CHEM. CLASS	REMARKS	REFERENCES
1 -1.5	STERNBERGITE	ORTH.	4.10-4.22	SLD	OPQ.VIOLET-BLU TARN. BLK STRK. PLTS, ROSETTES.1 PERF CLVG. AG-ORES	2.6.7.
1 -2	GRAPHITE	HEXA.	2.09-2.23	ELE	SCALES. 1 PERF CLVG TO INELASTIC PLTS. META OF CARBONACEOUS RKS	1.2.4.2
1.5-2	COVELLITE	HEXA.	4.6 -4.76	SLD	SOME IRID. GRAY STRK. 1 PERF CLVG-FLEX PLTS. ALTER OF CU-SULFIDES	2.6.8.1
2 -6.5	PYROLUSITE	TETR.	4.4 -5.08	OXD	OPQ. FIBR, GRNLR. PRISM CLVG. BOG + RESIDUAL MN-DEPS - COMMON	4.5.1.2
2.5	CHALCOPHANITE	HEXA.	3.90-4.10	OXD	BRWN STRK. TBLR, DRUSES. MSV. 1 CLVG(FLEX). SOL-HCL. SCNDRY OXIDE DEPOSITS	7.6.1.
2.5	FULOPPITE	MNCL.	5.23	SST	OPQ. PLAGIONITE GRP. S.G.-DEFINITIVE IN GRP. TARN-STL BLUE, BRONZY. STRK-REDDISH GRAY. STRIATED PRISM. RARE-WITH SPHAL(NAGYAG,ROUMANIA)	3.7.4.1
2.5	CANFIELDITE	ISOM.	6.28	SST	OPQ. CHEM DISTINGUISHED FROM ARGYRODITE. BLUISH-PURPLISH BLK, RED-DISH STL-GRAY. OCTAH + DODEC.MSV.RADIATING AGGS. LO-T VEINS	3.1.3.2
2.5	ARGYRODITE	ISOM.	6.26-6.29	SST	OPQ. CHEM DISTINGUISHED FROM CANFIELDITE. BLUISH-PURPLISH BLK, RED-DISH STL-GRAY. OCTAH + DODEC.MSV.RADIATING AGGS. LO-T VEINS	3.1.3.1
2.5	GEOCRONITE	ORTH.	6.3 -6.5	SST	OPQ. GRAYISH BLUE. TBLR, MSV. SOL-HOT HCL(H2S↑). SULFIDE VEINS	3.3.5.
2.5-3	DIGENITE	ISOM.	2.55-2.71	SLD	OPQ. MSV. CONCH FRACT. BRTL. SOL-HNO3. WITH OTHER COPPER SULFIDES	2.3.1.4
2.5-3	BOULANGERITE	MNCL.	6.23	SST	OPQ. YEL-SPOTTED BLUISH PB-GRAY. STRK-BRWN. PRISM-ACICULAR(STRIA-TED),FIBR(FLEX). 1 GOOD CLVG. SOL-HOT HCL(H2S↑). LO- TO MOD-T VNS	3.5.1.
2.5-3	STROMEYERITE	ORTH.	6.2 -6.3	SLD	OPQ. STL-GRAY. SUBCONCH. SOL-HNO3. WITH TETRAHEDRITE, BORNITE VNS	2.3.2.2
2.5-3	CLAUSTHALITE	ISOM.	7.8	SLD	OPQ.BLUISH PB-GRAY.MSV. GOOD CUBIC CLVG. SOL-HNO3. WITH SELENIDES	2.6.1.2
2.5-3.5	GUANAJUATITE	ORTH.	6.25-6.98	SLD	OPQ. BLUISH GRAY. STRIATED XLS. FIBR. SOL-WM AQ-REG. NEEDS CONFIRM	2.8.2.3
3	KLOCKMANNITE	HEXA.	>5.	SLD	SLATE-GRAY, TARN BLU-BLK. GRNLR. 1 PERF CLVG. WITH OTHER SELENIDES	2.6.8.2
3	UMANGITE	UNKN.	5.62	SLD	OPQ.RED-VIOLET TINT. BLK STRK. MSV. SOL-HNO3.WITH OTHER SELENIDES	2.4.2.
3	GUITERMANITE	UNKN.	5.94	SST	OPQ. POSSIBLY A MIXTURE RATHER THAN A DISTINCT SPECIES	3.3.9.
3	WEISSITE	UNKN.	6.	SLD	OPQ.BLUISH BLK, BLK TARN, BLK STRK. MSV. WITH PYRITE + TELLURIDES	2.5.3.
3 -4	CHALCOSTIBITE	ORTH.	4.90-5.00	SST	OPQ. BLDS, MSV. 3 GOOD CLVGS. DCMP-HNO3. QTZ-PYRITE VNS-(COMMONLY ALTERED ON EXPOSED SURFACES TO MALACHITE AND AZURITE)	3.5.9.1
3.5	EPIGENITE	ORTH.	4.5	SST	OPQ. STL-GRAY. TARN-BLK + BLUE. INCRUST. SOL-HNO3. VNS-ON BARITE	3.1.5.
3.5	VRBAITE	ORTH.	5.27-5.33	SST	SUBMET.DK BLUISH GRAY,DK RED WITH TRANSMITTED LIGHT. STRK-RED WITH YEL TINGE, TBLR,PYRAMIDAL XLS. SOL-HNO3,H2SO4. WITH REALGAR.	3.8.13.
4	STANNITE	TETR.	4.3 -4.5	SLD	OPQ. STL-GRAY, FE-BLK. MOST XLS STRIATED. DCMP-HNO3.SN-BEARING VNS	2.6.3.2
5.5-6	ANATASE	TETR.	3.90	OXD	SUBMET. PALE YEL STRK. ACUTE PYRAMIDAL. ADMN. BASAL + PYRAMIDAL CLVG. CONVERTS TO RUTILE ON HEATING. VNS,ACCESS-IG RKS,DETRITAL	4.5.2.
5.5-6.5	FRANKLINITE	ISOM.	5.07-5.22	OXD	SUBMET, MAGNETITE SERIES OF SPINEL GROUP. RED-BRWN STRK. OCTAH. OCTAH PARTING. SOL-HCL, WEAKLY MAGNETIC. RARE-FRANKLIN,N.J.	7.2.1.7
6 -6.5	RUTILE	TETR.	4.21-4.25	OXD	SUBMET. PRISM-STRIATED.ADMN.POOR CLVGS. VNS, ACCESS-META + IG RKS	4.5.1.1
9.5	MOISSANITE	HEXA.	3.1 -3.21	ELE	SUBMET. TABULAR. CONCHOIDAL FRACT. METEOR. (ARTIF=CARBORUNDUM)	1.1.7.4

METALLIC PURPLE

HARDNESS	NAME	XL. SYS.	SPECIFIC GRAVITY	CHEM. CLASS	REMARKS	REFERENCES
1 -1.5	STERNBERGITE	ORTH.	4.10-4.22	SLD	OPQ.VIOLET-BLU TARN, BLK STRK, PLTS, ROSETTES.1 PERF CLVG. AG-ORES	2.6.7.
2 -2.5	LORANDITE	MNCL.	5.53	SST	RED, EXPOS-DK PB-GRAY COVERED BY YEL POWDER. STRK-RED. PRISM-STRI-ATED. ADMN. 3 CLVGS.FLEX-SEPARATING TO FIBERS, WITH REALGAR, ETC.	3.5.10.
2 -3	IANTHINITE	ORTH.		OXD	SUBMET. BRWN-VIOLET STRK. PLTY, TBLR, BASAL CLVG. ALTER AFTER URANINITE	5.3.5.
2 -3	MURMANITE	MNCL.	2.84	NSI	BRONZE TARN. RED STRK. TBLR. 1 PERF CLVG. SOL-H2SO4. ALK SYENITE	W480 N
2.5	CANFIELDITE	ISOM.	6.28	SST	OPQ. CHEM DISTINGUISHED FROM ARGYRODITE. BLUISH-PURPLISH BLK, RED-DISH STL-GRAY. OCTAH + DODEC.MSV.RADIATING AGGS. LO-T VEINS	3.1.3.2
2.5	ARGYRODITE	ISOM.	6.26-6.29	SST	OPQ. CHEM DISTINGUISHED FROM CANFIELDITE. BLUISH-PURPLISH BLK, RED-DISH STL-GRAY. OCTAH + DODEC.MSV.RADIATING AGGS. LO-T VEINS	3.1.3.1
● 3	BORNITE	ISOM.	5.06-5.08	SLD	OPQ.PURPLISH IRID TARN. GRAY-BLK STRK. MSV. SOL-HNO3. CU-DEPS	2.4.3.
3	UMANGITE	UNKN.	5.62	SLD	OPQ.RED-VIOLET TINT. BLK STRK. MSV. SOL-HNO3.WITH OTHER SELENIDES	2.4.2.
3.5	RICKARDITE	UNKN.	7.54	SLD	OPQ.FRESH-RESEMBLES TARN BORNITE. MSV. SOL-HNO3. WITH NATIVE TE	2.5.2.
4.5	PARAMELACONITE	TETR.	6.04	OXD	OPQ. STRK-BRWNISH. STRIATED PRISM. ADMN. SOL-DILUTE ACIDS, DILUTE NH4OH + NH4CL SOLS. RARE-SCNDRY WITH CUPRITE,ETC.-BISBEE, ARIZ.	4.2.4.
4.5-5.5	VIOLARITE	ISOM.	4.5 -4.8	SLD	OPQ.MSV, GRNLR, IMPERF CUBIC CLVG. UNEVEN-SUBCONCH FRACT. SOL-HNO3 WITH OTHER CU-NI, + FE SULFIDES IN VNS. MEMBER OF LINNAEITE SERIES	2.7.1.4
5.5	COBALTITE	ISOM.	6.33	SLD	OPQ. TYPICALLY REDDISH AG-WHT, ALSO VIOLET STL-GRAY + DK GRAY. HAB-IT-LIKE PYRITE. CUBIC CLVG. DCMP-HNO3. HI-T DISSEM DEPS + VNS	2.9.2.1
5.5	BREITHAUPTITE	HEXA.	8.23	SLD	OPQ. VIOLET CU-RED, RED-BRWN STRK. SOL-HNO3.AG-BEARING CALCITE VNS	2.6.5.4
● 5.5-6	ANATASE	TETR.	3.90	OXD	SUBMET. PALE YEL STRK, ACUTE PYRAMIDAL. ADMN. BASAL + PYRAMIDAL CLVG. CONVERTS TO RUTILE ON HEATING. VNS,ACCESS-IG ROCKS,DETRITAL	4.5.2.
● 6 -6.5	RUTILE	TETR.	4.21-4.25	OXD	SUBMET. PRISM-STRIATED.ADMN.POOR CLVGS. VNS, ACCESS-META + IG RKS	4.5.1.1

METALLIC COLORLESS

HARDNESS	NAME	XL. SYS.	SPECIFIC GRAVITY	CHEM. CLASS	REMARKS	REFERENCES
● 3.5-4	SPHALERITE	ISOM.	3.9 -4.1	SLD	RESIN. CURVED XL FACES COMMON. FLUO-OR. TRIBOLUM. SOL-HCL. VEINS	2.6.2.1
● 5.5-6	ANATASE	TETR.	3.90	OXD	SUBMET. PALE YEL STRK. ACUTE PYRAMIDAL. ADMN. BASAL + PYRAMIDAL CLVG. CONVERTS TO RUTILE ON HEATING. VNS, ACCESS-IG RKS, DETRITAL	4.5.2.
● 6 -7	CASSITERITE	TETR.	6.99	OXD	SUBMET. RADIAL CONCRETIONARY MASSES. ADMN-DULL. HI-T VNS, GREISENS	4.5.1.5

METALLIC WHITE

HARDNESS	NAME	XL. SYS.	SPECIFIC GRAVITY	CHEM. CLASS	REMARKS	REFERENCES
0.0	MERCURY	AMOR.	13.60	ELE	OPQ. LIQ DROPS. WITH CINNABAR-VEINS OF VOL + HOT SPRING REGIONS	1.1.2.
1 -1.5	MELONITE	HEXA.	7.35	SLD	OPQ. REDDISH WHT, TARN-BRWN. PLTS. BASAL CLVG. SOL-HNO3. LO-T VNS	2.9.8.
1.5	LEAD	ISOM.	11.35	ELE	OPQ. ROUNDED MASSES + THIN PLATES. MAL. SOL-HNO3. VEINS. RARE	1.1.1.5
1.5-2	SYLVANITE	MNCL.	8.16	SLD	OPQ. YELLOWISH STL-GRAY TO AG-WHT. PRISM(SOME SKELETAL). DCMP-HNO3 (RUSTY COLORED GOLD RESID). WITH CALAVERITE. LO-T VNS, ETC.	2.9.7.3
1.5-2	MALDONITE	ISOM. C	15.70	SLD	OPQ. PINKISH AG-WHT. CU-RED TO BLK TARN. GRNLR. VEINS + GREISEN	1.1.1.2
1.5-2.5	MEHRLITE	UNKN.	8.38-8.44	SLD	FOLIATED MSS RESEMBLING TETRADYMITE. CLVG PLTS-SLIGHTLY ELASTIC	2.1.1.5
2	EMPLECTITE	ORTH.	6.38	SST	OPQ. GRAY, SN-WHT. PRISM-STRIATED. 2 CLVGS. DCMP-HNO3.SULFIDE DEPS	3.5.9.2
2	BERZELIANITE	ISOM.	6.71	SLD	OPQ. AG-WHT, TARN DK. DENDRI. MAL. SOL-HNO3. WITH OTHER SELENIDES	2.3.1.5
2	BISMUTHINITE	ORTH.	6.75-6.81	SLD	OPQ. PB-GRAY SN-WHT, YEL-IRID TARN. FIBR, BLDD. SOL-HNO3. HI-T VNS	2.8.2.2
2	ZINC	HEXA.	6.9 -7.2	ELE	OPQ. 1 PERF CLVG. BRTL. REPORTED OCCURRENCES NEED VERIFICATION	1.1.10.
2	TIN	TETR. C	7.28	ELE	OPQ. TIN-WHITE. HACKLY FRACTURE. MAL, DUCT. SOL-HCL. PLACERS	1.1.9.
2 -2.5	PARKERITE	MNCL.		SLD	OPQ. NEEDS CONFIRMATION.	2.9.9.
2 -2.5	FREIESLEBENITE	MNCL.	6.04-6.23	SST	OPQ. STL-GRAY, AG-WHT. PRISM-STRIATED. 1 POOR CLVG. AG- PB-DEPS	3.4.4.
2 -2.5	BISMUTH	HEXA.	9.7 -9.83	ELE	OPQ.REDDISH AG-WHT, IRIDESCENT TARN. SECT. SOL-HNO3. VEINS + PEGS	1.2.1.5
2 -3	WITTICHENITE	ORTH.	4.3 -4.5	SST	OPQ. STL-GRAY, SN-WHT, TARN-PALE. TBLR, COLUMNAR, ACICULAR, MSV. CONCH. SOL-HCL(H2St),DCMP-HNO3. WITH BARITE + FLUORITE -- RARE	3.2.3.
2 -3	KRENNERITE	ORTH.	8.62	SLD	OPQ. AG-WHT, BRASS-YEL. STRIATED PRISMS.BASAL CLVG. TELLURIDE DEPS	2.9.7.1
2.5	MUTHMANNITE	UNKN.	5.60	SLD	OPQ.BRASS YEL, FRESH SURF + STRK GRAY. TABULAR, 1 PERF CLVG. SOL- HNO3 YIELDING GOLD RESIDU. WITH OTHER TELLURIDES	2.6.14.2
2.5	OWYHEEITE	UNKN.	6.03	SST	OPQ. STL-GRAY, AG-WHT,TARN-YEL,STRK-GRAY(PAPER), REDDISH-BRWN(POR- CELAIN). ACICULAR,FIBR. SOL-HOT HCL. LO-T AG-BEARING VEINS	3.5.2.
2.5	EUCAIRITE	UNKN.	7.6 -7.8	SLD	OPQ.AG-WHT, PB-GRAY. SECT. SOL-BOILING HNO3. WITH OTHER SELENIDES	2.3.1.7
2.5-3	CALAVERITE	MNCL.	9.22-9.26	SLD	OPQ. BRASS-YEL, AG-WHT. YEL- TO GREEN-GRAY STRK. STRIATED LATHS. LO-T VNS + LO-T PORTIONS OF HI- + MOD-T VEINS	2.9.7.2
2.5-3	SILVER	ISOM. P	10.50	ELE	OPQ. GRAY TO BLK TARN. ELONGATE FMS. MAL, DUCT. SOL-HNO3. VEINS	1.1.1.3
2.5-3	GOLD	ISOM.	19.31	ELE	OPAQUE. BRANCHING XLS, NUGGETS. MAL, DUCT. VEINS + PLACERS	1.1.1.1
2.5-3.5	GALENOBISMUTITE	ORTH.	7.04	SST	OPQ. TARN-YEL, IRID. STRK-BLK. MSV, FIBR, LATHLK XLS(COMMONLY BENT + STRIATED, 1 CLVG. SOL-HCL WITH SULFIDES IN QTZ-SULFIDE VEINS	3.8.3.
3	NIGGLIITE	UNKN.	4.	SLD	OPQ. AG-WHT. RARE-IN CONCENTRATE OF OXIDE ZONE, MINE DUMP, INSIZWA	2.10.2.
3	ALTAITE	ISOM.	8.15	SLD	OPQ.YELLOWISH SN-WHT, BRONZE TARN. MSV. PERF CUBIC CLVG. AU,S-VNS	2.6.1.3
3 -3.5	ANTIMONY	HEXA.	6.61-6.72	ELE	OPQ. SN-WHT, GRAY STRK. MASSIVE. 1 PERF CLVG. UNEVN FRACT. VEINS	1.2.1.4
3 -3.5	DOMEYKITE	ISOM.	7.2 -7.9	SLD	OPQ. SN-WHT, STL-GRAY, BROWN-IRID TARN. MSV. SOL-HNO3. CU-DEPS	2.1.3.2

METALLIC WHITE

HARDNESS	NAME	XL. SYS.	SPECIFIC GRAVITY	CHEM. CLASS	REMARKS	REFERENCES
3 -3.5	GOLD AMALGAM	ISOM.	15.47	ELE	OPQ. GRAINS + LUMPS. CONCHOIDAL FRACTURE. SOL-HNO3. WITH PT. RARE	1.1.4.
3 -4	ALLEMONTITE	HEXA.	5.8 -6.2	ELE	OPQ.SN-WHT, REDDISH GRAY. DK TARN. 1 PERF CLVG. WITH AS, SB VEINS	1.2.1.3
3.5	ARSENIC	HEXA.	5.63-5.78	ELE	OPQ. TIN-WHT, DARK GRAY TARN. 1 PERF CLVG. BRTL. SULFIDE VEINS	1.2.1.1
3.5	MOSCHELLANDSBERGITE	ISOM. C	13.5	ELE	OPQ. AG-WHT. GRNLR. DDDEC + CUBIC CLVGS. SOL-HNO3. RARE	1.1.3.
3.5	POTARITE	ISOM.	16.11	ELE	OPQ. FIBR NUGGETS. BRITTLE. SOL-HNO3 TO BRWN SOLUT. PLACERS. RARE	1.1.5.
3.5-4	SPHALERITE	ISOM.	3.9 -4.1	SLD	RESIN. CURVED XL FACES COMMON. FLUO-OR. TRIBOLUM. SOL-HCL. VEINS	2.6.2.1
3.5-4	DYSCRASITE	ORTH.	9.6 -9.81	SLD	OPQ. AG-WHT, TARN GRAY, YEL, BLK. MSV. SECT. DECOMP-HNO3. AG-DEPS	2.2.1.
4	ALGODONITE	HEXA.	8.38	SLD	OPQ. STL-GRAY, AG-WHT, DULL TARN. INCRUST, MSV. SUBCONCH. AG-DEPS	2.1.3.1
4 -4.5	PLATINUM	ISOM.	19.	ELE	OPQ. GRAINS + SCALES. MAL, DUCT. ULTRABASIC IGNEOUS RKS + PLACERS	1.1.6.1
4 -5	HORSFORDITE	UNKN.	8.81	SLD	OPQ.AG-WHT, TARN EASILY. MSV. BRTL. UNEVN FRACT. RARE(I DEP, ASIA)	2.1.3.3
4 -5	STIBIOPALLADINITE	ISOM.	9.5	SLD	OPQ. AG-WHT, STL-GRAY. UNEVN FRACT. SOL-HOT AQUA REGIA. PT-DEPS	2.2.2.
4.5-5	WOLFACHITE	ORTH.	6.37	SLD	OPQ. AG- TO SN-WHT. BLK STRK. COLUMNAR RADIATING AGGS. DCMP-HNO3. RARE --AT WOLFACH(BADEN) ON NICCOLITE IN CALCITE	2.9.5.4
4.5-5	SAFFLORITE	ORTH.	6.95-7.45	SLD	OPQ. SN-WHT,TARN DK. MSV, FIBR. 1 CLVG. WITH CO-NI MINS-MESO-T VNS	2.9.3.2
4.5-5	PALLADIUM	ISOM.	11.9	ELE	OPQ.GRNS-SOME RADIALLY FIBROUS. MAL, DUCT. SOL-HNO3. WITH PT,RARE	1.1.6.2
5	GLAUCODOT	ORTH.	6.28-6.52	SLD	OPQ. GRAYISH SN-WHT, REDDISH AG-WHT. BLK STRK. PRISM. STRIATED ON PRISM ZONE FACES. DCMP-HNO3(PINK SOLUTION). WITH COBALTITE	2.9.5.2
5	PARARAMMELSBERGITE	ORTH.	7.12	SLD	OPQ. SN-WHT. MSV, TBLR. BASAL CLVG. ALTERS TO ERYTHRITE.MESO-T VNS	2.9.3.4
5	NICKEL IRON	ISOM.	7.8 -8.22	ELE	OPQ.AG-WHT, GRAY. MAL. MAG. SOL-HCL,CH3COOH(SLOW). METEOR + BASALT	1.1.7.2
5 -5.5	ULLMANNITE	ISOM.	6.61-6.69	SLD	OPQ. STL-GRAY. CUBIC. CUBIC CLVG. UNEVN. DCMP-HNO3. NI-BEARING VNS	2.9.2.3
5 -5.5	LOELLINGITE	ORTH.	7.39-7.41	SLD	OPQ. STL-GRAY. PRISMATIC, MSV. 1 CLVG. MESO-T,S-VNS + AG-AU DRES	2.9.3.1

METALLIC WHITE

Hardness	Mineral	System	S.G.		Description	Index
5.5	GERSDORFFITE	ISOM.	5.9	SLD	OPQ. AG-WHT, LK PYRITE. CUBIC CLVG. DCMP-WARM HNO3. WITH NI MINS	2.9.2.2
5.5	COBALTITE	ISOM.	6.33	SLD	OPQ. TYPICALLY REDDISH AG-WHT, ALSO VIOLET STL-GRAY + DK GRAY. HAB-IT-LIKE PYRITE. CUBIC CLVG. DCMP-HNO3. HI-T DISSEM DEPS + VEINS	2.9.2.1
5.5-6	ARSENOPYRITE	MNCL.	6.55-6.85	SLD	OPQ. AG-WHT, STL-GRAY. PRISM. DCMP-HNO3. WDSPRD-E.G.,ORE DEPS,PEGS	2.9.5.1
5.5-6	SKUTTERUDITE	ISOM.	6.1 -6.9	SLD	OPQ. SKUTTERUDITE, SMALTITE, NICKEL-SKUTTERUDITE, + CHLOANTHITE	2.10.1.1
5.5-6	SMALTITE	ISOM.	6.1 -6.9	SLD	CONSTITUTE A SERIES. MEMBERS INDISTINGUISHABLE MEGASCOPICALLY	2.10.1.2
5.5-6	NICKEL SKUTTERUDITE	ISOM.	6.1 -6.9	SLD	SN-WHT, AG-GRAY. TARN-GRAY OR IRID. CUBIC, OCTAH, GRNLR.CUBIC	2.10.1.3
5.5-6	CHLOANTHITE	ISOM.	6.1 -6.9	SLD	+ OCTAH CLVGS. SOL-HNO3.WITH OTHER CO +NI MINS,ETC. MOD-T VNS	2.10.1.4
5.5-6	RAMMELSBERGITE	ORTH.	7.0 -7.2	SLD	OPQ. REDDISH SN-WHT. GRAY STRK. MSV. WITH CO-NI MINS-MESO-T VNS	2.9.3.3
5.5-6	COHENITE	ORTH.	7.20-7.65	ELE	OPQ. BRONZY-GOLD ON EXPOS. BRTL. MAG. SOL-CU(NH3)6CL2. METEOR + TERRES FE	1.1.7.3
6	GUDMUNDITE	MNCL.	6.72	SLD	OPQ. AG-WHT, STL-GRAY. PRISM. HYDROTHERMAL-LATE IN SULFIDE DEPS	2.9.5.3
6 -6.5	MARCASITE	ORTH.	4.89	SLD	OPQ. SN-WHT. BRONZE-YEL. TBLR, COCKSCOMB-LK. LO-T, ACID DEPOSITION	2.9.4.
6 -7	CASSITERITE	TETR.	6.99	OXD	SUBMET. RADIAL CONCRETIONARY MASSES. ADMN-DULL. HI-T VNS, GREISENS	4.5.1.5
6 -7	SPERRYLITE	ISOM.	10.58	SLD	OPQ. SN-WHT, BLK STRK. MODIFIED CUBES. CONCH. BRTL. NORITE,PLACERS	2.9.1.4
6 -7	IRIDOSMINE	HEXA.	21.	ELE	OPQ. SN-WHT. FLAT GRNS. CLVG FLKS. MAL-BRTL. 1 PERF CLVG. PLACERS	1.1.6.5
6 -7	PLATINIRIDIUM	ISOM.	22.84	ELE	OPQ. YELLOWISH AG-WHT, FRACTURES GRAY. SOMEWHAT MAL. PLACERS.RARE	1.1.6.3
6.5-7	SCHREIBERSITE	TETR.	7.0 -7.3	ELE	OPQ.BRASS-YELLOW OR BROWN TARNISH. BRITTLE. MAG. DIFFICULTLY SOL-HCL + HNO3. METEORITES, COMBUSTION PRODUCT IN COAL MINES (FRANCE)	1.1.7.6
7	AUROSMIRIDIUM	ISOM.	20.	ELE	OPQ. AG-WHT. BRITTLE. IRREGULAR FRACTURE. WITH PT. PLACERS. RARE	1.1.6.4

METALLIC GRAY

HARDNESS	NAME	XL. SYS.	SPECIFIC GRAVITY	CHEM. CLASS	REMARKS	REFERENCES
0.0	MERCURY	AMOR.	13.60	ELE	OPQ. LIQ DROPS. WITH CINNABAR-VEINS OF VOL + HOT SPRING REGIONS	1.1.2.
1	LENGENBACHITE	TRCL.	5.80-5.85	SST	OPQ. STL-GRAY. TARN-IRID. STRIATED BLDS(FLEX, INELAST) SOME CURLED. ON JORDANITE. WITH PYRITE, IN CAVITIES IN DOLOMITE (SWITZERLAND)	3.3.7.
1 -1.5	MOLYBDENITE	HEXA.	4.62-4.73	SLD	PB-GRAY. STRK-GREEN(PORCELAIN), BLUE-BLK(PAPER). TBLR. FOL MSV. SCALES. DCMP-HNO3. ACCESS IN GRANITES,PEGS,VNS.CNTCT META RKS	2.9.6.1
1 -1.5	NAGYAGITE	MNCL.	7.36-7.46	SLD	OPQ. RESEMBLES TETRADYMITE. SOL-HNO3 GIVING AU-RESID. AU-TE-S VNS	2.1.2.
1 -2	GRAPHITE	HEXA.	2.09-2.23	ELE	SCALES. 1 PERF CLVG TO INELASTIC PLTS. META OF CARBONACEOUS RKS	1.2.4.2
1.5	TEALLITE	ORTH.	6.36	SST	OPQ. TARN-DULL OR IRID. TBLR-SQ, BENT PLTS. 1 CLVG. SOMEWHAT MAL. SOL-HOT HCL,HNO3. IN SILVER-TIN VEINS	3.5.11.
1.5	LEAD	ISOM.	11.35	ELE	OPQ. ROUNDED MASSES + THIN PLATES. MAL. SOL-HNO3. VEINS, RARE	1.1.1.5
1.5-2	TETRADYMITE	HEXA.	7.1 -7.5	SLD	OPQ.STL-GRAY, IRID-DULL TARN. GRNLR, FOL MSS. 1 PERF CLVG-INELAST PLATES. GOLD-QUARTZ VEINS AND CONTACT METAMORPHIC DEPOSITS	2.1.1.2
1.5-2	TELLUROBISMUTHITE	HEXA.	7.67-7.97	SLD	OPQ. PB-GRAY, PLTS, FOL MSS. 1 PERF CLVG-INELAST PLTS. AU-QTZ VNS	2.1.1.1
1.5-2	SYLVANITE	MNCL.	8.16	SLD	OPQ. YELLOWISH STL-GRAY TO AG-WHT. PRISM(SOME SKELETAL). DCMP-HNO3 (RUSTY COLORED GOLD RESID). WITH CALAVERITE, LO-T VEINS, ETC.	2.9.7.3
1.5-2	MALDONITE	ISOM. C	15.70	ELE	OPQ. PINKISH AG-WHT. CU-RED TO BLK TARN. GRNLR. VEINS + GREISEN	1.1.1.2
1.5-2.5	WEHRLITE	UNKN.	8.38-8.44	SLD	FOLIATED MSS RESEMBLING TETRADYMITE. CLVG PLTS-SLIGHTLY ELASTIC	2.1.1.5
2	FIZELYITE	UNKN.		SST	OPQ. STRIATED PRISMS. 1 CLVG. RARE-WITH SULFIDES-KISBANYA, HUNGARY	3.6.4.
2	STIBNITE	ORTH.	4.61-4.65	SLD	OPQ. PB-GRAY, BLK-IRID TARN. STRIATED BENT BLDS. SOL-HCL. LO-T VNS	2.8.2.1
2	SELENIUM	HEXA.	4.80	ELE	HOLLOW NEEDLES. RED STRK. ARTIF-INDUCED FUMAROLIC CONDITIONS PROD	1.2.2.1
2	LIVINGSTONITE	MNCL.	5.00	SST	SUBTRANSLUCENT. STRK-RED. NEEDLES. ADMN. 1 PERF,2 POOR CLVGS. SOL-WARM HNO3. WITH CINNIBAR, SULFUR,GYPSUM,ETC. (E.G.,SAN LUIS,POTOSI)	3.9.1.
2	RAMDOHRITE	UNKN.	5.43	SST	OPQ. PRISM, LANCE-SHPD XLS. IN QTZ, WITH SULFIDES-POTOSI, BOLIVIA	3.6.5.
2	ARSENOLAMPRITE	UNKN.	5.3 -5.5	ELE	OPQ.PB-GRAY, BLK STRK. FOLIATED MASSES. 1 CLVG. BRILLIANT. VEINS	1.2.1.2
2	EMPLECTITE	ORTH.	6.38	SST	OPQ. GRAY, SN-WHT. PRISM-STRIATED. 2 CLVGS. DCMP-HNO3.SULFIDE DEPS	3.5.9.2
2	BERZELIANITE	ISOM.	6.71	SLD	OPQ. AG-WHT, TARN DK. DENDRI. MAL. SOL-HNO3. WITH OTHER SELENIDES	2.3.1.5
2	SCHIRMERITE	UNKN.	6.74	SST	OPQ. PB-GRAY, FE-BLK. MSV, GRNLR. RARE-QTZ VEINS (3 LOCALES, COLO).	3.5.3.
2	BISMUTHINITE	ORTH.	6.75-6.81	SLD	OPQ. PB-GRAY SN-WHT, YEL-IRID TARN. FIBR, BLDD. SOL-HNO3. HI-T VNS	2.8.2.2
2	ALASKAITE (MIXTURE?)	UNKN.	6.78-6.88	SST	OPQ. TARN-BRONZY. MSV-FOL. SOL-HOT HCL.WITH CHALCOPYRITE-QTZ,S-VNS	3.8.7.
2	ZINC	HEXA.	6.9 -7.2	ELE	OPQ. 1 PERF CLVG. BRTL. REPORTED OCCURRENCES NEED VERIFICATION	1.1.10.
2	TIN	TETR. C	7.28	ELE	OPQ. TIN-WHITE. HACKLY FRACTURE. MAL. DUCT. SOL-HCL. PLACERS	1.1.9.
2	GRUENLINGITE	UNKN.	8.08	SLD	OPQ.STL-GRAY, IRID-DK TARN. 1 PERF CLVG. WITH BISMUTH MINERALS	2.1.1.3
2	JOSEITE	UNKN.	8.18	SLD	MEGASCOPICALLY INDISTINGUISHABLE FROM GRUENLINGITE. SAME OCCUR	2.1.1.4

METALLIC GRAY

H	SG	Crystal	Description	Name	Ref
2 -2.5	5.53	MNCL.	RED, EXPOS-DK PB-GRAY COVERED BY YEL POWDER, STRK-RED, PRISM-STRI-ATED. ADMN. 3 CLVGS,FLEX-SEPARATING TO FIBERS. WITH REALGAR, ETC.	LORANDITE	3.5.10.
2 -2.5	6.04-6.23	MNCL.	SST OPQ. STL-GRAY, AG-WHT. PRISM-STRIATED. 1 POOR CLVG. AG-PB-DEPS	FREIESLEBENITE	3.4.4.
2 -2.5	4.8 -6.3	HEXA.	ELE OPQ. BLK STRK. MASSIVE. PERF HEX PRISM CLVG. GANGUE OF AG-VEINS	SELEN-TELLURIUM	1.2.2.2
2 -2.5	7.06-7.08	ORTH.	SST OPQ. PB-GRAY, TARN-BRWN, CU-RED, COMMONLY WITH YELLOWISH GREEN COATING. PRISM-ACICULAR, STRIATED. IN GOLD-QUARTZ VEINS	AIKINITE	3.4.1.3
2 -2.5	7.12	UNKN.	SST OPQ. BLK STRK. RESEMBLES MOLYBDENITE. STRIAE AT 57° ON CLVG. WITH QUARTZ, MAGNETITE, + CORDIERITE IN AMPHIBOLITE-KOPPARBERG,SWEDEN	WITTITE	3.6.6.
2 -2.5	7.2 -7.4	ISOM.	SLD OPQ. BLKISH PB-GRAY. BRANCHING FMS, MSV. NOTABLY SECT. LO-T,S-VNS	ARGENTITE	2.3.1.1
2 -2.5	8.09	HEXA.	SLD SCARLET STRK. ADMN-DULL. NEAR VOLS + HOT SPRINGS-VNS + INCRUSTS	CINNABAR	2.6.9.
2 -2.5	9.7 -9.83	HEXA.	ELE OPQ.REDDISH AG-WHT, IRIDESCENT TARN. SECT. SOL-HNO3. VEINS + PEGS	BISMUTH	1.2.1.5
2 -3	4.3 -4.5	ORTH.	SST OPQ. STL-GRAY, SN-WHT, TARN-PALE. TBLR, COLUMNAR, ACICULAR, MSV. CONCH. SOL-HCL(H2S↑),DCMP-HNO3. WITH BARITE + FLUORITE -- RARE	WITTICHENITE	3.2.3.
2 -3	4.64	ORTH.	SST OPQ. TARN-IRID. BRWN. STRK-BROWNISH GRAY. STRIATED PRISM,FIBR MSV. POOR PRISMATIC CLEAVAGE. WITH QTZ & STIBNITE, LO-T HYDROTHERMAL VNS	BERTHIERITE	3.8.10.
2 -3	6.96	ORTH.	SST OPQ. STRK-BLK. PRISM. 1 GOOD, 1 FAIR CLVG. RARE-WITH QTZ + PB-SB SULFIDES (GLADHAMMAR, SWEDEN)	GLADITE	3.8.12.
2 -3	6.97	UNKN.	SST OPQ. NEEDS VERIFICATION-MAY BE MIXTURE	WEIBULLITE	3.8.4.
2 -3	6.92-7.15	UNKN.	SST OPQ. NEEDS CONFIRMATION-MAY BE MIXTURE	CHIVIATITE	3.8.6.
2 -3	7.0 -7.2	ORTH.	SST OPQ. NEEDS VERIFICATION -MAY BE MIXTURE.	LILLIANITE	3.3.12.
2 -3	7.98	HEXA.	SST OPQ.PB- TO STL-GRAY. THIN PLTS.GOOD BASAL,FAIR RHOMB CLVGS. IN QTZ(FAHLUN,SWEDEN)	PLATYNITE	3.8.5.
2 -3	8.24-8.45	MNCL.	SLD OPQ.PB- TO STL-GRAY. MSV. CUBIC CLVG. SECT. TELLURIDE,AU,TE-VEINS	HESSITE	2.3.1.8
2 -3	8.62	ORTH.	SLD OPQ. AG-WHT, BRASS-YEL. STRIATED PRISMS.BASAL CLVG. TELLURIDE DEPS	KRENNERITE	2.9.7.1
2 -6.5	4.4 -5.08	TETR.	OXD OPQ. FIBR, GRNLR. PRISM CLVG. BOG + RESIDUAL MN-DEPS - COMMON	PYROLUSITE	4.5.1.2
2.5	5.23	MNCL.	SST OPQ. PLAGIONITE GRP. S.G.-DEFINITIVE IN GRP. TARN-STL BLUE, BRONZY STRK-REDDISH GRAY. STRIATED PRISM. RARE-WITH SPHAL(NAGYAG,ROUMANIA)	FULOPPITE	3.7.4.1
2.5	5.20-5.30	MNCL.	SST SUBTRANSLUCENT. STRK-RED, TBLR, MSV. ADMN. 3 POOR CLVGS. DCMP-HNO3 WITH AG-MINS + SULFIDES IN LO-T HYDROTHERMAL VEINS	MIARGYRITE	3.5.4.
2.5	5.43-5.49	UNKN.	SST OPQ. MSV. CYLINDRICAL FMS. SOL-HOT HCL,HOT HNO3. WITH FRANKEITE-VN	CYLINDRITE	3.8.11.
2.5	5.51	MNCL.	SST SUBTRANSLUCENT. DK RED STRK. PRISM-STRIATED. SULFIDE DEP-IN VUGS	SAMSONITE	3.3.4.
2.5	5.54-5.58	MNCL.	SST OPQ. PLAGIONITE GRP. S.G.-DEFINITIVE IN GRP. STRIATED PRISM,1 CLVG. SOL-HOT HCL(H2S↑). WITH SULFOSALTS,SULFIDES, + NATIVE SB-ORE DEPS	PLAGIONITE	3.7.4.2
2.5	5.60	UNKN.	SLD OPQ.BRASS YEL, FRESH SURF + STRK GRAY. TABULAR. 1 PERF CLVG. SOL-HNO3 YIELDING GOD RESIDUE. WITH OTHER TELLURIDES	MUTHMANNITE	2.6.14.2
2.5	5.63	MNCL.	SST OPQ. TARN-IRID, STRIATED FIBR. 1 GOOD, 2 POOR CLVGS. DCMP-HNO3. WITH PB-SULFOSALTS + SULFIDES IN LO- TO MOD-T HYDROTHERMAL VEINS	JAMESONITE	3.6.7.
2.5	6.01	ORTH.	SST OPQ. STL-GRAY. TARN-BRASSY. MSV. SOL-HCL, HNO3. WITH OTHER BISMUTH MINERALS IN SEVERAL DEPOSITS	KLAPROTHITE (MIXTURE?)	3.4.5.
2.5	6.03	UNKN.	SST OPQ. STL-GRAY, AG-WHT,TARN-YEL.STRK-GRAY(PAPER), REDDISH-BRWN(POR-CELAIN). ACICULAR,FIBR. SOL-HOT HCL. LO-T AG-BEARING VEINS	OWYHEEITE	3.5.2.

METALLIC GRAY

HARDNESS	NAME	XL. SYS.	SPECIFIC GRAVITY	CHEM. CLASS	REMARKS	REFERENCES
2.5	SEMSEYITE	MNCL.	6.08	SST	OPQ. PLAGIONITE GRP. S.G.-DEFINITIVE IN GRP. TARN-DULL. PRISM. 1 CLVG. SEVERAL LOCALITIES WITH CORRODED GALENA, OTHER SULFIDES, ETC.	3.7.4.4
2.5	COCINERITE	UNKN.	6.14	SLD	OPQ. AG-GRAY, BLK TARN. MSV. WITH OXIDIZED CU-ORE. NEEDS CONFIRM	2.1.3.4
2.5	GRATONITE	HEXA.	6.20-6.24	SST	OPQ. PRISM. MSV. RARE-VUGS IN PYRITIC ORE, ASSOC REALGAR,ETC.-PERU	3.6.6.
2.5	CANFIELDITE	ISOM.	6.28	SST	OPQ. CHEM DISTINGUISHED FROM ARGYRODITE. BLUISH-PURPLISH BLK, RED-DISH STL-GRAY. OCTAH + DODEC,MSV,RADIATING AGGS, LO-T VEINS	3.1.3.2
2.5	ARGYRODITE	ISOM.	6.26-6.29	SST	OPQ. CHEM DISTINGUISHED FROM CANFIELDITE. BLUISH-PURPLISH BLK, RED-DISH STL-GRAY. OCTAH + DODEC,MSV,RADIATING AGGS, LO-T VEINS	3.1.3.1
2.5	MENEGHINITE	ORTH.	6.35-6.37	SST	OPQ. STRIATED PRISM, MSV, FIBR. 1 CLVG. OXIDIZED-HNO3. SULFIDE VNS	3.3.11.
2.5	REZBANYITE	UNKN.	6.09-6.39	SST	OPQ. TARN-DULL. MSV, GRNLR. NOT WELL DEFINED. RARE-2 LOCALITIES	3.8.2.
2.5	GEOCRONITE	ORTH.	6.3 -6.5	SST	OPQ. GRAYISH BLUE. TBLR, MSV. SOL-HOT HCL(H2S↑). SULFIDE VEINS	3.3.5.
2.5	MATILDITE	ORTH.	6.9	SST	OPQ. STRIATED PRISM., GRNLR. SOL-HNO3. MOD- TO HI-T SULFIDE VEINS	3.5.6.
2.5	POLYARGYRITE	ISOM.	6.97	SLD	OPQ. NEEDS CONFIRMATION.	3.1.2.
2.5	TUNGSTENITE	HEXA.	7.4	SLD	OPQ. DK PB-GRAY. SOILS FINGERS. MSV. SOL-AQ-REG. RPLCMNT DEP IN LS	2.9.6.2
2.5	EUCAIRITE	UNKN.	7.6 -7.8	SLD	OPQ.AG-WHT, PB-GRAY. SECT. SOL-BOILING HNO3. WITH OTHER SELENIDES	2.3.1.7
2.5	COLORADOITE	ISOM.	8.04	SLD	OPQ.MSV.GRNLR. FRIABLE-SUBCONCH. SOL-HNO3.WITH AU-+ AG-TELLURIDES	2.6.2.4
2.5	TIEMANNITE	ISOM.	8.19-8.47	SLD	OPQ.STL-GRAY, PB-GRAY. TETRAHEDRAL, MSV. UNEVEN-CONCH. VNS IN LS	2.6.2.3
2.5-3	CHALCOCITE	ORTH.	5.5 -5.8	SLD	OPQ.MSV. CONCH. SECT. SOL-HNO3. ENRICHED ZONES ATOP SULFIDE DEPS	2.3.2.1
2.5-3	BOURNONITE	ORTH.	5.80-5.86	SST	OPQ. STL-GRAY. PRISM-STRIATED, SUBPARALLEL AGGS. 3 POOR CLVGS. DCMP-HNO3 TO BLUE SOLUT + S + WHITE POWDER. MOD-T SULFIDE VEINS	3.4.1.1
2.5-3	FRANCKEITE	ORTH.	5.90	SST	OPQ. TARN-IRID. MSV,TBLR-STRIATED. DCMP-HNO3,SOL-AQ-REG. AG-SN-VNS	3.6.3.
2.5-3	DIAPHORITE	ORTH.	6.04	SST	OPQ. STL-GRAY. PRISM-STRIATED. SOL-HNO3. LO-T SULFIDE VEINS	3.4.3.
2.5-3	BOULANGERITE	MNCL.	6.23	SST	OPQ. YEL-SPOTTED BLUISH PB-GRAY. STRK-BRWN. PRISM-ACICULAR(STRIA-TED),FIBR(FLEX). 1 GOOD CLVG. SOL-HOT HCL(H2S↑). LO- TO MOD-T VEINS	3.5.1.
2.5-3	STROMEYERITE	ORTH.	6.2 -6.3	SLD	OPQ. STL-GRAY. SUBCONCH. SOL-HNO3. WITH TETRAHEDRITE, BORNITE VNS	2.3.2.2
2.5-3	KOBELLITE	UNKN.	6.33	SST	OPQ. NEEDS CONFIRMATION.	3.6.2.2
2.5-3	COSALITE	ORTH.	6.76	SST	OPQ. PRISM + HAIRLK, MSV. SOL-HNO3. MOD-T VNS, CNTCT META, PEGS	3.6.2.1
2.5-3	CROOKESITE	UNKN.	6.90	SLD	OPQ. PB-GRAY. MSV COMPACT. BRTL. SOL-HNO3. WITH OTHER SELENIDES	2.3.1.6
2.5-3	PENROSEITE	ISOM.	6.7 -7.1	SLD	OPQ.PB-GRAY, BLK STRK, TARN. RENIFM.CUBIC CLVG. SOL-HNO3(EFF). VNS	2.9.1.6
2.5-3	GALENA	ISOM.	7.57-7.59	SLD	OPQ.PB-GRAY. CUBIC, MSV. PERF CUBIC CLVG. DCMP-HNO3. SULFIDE VNS	2.6.1.1
2.5-3	CLAUSTHALITE	ISOM.	7.8	SLD	OPQ.BLUISH PB-GRAY.MSV. GOOD CUBIC CLVG. SOL-HNO3. WITH SELENIDES	2.6.1.2
2.5-3	PETZITE	UNKN.	8.7 -9.02	SLD	OPQ.STL-GRAY, FE-BLK, TARN. CUBIC CLVG. DCMP-HNO3. TELLURIDE VNS	2.3.1.9

METALLIC GRAY

Hardness	Mineral	Crystal System	Density	State	Description	Index
2.5-3	CALAVERITE	MNCL.	9.22-9.26	SLD	OPQ. BRASS-YEL, AG-WHT, YEL- TO GREEN-GRAY STRK. STRIATED LATHS. LO-T VNS + LO-T PORTIONS OF HI- + MOD-T VEINS	2.9.7.2
2.5-3	SILVER	ISOM.	P 10.50	ELE	OPQ. GRAY TO BLK TARN. ELONGATE FMS. MAL, DUCT. SOL-HNO3. VEINS	1.1.1.3
2.5-3.5	GUANAJUATITE	ORTH.	6.25-6.98	SLD	OPQ. BLUISH GRAY. STRIATED XLS, FIBR. SOL-WM AQ-REG. NEEDS CONFIRM	2.8.2.3
2.5-3.5	GALENOBISMUTITE	ORTH.	7.04	SST	OPQ. TARN-YEL, IRID, STRK-BLK. MSV, FIBR, LATHLK XLS(COMMONLY BENT + STRIATED). 1 CLVG. SOL-HCL. WITH SULFIDES IN QTZ-SULFIDE VEINS	3.8.3.
3	NIGGLIITE	UNKN.	4.	SLD	OPQ. AG-WHT. RARE-IN CONCENTRATE OF OXIDE ZONE, MINE DUMP, INSIZWA	2.10.2.
3	ENARGITE	ORTH.	4.40-4.50	SST	OPQ. TBLR, MSV, GRNLR. PRISM ZONE-STRIATED. 1 POOR, 3 GOOD CLVGS. SOL-AQ-REG. MOD-T TO LO-T VEIN + REPLACEMENT SULFIDE DEPOSITS	3.3.2.2
3	KLOCKMANNITE	HEXA.	>5.	SLD	SLATE-GRAY, TARN BLU-BLK. GRNLR. 1 PERF CLVG. WITH OTHER SELENIDES	2.6.8.2
3	SARTORITE	MNCL.	5.08-5.12	SST	OPQ. STRK CHOC-BRWN. PRISM-STRIATED. 1 FAIR CLVG. WITH SULFIDE IN VUGS IN DOLOMITE (VALAIS, SWITZERLAND)	3.8.9.
3	BAUMHAUERITE	MNCL.	5.33	SST	OPQ. TARN-IRID. STRK-CHOCOLATE BRWN. STRIATED PRISM. 1 PERF CLVG. RARE-WITH OTHER RARE SULFARSENIDES IN DOLO-VALAIS, SWITZERLAND	3.7.2.
3	RATHITE	MNCL.	5.33-5.41	SST	OPQ. TARN-IRID. STRK-CHOCOLATE BRWN. STRIATED PRISM. 1 PERF CLVG. RARE-WITH OTHER RARE SULFARSENIDES IN DOLO-VALAIS, SWITZERLAND	3.6.8.
3	SELIGMANNITE	ORTH.	5.38-5.44	SST	OPQ. STRK-CHOCOLATE-BRWN TO PURPLE-BLK. EQUANT. CAVITIES IN DOLO	3.4.1.2
3	DUFRENOYSITE	MNCL.	5.50-5.56	SST	SUBTRANSLUCENT. STRK-REDDISH BRWN. TBLR, ELONGATE + STRIATED. 1 PERF CLVG. DRUSES IN DOLO(VALAIS,SWITZ),SULFIDE DEPS, ETC.	3.6.1.
3	GUITERMANITE	UNKN.	5.94	SST	OPQ. POSSIBLY A MIXTURE RATHER THAN A DISTINCT SPECIES	3.3.9.
3	JORDANITE	MNCL.	6.33-6.43	SST	OPQ. PB-GRAY, TARN-IRID. TBLR. 1 PERF CLVG. DCMP-HNO3. CVTS-DOLO	3.3.8.
3	GOONGARRITE (MIXTURE?)	MNCL.	7.29	SST	OPQ. PLTY, RADIATING FIBERS. GOOD CLVG. RARE-WITH AU IN QTZ VEIN	3.3.10.
3	METACINNABAR	ISOM.	7.65	SLD	OPQ.GRAY-BLK. BLK STRK. TETRA-ROUGH FACES, MSV. WITH CINNABAR,VNS	2.6.2.2
3	GOLDFIELDITE	UNKN.			OPQ. SPECIES OF DOUBTFUL VALIDITY.	3.2.5.
3 -3.5	LAUTITE	ORTH.	4.8 -5.0	SLD	SUBMET. OPQ. BLK-GRAY(RED TINGE).SOL-HNO3. VNS-WITH NATIVE AS,ETC.	2.9.5.5
3 -3.5	ZINKENITE	HEXA.	5.25-5.35	SST	OPQ. TARN-IRID,COLUMNAR, RADIAL FIBR AGGS.SOL-HOT HCL(H2St). S-VNS	3.8.8.
3 -3.5	ANDORITE	ORTH.	5.33-5.37	SST	OPQ. BLK STRK. TARN-YEL,IRID. STRIATED PRISM, MSV.SOL-HCL. S-VNS	3.7.1.1
3 -3.5	LINDSTROMITE	ORTH.	7.01	SST	OPQ. BLK STRK. STRIATED PRISM. RARE-ON QTZ AT GLADHAMMAR,SWEDEN	3.7.1.2
3 -3.5	DOMEYKITE	ISOM.	7.2 -7.9	SLD	OPQ. SN-WHT, STL-GRAY, BROWN-IRID TARN. MSV. SOL-HNO3. CU-DEPS	2.1.3.2
3 -4	HAMMARITE	MNCL.			OPQ. REDDISH STL-GRAY. BLK STRK. PRISM-CURVED FACES, ACICULAR. RARE - ON DRUSY QUARTZ (GLADHAMMAR, SWEDEN).	3.5.13.
3 -4	CHALCOSTIBITE	ORTH.	4.90-5.00	SST	OPQ. BLDS, MSV. 3 GOOD CLVGS. DCMP-HNO3. QTZ-PYRITE VNS-(COMMONLY ALTERED ON EXPOSED SURFACES TO MALACHITE AND AZURITE)	3.5.9.1
3 -4	ALLEMONTITE	HEXA.	5.8 -6.2	ELE	OPQ.SN-WHT, REDDISH GRAY, DK TARN. 1 PERF CLVG. WITH AS, SB,VEINS	1.2.1.3
3 -4.5	TENNANTITE	ISOM.	4.62	SST	SOFTER + LIGHTER THAN TETRAHEDRITE. TETRA, MSV. DCMP-HNO3. VNS	3.2.4.2
3 -4.5	TETRAHEDRITE	ISOM.	4.99	SST	HARDER + HEAVIER THAN TENNANTITE. TETRA, MSV. DCMP-HNO3.WDSPRD-VNS	3.2.4.1

METALLIC GRAY

HARDNESS	NAME	XL. SYS.	SPECIFIC GRAVITY	CHEM. CLASS	REMARKS	REFERENCES
3.5	EPIGENITE	ORTH.	4.5	SST	OPQ. STL-GRAY. TARN-BLK + BLUE. INCRUST. SOL-HNO3. VNS-ON BARITE	3.1.5.
3.5	VRBAITE	ORTH.	5.27-5.33	SST	SUBMET.DK BLUISH GRAY,DK RED WITH TRANSMITTED LIGHT. STRK-RED WITH YEL TINGE. TBLR,PYRAMIDAL. 1 CLVG. SOL-HNO3.H2SO4. WITH REALGAR	3.8.13.
3.5	BENJAMINITE	UNKN.	6.34	SST	OPQ. TARN-DULL, YEL, CU-RED. GRNL. 1 CLVG. SOL-HCL(HOT), HNO3. RARE-QTZ,MUSCOVITE,FLUORITE,SULFIDE VN,NORTH OF MANHATTAN,NEVADA	3.5.12.
3.5	TENORITE	MNCL.	5.8 -6.4	OXD	BRWN-TRANSMITTED LIGHT. PLTY, EARTHY-MSV. SOL-DILUTE HCL OR HNO3. OXIDIZED PARTS OF CU-DEPS; AS SUBLIMATED XLS ON LAVAS	4.2.3.
3.5	MOSCHELLANDSBERGITE	ISOM.	13.5	ELE	OPQ. AG-WHT. GRNLR. DODEC + CUBIC CLVGS. SOL-HNO3. RARE	1.1.3.
3.5	POTARITE	ISOM.	16.11	ELE	OPQ. FIBR NUGGETS. BRITTLE. SOL-HNO3 TO BRWN SOLUT. PLACERS, RARE	1.1.5.
3.5-4	FAMATINITE	UNKN.	4.47-4.57	SST	OPQ. GRAY-RED TINGE. CRUSTS, MSV. WITH ENARGITE,ETC.-SULFIDE DEPS	3.3.2.1
3.5-4	DYSCRASITE	ORTH.	9.6 -9.81	SLD	OPQ. AG-WHT, TARN GRAY, YEL, BLK. MSV. SECT. DECOMP-HNO3. AG-DEPS	2.2.1.
4	MANGANITE	MNCL.	4.32-4.34	OXD	SUBMET. RED-BRWN STRK. STRIATED PRISMS. 3 CLVGS.SOL-HCL. LO-T VNS, SECONDARY, ETC.	6.1.3.
4	STANNITE	TETR.	4.3 -4.5	SLD	OPQ. STL-GRAY, FE-BLK. MOST XLS STRIATED. DCMP-HNO3.SN-BEARING VNS	2.6.3.2
4	GERMANITE	ISOM.	4.46-4.59	SST	OPQ. REDDISH GRAY. DK GRAY STRK. SOL-HNO3.WITH OTHER SULFIDES-DEPS	3.3.1.2
4	IRON	ISOM.	7.3 -7.87	ELE	OPQ.STL-GRAY, FE-BLK. MAL. MAG. SOL-HCL + CH3COOH.METEOR + BASALT	1.1.7.1
4	ALGODONITE	HEXA.	8.38	SLD	OPQ. STL-GRAY, AG-WHT, DULL TARN. INCRUST, MSV. SUBCONCH. AG-DEPS	2.1.3.1
4 -4.5	WOLFRAMITE	MNCL.	7.31	WOS	SUBMET. TARN-IRID. PRISM-EQUANT--STRIATED.MSV-GRNLR,ETC. 1 CLVG. DCMP-AQ-REG,H2SO4 OR HCL(SLOWLY). DIVERSE-VNS,CNTCT META,PLACERS	48.1.1.2
4 -4.5	PLATINUM	ISOM.	19.	ELE	OPQ. GRAINS + SCALES. MAL, DUCT. ULTRABASIC IGNEOUS RKS + PLACERS	1.1.6.1
4 -5	STAINIERITE	UNKN.	4.13-4.47	OXD	SUBMET. OPQ. MAMMILLARY. UNEVN. SOL-HCL. OXIDATION PROD OF CO-MINS	6.1.4.
4 -5	BERTHONITE	UNKN.	5.49	SST	OPQ. MSV, GRNLR. DCMP-HNO3. RARE-WITH GALENA IN VNS, FE-DEP(TUNIS)	3.4.2.
4 -5	HORSFORDITE	UNKN.	8.81	SLD	OPQ. AG-WHT, TARN EASILY. MSV. BRTL. UNEVN FRACT. RARE(I DEP, ASIA)	2.1.3.3
4 -5	COOPERITE	TETR.	9.5	SLD	OPQ. STL-GRAY. IRREGULAR GRNS. CONCH. IN PLATINUM-BEARING NORITES	2.6.11.
4 -5	STIBIOPALLADINITE	ISOM.	9.5	SLD	OPQ. AG-WHT, STL-GRAY. UNEVN FRACT. SOL-HOT AQUA REGIA. PT-DEPS	2.2.2.
4.5	CESAROLITE	UNKN.	5.29	OXD	SUBMET. STL-GRAY, FRIABLE MASSES(RESEMBLES COKE).SOL-HCL. RARE-CAVITIES IN GALENA (SIDI-AMOR-BEN-SALEM, TUNIS)	7.7.1.3
4.5	CORONADITE	TETR.	5.44	OXD	SUBMET. OPQ. BRWN-GRAY STRK. MSV-BOTRYOIDAL. OXIDIZED ZONE-MN DEPS	7.7.1.1
4.5-5	WOLFACHITE	ORTH.	6.37	SLD	OPQ. AG- TO SN-WHT. BLK STRK. COLUMNAR RADIATING AGGS. DCMP-HNO3. RARE - AT WOLFACH(BADEN) ON NICCOLITE IN CALCITE	2.9.5.4
4.5-5	SAFFLORITE	ORTH.	6.95-7.45	SLD	OPQ. SN-WHT,TARN DK. MSV, FIBR. 1 CLVG. WITH CO-NI MINS-MESO-T VNS	2.9.3.2
4.5-5	PALLADIUM	ISOM.	11.9	ELE	OPQ.GRNS-SOME RADIALLY FIBROUS. MAL, DUCT. SOL-HNO3. WITH PT,RARE	1.1.6.2
4.5-5.5	LINNAEITE	ISOM.	4.5 -4.8	SLD	OPQ. MINERALS OF LINNAEITE SERIES RANGE IN COMPOSITION BETWEEN LINNAEITE AND POLYDYMITE. LIGHT GRAY TO STEEL GRAY (VIOLARITE-	2.7.1.1
4.5-5.5	SIEGENITE	ISOM.	4.5 -4.8	SLD		2.7.1.2

METALLIC GRAY

Hardness	Mineral	Crystal System	S.G.	State	Description	Ref
4.5-5.5	CARROLLITE	ISOM.	4.5-4.8	SLD	VIOLET GRAY), TARNISH TO COPPER-RED OR VIOLET GRAY. OCTA XLS.	2.7.1.3
4.5-5.5	VIOLARITE	ISOM.	4.5-4.8	SLD	MSV, GRNLR. IMPERF CUBIC CLVG. UNEVEN-SUBCONCH FRACT. SOL-HNO3	2.7.1.4
4.5-5.5	POLYDYMITE	ISOM.	4.5-4.8	SLD	YIELDING SULFUR. RARE, ASSOC WITH OTHER CU,NI+FE-SULFIDES, VNS	2.7.1.5
5	GLAUCODOT	ORTH.	6.28-6.52	SLD	OPQ. GRAYISH SN-WHT, REDDISH AG-WHT. BLK STRK. PRISM. STRIATED ON PRISM ZONE FACES. DCMP-HNO3(PINK SOLUTION). WITH COBALTITE	2.9.5.2
5	MAUCHERITE	TETR.	8.0	SLD	OPQ.REDDISH GRAY. UNEVN. SOL-HNO3,HCL,H2SO4 WITH NICCOLITE, VEINS	2.4.1.
5	NICKEL IRON	ISOM.	7.8-8.22	ELE	OPQ.AG-WHT, GRAY. MAL. MAG. SOL-HCL, CH3CUOH(SLOW).METEOR + BASALT	1.1.7.2
5	YTTROTANTALITE	ORTH.	5.5-5.9	OXD	SUBMET. GRAY STRK. PRISM. VITR. METAMICT. RARE-PEGS	8.1.4.
5-5.5	ULLMANNITE	ISOM.	6.61-6.69	SLD	OPQ. STL-GRAY. CUBIC. CUBIC CLVG. UNEVN. DCMP-HNO3. NI-BEARING VNS	2.9.2.3
5-5.5	LOELLINGITE	ORTH.	7.39-7.41	SLD	OPQ. STL-GRAY. PRISMATIC, MSV. 1 CLVG. MESO-T,S-VNS + AG-AU ORES	2.9.3.1
5-5.5	NICCOLITE	HEXA.	7.78	SLD	OPQ. PALE CU-RED,GRAY TARN, BRWNISH STRK. SOL-AQUA REGIA. AS,S-VNS	2.6.5.3
5-6	PSILOMELANE	ORTH.	4.70-4.72	OXD	SUBMET. OPQ. MSV, MAMMILLARY. SOL-HCL. SCNDRYIE.G.=RESIDUAL DEPS)	6.1.8.
5-6	ILMENITE	HEXA.	4.68-4.76	OXD	SUBMET. OPQ. TBLR. VNS + DISSEM DEPS ASSOC WITH GABBROS + DIORITES	4.4.1.3
5-6	HOLLANDITE	TETR.	4.95	OXD	AG-GRAY TO BLK, BLK STRK. PRISM, FIBR. PRISM CLVG. RARE-QTZ VNS TRANSECTING MANGANESE ORE (CENTRAL INDIA)	7.7.1.2
5-6	HEMATITE	HEXA.	5.26	OXD	MET-EARTHY. STRK-RED. PLTY,GRNLR. SOL-HCL. SED FE-FMS, RARE IN VNS	4.4.1.2
5-6	PEROVSKITE	MNCL.	3.97-4.05	OXD	SUBMET. CUBIC, OCTAH, GRNLR. ADMN. 1 CLVG. DCMP-HF, H2SO4.BASIC IG	7.4.2.1
5.5	ARIZONITE (MIXTURE ?)	MNCL.	4.25	OXD	SUBMET. STL-GRAY. BRWN STRK. MSV. DCMP-HCL. RARE-PEG(25 MI SE OF HACKBERRY, ARIZONA)	8.2.1.
5.5	GERSDORFFITE	ISOM.	5.9	SLD	OPQ. AG-WHT. LK PYRITE. CUBIC CLVG. DCMP-WARM HNO3. WITH NI MINS	2.9.2.2
5.5	COBALTITE	ISOM.	6.33	SLD	OPQ. TYPICALLY REDDISH AG-WHT, ALSO VIOLET STL-GRAY + DK GRAY. HAB-IT-LIKE PYRITE. CUBIC CLVG. DCMP-HNO3. HI-T DISSEM DEPS + VEINS	2.9.2.1
5.5-6	ANATASE	TETR.	3.90	OXD	SUBMET. PALE YEL STRK. ACUTE PYRAMIDAL. ADMN. BASAL + PYRAMIDAL CLVG. CONVERTS TO RUTILE ON HEATING. VNS,ACCESS-IG RKS, DETRITAL	4.5.2.
5.5-6	BRAVOITE	ISOM.	4.28-4.66	SLD	OPQ. STL-GRAY. CRUSTS, NODULES. CONCH, CUBIC CLVG. SULFIDE VEINS	2.9.1.2
5.5-6	ARSENOPYRITE	MNCL.	6.55-6.85	SLD	OPQ. AG-WHT, STL-GRAY. PRISM. DCMP-HNO3. WDSPRD-E.G.,ORE DEPS,PEGS	2.9.5.1
5.5-6	SKUTTERUDITE	ISOM.	6.1-6.9	SLD	OPQ. SKUTTERUDITE, SMALTITE, NICKEL-SKUTTERUDITE, + CHLOANTHITE	2.10.1.1
5.5-6	SMALTITE	ISOM.	6.1-6.9	SLD	CONSTITUTE A SERIES. MEMBERS INDISTINGUISHABLE MEGASCOPICALLY	2.10.1.2
5.5-6	NICKEL SKUTTERUDITE	ISOM.	6.1-6.9	SLD	SN-WHT, AG-GRAY. TARN-GRAY OR IRID. CUBIC, OCTAH, GRNLR,CUBIC	2.10.1.3
5.5-6	CHLOANTHITE	ISOM.	6.1-6.9	SLD	+ OCTAH CLVGS. SOL-HNO3.WITH OTHER CO +NI MINS,ETC. MOD-T VNS	2.10.1.4
5.5-6.5	FERGUSONITE	TETR. C	5.38	OXD	SUBMET. IN SERIES WITH FORMANITE. PRISM. COMMONLY HEMIHEDRAL. IR-REGULAR MASSES. VITR. METAMICT. DCMP-HF. GRANITIC PEGMATITES	8.1.3.1
5.5-6.5	FORMANITE	TETR. C	7.03	OXD	SUBMET.IN SERIES WITH FERGUSONITE,WHICH SEE. PLACERS(W. AUSTRALIA)	8.1.3.2

METALLIC GRAY

HARDNESS	NAME	XL. SYS.	SPECIFIC GRAVITY	CHEM. CLASS	REMARKS	REFERENCES
6	MAGNETOPLUMBITE	HEXA.	5.52	OXD	SUBMET. OPQ. BRWN STRK. PYRAMIDAL. BASAL CLVG. MAG. SOL-HCL(SLOW). RARE-WITH MANGANOPHYLLITE (LANGBAN, SWEDEN)	7.3.3.
6	GUDMUNDITE	MNCL.	6.72	SLD	OPQ. AG-WHT, STL-GRAY. PRISM. HYDROTHERMAL-LATE IN SULFIDE DEPS	2.9.5.3
6 -6.5	BRAUNITE	TETR.	4.72-4.83	NSI	SUBMET.-OPQ. PYRAMIDAL, GRNLR. PYRAMIDAL CLVG. GELAT. MAG. WITH OTHER MN MINERALS -- VEINS, WEATHERING PRODUCT DEPOSITS, ETC.	W60 N
6 -7	CASSITERITE	TETR.	6.99	OXD	SUBMET. RADIAL CONCRETIONARY MASSES. ADMN-DULL. HI-T VNS, GREISENS	4.5.1.5
6 -7	TANTALUM	ISOM. C	16.72	ELE	OPQ. GRAYISH YELLOW. MINUTE XLS + GRAINS. IN GOLD WASHINGS	1.1.8.
6 -7	IRIDOSMINE	HEXA.	21.	ELE	OPQ. SN-WHT. FLAT GRNS, CLVG FLKS. MAL-BRTL. 1 PERF CLVG. PLACERS	1.1.6.5
6 -7	SISERSKITE	HEXA.	21.	ELE	OPQ.STL-GRAY. FLAT GRNS, CLVG FLKS. MAL-BRTL. 1 PERF CLVG.PLACERS	1.1.6.6
6 -7	PLATINIRIDIUM	ISOM.	22.84	ELE	OPQ. YELLOWISH AG-WHT, FRACTURES GRAY. SOMEWHAT MAL. PLACERS.RARE	1.1.6.3
6.5	THORIANITE	ISOM.	9.7	OXD	SUBMET. CUBIC. HORNLK. POOR BASAL CLVG. RDACTV. PLACERS, IN SERPEN-TINE OF CONTACT METAMORPHOSED ZONES	5.1.2.2
6.5-7	SCHREIBERSITE	TETR.	7.0 -7.3	ELE	OPQ.BRASS-YELLOW OR BROWN TARNISH. BRITTLE. MAG. DIFFICULTLY SOL-HCL + HNO3. METEORITES, COMBUSTION PRODUCT IN COAL MINES (FRANCE)	1.1.7.6
7	AUROSMIRIDIUM	ISOM.	20.	ELE	OPQ. AG-WHT. BRITTLE. IRREGULAR FRACTURE. WITH PT. PLACERS. RARE	1.1.6.4
7.5	LAURITE	ISOM.	6.0 -6.99	SLD	OPQ. DK GRAY STRK. SUBCONCH. OCTA CLVG. BRTL. PT-BEARING SANDS	2.9.1.3

METALLIC BLACK

HARDNESS	NAME	XL. SYS.	SPECIFIC GRAVITY	CHEM. CLASS	REMARKS	REFERENCES
1	TODOROKITE	MNCL.	3.67	OXD	AGGS OF LATHLK XLS- 2 CLVGS. SOL-HCL. RARE-ALTER AFTER INESITE	4.5.1.4
1 -1.5	NAGYAGITE	MNCL.	7.36-7.46	SLD	OPQ. RESEMBLES TETRADYMITE. SOL-HNO3 GIVING AU-RESID. AU-TE-S VNS	2.1.2.
1.5	TEALLITE	ORTH.	6.36	SST	OPQ. TARN-DULL OR IRID. TBLR-SQ, BENT PLTS. 1 CLVG. SOMEWHAT MAL. SOL-HOT HCL, HNO3. IN SILVER-TIN VEINS	3.5.11.
1.5-2	MALDONITE	ISOM. C	15.70	ELE	OPQ. PINKISH AG-WHT. CU-RED TO BLK TARN. GRNLR. VEINS + GREISEN	1.1.1.2
2	LIVINGSTONITE	MNCL.	5.00	SST	SUBTRANSLUCENT. STRK-RED. NEEDLES. ADMN. 1 PERF,2 POOR CLVGS. SOL-WARM HNO3. WITH CINNIBAR,SULFUR,GYPSUM,ETC.(E.G.,SAN LUIS,POTOSI)	3.9.1.
2	RAMDOHRITE	UNKN.	5.43	SST	OPQ. PRISM. LANCE-SHPD XLS. IN QTZ, WITH SULFIDES-POTOSI, BOLIVIA	3.6.5.
2	SCHIRMERITE	UNKN.	6.74	SST	OPQ. PB-GRAY, FE-BLK. MSV. GRNLR. RARE-QTZ VEINS (3 LOCALES, COLO).	3.5.3.
2	JOSEITE	UNKN.	8.18	SLD	MEGASCOPICALLY INDISTINGUISHABLE FROM GRUENLINGITE. SAME OCCUR	2.1.1.4
2 -2.5	DAUBREELITE	ISOM.	3.80-3.82	SLD	OPQ. MSV. SCALY, PLATY. UNEVN FRACT. 1 CLVG. SOL-HNO3. METEORITES	2.7.2.
2 -2.5	STEPHANITE	ORTH.	6.22-6.28	SST	OPQ. FE-BLK. PRISM-TBLR. OXIDIZED-HNO3. OF LATE FM IN MANY AG-DEPS	3.1.4.
2 -2.5	SELEN-TELLURIUM	HEXA.	4.8 -6.3	ELE	OPQ. BLK STRK. MASSIVE. PERF HEX PRISM CLVG. GANGUE OF AG-VEINS	1.2.2.2
2 -2.5	AIKINITE	ORTH.	7.06-7.08	SST	OPQ. PB-GRAY, TARN-BRWN, CU-RED, COMMONLY WITH YELLOWISH GREEN COATING. PRISM-ACICULAR. STRIATED. IN GOLD-QUARTZ VEINS	3.4.1.3
2 -2.5	ARGENTITE	ISOM.	7.2 -7.4	SLD	OPQ. BLKISH PB-GRAY. BRANCHING FMS, MSV. NOTABLY SECT. LO-T,S-VNS	2.3.1.1
2 -3	IANTHINITE	ORTH.		OXD	SUBMET. BRWN-VIOLET STRK. PLTY, TBLR. BASAL CLVG. ALTER AFTER URANINITE	5.3.5.
2 -3	POLYBASITE	MNCL.	6.0 -6.2	SST	FE-BLK. TBLR, MSV. STRIAE ON BASE. DCMP-HNO3. LO- TO MOD-T, AG-VNS	3.1.1.1
2 -3	PLATYNITE	HEXA.	7.98	SLD	OPQ. THIN PLTS.GOOD BASAL,FAIR RHOMB CLVGS. IN QTZ(FAHLUN,SWEDEN)	3.8.5.
2 -6.5	PYROLUSITE	TETR.	4.4 -5.08	OXD	OPQ. FIBR, GRNLR. PRISM CLVG. BOG + RESIDUAL MN-DEPS - COMMON	4.5.1.2
2.5	ACANTHITE	ORTH.	7.2 -7.3	SLD	OPQ. FE-BLK. MSV. UNEVN FRACT. SECT. WITH ARGENTITE, LO-T,S-VNS	2.2.2.3
2.5	CHALCOPHANITE	HEXA.	3.90-4.10	OXD	BRWN STRK. TBLR, DRUSES, MSV. 1 CLVG(FLEXI). SOL-HCL. SCNDRY OXIDE	7.6.1.
2.5	MIARGYRITE	MNCL.	5.20-5.30	SST	SUBTRANSLUCENT. STRK-RED. TBLR, MSV. ADMN. 3 POOR CLVGS. DCMP-HNO3. WITH AG-MINS + SULFIDES IN LO-T HYDROTHERMAL VEINS	3.5.4.
2.5	CYLINDRITE	UNKN.	5.43-5.49	SST	OPQ. MSV, CYLINDRICAL FMS. SOL-HOT HCL,HOT HNO3. WITH FRANKEITE-VN	3.8.11.
2.5	SAMSONITE	MNCL.	5.51	SST	SUBTRANSLUCENT. DK RED STRK. PRISM-STRIATED. SULFIDE DEP-IN VUGS	3.3.4.
2.5	PLAGIONITE	MNCL.	5.54-5.58	SST	OPQ. PLAGIONITE GRP. S.G.-DEFINITIVE IN GRP. STRIATED PRISM.1 CLVG. SOL-HOT HCL(H2S↑). WITH SULFOSALTS,SULFIDES.NATIVE SB-ORE DEPS	3.7.4.2
2.5	ARAMAYOITE	TRCL.	5.60	SST	SUBTRANSLUCENT. STRIATED PLTS. 1 PERF, 2 OTHER CLVGS. PLIABLE BUT INELASTIC. SECTILE. RARE-WITH PYRITE,ETC.(POTOSI,BOLIVIA)	3.5.5.
2.5	JAMESONITE	MNCL.	5.63	SST	OPQ. TARN-IRID. STRIATED FIBR. 1 GOOD, 2 POOR CLVGS. DCMP-HNO3. WITH PB-SULFOSALTS + SULFIDES IN LO- TO MOD-T HYDROTHERMAL VEINS	3.6.7.
2.5	SEMSEYITE	MNCL.	6.08	SST	OPQ. PLAGIONITE GRP. S.G.-DEFINITIVE IN GRP. TARN-DULL. PRISM. 1 CLVG. SEVERAL LOCALITIES WITH CORRODED GALENA, OTHER SULFIDES, ETC.	3.7.4.4
2.5	COCINERITE	UNKN.	6.14	SLD	OPQ. AG-GRAY, BLK TARN. MSV. WITH OXIDIZED CU-ORE. NEEDS CONFIRM	2.1.3.4

METALLIC BLACK

HARDNESS	NAME	XL. SYS.	SPECIFIC GRAVITY	CHEM. CLASS	REMARKS	REFERENCES
2.5	CANFIELDITE	ISOM.	6.28	SST	OPQ. CHEM DISTINGUISHED FROM ARGYRODITE. BLUISH-PURPLISH BLK, RED-DISH STL-GRAY. OCTAH + DODEC,MSV.RADIATING AGGS. LO-T VEINS	3.1.3.2
2.5	ARGYRODITE	ISOM.	6.26-6.29	SST	OPQ. CHEM DISTINGUISHED FROM CANFIELDITE. BLUISH-PURPLISH BLK, RED DISH STL-GRAY. OCTAH + DODEC.MSV.RADIATING AGGS. LO-T VEINS	3.1.3.1
2.5	MENEGHINITE	ORTH.	6.35-6.37	SST	OPQ. STRIATED PRISM, MSV. FIBR. 1 CLVG. OXIDIZED-HNO3. SULFIDE VNS	3.3.11.
2.5	QUENSELITE	MNCL.	6.84	OXD	SUBMET. OPQ. BRWNISH STRK. TBLR-STRIATED, 1 CLVG(FLEX). SOL-DILUTE ACIDS. RARE-WITH CALCITE + BARITE IN CREVICES (LANGBAN, SWEDEN)	7.4.1.
2.5	MATILDITE	ORTH.	6.9	SST	OPQ. STRIATED PRISM. GRNLR. SOL-HNO3. MOD- TO HI-T SULFIDE VEINS	3.5.6.
2.5	POLYARGYRITE	ISOM.	6.97	SST	OPQ. NEEDS CONFIRMATION.	3.1.2.
2.5	AGUILARITE	ISOM.	7.59	SLD	OPQ. FE-BLK. SKELETAL DODECAHEDRA, MSV. SECT. WITH ARGENTITE + AG	2.3.1.2
2.5	NAUMANNITE	ISOM. C	7.87	SLD	OPQ. FE-BLK. CUBES, GRNLR, PLTY. MAL. SECT. WITH OTHER SELENIDES	2.3.1.3
2.5	COLORADOITE	ISOM.	8.04	SLD	OPQ.MSV.GRNLR. FRIABLE-SUBCONCH. SOL-HNO3.WITH AU-+ AG-TELLURIDES	2.6.2.4
2.5	TIEMANNITE	ISOM.	8.19-8.47	SLD	OPQ.STL-GRAY, PB-GRAY. TETRAHEDRAL, MSV. UNEVEN-CONCH. VNS IN LS	2.6.2.3
2.5	DIGENITE	ISOM.	2.55-2.71	SLD	OPQ. MSV. CONCH FRACT. BRTL. SOL-HNO3. WITH OTHER COPPER SULFIDES	2.3.1.4
2.5-3	HETEROMORPHITE	MNCL.	5.73	SST	OPQ. PLAGIONITE GRP. S.G.-DEFINITIVE IN GRP. STRIATED PYRAMIDAL. 1 CLVG. WITH SPHALERITE IN CAVITIES AT STIBNITE MINES-WESTPHALIA	3.7.4.3
2.5-3	CHALCOCITE	ORTH.	5.5 -5.8	SLD	OPQ.MSV. CONCH. SECT. SOL-HNO3. ENRICHED ZONES ATOP SULFIDE DEPS	2.3.2.1
2.5-3	BOURNONITE	ORTH.	5.80-5.86	SST	OPQ. STL-GRAY. PRISM-STRIATED, SUBPARALLEL AGGS. 3 POOR CLVGS. DCMP-HNO3 TO BLUE SOLUT + S + WHITE POWDER. MOD-T SULFIDE VEINS	3.4.1.1
2.5-3	FRANCKEITE	ORTH.	5.90	SST	OPQ. TARN-IRID. MSV.TBLR-STRIATED. DCMP-HNO3,SOL-AQ-REG. AG-SN-VNS	3.6.3.
2.5-3	KOBELLITE	UNKN.	6.33	SST	OPQ. NEEDS CONFIRMATION.	3.6.2.2
2.5-3	PETZITE	UNKN.	8.7 -9.02	SLD	OPQ.STL-GRAY, FE-BLK, TARN. CUBIC CLVG. DCMP-HNO3. TELLURIDE VNS	2.3.1.9
2.5-3	SILVER	ISOM. P	10.50	ELE	OPQ. GRAY TO BLK TARN. ELONGATE FMS. MAL. DUCT. SOL-HNO3. VEINS	1.1.1.3
2.5-3.5	HULSITE	ORTH.	4.28	BOR	SUBMET-VITR. TBLR. PRISM CLVG. SOL-ACIDS. RARE-CNTCT META LS	24.1.3.
3	ENARGITE	ORTH.	4.40-4.50	SST	OPQ. TBLR, MSV, GRNLR. PRISM ZONE-STRIATED. 1 POOR, 3 GOOD CLVGS. SOL-AQ-REG. MOD-T TO LO-T VEIN + REPLACEMENT SULFIDE DEPOSITS	3.3.2.2
3	KLOCKMANNITE	HEXA.	>5.	SLD	SLATE-GRAY, TARN BLU-BLK. GRNLR. 1 PERF CLVG. WITH OTHER SELENIDES	2.6.8.2
3	STYLOTYPITE	MNCL.	4.79-5.18	SST	OPQ. PRISM-SQ. TWINNED-INDIVIDUALS AT 90°. RARE(COPIAPO, CHILE)	3.2.2.2
3	SELIGMANNITE	ORTH.	5.38-5.44	SST	OPQ. STRK-CHOCOLATE-BRWN TO PURPLE-BLK. EQUANT. CAVITIES IN DOLO	3.4.1.2
3	WEISSITE	UNKN.	6.	SLD	OPQ.BLUISH BLK, BLK TARN, BLK STRK. MSV. WITH PYRITE + TELLURIDES	2.5.3.
3	PEARCEITE	MNCL.	6.13-6.17	SST	OPQ. SHORT TBLR 6-SIDED PRISMS. STRIAE ON BASE. DCMP-HNO3. LO- TO MODERATE-TEMPERATURE SILVER VEINS	3.1.1.2
3	METACINNABAR	ISOM.	7.65	SLD	OPQ.GRAY-BLK. BLK STRK. TETRA-ROUGH FACES. MSV. WITH CINNABAR,VNS	2.6.2.2
3 -3.5	LAUTITE	ORTH.	4.8 -5.0	SLD	SUBMET. OPQ. BLK-GRAY(RED TINGE).SOL-HNO3. VNS-WITH NATIVE AS,ETC.	2.9.5.5

METALLIC BLACK

H		Name	Crystal	SG	Class	Description	Ref
3	-4.5	TENNANTITE	ISOM.	4.62	SST	SOFTER + LIGHTER THAN TETRAHEDRITE. TETRA. MSV. DCMP-HNO3. VNS	3.2.4.2
3	-4.5	TETRAHEDRITE	ISOM.	4.99	SST	HARDER + HEAVIER THAN TENNANTITE. TETRA. MSV. DCMP-HNO3,WDSPRD-VNS	3.2.4.1
3.5		HEMATOLITE	HEXA.	3.49	ARS	SUBMET.,TBLR,RHOMB,STRIATED, VITR. 1 CLVG. SOL-ACIDS. WITH BARITE, JACOBSITE, ETC. IN VNS IN LIMESTONE (NORDMARK, SWEDEN)	41.1.3.
3.5		KALKOWSKITE	UNKN.	3.98-4.04	OXD	SUBMET. REDDISH STRK. FIBR PLTS.WAXY. DCMP-HCL.RARE-SCHIST(BRAZIL)	8.2.2.
3.5		EPIGENITE	ORTH.	4.5	SST	OPQ. STL-GRAY. TARN-BLK + BLUE. INCRUST. SOL-HNO3. VNS-ON BARITE	3.1.5.
3.5		TENORITE	MNCL.	5.8 -6.4	OXD	BRWN-TRANSMITTED LIGHT. PLTY, EARTHY-MSV. SOL-DILUTE HCL OR HNO3. OXIDIZED PARTS OF CU-DEPS, AS SUBLIMATED XLS ON LAVAS	4.2.3.
3.5-4		WARWICKITE	ORTH.	3.34-3.36	BOR	SUBMET-PEARLY. BLUE-BLK STRK. SLENDER PRISMS, ROUNDED ENDS. DCMP-H2SO4. RARE - CONTACT METAMORPHOSED LIMESTONE	24.1.4.
3.5-4		ALABANDITE	ISOM.	3.9 -4.04	SLD	SUBMET.GRN STRK, TARN-BRWN. PERF CUBIC CLVG. SOL-HCL. SULFIDE VNS	2.6.1.4
3.5-4		SPHALERITE	ISOM.	3.9 -4.1	SLD	RESIN. CURVED XL FACES COMMON. FLUO-OR. TRIBOLUM. SOL-HCL. VEINS	2.6.2.1
3.5-4		CUPRITE	ISOM.	6.14	OXD	SUBMET(SOME). OCTAH, HAIRLK, MSV. COMMON-OXIDIZED ZONES OF CU-DEPS	4.1.1.
4		HAUERITE	ISOM.	3.46	SLD	REDDISH BRWN-BLK. SUBCONCH. CUBIC CLVG. SOL-HCL. WITH GYP. SULFUR	2.9.1.5
4		MANGANITE	MNCL.	4.32-4.34	OXD	SUBMET. RED-BRWN STRK. STRIATED PRISMS. 3 CLVGS.SOL-HCL. LO-T VNS, SECONDARY, ETC.	6.1.3.
4		STANNITE	TETR.	4.3 -4.5	SLD	OPQ. STL-GRAY, FE-BLK. MOST XLS STRIATED. DCMP-HNO3.SN-BEARING VNS	2.6.3.2
4		CREDNERITE	MNCL.	4.99-5.03	OXD	OPQ. BRWN-BLK STRK. PLTY, RADIATING GRPS. 1 CLVG. SOL-HCL. SCNDRY	7.2.4.
4		IRON	ISOM.	7.3 -7.87	ELE	OPQ.STL-GRAY, FE-BLK. MAL. MAG. SOL-HCL + CH3COOH.METEOR + BASALT	1.1.7.1
4	-4.5	WOLFRAMITE	MNCL.	7.31	WOS	SUBMET. TARN-IRID. PRISM-EQUANT--STRIATED.MSV-GRNLR,ETC. 1 CLVG. DCMP-AQ-REG,H2SO4 OR HCL(SLOWLY).DIVERSE-VNS,CONTACT META,PLACERS	48.1.1.2
4	-4.5	PLATINUM	ISOM.	19.	ELE	OPQ. GRAINS + SCALES. MAL, DUCT. ULTRABASIC IGNEOUS RKS + PLACERS	1.1.6.1
4	-5	STAINIERITE	UNKN.	4.13-4.47	OXD	SUBMET. OPQ. MAMMILLARY. UNEVN. SOL-HCL. OXIDATION PROD OF CO-MINS	6.1.4.
4	-5	BERTHONITE	UNKN.	5.49	SST	OPQ. MSV, GRNLR. DCMP-HNO3. RARE-WITH GALENA IN VNS, FE-DEP(TUNIS)	3.4.2.
4	-5.5	BETAFITE	ISOM.	3.7 -5.	OXD	SUBMET. OCTAH. WAXY. CONCH. METAMICT. DCMP-ACIDS. GRANITIC PEGS	8.4.1.
4.5		PARAMELACONITE	TETR.	6.04	OXD	OPQ. STRK-BRWNISH. STRIATED PRISM. ADMN. SOL-DILUTE ACIDS, DILUTE NH4OH + NH4CL SOLS. RARE-SCNDRY WITH CUPRITE,ETC.-BISBEE,ARIZONA	4.2.4.
4.5		FERBERITE	MNCL.	7.51	WOS	SUBMET. TARN-IRID. ELONGATE-STRIATED,WEDGE-SHAPED,BLDS.MSV.1 CLVG. DCMP-AQ-REG,H2SO4 OR HCL(SLOWLY).DIVERSE-VNS, CNTCT META,PLACERS	48.1.1.3
4.5-5		CORONADITE	TETR.	5.44	OXD	SUBMET. OPQ. BRWN-GRAY STRK. MSV-BOTRYOIDAL. OXIDIZED ZONE-MN DEPS	7.7.1.1
4.5-5		WOLFACHITE	ORTH.	6.37	SLD	OPQ. AG- TO SN-WHT. BLK STRK. COLUMNAR RADIATING AGGS. DCMP-HNO3. RARE - AT WOLFACH(BADEN) ON NICCOLITE IN CALCITE	2.9.5.4
5		LORANSKITE	UNKN.	4.6	OXD	SUBMET. RESEMBLES SAMARSKITE. WITH WIIKITE-PEG(PITKARANTA,FINLAND) ILL-DEFINED	8.1.7.
5		TREVORITE	ISOM.	5.16	OXD	SUBMET. MAGNETITE SERIES OF SPINEL GROUP. BRWN STRK,GRNLR, MSV. DIFFICULTLY SOL-HCL. STRONGLY MAGNETIC.RARE-NI DEPS, TRANSVAAL	7.2.1.9
5		GLAUCODOT	ORTH.	6.28-6.52	SLD	OPQ. GRAYISH SN-WHT. REDDISH AG-WHT. BLK STRK. PRISM. STRIATED ON PRISM ZONE FACES. DCMP-HNO3(PINK SOLUTION). WITH COBALTITE	2.9.5.2

METALLIC BLACK

HARDNESS	NAME	XL. SYS.	SPECIFIC GRAVITY	CHEM. CLASS	REMARKS	REFERENCES
5	BISMUTOTANTALITE	ORTH.	8.26	OXD	SUBMET. EXPOS-PINKISH YEL. PRISM. RARE-PEGMATITE(SW UGANDA)	8.1.9.
5 -5.5	GOETHITE	ORTH.	3.3 -4.29	OXD	SUBMET. YELLOWISH STRK. MSV, FIBR, ETC. SILKY-ADMN. SOL-HCL. WXING PRODUCT	7.1.2.2
5 -5.5	NICCOLITE	HEXA.	7.78	SLD	OPQ. PALE CU-RED,GRAY TARN, BRWNISH STRK. SOL-AQUA REGIA. AS,S-VNS	2.6.5.3
5 -6	GEIKIELITE	HEXA.	4.05	OXD	SUBMET.BRWN-RED STRK. TBLR. RHOMB CLVG. RARE-GEM GRAVELS(CEYLON)	4.4.1.4
5 -6	HYDROHETAEROLITE	TETR.	4.6	OXD	SUBMET. BOTRYOIDAL, FIBR. CLVG // FIBERS. SOL-HCL. WITH CHALCOPHAN-ITE, ETC.	7.2.2.3
5 -6	PSILOMELANE	ORTH.	4.70-4.72	OXD	SUBMET. OPQ. MSV, MAMMILLARY. SOL-HCL. SCNDRY(E.G.,RESIDUAL DEPS)	6.1.8.
5 -6	ILMENITE	HEXA.	4.68-4.76	OXD	SUBMET. OPQ. TBLR. VNS + DISSEM DEPS ASSOC WITH GABBROS + DIORITES	4.4.1.3
5 -6	HOLLANDITE	TETR.	4.95	OXD	AG-GRAY TO BLK. BLK STRK. PRISM, FIBR, PRISM CLVG. RARE-QTZ VNS TRANSECTING MANGANESE ORE (CENTRAL INDIA)	7.7.1.2
5 -6	PRIORITE	ORTH.	4.85-5.05	OXD	SUBMET. IN SERIES WITH ESCHYNITE(SEE). RED-YEL STRK. GRANITIC PEGS	8.3.5.2
5 -6	ESCHYNITE	ORTH.	5.14-5.24	OXD	SUBMET. IN SERIES WITH PRIORITE. BLK-BRWN STRK. PRISM, MSV. WAXY.	8.3.5.1
5 -6	"SAMARSKITE"	UNKN.	5.67	OXD	SUBMET. DISTINGUISHED FROM SAMARSKITE ONLY BY X-RAY ANALYSIS	8.3.6.2
5 -6	SAMARSKITE	ORTH.	5.69	OXD	SUBMET. RED-BRWN TO BLK STRK. PRISM, MSV. VITR. RDACTV. PEGS	8.3.6.1
5 -6	URANINITE	ISOM.	10.63	OXD	SUBMET. OPQ. OCTAH,ETC., MSV.GRSY-DULL. RDACTV. PEGS, HI+MOD-T VNS	5.1.2.1
5.5	PEROVSKITE	MNCL.	3.97-4.05	OXD	SUBMET. CUBIC, OCTAH, GRNLR. ADMN. 1 CLVG. DCMP-HF, H2SO4.BASIC IG	7.4.2.1
5.5	UHLIGITE	MNCL.	4.05-4.25	OXD	BRWN-GRAY STRK. OCTAH-STRIATED. 1 CLVG.POSSIBLY VAR OF PEROVSKITE	7.4.2.2
5.5	MAGNESIOCHROMITE	ISOM.	4.1 -4.3	OXD	CHROMITE SERIES OF SPINEL GROUP. MSV., GRNLR. OCTAH. BRWN STRK. MAG (FEEBLY). WITH OLIVINE-RICH(COMMONLY SERPENTINIZED) IGNEOUS ROCKS	7.2.1.11
5.5	CATOPTRITE	MNCL.	4.5	ANT	SUBTRANSLUCENT. TBLR,EQUANT,ANHEDRA. WITH MAGNETITE IN XLINE LS-RARE	44.2.3.
5.5	CHROMITE	ISOM.	4.5 -4.8	OXD	CHROMITE SERIES OF SPINEL GROUP. MSV., GRNLR. MAG(FEEBLY). WITH OLIVINE-RICH(COMMONLY SERPENTINIZED)IGNEOUS ROCKS	7.2.1.12
5.5	HAUSMANNITE	TETR.	4.83-4.85	OXD	SUBMET. PSEUDO-OCTAH. MSV. BASAL CLVG. SOL-HOT HCL. HI-T MN DEPS	7.2.2.1
5.5	DELAFOSSITE	HEXA.	5.41	OXD	OPQ. TBLR, BOTRYOIDAL. RHOMB CLVG. SOL-HCL, HNO3. SCNDRY- CU-DEPS	7.1.1.
5.5	GERSDORFFITE	ISOM.	5.9	SLD	OPQ. AG-WHT. LK PYRITE. CUBIC CLVG. DCMP-WARM HNO3. WITH NI MINS	2.9.2.2
5.5	PLATTNERITE	TETR.	9.40-9.44	OXD	OPQ. BRWN STRK. PRISM, NODULAR. ADMN, TARN-DULL. SOL-HCL. WITH PY-ROMORPHITE	4.5.1.6
5.5 -6	ANATASE	TETR.	3.90	OXD	SUBMET. PALE YEL STRK. ACUTE PYRAMIDAL. ADMN. BASAL + PYRAMIDAL CLVG. CONVERTS TO RUTILE ON HEATING. VNS,ACCESS-IG RKS, DETRITAL	4.5.2.
5.5 -6	YTTROCRASITE	ORTH.	4.80	OXD	BLK. EXPOS-BRWN. XLS. UNEVN. RDACTV. DCMP-HF.RARE-PEGILLAND,TEXAS)	8.3.4.
5.5 -6.5	MAGNESIOFERRITE	ISOM.	4.56-4.65	OXD	SUBMET. MAGNETITE SERIES OF SPINEL GROUP. BLK STRK. GRNLR, MSV. DIFFICULTLY SOL-HCL. STRONGLY MAGNETIC. FUMAROLE DEPOSITS	7.2.1.5
5.5 -6.5	JACOBSITE	ISOM.	4.76	OXD	SUBMET. MAGNETITE SERIES OF SPINEL GROUP. BRWN STRK.GRNLR, MSV. SOL-HCL.WEAKLY MAGNETIC. MN DEPS, E.G., LANGBAN, SWEDEN	7.2.1.8
5.5 -6.5	MAGNETITE	ISOM.	5.18	OXD	SUBMET. MAGNETITE SERIES OF SPINEL GROUP. BLK STRK. OCTAH, GRNLR. OCTAH PARTING. DIFFICULTLY SOL-HCL. STRONGLY MAGNETIC. WIDESPREAD	7.2.1.6

METALLIC BLACK

Hardness	Mineral	System	S.G.	Class	Description	Ref.
5.5-6.5	FRANKLINITE	ISOM.	5.07-5.22	OXD	SUBMET, MAGNETITE SERIES OF SPINEL GROUP. RED-BRWN STRK. OCTAH. OCTAH PARTING. SOL-HCL. WEAKLY MAGNETIC. RARE-FRANKLIN, N.J.	7.2.1.7
5.5-6.5	FERGUSONITE	TETR.	C 5.38	OXD	SUBMET. IN SERIES WITH FORMANITE, PRISM, COMMONLY HEMIHEDRAL, IR-REGULAR MASSES. VITR. METAMICT. DCMP-HF. GRANITIC PEGMATITES	8.1.3.1
5.5-6.5	EUXENITE	ORTH.	4.9-5.9	OXD	SUBMET. IN SERIES WITH POLYCRASE. STRK YEL,GRAY,RED-BRWN. PRISM, MSV.GRSY.CONCH. DCMP-HOT HCL,HF,H2SO4. GRANITIC PEGMATITES	8.3.3.1
5.5-6.5	POLYCRASE	ORTH.	4.9-5.9	OXD	SUBMET. IN SERIES WITH EUXENITE, WHICH SEE. PEGMATITES	8.3.3.2
5.5-6.5	FORMANITE	TETR.	C 7.03	OXD	SUBMET. IN SERIES WITH FERGUSONITE,WHICH SEE. PLACERS(W. AUSTRALIA)	8.1.3.2
6	PINAKIOLITE	ORTH.	3.88	BOR	SUBTRANSLUCENT.BRWNISH STRK. RECTANGULAR TBLR-SOME BENT OR BROKEN. SOL-HCL. RARE-LANGBAN, SWEDEN	24.1.2.
6	PSEUDOBROOKITE	ORTH.	4.33-4.39	OXD	SUBMET. TBLR-STRIATED. GRSY. 1 CLVG. VOLCANIC AREAS FUMAROLIC,ETC.	7.5.1.
6	HETAEROLITE	TETR.	5.18	OXD	SUBMET. BRWN STRK. PSEUDO-OCTAH. MSV. BASAL CLVG. RARE-FRNKLN,N.J.	7.2.2.2
6	COLUMBITE	ORTH.	5.15-5.25	OXD	SUBMET. IN SERIES WITH TANTALITE. RED-BLK STRK. TARN-IRID. TBLR. PRISM. 2 CLEAVAGES. BRITTLE. GRANITIC PEGMATITES	8.3.2.1
6	MAGNETOPLUMBITE	HEXA.	5.52	OXD	SUBMET. OPQ. BRWN STRK. PYRAMIDAL. BASAL CLVG. MAG. SOL-HCL(SLOW). RARE-WITH MANGANOPHYLLITE (LANGBAN, SWEDEN)	7.3.3.
6 -6.5	RUTILE	TETR.	4.21-4.25	OXD	SUBMET. PRISM-STRIATED.ADMN.POOR CLVGS. VNS, ACCESS-META + IG RKS	4.5.1.1
6 -6.5	BRAUNITE	TETR.	4.72-4.83	NSI	SUBMET. OPQ. PYRAMIDAL, GRNLR. PYRAMIDAL CLVG. GELAT. MAG. WITH OTHER MN MINERALS -- VEINS, WEATHERING PRODUCT DEPOSITS, ETC.	W60 N
6 -6.5	BIXBYITE	ISOM.	4.95	OXD	SUBMET. OPQ. CUBIC. OCTAH CLVG. SOL-HCL. IN CAVITIES IN RHYOLITE	4.4.5.
6 -6.5	SENAITE	HEXA.	5.30	OXD	SUBMET. BRWN-BLK STRK. ROUGH XLS. CONCH. DCMP-HF, H2SO4. PLACERS	4.4.1.6
6 -6.5	TAPIOLITE	TETR.	7.85-7.95	OXD	SUBMET. STRK-BRWNISH. PRISM. UNEVN. PEGMATITES, PLACERS	8.3.1.1
6 -6.5	MOSSITE	TETR.	7.85-7.95	OXD	SUBMET. POSSIBLE CB END-MEMBER OF TAPIOLITE SERIES(SEE). RARE-PEGS	8.3.1.2
6 -6.5	TANTALITE	ORTH.	7.90-8.00	OXD	SUBMET. IN SERIES WITH COLUMBITE, WHICH SEE. PEGMATITES	8.3.2.2
6 -7	CASSITERITE	TETR.	6.99	OXD	SUBMET. RADIAL CONCRETIONARY MASSES. ADMN-DULL. HI-T VNS, GREISENS	4.5.1.5
6.5	HOEGBOMITE	HEXA.	3.81	OXD	TBLR. BASAL CLVG. CONCH. MAG(SLIGHTLY).WITH CORUNDUM IN EMERY DEPS	7.3.1.1
6.5	POLYMIGNYTE	ORTH.	4.77-4.85	OXD	SUBMET. BRWN STRK. PRISM-STRIATED. 2 CLVGS. ALKALIC PEGMATITES	8.1.5.
6.5	THORIANITE	ISOM.	9.7	OXD	SUBMET. CUBIC. HORNLK. POOR BASAL CLVG. RDAGTV. PLACERS, IN SERPEN-TINE OF CONTACT METAMORPHOSED ZONES	5.1.2.2
7.5	LAURITE	ISOM.	6.0 -6.99	SLD	SUBMET. OPQ. DK GRAY STRK. SUBCONCH. OCTA CLVG. BRTL. PT-BEARING SANDS	2.9.1.3
9.5	MOISSANITE	HEXA.	3.1 -3.21	ELE	SUBMET. TABULAR. CONCHOIDAL FRACT. METEOR. (ARTIF=CARBORUNDUM)	1.1.7.4

METALLIC BROWN

HARDNESS	NAME	XL. SYS.	SPECIFIC GRAVITY	CHEM. CLASS	REMARKS	REFERENCES
1 -1.5	STERNBERGITE	ORTH.	4.10-4.22	SLD	OPQ.VIOLET-BLU TARN, BLK STRK, PLTS, ROSETTES.1 PERF CLVG, AG-ORES	2.6.7.
1 -1.5	MELONITE	HEXA.	7.35	SLD	OPQ. REDDISH WHT, TARN-BRWN. PLTS. BASAL CLVG. SOL-HNO3. LO-T VNS	2.9.8.
1.5-2	ORPIMENT	MNCL.	3.49	SLD	FOL, FIBR, GRNLR. RESIN. 1 CLVG-INELAST. LO-T VNS, ALTER PROD.	2.8.1.
2	ALASKAITE (MIXTURE ?)	UNKN.	6.78-6.88	SST	OPQ. TARN-BRONZY. MSV-FOL. SOL-HOT HCL.WITH CHALCOPYRITE-QTZ,S-VNS	3.8.7.
2 -2.5	PARKERITE	MNCL.		SLD	OPQ. NEEDS CONFIKMATION.	2.9.9.
2 -2.5	CINNABAR	HEXA.	8.09	SLD	SCARLET STRK. ADMN-DULL. NEAR VOLS + HOT SPRINGS-VNS + INCRUSTS	2.6.9.
2 -3	HEMATOPHANITE	TETR.	7.70	OXD	SUBMET. YEL-RED STRK. TBLR, LAMELLAR AGGS. 1 CLVG. SOL-HCL, HNO3 RARE -- WITH PLUMBOFERRITE, ETC. (VERMLAND, SWEDEN)	7.3.4.
2.5	FULOPPITE	MNCL.	5.23	SST	OPQ. PLAGIONITE GRP. S.G.-DEFINITIVE IN GRP. TARN-STL BLUE, BRONZY. STRK-REDDISH GRAY. STRIATED PRISM.RARE--WITH SPHAL(NAGYAG,ROUMANIA)	3.7.4.1
2.5-3	COPPER	ISOM. C	8.94	ELE	OPQ.BRANCHING XLS. MAL, DUCT. SOL-HNO3. BASALTIC VOLS. VEINS	1.1.1.4
3	ASTROPHYLLITE	TRCL.	3.3 -3.4	SSI	SUBMET-PEARLY. BRONZY-GLDN YEL. CLVG-BRTL FLKS. DCMP-HCL. ALK RKS	W480 N
3	BORNITE	ISOM.	5.06-5.08	SLD	OPQ.PURPLISH IRID TARN. GRAY-BLK STRK. MSV. SOL-HNO3. CU-DEPS	2.4.3.
3 -3.5	EMPRESSITE	UNKN.	7.51	SLD	OPQ.BRONZY, GRAY-BLK STRK. MSV. FRIABLE. SOL-HNO3.WITH GALENA + TE	2.6.14.1
3 -4	COLUSITE	ISOM.	4.50	SST	OPQ. BRONZE. BLK STRK. MSV. WITH PYRITE, CU-MINS,ETC.-SULFIDE VNS	3.3.1.3
3.5	HEMATOLITE	HEXA.	3.49	ARS	SUBMET. TBLR,RHOMB,STRIATED. VITR. 1 CLVG. SOL-ACIDS. WITH BARITE, JACOBSITE,ETC. IN VNS IN LIMESTONE (NORDMARK, SWEDEN)	41.1.3.
3.5	SULVANITE	ISOM.	3.86-4.00	SST	OPQ. BRONZE-YEL. BLK STRK. CUBIC,MSV. WITH MALACHITE,ETC.-ORE DEPS	3.3.1.1
3.5	KALKOWSKITE	UNKN.	3.98-4.04	OXD	SUBMET. REDDISH STRK. FIBR PLTS.WAXY. DCMP-HCL.RARE-SCHIST(BRAZIL)	8.2.2.
3.5-4	WARWICKITE	ORTH.	3.34-3.36	BOR	SUBMET-PEARLY. BLUE-BLK STRK. SLENDER PRISMS, ROUNDED ENDS. DCMP-H2SO4. RARE -- CONTACT METAMORPHOSED LIMESTONE	24.1.4.
3.5-4	ALABANDITE	ISOM.	3.9 -4.04	SLD	SUBMET.GRN STRK, TARN-BRWN. PERF CUBIC CLVG. SOL-HCL. SULFIDE VNS	2.6.1.4
3.5-4	SPHALERITE	ISOM.	3.9 -4.1	SLD	RESIN. CURVED XL FACES COMMON. FLUO-OR. TRIBOLUM. SOL-HCL. VEINS	2.6.2.1
3.5-4.5	PYRRHOTITE	HEXA.	4.58-4.79	SLD	OPQ. BRONZY, TARN. MSV. UNEVN. MAG. DCMP-HCL. BASALTS. S-VEINS	2.6.5.1
4	OLDHAMITE	ISOM.	2.58	SLD	SMALL SPHERULES. CUBIC CLVG. SOL-HCL, DCMP-BOILING H2O. METEORIC	2.6.1.5
4	HAUERITE	ISOM.	3.46	SLD	REDDISH BRWN-BLK. SUBCONCH. CUBIC CLVG. SOL-HCL. WITH GYP, SULFUR	2.9.1.5
4	HUEBNERITE	MNCL.	7.12	WOS	SUBMET. TARN-IRID. PRISM-STRIATED. –RADIATING XL GRPS. 1 CLVG. DCMP-AQ-REG,H2SO4 OR HCL(SLOWLY). DIVERSE-VNS,CNTCT META, PLACERS	48.1.1.1
4 -4.5	WOLFRAMITE	MNCL.	7.31	WOS	SUBMET. TARN-IRID. PRISM-EQUANT--STRIATED,MSV-GRNLR,ETC. 1 CLVG. DCMP-AQ-REG,H2SO4 OR HCL(SLOWLY). DIVERSE-VNS,CNTCT META, PLACERS	48.1.1.2
4 -5.5	BETAFITE	ISOM.	3.7 -5.	OXD	SUBMET. OCTAH. WAXY. CONCH. METAMICT. DCMP-ACIDS. GRANITIC PEGS	8.4.1.

METALLIC BROWN

H	Mineral	System	G	Streak/Color	Remarks	Ref
5	LEPIDOCROCITE	ORTH.	4.05-4.13	OXD	SUBMET. ORANGE STRK. SCALY, FIBR MSV. 3 CLVGS. WITH GOETHITE, ETC.	6.1.2.1
5	TREVORITE	ISOM.	5.16	OXD	SUBMET. MAGNETITE SERIES OF SPINEL GROUP. BRWN STRK.GRNLR, MSV. DIFFICULTY SOL-HCL. STRONGLY MAGNETIC. RARE-NI DEPS,TRANSVAAL	7.2.1.9
5 -5.5	GOETHITE	ORTH.	3.3 -4.29	OXD	SUBMET. YELLOWISH STRK. MSV, FIBR, ETC. SILKY-ADMN. SOL-HCL. WXING PRODUCT	7.1.2.2
5 -6	BRONZITE (ORTHOPYROXENE)	ORTH.	3.3 -3.45	ISI	SUBMET.(BRONZY). FS12-30. IN BASIC + ULTRABSIC IG RKS, SOME METAS	W405 2-9
5 -6	GEIKIELITE	HEXA.	4.05	OXD	SUBMET.BRWN-RED STRK. TBLR. RHOMB CLVG. RAKE-GEM GRAVELS(CEYLON)	4.4.1.4
5 -6	HYDROHETAEROLITE	TETR.	4.6	OXD	SUBMET. BOTRYOIDAL, FIBR. CLVG FIBERS. SOL-HCL. WITH CHALCOPHAN-ITE, ETC.	7.2.2.3
5 -6	PRIORITE	ORTH.	4.85-5.05	OXD	SUBMET. IN SERIES WITH ESCHYNITE(SEE). RED-YEL STRK. GRANITIC PEGS	8.3.5.2
5 -6	ESCHYNITE	ORTH.	5.14-5.24	OXD	SUBMET. IN SERIES WITH PRIORITE. BLK-BRWN STRK. PRISM, MSV. WAXY,	8.3.5.1
5 -6	"SAMARSKITE"	UNKN.	5.67	OXD	SUBMET. DISTINGUISHED FROM SAMARSKITE ONLY BY X-RAY ANALYSIS	8.3.6.2
5 -6	SAMARSKITE	ORTH.	5.69	OXD	SUBMET. RED-BRWN TO BLK STRK. PRISM, MSV. VITR. RDACTV. PEGS	8.3.6.1
5 -6	URANINITE	ISOM.	10.63	OXD	SUBMET. OPQ. OCTAH,ETC. MSV.GRSY-DULL. RDACTV. PEGS, HI+MOD-T VNS	5.1.2.1
5.5	PEROVSKITE	MNCL.	3.97-4.05	OXD	SUBMET. CUBIC, OCTAH, GRNLR, ADMN. 1 CLVG. DCMP-HF, H2SO4,BASIC IG	7.4.2.1
5.5	MAGNESIOCHROMITE	ISOM.	4.1 -4.3	OXD	CHROMITE SERIES OF SPINEL GROUP. MSV, GRNLR, OCTAH. BRWN STRK. MAG (FEEBLY), WITH OLIVINE-RICH(COMMONLY SERPENTINIZED) IGNEOUS ROCKS	7.2.1.11
5.5	HAUSMANNITE	TETR.	4.83-4.85	OXD	SUBMET. PSEUDO-OCTAH, MSV. BASAL CLVG. SOL-HOT HCL. HI-T MN DEPS	7.2.2.1
5.5	PLATTNERITE	TETR.	9.40-9.44	OXD	OPQ. BRWN STRK. PRISM, NODULAR, ADMN, TARN-DULL. SOL-HCL. WITH PY- ROMORPHITE	4.5.1.6
5.5-6	ANATASE	TETR.	3.90	OXD	SUBMET. PALE YEL STRK. ACUTE PYRAMIDAL, ADMN. BASAL + PYRAMIDAL CLVG. CONVERTS TO RUTILE ON HEATING. VNS, ACCESS - IG RKS,DETRITAL	4.5.2.
5.5-6	YTTROCRASITE	ORTH.	4.80	OXD	BLK. EXPOS-BRWN. XLS, UNEVN. RDACTV. DCMP-HF.RARE-PEGS(LLANO,TEXAS)	8.3.4.
5.5-6.5	MAGNESIOFERRITE	ISOM.	4.56-4.65	OXD	SUBMET. MAGNETITE SERIES OF SPINEL GROUP. BLK STRK; GRNLR, MSV. DIFFICULTLY SOL-HCL. STRONGLY MAGNETIC. WIDESPREAD	7.2.1.5
5.5-6.5	JACOBSITE	ISOM.	4.76	OXD	SUBMET. MAGNETITE SERIES OF SPINEL GROUP. BRWN STRK, GRNLR, MSV. SOL-HCL. WEAKLY MAGNETIC. MN DEPS, E.G., LANGBAN, SWEDEN	7.2.1.8
5.5-6.5	MAGNETITE	ISOM.	5.18	OXD	SUBMET. MAGNETITE SERIES OF SPINEL GROUP. BLK STRK. OCTAH, GRNLR, OCTAH PARTING. DIFFICULTLY SOL-HCL. STRONGLY MAGNETIC. WIDESPREAD	7.2.1.6
5.5-6.5	FRANKLINITE	ISOM.	5.07-5.22	OXD	SUBMET. MAGNETITE SERIES OF SPINEL GROUP. RED-BRWN STRK. OCTAH. OCTAH PARTING. SOL-HCL. WEAKLY MAGNETIC. RARE-FRANKLIN, N.J.	7.2.1.7
5.5-6.5	FERGUSONITE	TETR. C	5.38	OXD	SUBMET. IN SERIES WITH FORMANITE. PRISM, COMMONLY HEMIHEDRAL, IR-REGULAR MASSES. VITR. METAMICT. DCMP-HF. GRANITIC PEGMATITES	8.1.3.1
5.5-6.5	EUXENITE	ORTH.	4.9 -5.9	OXD	SUBMET. IN SERIES WITH POLYCRASE. STRK YEL,GRAY,RED-BRWN. PRISM, MSV.GRSY,CONCH. DCMP-HOT HCL,HF,H2SO4. GRANITIC PEGMATITES	8.3.3.1
5.5-6.5	POLYCRASE	ORTH.	4.9 -5.9	OXD	SUBMET. IN SERIES WITH EUXENITE, WHICH SEE. PEGMATITES	8.3.3.2
5.5-6.5	FORMANITE	TETR. C	7.03	OXD	SUBMET. IN SERIES WITH FERGUSONITE, GRSY. 1 CLVG. VOLCANIC AREAS FUMAROLIC,ETC. PLACERS(W. AUSTRALIA)	8.1.3.2
6	PSEUDOBROOKITE	ORTH.	4.33-4.39	OXD	SUBMET. TBLR-STRIATED. GRSY. 1 CLVG. VOLCANIC AREAS FUMAROLIC,ETC.	7.5.1.

METALLIC BROWN

HARDNESS	NAME	XL. SYS.	SPECIFIC GRAVITY	CHEM. CLASS	REMARKS	REFERENCES
6	COLUMBITE	ORTH.	5.15-5.25	OXD	SUBMET. IN SERIES WITH TANTALITE. RED-BLK STRK. TARN-IRID. TBLR. PRISMATIC. 2 CLEAVAGES. BRITTLE. GRANITIC PEGMATITES	8.3.2.1
6 -6.5	RUTILE	TETR.	4.21-4.25	OXD	SUBMET. PRISM-STRIATED. ADMN. POOR CLVGS. VNS. ACCESS-META + IG RKS	4.5.1.1
6 -6.5	BRAUNITE	TETR.	4.72-4.83	NSI	SUBMET. OPQ. PYRAMIDAL, GRNLR. PYRAMIDAL CLVG. GELAT. MAG. WITH OTHER MN MINERALS -- VEINS, WEATHERING PRODUCT DEPOSITS, ETC.	W60 N
6 -6.5	TAPIOLITE	TETR.	7.85-7.95	OXD	SUBMET. STRK-BRWNISH. PRISM. UNEVN. PEGMATITES, PLACERS	8.3.1.1
6 -6.5	MOSSITE	TETR.	7.85-7.95	OXD	SUBMET. POSSIBLE CB END-MEMBER OF TAPIOLITE SERIES(SEE). RARE-PEGS	8.3.1.2
6 -6.5	TANTALITE	ORTH.	7.90-8.00	OXD	SUBMET. IN SERIES WITH COLUMBITE, WHICH SEE. PEGMATITES	8.3.2.2
6 -7	CASSITERITE	TETR.	6.99	OXD	SUBMET. RADIAL CONCRETIONARY MASSES. ADMN-DULL. HI-T VNS, GREISENS	4.5.1.5
6.5-7	SCHREIBERSITE	TETR.	7.0 -7.3	ELE	OPQ. BRASS-YELLOW OR BROWN TARNISH. BRITTLE. MAG. DIFFICULTLY SOL-HCL + HNO3. METEORITES. COMBUSTION PRODUCT IN COAL MINES (FRANCE)	1.1.7.6

Table II.

MINERALS

with

NONMETALLIC LUSTER

NONMETALLIC RED

HARDNESS	NAME	XL. SYS.	SPECIFIC GRAVITY	CHEM. CLASS	REMARKS	REFERENCES
1	MOLYSITE	HEXA. A	2.90	HAL	MSV. BASAL CLVG. DLQSNT. VOL SUBLIMATE	9.3.1.
1	SCACCHITE	HEXA.	2.98	HAL	MSV. BASAL CLVG. SOL-H2O. DLQSNT. WITH OTHER SALTS-VOL SUBLIMATE	9.2.3.2
1	-1.5	KERNESITE	MNCL.	4.68	SLD	LATHS, HAIRLK TUFTS. ADMN. 2 CLVGS. SECT. ALTER PROD OF STIBNITE	2.8.3.
1	-2	KOENENITE	HEXA.	1.98	HAL	CRUSTS, PEARLY. BASAL CLVG(FLEX). DCMP-HOT H2O. K-DEPSIN. GERMANY)	10.2.8.
1	-2	MONTMORILLONITE (CLAY)	MNCL.	2.2 -2.7	PSI	IRREG+HEX-SHAPED PLTS. "SWELLING CLAY". ALTER-HI MG +CA,LO K RKS (ALKALINE CONDITIONS). SOILS, BENTONITES, ETC.	W398 3-226
1	-2	SAUCONITE (CLAY)	MNCL. 2	3.	PSI	FINE-GRAINED AGGS, VERMIFM, LAMELLAR,ETC. "SWELLING CLAY". ALTER- HI MG +CA,LO K RKS (ALK CONDITIONS). HYDROTHERMAL, SOILS, BENTONITE	W400 3-226
1	-3	CLAY	MNCL.	1.85-3.0	PSI	NAME GIVEN SEVERAL HYDROUS AL-SILICATES WITH ALKALIES OR ALKALINE EARTHS + MG OR FE. NAMES UNSETTLED. MOST DISTINGUISHED BY NONMEGASCOPIC MEANS. FINE-GRAINED AGGS(COMPACT,MEALY,ETC.). EARTHY-UNCTUOUS,PLASTIC WHEN WET,DEHYDRATED ON HEATING. SOME ARE ABSORBENT. WXNG PRODS,HYDROTHERMAL ALTER, DETRITAL, DIAGENIC	W398 3-191
1.5	APJOHNITE	MNCL.	1.78	SUF	FIBR. SILKY. SOL-H2O. EFFLORESCENCES IN SHELTERED PLACES	29.7.3.3
1.5	PICKERINGITE	MNCL.	1.73-1.79	SUF	FIBR. VITR. ASTRINGENT. EFFLORESCENCES IN SHELTERED PLACES	29.7.3.1
1.5	SZMIKITE	MNCL.	3.15	SUF	STALAC. SPLINTERY. RARE-EFFLOR(FELSOBANYA, ROUMANIA)	29.6.2.3
1.5	ARSENOLITE	ISOM.	3.86-3.88	OXD	WHITE-COLOR TINTED. OCTAH, CRUSTS, VITR-SILKY, OCTAH CLVG. ASTRIN- GENT-SWEETISH TASTE. SCNDRY AFTER ARESNOPYRITE + OTHER AS-MINS	4.4.2.1
1.5	ALUNOGEN	TRCL.	1.77	SUF	FIBR MASSES. VITR-SILKY. SHARP, ACID TASTE. EFFLOR IN FES2-BEARING ROCKS	29.8.6.
1.5-2	BARBERTONITE	HEXA.	2.05-2.15	OXD	PLTY-,FIBR-MSV.WAXY,PEARLY.BASAL CLVG(FLEX). VNLETS IN SERPENTINE	6.1.6.2
1.5-2	STICHTITE	HEXA.	2.16	OXD	PLTY, FIBR-MSV. WAXY, BASAL CLVG(FLEX). EFF-HCL. VNS IN SERPENTINE	6.1.5.2
1.5-2	SODA-NITER	HEXA.	2.24-2.29	NIT	INCRUST, RHOMB. VITR. RHOMB + OTHER CLVGS. COOLING TASTE. IN ARID SOILS IN PROTECTED LOCATIONS	18.1.1.
1.5-2	REALGAR	MNCL.	3.48-3.56	SLD	RED, DR-YEL. RESIN-GRSY. DCMP-HNO3. WITH AS-MINS IN PB,AG,AU-VEINS	2.6.10.
1.5-2	TRECHMANNITE	HEXA.	4.7	SST	EQUANT. ADMN. RHOMB +BASAL CLVGS. ON TENNANTITE(VALAIS,SWITZERLAND)	3.5.8.
1.5-2	COVELLITE	HEXA.	4.6 -4.76	SLD	SOME IRID. GRAY STRK. 1 PERF CLVG-FLEX PLTS. ALTER OF CU-SULFIDES	2.6.8.1
1.5-2	SMITHITE	MNCL.	4.88	SST	EQUANT, TBLR. ADMN. 1 PERF CLVG. RARE-WITH SULFIDES(VALAIS,SWITZER-LAND)	3.5.7.
1.5-2.5	MALLARDITE	MNCL. A	1.85	SUF	FIBR CRUSTS. EXPOS-BECOMES MEALY. VITR.CLVGS.SOL-H2O. RARE EFFLOR	29.6.8.6
1.5-2.5	BIEBERITE	MNCL.	1.96	SUF	CRUSTS,STALAC. VITR. CLVGS. SOL-H2O. EFFLOR-AT CO-SULFIDE DEPS	29.6.8.5
1.5-2.5	SULFUR	ORTH.	2.07	ELE	CONCH-UNEVN. BRTL. BURNS-BLU FL + ACRID ODOR. SOL-CS2. VOLS + SED	1.2.3.1
1.5-2.5	ERYTHRITE	MNCL. C	3.18	ARS	RADIAL-STELLATE GROUPS,GLBL, ADMN-DULL. CLVGS. SOL-ACIDS. OXIDA- TION PRODUCT OF COBALT AND NICKEL ARSENIDES	40.2.15.2
1.5-4.5	ARSENIOSIDERITE	TETR.	3.60	ARS	STRK-YEL. SUBTRANSLUCENT. RADIAL FIBR AGGS. SUBMET-SILKY. 1 CLVG. SOL-HOT ACIDS. SECONDARY — DIVERSE OCCURRENCES	42.6.4.
2	BOUSSINGAULTITE	MNCL. A	1.72	SUF	CRUSTS, STALAC, PRISM. CLVG SOL-H2O. FUMAROLIC, GEYSER DEPOSITS	29.3.7.3
2	SYLVITE	ISOM.	1.98-2.00	HAL	CUBIC, GRNLR. VITR. CUBIC CLVG. BITTER TASTE. SED EVAPORITE DEPS	9.1.1.2

NONMETALLIC RED

H	Mineral	Cryst.	S.G.	Type	Description	Ref
2	BRUGNATELLITE	HEXA.	2.14	OXD	FLKS, FOL-MSV. PEARLY. BASAL CLVG. EFF-HCL. INCRUST-CRACKS IN SERPENTINE	6.1.7.
2	HALITE	ISOM.	2.17	HAL	CUBIC, GRNLR. VITR. CUBIC CLVG. SALTY TASTE. SED EVAPORITE DEPS	9.1.1.1
2	THOMSENOLITE	MNCL.	2.98	HAL	PRISM, OPALINE CRUSTS. VITR. BASAL,PYRAMIDAL CLVGS. SOL-H2SO4. AT CRYOLITE LOCALITIES	11.5.4.
2	PYROSTILPNITE	MNCL.	5.94	SST	STRK-ORANGE-YEL. TBLR, SHEAFLK AGGS. ADMN. 1 PERF CLVG. THIN PLTS SLIGHTLY FLEXIBLE. DCMP-HNO3. WITH RUBY SILVERS IN LO-T AG DEPS	3.2.2.1
2	LITHARGE	TETR. A	9.14	OXD	CRUSTS, GRSY-DULL. PRISM CLVG. SOL-HCL,HNO3,ALKALIES. OXIDIZED PB- DEPOSITS	4.2.6.
2	MASSICOT	ORTH. A	9.56	OXD	MSV, SCALY, GRSY-DULL. 2 CLVGS.SOL-HCL,HNO3,ALKALIES. OXIDIZED PB- DEPOSITS	4.2.7.
2	-2.5 MELLITE	TETR.	1.64	CCC	MSV,GRNLR,CRUSTS,NODULAR. RESIN,PRISM CLVG. CONCH. SOL-HNO3. FLUO- BLUE(SW). SECONDARY IN OPEN SPACES IN LIGNITE	50.2.1.
2	-2.5 EPSOMITE	ORTH. A	1.68	SUF	CRUSTS. VITR-EARTHY. CLVGS. BITTER, SALTY. METALLIC. COMMON EFFLOR	29.6.9.1
2	-2.5 SEPIOLITE (CLAY)	ORTH.	2.08	PSI	POROUS(FLOATS ON H2O WHEN DRY)-COMPACT. SMOOTH FEEL. GELAT. WITH SERP OR OPAL; DESERT SOILS, CLAYEY SEDS WITH SALINES, LAKE SEDS. MEERSCHAUM IS A VARIETY	W444 N
2	-2.5 BOTRYOGEN	MNCL.	2.14	SUF	AGGS-RADIATING,WITH XLINE SURF. VITR. CLVGS, CONCH. SOL-HCL. SCNDRY-IN SULFATE CAPPINGS ATOP PYRITE DEPS-ESP IN ARID REGIONS	31.6.7.
2	-2.5 KAOLINITE (CLAY)	TRCL.	2.60-2.68	PSI	ALTER(ACID CONDITIONS) OF FELDSPARS, ETC. HYDROTHERMAL + WXING PRODUCTS	W362 3-194
2	-2.5 VILLIAUMITE	ISOM.	2.79	HAL	MSV, GRNLR. VITR. CUBIC CLVG. SOL-H2O. CAVITIES IN ALK IG RKS	9.1.1.3
2	-2.5 PHLOGOPITE	MNCL.	2.76-2.90	PSI	MG-MICA. ATTKD-HCL. CNTCT META-LS. ULTRABASIC IGNEOUS ROCKS	W373 3-42
2	-2.5 HYDROZINCITE	MNCL.	3.5 -4.0	CBT	MSV,INCRUST. SILKY-DULL,1 CLVG. FLUO-BLUE,LILAC. SOL-ACIDS.SCNDRY WITH OTHER ZINC MINERALS	16.1.3.
2	-2.5 PROUSTITE	HEXA.	5.57	SST	STRK-VERMILLION. PRISM, RHOMB, MSV. ADMN. DCMP-HNO3. LO-T AG-DEPS	3.2.1.2
2	-2.5 SCHWARTZEMBERGITE	TETR.	7.39	IOD	ROUNDED FLAT PYRAMIDAL XLS, CRUSTS. ADMN. 1 CLVG. SOL-HCL. PB DEPS	22.1.2.
2	-2.5 CINNABAR	HEXA.	8.09	SLD	SCARLET STRK. ADMN-DULL. NEAR VOLS + HOT SPRINGS-VNS + INCRUSTS	2.6.9.
2	-2.5 KAEMMERERITE (CHLORITE)	MNCL.		PSI	SMALL HEX-SHAPED PYRAMIDAL XLS IN ULTRABASIC RKS + CHROMIUM DEPS	W386 3-146
2	-3 PASCOITE	TRCL.	1.87	VAS	STRK-YEL. GRNLR CRUSTS OF LATHLK XLS. VITR.1 CLVG. CONCH. SOL-H2O. EFFLORESCENCES-VANADIUM DEPOSITS, WITH CARNOTITE IN SANDSTONES	47.1.13.
2	-3 RICHELLITE	UNKN.	2.	PHO	MSV COMPACT-FOL,RADIAL FIBR GLBLS, GRSY-HORNLK. SOL-ACIDS. RARE- WITH HALLOYSITE, ETC. (RICHELLE, BELGIUM)	42.6.7.
2	-3 PITTICITE	UNKN.	2.2 -2.5	ARS	MSV,BOTRYOIDAL,OPALINE CRUSTS,ETC. DULL-GRSY. SOL-ACIDS. OXIDATION ZONES OF AS-BEARING DEPS, DEPOSITED FROM MINE AND SPRING WATERS	43.2.6.
2	-3 COOKEITE (CHLORITE)	MNCL.	2.69	PSI	LI-BEARING CHLOR. WITH TOURMALINE IN PEGS, PROBABLY AS ALTER PROD	W378 3-147
2	-3 DELESSITE (CHLORITE)	MNCL.	2.7 -2.9	PSI	COMMONLY AS REPLACEMENT OF BIOTITE + OTHER FERROMAGNESIAN MINERALS	W383 3-137
2	-3 CLINOCHLORE (CHLORITE)	MNCL.	2.6 -3.0	PSI	HEX-OUTLINE (COMMONLY SLENDER)XLS,MSV. VERY COMMON,E.G.IN TALC RKS	W383 3-137
2	-3 PENNANTITE (CHLORITE)	MNCL.	3.06	PSI	TINY FLAKES. IN MANGANESE ORES + VEINS TRANSECTING THE ORES	W389 3-146
2	-3 PENNINITE (CHLORITE)	MNCL.	2.6 -3.1	PSI	PSEUDORHOMB XLS. WITH MAGNETITE + CHONDRODITE(SERP RKS), SCHISTS	W383 3-137
2	-3 CHLORITE	MNCL.	2.6 -3.3	PSI	GROUP NAME. MEMBER NAMES UNSETTLED. ONLY FEW RARE VARIETIES MEGASCOPICALLY DISTINGUISHABLE. DISSEM FLKS, PEARLY, 1 PERF CLVG TO FLEX,INELAST FOLIA. DCMP-IN BOILING H2SO4. SCNDRY-HYDROTHERMAL ALTER, LO-GRADE META, DETRITAL, AUTHIGENIC	W381 3-131

NONMETALLIC RED

HARDNESS	NAME	XL. SYS.	SPECIFIC GRAVITY	CHEM. CLASS	REMARKS	REFERENCES
2 -3	XANTHOCONITE	MNCL.	5.40-5.68	SST	STRK-ORANGE-YEL. TBLR. BLDR. LATHLK. ADMN. 1 CLVG. WITH RUBY SILVERS	3.2.2.3
2 -3	PHOSGENITE	TETR.	6.13	CBT	PRISM. MSV. ADMN. CLVGS(PERCUSSION FIGURE ON BASAL PLANE) FLUO-YEL (LW ULTRAVIOLET,CATHODE + X-RAYS). EFF-HNO3. SCNDRY AFTER PB MINS	16.1.7.
2 -3	URANOSPHAERITE	ORTH.	6.36	OXD	YEL STRK. GLBL FMS. GRSY. 1 CLVG. OXIDATION PROD OF PITCHBLENDE	5.3.3.
2 -3	HEMATOPHANITE	TETR.	7.70	OXD	SUBMET. YEL-RED STRK. TBLR. LAMELLAR AGGS. 1 CLVG. SOL-HCL, HNO3. RARE-WITH PLUMBOFERRITE, ETC. (VERMLAND, SWEDEN)	7.3.4.
2 -4	MICA	MNCL.	2.4 -3.4	PSI	GROUP NAME. SOME SPECIES DISTINGUISHABLE ON BASIS OF COLOR. SERIES ARE NOT COM-PLETE. HEX-SHAPED XLS,DISSEM. PERF CLVG TO FLEX-ELAST FLKS. PERCUSSION + PRES-SURE FIGURES MAY BE IMPOSED ON FLKS. CHIEF OCCURRENCES ARE NOTED UNDER SPECIES	W365 3-1
2 -4	BAUXITE		2	3.5	OXD	FIELD TERM FOR MATERIALS RICH IN HYDROUS ALUMINUM OXIDES.	6.2.3.
2.5	CARNALLITE	ORTH.	1.60	HAL	GRNLR. PSEUDOHEX PRISM. GRSY. CONCH. DLQSNT. BITTER. SED SALT DEPS	11.1.2.
2.5	PICROMERITE	MNCL. A	2.03	SUF	PRISM. CRUSTS. VITR. BITTER. FUMAROLIC. MARINE SALINE DEPS-AS BEDS	29.3.7.1
2.5	QUENSTEDTITE	TRCL.	2.15	SUF	AGGS TBLR XLS. 2 CLVGS. SOL-H2O. RARE-WITH COQUIMBITE	29.8.5.
2.5	AMARANTITE	TRCL.	2.19-2.29	SUF	STRK-YEL. ELONGATE,RADIATED AGGS. VITR. SOL-HCL,DCMP-H2O. SCNDRY-WITH COPIAPITE, ETC.	31.6.4.
2.5	LOPEZITE	TRCL.	2.69	CHR	SHOT-LK AGGS. SOL-H2O.(ORANGE SOLUTION). VUGS IN MSV NITRATE ROCK	35.2.1.
2.5	PHARMACOSIDERITE	TETR.	2.80	ARS	STRIATED CUBES,TETRA,GRNLR. ADMN.1 CLVG. UNEVN. SECTILE. OXIDATION PRODUCT OF OTHER ARSENIC MINERALS	42.9.1.
2.5	CRYOLITE	MNCL.	2.96-2.98	HAL	MSV, GRNLR. GRSY. UNEVN. THEMOLUM. SOL-H2O(SLIGHTLY). RARE-PEGS	11.5.1.
2.5	SZOMOLNOKITE	MNCL.	3.03-3.07	SUF	BIPYRAMIDAL,STALAC.VITR. CONCH.SOL-H2O(SLOW). EFFLOR-ACID SOLUTION	29.6.2.2
2.5	BLAKEITE	UNKN.	3.1	TLI	STRK-YELLOWISH. MICROXLINE CRUSTS. DULL. FRIABLE. OXIDE ZONE OF VN WITH NATIVE TELLURIUM AND PYRITE	34.2.3.
2.5	MARSHITE	ISOM.	5.68	HAL	YEL STRK. TETRA-STRIATED. ADMN. FLUO-RED. RARE-CU DEPS	9.1.3.3
2.5	PYRARGYRITE	HEXA.	5.85	SST	STRK-PURPLISH. HEMIMORPHIC PRISM,MSV. ADMN. DCMP-HNO3.LO-T AG-DEPS	3.2.1.1
2.5	MENDIPITE	ORTH.	7.24	HAL	FIBR, RADIATED. SILKY,RESIN. PRISM + OTHER CLVGS. SOL-HNO3. SCNDRY WITH CERUSSITE, ETC.	10.1.4.
2.5	MINIUM	UNKN.	8.9 -9.2	OXD	MSV, EARTHY.GRSY-DULL. SOL-HCL(CL↑).DCMP-HNO3.ALTER PROD OF GALENA DEPOSITS	4.3.1.
2.5	MONTROYDITE	ORTH.	11.23	OXD	YEL-BRWN STRK. BENT, ELONGATE-EQUANT XLS, WORMLK-SPHERICAL AGGS. VITR. 1 CLVG. SECTILE. SOL-HCL, HNO3,KI,ETC. WITH OTHER HG MINS	4.2.5.
2.5	KAINITE	MNCL.	2.15	SUF	GRNLR, TBLR. VITR. 1 CLVG. SALTY,BITTER. IN MARINE POTASH DEPS	31.4.1.
2.5-3	BLOEDITE	MNCL.	2.22-2.28	SUF	PRISM, GRNLR. VITR. CONCH. FAINTLY SALTY + BITTER. EVAPORITE DEPS	29.3.5.1
2.5-3	LOEWEITE	UNKN.	2.37-2.42	SUF	ANHEDRAL GRAINS. VITR. CONCH. BITTER. DISSEM IN MARINE SALINE DEPS	29.3.4.
2.5-3	THENARDITE	ORTH.	2.66	SUF	DIPYRAMIDAL, CRUSTS. VITR. CLVGS. FAINTLY SALTY. SALINE LAKE DEPS, EFFLORESCENCES, FUMAROLIC	28.2.4.
2.5-3	LANTHANITE	ORTH.	2.61-2.74	CBT	PLTY-TBLR, GRNLR. PEARLY. MICACEOUS CLVG. SOL-ACIDS. COATING ON CERITE, ALLANITE, ETC.	15.2.10.
2.5-3	GLAUBERITE	MNCL.	2.75-2.85	SUF	TBLR, PRISM. DIPYRAMIDAL-STRIATED. VITR,PEARLY. CLVGS. CONCH. BRTL. SALTY TASTE. SOL-HCL. SALINE DEPS, DISSEM IN CLASTIC SEDS, FUMAROLIC	28.4.2.
2.5-3	BIOTITE	MNCL.	2.7 -3.3	PSI	MG-,FE-MICA. BLEACHED + ATTKD-H2SO4.COMMON-E.G.,ACID IG,PEGS, META	W373 3-55

NONMETALLIC RED

2.5-3	KOETTIGITE	MNCL.	3.33	ARS	PRISM,MSV,CRUSTS,FIBR. SILKY, 1 CLVG. SOL-ACIDS. ALTER PROD OF SMALTITE AND SPHALERITE	40.2.15.4
2.5-3	SIDEROPHYLLITE	MNCL. 2.8 -3.4		PSI	FE-MICA. RARE IN PEGMATITES	W373 3-65
2.5-3	VALENTINITE	ORTH.	5.76	OXD	PRISM, TBLR, MSV. ADMN. PRISM CLVG. SOL-HCL. SCNDRY AFTER SB-MINS	4.4.4.
2.5-3	CROCOITE	MNCL.	5.96-6.02	CHR	PRISM,MSV,ETC. ADMN-VITR. CLVGS. CONCH. SECTILE. SCNDRY-IN GOSSANS	35.3.1.
2.5-3	VANADINITE	HEXA.	6.88	VAN	PRISM,HAIRLK. RESIN. CONCH. SOME SHOW CONCENTRIC ZONING. SOL-HNO3 (YEL), HCL(GREEN). OXIDIZED ZONES OF LEAD DEPOSITS	41.7.2.3
2.5-3	WULFENITE	TETR. 6.5 -7.0		MBS	TBLR,DIPYRAMIDAL,MSV,GRNLR. RESIN-ADMN. CLVGS-UNEVN. DCMP-HCL, HNO3, SOL-H2SO4. ALKS, SCNDRY-OXIDIZED ZONES OF PB + MO-BEARING RKS	48.1.4.1
2.5-3	LORETTOITE	TETR. 7.39-7.95		HAL	YEL-STRK. MSV- BLDD, FIBR. ADMN. 1 CLVG. SOL-ACIDS. RARE-PB DEPS	10.1.3.
2.5-3	STOLZITE	TETR. 7.9 -8.34		WOS	DIPYRAMIDAL-STRIATED, RESIN. CLVGS. CONCH. SOL-HCL(YEL). WITH LIMONITE, ETC.-OXIDIZED DEPOSITS WITH TUNGSTEN MINERALS	48.1.4.2
2.5-3.5	GIBBSITE	MNCL. 2.3 -2.42		OXD	TBLR, MAMMILLARY, PEARLY-VITR. 1 CLVG. CLAYLK ODOR WHEN DAMP. WX-ING PRODUCT. BAUXITE DEPOSITS. LOW TEMPERATURE VEINS	6.2.2.
2.5-4	DELVAUXITE	UNKN. 1.99-2.83		PHO	NODULES,BOTRYOIDAL CRUSTS. WAXY-VITR. DATA LACKING.	42.4.7.
2.5-4	MUSCOVITE	MNCL. 2.77-2.88		PSI	K-MICA. COMMON RK-FORMING MIN-E.G., ACID IG, PEGS, GREISENS, CNTCT METASOMATIZED(FLUORINE) ZONES, REGIONAL META-ARGILLACEOUS ROCKS	W367 3-11
2.5-4	LEPIDOLITE	MNCL. 2.80-2.90		PSI	LI-MICA. TABULAR XLS, AGGS OF SMALL FLAKES, DISTINGUISHED FROM ROSE-COLORED MUSCOVITE BY CHEM OR X-RAY ANAL. RARE,LI-BEARING PEGS	W370 3-85
2.5-5	GUMMITE		3.9 -6.4	OXD	FIELD TERM FOR HYDROUS OXIDES OF URANIUM OF UNKNOWN IDENTITY	5.2.1.
3	INDERITE	TRCL.	1.86	BOR	ACICULAR-NODULAR AGGS. VITR,DULL. CLVGS. SOL-HCL. EVAPORITE DEPS	25.1.15.
3	HOHMANNITE	TRCL.	2.2	SUF	GRNLR, PRISM. VITR. CLVGS. EXPOS-DEHYDRATES TO METAHOHMANNITE. SOL-HCL,DCMP-HOT H2O. SECONDARY -- COMMONLY WITH COPIAPITE	31.6.5.
3	RINNEITE	HEXA.	2.35	HAL	EXPOS-BRWN. GRNLR. SILKY, CONCH. ASTRINGENT. SCNDRY-SED SALT DEPS	11.4.3.
3	CALCITE	HEXA.	2.71	CBT	VARIOUS TRIGONAL XL HABITS, MSV. VITR-PEARLY, RHOMB CLVG. SOME IS FLUO, PHOSPHO, THERMOLUM. EFF-DILUTE HCL.WDSPRD-VNS,DIVERSE RKS	14.1.1.1
3	APHTHITALITE	HEXA. 2.65-2.71		SUF	TRIGONAL-TBLR, MAMMILLARY,MSV. VITR. CLVGS. CONCH. SOL-H2O, ACIDS. FUMAROLIC, OCEANIC + LACUSTRINE SALINE DEPOSITS	28.2.2.
3	CHURCHITE	MNCL.	3.14	PHO	FAN-SHAPED GROUPS OF COLUMNAR XLS. VITR,PEARLY. 1 CLVG. CONCH. RARE -- COATINGS AT COPPER DEPOSIT	40.3.4.
3	GILLESPITE	TETR.	3.4	PSI	1 PERF, 1 POOR CLVG. LEACHED-HCL(SIO2 + H2O RESIDUE). WITH QTZ, CELSIAN, ETC.	W357 N
3	HEMAFIBRITE	ORTH.	3.65	ARS	PRISM,RADIAL FIBR AGGS.VITR.CLVGS. SOL-ACIDS. WITH SCNDRY MN-MINS	42.2.3.
3 -3.5	LAUMONTITE (ZEOLITE)	MNCL. 2.2 -2.3		TSI	PERF CLVGS-3 DIRECTIONS IN 1 ZONE. GELAT. CAVITIES IN SEVERAL RKS	W342 4-401
3 -3.5	HISINGERITE	AMOR.	3.	PSI	RESINOUS. CONCH. 1 POOR CLVG. DCMP-HCL. SCNDRY(E.G.,AFTER OLIVINE)	W400 N
3 -3.5	REDDINGITE	ORTH.	3.0 -3.2	PHO	MSV,GRNLR,FIBR,TBLR. VITR. 1 CLVG. UNEVN. SOL-ACIDS. PEGMATITES	40.2.5.1
3 -3.5	CELESTITE	ORTH.	3.96-3.98	SUF	TYPICALLY COLORLESS-BLUISH. TBLR,ETC. VITR-PEARLY. CLVGS. SOME FLUO,THERMOLUM. SOL-H2O(SLIGHTLY). SED RKS(ESP GYPSUM + LS), VNS	28.3.1.2
3 -3.5	BARITE	ORTH.	4.50	SUF	TBLR, MSV,ETC. VITR-PEARLY. CLVGS. SOME FLUO, PHOSPHO, THERMOLUM. SOME FETID WHEN RUBBED. COMMON-VNS, SED RKS, CAVITIES IN IG RKS	28.3.1.1
3 -3.5	LAUTITE	ORTH.	4.8 -5.0	SLD	SUBMET. OPQ. BLK-GRAYIRED TINGE).SOL-HNO3. VNS-WITH NATIVE AS,ETC.	2.9.5.5

41

NONMETALLIC RED

HARDNESS	NAME	XL. SYS.	SPECIFIC GRAVITY	CHEM. CLASS	REMARKS	REFERENCES
3 -3.5	PHOENICOCHROITE	ORTH.	5.75	CHR	TBLR,MSV,COATINGS. RESIN. 1 CLVG SOL-HCL GOSSANS WITH CROCOITE	35.3.2.
3 -3.5	DESCLOIZITE	ORTH.	6.1 -6.3	VAN	VARIABLE-XLS TO COMPACT MSV. GRSY.COMMONLY COLOR ZONED. SOL-ACIDS. OXIDIZED ZONES OF VANADIUM-BEARING ORE DEPOSITS	41.5.2.1
3 -4	ARSENIOPLEITE	HEXA.		PHO	SUBTRANSLUCENT. YEL-BRWN STRK. MSV,GRNLR. RHOMB CLVG. CONCH. SOL-HCL, HNO3. VNS + NODULES WITH RHODONITE (UKERO,SWEDEN)	41.5.12.
3 -4	MORDENITE (ZEOLITE)	ORTH.	2.12-2.15	TSI	STRIATED PRISMS, ACICULAR, COTTON-LK AGGS. ZEOL-ASSOC, AUTHIGENIC	W339 4-401
3 -4	EVANSITE	UNKN.	1.8 -2.2	PHO	MSV,COLLOFM COATINGS. VITR. CONCH. BRTL. SOL-ACIDS. WITH LIMONITE	42.2.6.
3 -4	GISMONDINE (ZEOLITE)	ORTH.	2.1 -2.3	TSI	PSEUDOTETRAG BIPYRAMIDS. 1 CLVG. GELAT. LEUCITE RKS, ALTER OF PLAG	W339 4-401
3 -4	CACOXENITE	HEXA.	2.2 -2.4	PHO	ACICULAR,TUFTS,RADIAL AGGS,FIBR CRUSTS. SILKY. SOL-ACIDS. SCNDRY-DIVERSE OCCURRENCES (WITH DUFRENITE) NAVELLITE, ETC.)	42.9.2.
3 -4	DIADOCHITE	TRCL.	2.0 -2.4	PHO	NODULAR,PLTY,GEL-LK,COLLOFM. DULL-WAXY. UNEVN-CONCH. SOL-ACIDS. IN GOSSANS AND AS EFFLORESCENCES	43.2.4.
3 -4	ASHCROFTINE (ZEOLITE)	TETR.	2.61	TSI	FIBR PRISM. 2 PERF CLVGS. RARE-PINK POWDER IN AUGITE SYENITE IN GREENLAND	W334 4-404
3 -4	STILPNOMELANE	MNCL.	2.59-2.96	PSI	MICACEOUS(BUT 2ND CLVG ⊥ 1ST). SOL-HF+H2SO4(1/1). LO-GRD META RKS	W390 3-103
3 -4	ST! PNOCHLORANE	UNKN.	2.5 -3.0	PSI	YELLOWISH TO BRONZY RED. SCALES.	WN N
3 -4	FOURMARIERITE	ORTH.	6.05	OXD	TBLR-STRIATED. ADMN. BASAL CLVG. SOL-ACIDS. SCNDRY URANIUM MINERAL	5.3.1.
3 -5.5	ZEOLITE		1.9 -2.45	TSI	GROUP OF MIN FAMILIES WITH SOMEWHAT RELATED APPEARANCES, OCCURRENCES, AND COMPOSITIONS. MANY ARE INDISTINGUISHABLE MEGASCOPICALLY. DIVERSE HABITS(E.G., BUNDLES, FIBROUS). PEARLY-GLASSY. SUBCONCH. SOME FLUO-ORANGE OR YEL-GREEN. CONTINUOUS, IN PART REVERSIBLE, DEHYDRATION(I.E.,WATER CAN BE EXPELLED WITH-OUT DESTROYING XL STRUCTURE), SCNDRY IN OPEN SPACES; COMMONLY TWO OR MORE OCCUR TOGETHER IN CAVITIES IN BASALTIC RKS — THIS OCCURRENCE IS REFERRED TO HEREIN AS THE ZEOLITE ASSOCIATION (ABBREVIATED "ZEOL-ASSOC").	W330 4-351
3.5	GORDONITE	TRCL.	2.23	PHO	SHEAFLK AGGS. VITR,PEARLY. CLVGS. CONCH. SOL-ACIDS. IN VARISCITE NODULES	42.8.8.
3.5	BORICKITE	UNKN.	2.69-2.71	PHO	SUBTRANSLUCENT. COMPACT-SOME OPALINE. WAXY. SOL-HCL. RARE-GOSSAN	42.1.1.
3.5	POLYHALITE	TRCL.	2.78	SUF	MSV-FIBR, FOL. VITR. 1 CLVG. DCMP-H2O(4 GYP). MARINE SALINE DEPS	29.4.2.
3.5	BERMANITE	ORTH.	2.84	PHO	TBLR,FANLK + ROSETTELK AGGS, LAMELLAR MASSES.VITR.CLVGS.SOL-HNO3. SECONDARY - IN TRIPLITE (PEGMATITES)	42.8.1.
3.5	ANHYDRITE	ORTH.	2.98	SUF	MSV, GRNLR, FIBR. VITR-PEARLY. 2 CLVGS. SOL-ACIDS. SED ROCKS	28.3.2.
3.5	AKROCHORDITE	MNCL.	3.19	ARS	WARTLK AGGS. 2 CLVGS. SOL-H2SO4 (PURPLE). RARE-LANGBAN, SWEDEN	42.3.3.
3.5	HUREAULITE	MNCL.	3.15-3.20	PHO	TBLR,MSV,SCALY,FIBR. VITR. 1 CLVG. SOL-ACIDS. WITH ALTERED LITHI-OPHILITE AND/OR TRIPHYLITE	39.1.3.
3.5	CHLOROPHOENICITE	MNCL.	3.46	ARS	LONG PRISM, STRIATED. VITR,PEARLY. 1 CLVG. SOL-ACIDS. FRNKLN, N.J.	41.1.4.1
3.5	HEMATOLITE	HEXA.	3.49	ARS	SUBMET. TBLR,RHOMB,STRIATED. VITR. 1 CLVG. SOL-ACIDS. WITH BARITE, JACOBSITE,ETC. IN VNS IN LIMESTONE (NORDMARK, SWEDEN)	41.1.3.
3.5	ROSELITE	MNCL.	3.50-3.74	ARS	SHORT PRISM,SPHERICAL AGGS. VITR. 1 CLVG. SOL-ACIDS. IN QTZ DRUSES	40.2.4.1

NONMETALLIC RED

Hardness	Mineral	Crystal	SG	Class	Description	Code
3.5	STRONTIANITE	ORTH.	3.64-3.78	CBT	SPEAR-LK, MSV. VITR. CLVGS. FLUO(ETC.)-LK ARAGONITE. SOL-HCL. VNS	14.1.3.3
3.5	RHABDOPHANE	HEXA.	3.94-4.01	PHO	RADIATED FIBR MAMMILLARY CRUSTS, GRSY. UNEVN. SOL-HCL. WITH LIMONITE	40.3.5.
3.5	SCHAFARZIKITE	TETR.	4.3	ATI	BRWN STRK. SUBTRANSLUCENT. PRISM-STRIATED. CLVGS. RARE-SB-DEPOSITS	45.1.4.
3.5	ADAMITE	ORTH.	4.32-4.48	ARS	ELONGATE,CRUSTS,RADIAL AGGS, VITR. CLVGS. UNEVN. SOL-DILUTE ACIDS. SOME FLUO YEL.OXIDIZED ZONES OF DEPS WITH ZN- + AS-RICH MINERALS	41.6.5.3
3.5	CARMINITE	ORTH.	4.10-5.22	ARS	LATHLK,FIBR AGGS,ETC.VITR,PEARLY.SOL-HNO3,DCMP-HCL. SCNDRY-VN DEPS	41.8.3.
3.5	PYROBELONITE	ORTH.	5.38	VAN	ACICULAR. ADMN. CONCH. RARE-WITH HAUSMANNITE,ETC.(LANGBAN, SWEDEN)	41.5.2.3
3.5-4	HEULANDITE (ZEOLITE)	MNCL.	2.1 -2.2	TSI	PLTY GRP. TBLR-BROAD CENTRAL PORTIONS(COFFIN-LIKE). ZEOL-ASSOC	W347 4-377
3.5-4	STILBITE (ZEOLITE)	MNCL.	2.1 -2.2	TSI	PLTY GRP. SHEAF-LIKE AGGS. 1 GOOD CLVG. DCMP-HCL. ZEOL-ASSOC	W345 4-377
3.5-4	METASTRENGITE	MNCL.	2.76	PHO	TBLR,BOTRYOIDAL,RADIAL FIBR CRUSTS. VITR. 2 CLVGS. SOL-HCL. WITH STRENGITE	40.3.2.2
3.5-4	LANGBEINITE	ISOM.	2.83	SUF	NODULAR, DISSEM. VITR.CONCH. SOL-H2O(SLOWLY). MARINE SALTS	28.4.3.1
3.5-4	DOLOMITE	HEXA.	2.84-2.86	CBT	RHOMB-CURVED, GRNLR. VITR-PEARLY. RHOMB CLVG. SOME FLUO. EFF-WARM HCL IF POWDERED. SED RKS, XLS IN CAVITIES, ETC.	14.2.1.1
3.5-4	STRENGITE	ORTH.	2.5 -2.87	PHO	TBLR,PSEUDO-OCTAH,BOTRYOIDAL AGGS,ETC. MAXY. 2 CLVGS. SOL-HCL. ALTERATION PRODUCT OF FE-BEARING PHOSPHATES	40.3.1.2
3.5-4	ALUNITE	HEXA.	2.6 -2.9	SUF	GRNLR, MSV. VITR,PEARLY. CLVGS. SOL-H2SO4. "ALUNITIZED" ROCKS	30.2.4.1
3.5-4	NATROALUNITE	HEXA.	2.6 -2.9	SUF	GRNLR, MSV. VITR,PEARLY. CLVGS. SOL-H2SO4. RARE, BUT VARIED OCCUR.	30.2.4.2
3.5-4	ARAGONITE	ORTH.	2.94-2.95	CBT	ACICULAR, RADIAL FIBR, STALACTITIC. VITR. CLVGS. FLUO(ULTRA-VIOLET, X-RAYS, ELECTRON BEAMS). THERMOLUM. EFF-ACIDS. LO-T DEPOSITION	14.1.3.1
3.5-4	KUTNAHORITE	HEXA.	3.0	CBT	DISTINGUISHED FROM DOLOMITE(SEE) BY NONMEGASCOPIC METHODS. RARE	14.2.1.3
3.5-4	ANKERITE	HEXA.	3.01-3.03	CBT	DISTINGUISHED FROM DOLOMITE(SEE) BY NONMEGASCOPIC METHODS. VNS	14.2.1.2
3.5-4	BERAUNITE	MNCL.	2.8 -3.08	PHO	STRK-YEL,OLIVE. RADIATED FOL GLBLS,CRUSTS,ETC. VITR. 1 CLVG. SOL-ACIDS. SCNDRY-WITH OTHER PHOSPHATES + LIMONITE--FE DEPS + PEGS	42.7.2.
3.5-4	WARWICKITE	ORTH.	3.34-3.36	BOR	SUBMET-PEARLY. BLUE-BLK STRK. SLENDER PRISMS, ROUNDED ENDS. DCMP-H2SO4. RARE -- CONTACT METAMORPHOSED LIMESTONE	24.1.4.
3.5-4	RHODOCHROSITE	HEXA.	3.70	CBT	GRNLR, BOTRYOIDAL. VITR-PEARLY. RHOMB CLVG. EFF-WARM ACID. VNS	14.1.1.4
3.5-4	TARBUTTITE	TRCL.	4.12	PHO	CRUSTS,SHEAFLK AGGS OF ROUNDED,STRIATED,EQUANT XLS. VITR,PEARLY. 1 CLEAVAGE. OXIDIZED ZONES OF ZINC DEPOSITS	41.6.7.
3.5-4	CUPRITE	ISOM.	6.14	OXD	SUBMET(SOME). OCTAH, HAIRLK, MSV. COMMON-OXIDIZED ZONES OF CU-DEPS	4.1.1.
3.5-4	PYROMORPHITE	HEXA.	7.00-7.08	PHO	PRISM,GLBL,ETC. RESIN. RHOMB CLVG. UNEVN. SOL-HNO3. OXIDIZED ZONES OF LEAD DEPOSITS	41.7.2.1
3.5-4	KLEINITE	HEXA.	8.0	HAL	REDDISH-DAYLIGHT, YEL TO ORANGE-OTHERWISE. PRISM. ADMN-GRSY. BASAL + RHOMB CLVGS. SOL-HCL,HNO3,NH4BR. RARE--WITH OTHER HG MINS (TEXAS)	10.2.10.
3.5-4.5	MORINITE	MNCL.	2.95	PHO	LAMELLAR. VITR-PEARLY. CLVG. ALTER PROD OF AMBLYGONITE-RARE	41.2.2.
3.5-4.5	DUFRENITE	MNCL.	3.1 -3.34	PHO	BOTRYOIDAL-RADIAL,CRUSTS. VITR-SILKY. CLVGS. SOL-DILUTE ACIDS. WITH LIMONITE IN GOSSANS	41.6.9.
3.5-4.5	SIDERITE	HEXA.	3.95-3.97	CBT	TARN-IRID. MSV. RHOMB,ETC. VITR.RHOMB CLVG. SOL-HOT ACID. SED, VNS	14.1.1.3
3.5-7	BRITTLE MICA	MNCL.	2.6 -3.2	PSI	GROUP OF MORE OR LESS RELATED MINS THAT RESEMBLE MICAS BUT CLEAVE TO BRTL FLKS AND HAVE HARDNESSES OF 3.5-4 VS. 2-3(MICA) ON CLVG FLKS AND 6 VS. 4 ⊥ TO FLKS. WITH CORUNDUM + DIASPORE IN EMERY DEPS, CHLORITE + MICA SCHISTS, SKARN	W392 3-95

NONMETALLIC RED

HARDNESS	NAME	XL. SYS.	SPECIFIC GRAVITY	CHEM. CLASS	REMARKS	REFERENCES
3.5-6	MARGARITE (BRITTLE MICA)	MNCL.	3	PSI	AGGS OF LAMELLAE. ATTKD-H2SO4. WITH DIASPORE + CORUNDUM-EMERY DEPS	W392 3-95
3.5-6	CLINTONITE (BRITTLE MICA)	MNCL.	3	PSI	WITH TALC-CHLORITE SCHIST, WITH SPINEL IN METASOMATIZED CALC RKS. DISTINGUISHED FROM XANTHOPHYLLITE BY OPTICAL OR X-RAY ANALYSES	W391 3-99
3.5-6	XANTHOPHYLLITE (BRITTLE MICA)	MNCL.	3	PSI	WITH TALC-CHLORITE SCHIST, WITH HUMITES IN METASOMATIZED CALC RKS. DISTINGUISHED FROM CLINTONITE BY OPTICAL OR X-RAY ANALYSES	W391 3-99
4	EPISTILBITE (ZEOLITE)	MNCL.	2.1 -2.3	TSI	PLTY GRP. SHEAF-LK + SPHERICAL AGGS. DCMP-HCL.PIEZOELECTRIC. ZEOL- ASSOCIATION	W346 4-377
4	MOSANDRITE	MNCL.	2.93-3.03	NSI	REDDISH BRWN. LONG PRISM XLS. 1 CLVG. ALKALIC PEGMATITES-RARE	W517 N
4	FLUORITE	ISOM.	3.18	HAL	CUBIC, MSV. VITR. OCTAH CLVG. FLUO-VIOLET, DCMP-H2SO4. WDSPRD-VNS, CAVITIES, DISSEMINATED -- IGNEOUS ROCKS, ETC.	9.2.1.
4	SARCOPSIDE	MNCL.	3.64-3.73	PHO	RED TO BROWNISH,EXPOS-DK BROWN,BLUE,VIOLET,GREEN.FIBR MASSES,SILKY. 2 CLVGS(1 ⊥ // FIBERS). SOL-ACIDS. RARE-PEGMATITES	41.6.4.
4	HOLDENITE	ORTH.	4.11	ARS	EQUANT. VITR. 1 POOR CLVG. SUBCONCH. VN CUTTING ORE(FRANKLIN,N.J.)	41.1.2.
4	SPHEROCOBALTITE	HEXA.	4.13	CBT	PINK STRK. MSV-RADIAL STRUCTURE. VITR. EFF-HOT ACID. VNS	14.1.1.5
4	ZINCITE	HEXA.	5.64-5.68	OXD	HEMIMORPHIC PYRAMIDAL,MSV,GRNLR. SOME FLUO. SOL-ACIDS. RARE-FRNKLN	4.2.2.1
4	PUCHERITE	ORTH.	C 6.57	VAS	TBLR,ACICULAR,MSV,CRUSTS, VITR-ADMN.1 CLVG. SUBCONCH.SOL-HCL(RED). SECONDARY, WITH BISMUTITE AND BEYERITE	47.1.8.
4	HUEBNERITE	MNCL.	7.12	WOS	SUBMET. TARN-IRID. PRISM-STRIATED, -RADIATING XL GRPS. 1 CLVG. DCMP-AQ-REG,H2SO4 OR HCL(SLOWLY). DIVERSE-VNS,CNTCT META,PLACERS	48.1.1.1
4	KALIBORITE	MNCL.	2.13	BOR	XL AGGS, GRNLR. VITR. CLVGS. SOL-ACIDS. SALINE EVAPORITE DEPOSITS	25.1.22.
-4.5	PHILLIPSITE (ZEOLITE)	MNCL.	2.2	TSI	PENETRATION TWINS, SPHERULITES, 2 GOOD CLVGS. GELAT. DISTINGUISHED FROM HARMOTOME BY OPTICAL METHODS. ZEOL-ASSOC,DEEP-SEA RED CLAYS	W343 4-386
-4.5	SULFOBORITE	ORTH.	2.38-2.45	BOR	PRISM. PRISM + BASAL CLVGS. BRTL. SOL-ACIDS,DCMP-H2O. SALT DEPS	27.1.3.
-4.5	JEZEKITE	MNCL.	2.94	PHO	PRISM,STRIATED,FIBR CRUSTS,AGGS. VITR. 2 CLVGS. DCMP-HOT H2SO4. DRUSES IN GRANITE (SAXONY, GERMANY)	41.2.3.
-4.5	HETEROSITE	ORTH.	3.2 -3.4	PHO	SUBTRANSLUCENT. EXPOS-BRWNISH, SATINY-DULL. 2 CLVGS(CURVED OR CRIN-KLED SURFACES). SOL-HCL. OXIDATION PROD OF TRIPHYLITE + LITHIOPHILITE	38.1.5.1
-4.5	PURPURITE	ORTH.	3.2 -3.4	PHO	SUBTRANSLUCENT. DISTINGUISHED BY CHEM ANAL FROM HETEROSITE	38.1.5.2
-4.5	ALSTONITE	ORTH.	3.67-3.71	CBT	PSEUDOHEX, STRIATED. VITR. FLUO-YEL(LW). SOL-HCL. LO-T DEPS	14.2.2.
-4.5	VOLTZITE	UNKN.	3.7 -3.8	SLD	SECONDARY INCRUSTS. VITR-GRSY. GIVES H2S IN HCL. WITH SULFIDES	2.6.4.3
-4.5	HODGKINSONITE	MNCL.	3.91	NSI	PYRAMIDAL XLS. BASAL CLVG. GELAT. WITH WILLEMITE(ETC.)-FRNKLN,N.J.	W515 N
-4.5	BASTNASITE	HEXA.	4.9 -5.2	CBT	TBLR-GROOVED, GRNLR. VITR,PEARLY. CLVGS. SOL-HOT ACIDS. CNTCT META	16.2.9.
-4.5	BINDHEIMITE	ISOM.	4.6 -5.6	ANT	DENSE-EARTHY CRUSTS,NODULAR-CONCENTRIC. SOME IS OPALINE. RESIN-DULL. DCMP-HNO3, HCL. OXIDATION ZONES OF ANTIMONY-BEARING LEAD DEPOSITS	44.1.1.1
-4.5	CLARKEITE	UNKN.	6.39	OXD	YEL-BRWN STRK. MSV. WAXY. CONCH. RDACTV. ALTER PROD OF URANINITE	5.2.2.
-5	FRIEDELITE	HEXA.	3.06-3.19	PSI	TBLR, MSV. BASAL CLVG. MN-RICH ORE DEPS (E.G., FRNKLN, N. J.)	W359 N
-5	LITHIOPHILITE	ORTH.	3.50-3.58	PHO	MSV, ANHEDRAL. VITR. CLVGS. UNEVN. SOL-ACIDS. PEGMATITES	38.1.1.2
-5	SARKINITE	MNCL.	4.08-4.18	ARS	TBLR,SPHERICAL GROUPS,GRNLR. GRSY.1 CLVG. UNEVN. SOL-DILUTE ACIDS. DIVERSE (E.G., VEINS)	41.6.3.3
-5	XENOTIME	TETR.	4.4 -5.1	PHO	PRISM + PYRAMIDAL. VITR. 1 CLVG. ACID + ALK IG RKS + PEGS,DETRITAL	38.4.1.

NONMETALLIC RED

H	Mineral	System	SG	Code	Description	Ref
4 -5	CURITE	ORTH.	7.26	OXD	MSV, SACCH, STRIATED PRISMS, ADMN. 1 CLVG. SOL-ACIDS. ALTER AFTER URANINITE	5.3.2.
4 -5.5	LIMONITE		2.7 -4.3	OXD	FIELD TERM FOR HYDROUS IRON OXIDES OF UNKNOWN IDENTITIES.	7.1.3.
4 -5.5	STIBICONITE	ISOM.	5.58	OXD	MSV, POWDER, CRUSTS. EARTHY-GLASSY. OXIDATION PROD OF SB-MINS	4.5.7.
4.5	CHABAZITE (ZEOLITE)	HEXA.	2.05-2.10	TSI	RHOMBS(NEARLY CUBES). RHOMB CLVG. DCMP-HCL.ZEOL-ASSOC, RK CAVITIES	W334 4-387
4.5	GMELINITE (ZEOLITE)	HEXA.	2.0 -2.2	TSI	PRISM CLVG. DCMP-HCL. ZEOL-ASSOC	W335 4-387
4.5	LEVYNITE (ZEOLITE)	ORTH.	2.0 -2.2	TSI	CHEM OR X-RAY ANAL DISTINGUISH CHABAZITE +LEVYNITE.RARE-ZEOL-ASSOC	W335 4-387
4.5	HARMOTOME (ZEOLITE)	MNCL.	2.41-2.47	TSI	HABIT LK PHILLIPSITE(ABOVE), + CRUCIFM TWINS., DCMP-HCL. ZEOL-ASSOC, ALSO WITH AL-WXING PRODS,IN GNEISSES, + WITH MN-MINERALIZATION	W344 4-386
4.5	BULTFONTEINITE	TRCL.	2.73	NSI	PINK. ACICULAR IN RADIAL SPHERULITES. PRISM CLVG. SOL-ACID. SCDRY	W518 N
4.5	WOODHOUSEITE	HEXA.	3.01	PHO	PSEUDOCUBIC,TBLR -CURVED,STRIATED. VITR,PEARLY. 1 CLVG. VUGS,ETC.	43.1.1.5
4.5	FILLOWITE	MNCL.	3.43	PHO	GRNLR,PSEUDORHOMB. GRSY. 1 CLVG. SOL-ACIDS. PEG-WITH TRIPLOIDITE, ETC.	40.2.2.
4.5	FRONDELITE	ORTH.	3.3 -3.49	PHO	SUBTRANSLUCENT- RADIAL FIBR BOTRYOIDAL(SOME HAVE CONCENTRIC COLOR BANDING). VITR-DULL.CLVGS.UNEVN.SOL-HCL.SCNDRY-PO4 MINS IN PEGS	41.6.6.1
4.5	ROCKBRIDGEITE	ORTH.	3.3 -3.49	PHO	IN SERIES WITH FRONDELITE(SEE FOR PROPS + OCCUR.) WITH LIMONITE	41.6.6.2
4.5	SYNADELPHITE	TRCL.	3.57-3.79	ARS	COMMONLY COLOR ZONED. PSEUDOPRISMS, MSV. VITR. 1 CLVG. UNEVN. SOL-ACIDS. RARE WITH OTHER MANGANESE MINERALS	41.1.5.
4.5	ALLACTITE	MNCL.	3.83	ARS	ELONGATE,ROSETTELK AGGS. VITR. 1 CLVG. SOL-HCL. VNS(E.G.,LANGBAN)	41.2.4.
4.5	LESSINGITE	HEXA.	4.69	NSI	REDDISH OR GREENISH YEL. VITR. RARE IN PLACERS WITH OTHER CE-MINS	W512 N
4.5-5	APOPHYLLITE	TETR.	2.33-2.37	PSI	VITR SQ PRISMS, PEARLY BASES. BASAL + PRISM CLVG. DCMP-HCL(SILICA RESIDUE). SECONDARY IN CAVITIES, COMMONLY WITH ZEOLITES, ETC.	W394 3-258
4.5-5	AUGELITE	MNCL.	2.70	PHO	TBLR,ACICULAR,MSV. VITR,PEARLY. CLVGS. SOL-HCL(SLOW). VNS,ETC.	41.6.8.
4.5-5	GOYAZITE	HEXA.	3.26	PHO	RHOMB-STRIATED,ROUNDED GRAINS. GRSY,PEARLY.1 CLVG.SOL-ACIDS(SLOW). DIAMOND SANDS, CREVICES	41.5.8.3
4.5-5	TRIPLOIDITE	MNCL.	3.66	PHO	//-DIVERGENT FIBR-COLUMNAR AGGS,STRIATED PRISM. VITR-ADMN. CLVGS. UNEVEN-SUBCONCH. SOL-ACIDS. PEGMATITES - PROBABLY SECONDARY	41.6.3.1
4.5-5	WOLFEITE	MNCL.	3.83	PHO	IN SERIES WITH TRIPLOIDITE(SEE FOR PROPERTIES + OCCURRENCE.)	41.6.3.2
4.5-5	PLUMBOGUMMITE	HEXA.	4.01	PHO	CONCENTRIC BOTRYOIDAL, GUM-LIKE LUSTER. UNEVN. SOL-HOT ACIDS. SECONDARY - IN LEAD DEPOSITS WITH PYROMORPHITE	41.5.8.1
4.5-5	MANGANBERZELIITE	ISOM.	4.46	ARS	DISTINGUISHED FROM BERZELIITE BY CHEMICAL ANALYSIS	38.2.1.2
4.5-5	SCHEELITE	TETR.	6.08-6.12	WOS	DIPYRAMIDAL,MSV,GRNLR,VITR.CLVGS.UNEVN.DCMP-HCL,HNO3,FLUO-BLUE(X-SW ULTRAVT-OLET-, + CATHODE-RAYS),THERMOLM. HI-T VNS,PEGS,GREISSENS,CNTCT META RKS,ETC.	48.1.3.1
4.5-5.5	ELLESTADITE	HEXA.	3.07	PHO	MSV,GRNLR. CLVGS. VNS IN CNTCT META LS(E.G., CRESTMORE, CALIF.)	41.7.8.
4.5-5.5	WILKEITE	HEXA.	3.1	PHO	ROUNDED XLS,GRNLR. SUBRESIN. 1 CLVG. GELAT. CNTCT META LS	41.7.7.
5	NATROLITE (ZEOLITE)	ORTH.	2.20-2.26	TSI	FIBR GRP. SQ NEEDLES. PRISM CLVG. GELAT. PYROELECTRIC. ZEOL-ASSOC	W340 4-358
5	MESOLITE (ZEOLITE)	MNCL.	2.25-2.27	TSI	FIBR GRP. TUFTS OF HAIR-LIKE XLS. GELAT. PYROELECTRIC. ZEOL-ASSOC	W341 4-358
5	GONNARDITE (ZEOLITE)	ORTH.	2.2 -2.4	TSI	FIBR GRP. RADIATED SPHERICAL MASSES. GELAT. RARE. ZEOL-ASSOCIATION	W337 4-359

NONMETALLIC RED

HARDNESS	NAME	XL. SYS.	SPECIFIC GRAVITY	CHEM. CLASS	REMARKS	REFERENCES
5	EOSPHORITE	ORTH.	3.04-3.08	PHO	PRISM-STRIATED,BOTRYOIDAL-FIBR. VITR. 1 CLVG. UNEVN. SOL-ACIDS. WITH RHODOCHROSITE, ETC. -- PEGMATITES	42.5.2.2
5	HYDROXYLAPATITE	HEXA.	2.9 -3.1	PHO	APATITE GROUP(SEE FOR PROPS.) RARE-TALC SCHIST	41.7.1.3
5	CARBONATE-APATITE	HEXA.	2.9 -3.1	PHO	APATITE GROUP(SEE FOR PROPS.)RARE-NODULES(PHOSPHATE RK),GUANO,ETC.	41.7.1.4
5	CHLORAPATITE	HEXA. A	3.17	PHO	APATITE GROUP(SEE FOR PROPS.) RARE-VNS IN GABBRO, METEORITES	41.7.1.2
5	FLUORAPATITE	HEXA. A	3.18	PHO	APATITE GROUP(SEE FOR PROPS.) COMMON-IGS, METAS, MARINE SEDS, ETC.	41.7.1.1
5	APATITE	HEXA.	2.9 -3.2	PHO	GROUP NAME, MOST TYPES INDISTINGUISHABLE MEGASCOPICALLY. HEX PRISMS-COMMONLY ROUNDED,MSV,GRNLR. VITR. SOL-ACIDS. SOME FLUO--YEL-ORANGE. IG RKS,VNS,PEGS	41.7.1.
5	SVANBERGITE	HEXA.	3.22	PHO	RHOMB,PSEUDOCUBIC,GRNLR,VITR-ADMN.1 CLVG. DIVERSE META RKS,PLACERS	43.1.1.4
5	KATOPHORITE (AMPHIBOLE)	MNCL.	3.20-3.50	ISI	RARE IN INTERMEDIATE TO BASIC ALKALIC IGNEOUS ROCKS	W440 2-359
5	MAGNESIOKATOPHORITE (AMPHIBOLE)	MNCL.	3.20-3.50	ISI	RARE IN INTERMEDIATE AND BASIC ALKALIC IGNEOUS ROCKS	WN 2-359
5	FERMORITE	HEXA.	3.52	ARS	MSV,GRNLR. GRSY. EVN FRACTURE. SOL-ACIDS. VNS IN MN ORE(INDIA)	41.7.6.
5	GRAFTONITE	MNCL.	3.67-3.79	PHO	PRISM, MSV. VITR. UNEVN. SOL-ACIDS. WITH TRIPHYLITE IN PEGMATITES	38.3.2.
5	DURANGITE	MNCL.	3.94-4.07	ARS	PYRAMIDAL. VITR. PRISM CLVG. SOL-H2SO4.WITH CASSITERITE(E.G.,PEGS)	41.5.7.
5	LEPIDOCROCITE	ORTH.	4.05-4.13	OXD	SUBMET. ORANGE STRK. SCALY, FIBR MSV. 3 CLVGS. WITH GOETHITE, ETC.	6.1.2.1
5	KENTROLITE	ORTH.	6.19	NSI	DK REDDISH BRWN. PRISM XLS + CLVG. SOL-HCL. MN-DEPS(SO.CHILE)-RARE	W523 N
5	BISMUTOTANTALITE	ORTH.	8.26	OXD	SUBMET. EXPOS-PINKISH YEL. PRISM. RARE-PEGMATITE(SW UGANDA)	8.1.9.
5 -5.5	THOMSONITE (ZEOLITE)	ORTH.	2.10-2.39	TSI	FIBR GRP. RADIATED SPHERICAL MASSES. GELAT. PYROELECT. ZEOL-ASSOC	W336 4-359
5 -5.5	EDINGTONITE (ZEOLITE)	TETR.	2.7 -2.8	TSI	FIBR GRP. PSEUDOTETRAGONAL. PRISM CLVG. GELAT. PYROELECTRIC. ZEOL-ASSOCIATION, ORE DEPOSITS	M338 4-359
5 -5.5	DATOLITE	MNCL.	2.96-3.00	NSI	TINTED WHT-COLORLESS. GLASSY. GELAT. SCNDRY WITH ZEOLITES-CAVITIES	W355 1-171
5 -5.5	SCHIZOLITE	TRCL.	2.97-3.13	ISI	(MANGANOAN PECTOLITE). ALTER-BRWN. RADIATING AGGS. ALK SYENITES	W462 2-177
5 -5.5	WAGNERITE	MNCL.	3.15	PHO	NEARLY OPQ-ALTERED. PRISM STRIATED,MSV. VITR. CLVGS. UNEVN-SPLINTERY. SOL-ACIDS. DIVERSE (E.G., VEINS AND VOLCANIC ROCKS)	41.6.1.
5 -5.5	KAINOSITE	ORTH.	3.34-3.61	CSI	SHORT PRISMS. 1 CLVG. EFF-HCL. ORE VEINS-RARE(E.G.,N. BURGESS,ONT)	W463 N
5 -5.5	TRIPLITE	MNCL.	3.5 -3.9	PHO	SUBTRANSLUCENT. MSV. VITR-RESIN. 3 CLVGS. UNEVN. SOL-ACIDS. PEGS	41.6.2.
5 -5.5	GOETHITE	ORTH.	3.3 -4.29	OXD	SUBMET. YELLOWISH STRK. MSV, FIBR, ETC. SILKY-ADMN. SOL-HCL. WXING PRODUCT	7.1.2.2
5 -5.5	PYROCHLORE	ISOM.	4.45	OXD	OCTAH. VITR. SUBCONCH. DCMP-H2SO4(SLOW). ALK PEGS + IG RKS	8.1.1.1
5 -5.5	MONAZITE	MNCL.	4.6 -5.4	PHO	TBLR-STRIATED. WAXY-ADMN. CLVGS. UNEVN. DCMP-ACIDS(SLOW). ACCESS IN GRANITIC IG + GNEISSIC RKS, PEGS, DETRITAL SED ROCKS, ETC.	38.4.2.
5 -5.5	MICROLITE	ISOM.	6.38-6.66	OXD	OCTAH. VITR. SUBCONCH. DCMP-H2SO4(SLOW).GRANITIC PEGS WITH TA MINS	8.1.1.2
5 -6	VISHNEVITE	HEXA.	2.32-2.42	TSI	SULFATE-RICH END MEMBER OF CANCRINITE FAMILY. OCCURRENCE LIKE CANCRINITE. CARBONATE-RICH MEMBERS EFFERVESCE AS WELL AS GELATINIZE WITH HCL	WN 4-310

NONMETALLIC RED

H	Mineral	System	S.G.		Description	Ref.	Occurrence
● 5 -6	CANCRINITE	HEXA.	2.42-2.51	TSI	MSV (RARE PRISM XLS). PRISM CLVG. EFF-HCL. NEPHELINE-BEARING RKS	M354 4-310	
5 -6	MARIALITE (SCAPOLITE)	TETR.	2.50-2.62	TSI	ME-0-20. PURE MARIALITE UNKNOWN IN NATURE	M352 4-321	
5 -6	DIPYRE (SCAPOLITE)	TETR.	2.57-2.69	TSI	ME-20-50.	M352 4-321	
5 -6	MIZZONITE (SCAPOLITE)	TETR.	2.67-2.74	TSI	ME-50-80.	M352 4-322	
● 5 -6	SCAPOLITE	TETR.	2.50-2.78	TSI	GROUP NAME. MOST TYPES INDISTINGUISHABLE MEGASCOPICALLY BUT S.G. MAY BE INDICATIVE. LONG,STRIATED PRISMS + AGGS OF COARSE XLS. SUBCONCH. POOR CLVG GIVING IRREGULAR, STRIATED-APPEARING SURFACES. MARIALITE IS INSOL; MEIONITE IS DCMP IN HCL. MOST FLUO--YEL-RED(ULTRAVIOLET,LW). CNTCT META CALC RKS,PEGS	M352 4-321	
5 -6	(WERNERITE) (SCAPOLITE)	TETR.	2.50-2.78	TSI	GROUP NAME NO LONGER USED	M352 4-322	
5 -6	MEIONITE (SCAPOLITE)	TETR.	2.73-2.78	TSI	ME-80-100. PURE MEIONITE UNKNOWN IN NATURE	M352 4-321	
5 -6	CUSPIDINE	MNCL.	2.95	SSI	SPEAR-SHAPED XLS. BASAL CLVG. SOL-HNO3. CONTACT METAMORPHIC ZONES	M480 N	
5 -6	EUDIALYTE	TETR.	2.8 -3.1	CSI	SCALENOHEDRAL. GOOD BASAL CLVG. GELAT. ALK RKS(E.G., NEPH SYENITE)	M453 N	
5 -6	NEPTUNITE	MNCL.	3.19	CSI	BLK,RED IN SPLINTERS. PRISM. PERF PRISM CLVG-AT 80°. ALK SYENITE	M463 N	
5 -6	RICHTERITE (AMPHIBOLE)	MNCL.	2.97-3.45	ISI	THERMALLY META LS + SKARNS. HYDROTHERMAL VNS IN ALK IGNEOUS ROCKS	M435 2-352	
5 -6	AMPHIBOLE		3.0 -3.57	ISI	GROUP NAME. SEVERAL END AND OTHER MEMBERS WITH NAMES UNSETTLED FOR SOME, MANY MEGASCOPICALLY INDISTINGUISHABLE BUT SOME COLORS AND OCCURRENCES ARE INDICATIVE. ORTH AND MNCL SERIES. LONG SLENDER XLS. PRISM CLVG AT 56° +124°. SOL-HF(SLOWLY). ALTERS TO BIOTITE +/OR CHLORITE. IGNEOUS AND METAMORPHIC RKS, VNS	M422 2-203	
5 -6	ARSENOCLASITE	ORTH.	4.16	ARS	MSV,GRNLR. 1 CLVG. RARE-ALONG FISSURES IN DOLO(LANGBAN, SWEDEN)	41.4.4.	
5 -6	PYROPHANITE	HEXA.	4.54	OXD	SUBMET. YEL STRK,SCALY,RHOMB CLVG. RARE-CAVITIES(VERMLAND, SWEDEN)	4.4.1.5	
● 5 -6	HEMATITE	HEXA.	5.26	OXD	MET-EARTHY. STRK-RED. PLTY,GRNLR. SOL-HCL. SED FE-FMS, RARE IN VNS	4.4.1.2	
5 -7	EPHESITE (BRITTLE MICA)	MNCL.	3.	PSI	NA-MARGARITE. RARE-E.G., WHERE PEG HAS INVADED AMPHIBOLITE	M393 3-97	
● 5 -7	PYROXENE		2.96-3.96	ISI	GROUP NAME. SEVERAL END AND OTHER MEMBERS WITH NAMES UNSETTLED FOR SOME, MANY MEGASCOPICALLY INDISTINGUISHABLE; OCCURRENCE MAY BE INDICATIVE. ORTH + MNCL SERIES.SHORT PRISMS,PRISM CLVG AT 87° + 92°. ALTERS TO AMPHIBOLE.IG + META RKS	M402 2-1	
● 5.5	ANALCIME (ZEOLITE)	ISOM.	2.24-2.29	TSI	TRAPEZOHEDRA, MSV, GRNLR. POOR CUBIC CLVG. GELAT. AMYGDULES-BASALT	M333 4-338	
5.5	HELLANDITE	MNCL.	3.35-3.70	NSI	TBLR. SOL-HCL. COMMONLY METAMICT. PEGS WITH ALLANITE, ETC.-RARE	M518 N	
5.5	TARAMELLITE	ORTH.	3.92	ISI	REDDISH BRWN. FIBR. CLVG // FIBERS. CNTCT METAMORPHOSED LIMESTONE	M401 N	
5.5	ALLEGHANYITE	MNCL.	4.02	NSI	SIMILAR TO OLIVINE + HUMITE MINS. WITH MN-MINS IN VNS IN META RKS	M516 1-35	
5.5	WILLEMITE	HEXA.	3.9 -4.1	NSI	PRISMS, MSV. MAY FLUO-GRN. SOME PHOS + TRIBOLUM. GELAT. ZINC DEPS	M497 N	
5.5	CERITE	MNCL.	4.65-4.91	SSI	SHORT PRISM, GRNLR. GELAT. WITH ALLANITE ETC. IN GNEISS + GRANITE	M507 N	
5.5	STIBIOCOLUMBITE	ORTH.	5.68	OXD	IN SERIES WITH STIBIOTANTALITE, WHICH SEE. RARE IN PEGMATITES	8.1.8.2	
5.5	STIBIOTANTALITE	ORTH.	7.34	OXD	IN SERIES,WITH STIBIOCOLUMBITE. YEL, YEL-BRWN STRK,PRISM-STRIATED. RESIN-ADMN. 1 GOOD,1 POOR-CLVGS. SOL-HF. PYROELECT. PEGS, PLACERS	8.1.8.1	

47

NONMETALLIC RED

HARDNESS	NAME	XL. SYS.	SPECIFIC GRAVITY	CHEM. CLASS	REMARKS	REFERENCES
5.5-6	SODALITE	ISOM.	2.27-2.33	TSI	DODEC, MSV. POOR DODEC CLVG. GELAT. SOME FLUO-YEL-ORANGE. DISSOLVE-HNO3; EVAPORATE SLOWLY, HALITE REMAINS. NEPHELINE-BEARING ROCKS	W348 4-289
5.5-6	MONTEBRASITE	TRCL.	2.98	PHO	AMBLYGONITE SERIES(SEE FOR PROPERTIES + OCCURRENCES.)	41.5.5.2
5.5-6	NATROMONTEBRASITE	TRCL.	3.04-3.1	PHO	AMBLYGONITE SERIES(SEE FOR PROPERTIES + OCCURRENCES.)	41.5.5.3
5.5-6	AMBLYGONITE	TRCL.	3.11	PHO	EQUANT, LARGE MASSES. VITR,PEARLY. CLVGS. SOL-ACIDS(DIFFICULTY). LI-, PO4-RICH PEGS	41.5.5.1
5.5-6	TITANAUGITE (CLINOPYROXENE)	MNCL.	3.3-3.6	ISI	AUGITE WITH NOTABLE TI-CONTENT. ALKALIC BASIC IGNEOUS ROCKS	W411 2-113
5.5-6	PYROXMANGITE	TRCL.	3.61-3.80	ISI	PINK, EXPOS-BRWN OR BLK. F-BLK MAG GLBL. META + METASOMATIC MN-RKS	W460 2-196
5.5-6	LEUCOPHENICITE	MNCL.	3.85	NSI	PURPLISH RED. STRIATED ELONGATE XLS. GELAT. RARE-FRNKLN, N.J.	W516 N
5.5-6	ANATASE	TETR.	3.90	OXD	SUBMET. PALE YEL STRK. ACUTE PYRAMIDAL. ADMN. BASAL + PYRAMIDAL CLVG. CONVERTS TO RUTILE ON HEATING. VNS,ACCESS-IG RKS,DETRITAL	4.5.2.
5.5-6	BROOKITE	ORTH.	4.08-4.20	OXD	GRAY TO YEL STRK.TBLR XLS. ADMN. POOR CLVGS. VNS, DISSEM, DETRITAL	4.5.3.
5.5-6.5	OPAL	TETR.	1.8-2.25	SIL	SUBMICROXLINE QTZ + H2O.RESIN-PEARLY, MSV.MAMILLARY CRUSTS,CONCH,SOL-HF +CAUS-TIC ALKS.SOME FLUO-YEL-GREEN. INCRUSTATIONS, IN CAVITIES(ESP IN VOLCANIC RKS)	D-111 287
5.5-6.5	BUSTAMITE	TRCL.	3.32-3.43	ISI	PINK, EXPOS-FADES. 1 PERF +OTHER CLVGS. MN-ORE DEPS(ESP META ONES)	W457 2-191
5.5-6.5	RHODONITE	TRCL.	3.57-3.76	ISI	TBLR(COMMONLY ROUGH), MSV. ALTERS READILY TO BLK MN-OXIDES	W459 2-182
5.5-6.5	ROMEITE	ISOM.	4.7-5.4	ANT	OCTAH,MSV. GRSY. OCTAH CLVG. DIVERSE(E.G.,MN DEPS)	44.1.1.2
6	SARCOLITE	TETR.	2.92	TSI	FLESH RED. NEARLY EQUANT XLS. RARE-VESUVIUS, ITALY	W496 N
6	INESITE	TRCL.	3.03	ISI	PRISM, FIBR, RADIATING. 1 PERF, 1 GOOD-CLVGS. DCMP-HCL. MN-DEPS	W421 N
6	HUMITE	ORTH.	3.20-3.32	NSI	GROUP NAME, ALSO APPLIED TO SPECIES OF GROUP. IRREGULAR-SHAPED GRAINS. VITR-RESIN. GELAT. SIMILAR TO OLIVINE. CONTACT METAMORPHOSED DOLOMITIC LIMESTONES	W514 1-49
6	ZOISITE	ORTH.	3.15-3.37	SSI	DISTINGUISHED FROM CLINOZOISITE BY OPTICAL OR X-RAY ANAL. IG,META(E.G.,AMPHIBOLITES)	W446 1-185
6	THULITE	ORTH.	3.15-3.37	SSI	MN-BEARING(LD TENOR) ZOISITE. PINK VS. DK REDDISH BRWN OF PIEMON-TITE	W447 N
6	DANALITE	ISOM.	3.28-3.44	ISI	HELVITE(SEE) GRP PROPS. FLESH-RED TO GRAY. OCTAH. GRANITES, SKARNS	W351 4-303
6	HELVITE	ISOM.	3.20-3.44	ISI	TETRA,MSV. POOR OCTAH CLVG.GELAT(H2S↑).PYROELECTRIC.BOIL IN HCL + AS2O3,DECANT, WASH,HELVITE IS STAINED CANARY YELLOW (AS2S3). RARE-GNEISS,GRANITE,CNTCT ZONES	M351 4-303
6	EPIDOTE	MNCL.	3.38-3.49	SSI	GROUP NAME. NO GENERALLY ACCEPTED NOMENCLATURE FOR GROUP MEMBERS. CLOSELY RE-LATED TO ZOISITE + ALLANITE. PISTACHIO-BLACKISH GREEN-BROWNISH. LONG THIN GROOVED XLS, GRNLR MASSES. VNS, OTHER CAVITIES, REGIONAL META + IG ROCKS	W448 1-193
6	PIEMONTITE	MNCL.	3.45-3.52	SSI	MN-BEARING EPIDOTE. COLOR GENERALLY DISTINCTIVE(DKR THAN THULITE)	W448 1-194
6	ACMITE (CLINOPYROXENE)	MNCL.	3.55-3.60	ISI	OCCURRENCES LIKE AEGIRENE + AEGIRINE-AUGITE (ALKALIC SYENITES,ETC.)	W411 2-80
6	PSEUDOBROOKITE	ORTH.	4.33-4.39	OXD	SUBMET. TBLR-STRIATED. GRSY. 1 CLVG. VOLCANIC AREAS FUMAROLIC,ETC.	7.5.1.
6	COLUMBITE	ORTH.	5.15-5.25	OXD	SUBMET. IN SERIES WITH TANTALITE. RED-BLK STRK. TARN-IRID. TBLR. PRISMATIC. 2 CLEAVAGES, BRITTLE. GRANITIC PEGMATITES	8.3.2.1
-6.5	MICROCLINE (K-FELDSPAR)	TRCL.	2.56-2.63	TSI	DISTINGUISHED FROM ORTHOCLASE OPTICALLY. PLUTONIC IG RKS, PEGS	M308 4-7
-6.5	ORTHOCLASE (K-FELDSPAR)	TRCL.	2.55-2.63	TSI	DISTINGUISHED FROM MICROCLINE OPTICALLY. PLUTONIC IG RKS, PEGS	M303 4-7

NONMETALLIC RED

Hardness	Mineral	Crystal System	S.G.	Class	Description	Reference
6 -6.5	PERTHITE (FELDSPARS)	TRCL.	2.56-2.65	TSI	MEGASCOPICALLY, MICROSCOPICALLY, OR SUBMICROSCOPICALLY INTERDIGITATED MICROCLINE OR ORTHOCLASE + PLAGIOCLASE (TYPICALLY ALBITE)	W229 4-7
6 -6.5	OLIGOCLASE (PLAGIOCLASE)	TRCL.	2.62-2.67	TSI	AN10-30. SEE FELDSPAR. PLUTONIC IG RKS, FEW PEGS. SUNSTONE IS VAR.	W312 4-95
6 -6.5	FELDSPAR		2.55-3.39	TSI	INCLUDES PLAGIOCLASE(NA-CA)SERIES, ALKALI(K-NA) AND BA FELDSPARS. MOST EASILY DISTINGUISHED BY NONMEGASCOPIC MEANS. MONOCLINIC ONES HAVE 2 CLVGS AT 90° AND SIMPLE(IF ANY) TWINNING; TRICLINIC ONES HAVE 2 CLVGS UP TO 4° OFF 90° AND TYPICALLY HAVE POLYSYNTHETIC TWINNING (GIVING STRIATED APPEARANCE ON SOME CLVG SURFACES OF PLAGIOCLASE). PEGMATITES IG, META, AND SOME SED ROCKS	W261 4-1
6 -6.5	RUTILE	TETR.	4.21-4.25	OXD	SUBMET. PRISM-STRIATED. ADMN. POOR CLVGS. VNS, ACCESS-META + IG RKS	4.5.1.1
6 -6.5	TANTALITE	ORTH.	7.90-8.00	OXD	SUBMET. IN SERIES WITH COLUMBITE, WHICH SEE. PEGMATITES	8.3.2.2
6 -7	BERLINITE	HEXA.	2.64	PHO	MSV, GRNLR. VITR. CONCH. RARE-WITH AUGELITE AT FE-MINE(KRISTIANSTAD, SWEDEN)	38.4.3.
6 -7	MULLITE	ORTH.	3.03-3.16	NSI	INDISTINGUISHABLE MEGASCOPICALLY FROM SILLIMANITE. HI-T META-RARE	W401 1-124
6 -7	VESUVIANITE	TETR.	3.33-3.43	SSI	PRISM. MSV. 1 POOR CLVG. SUBCONCH. ATTKD-HCL. META-ESP CNTCT LS	W508 1-113
6 -7	TRIMERITE	MNCL.	3.47	TSI	PSEUDOHEX PRISMS. BASAL CLVG. SOL-HCL. IN CALCITE OF MN-DEP(LANGBAN)	M260 N
6 -7	CASSITERITE	TETR.	6.99	OXD	SUBMET. RADIAL CONCRETIONARY MASSES. ADMN-DULL. HI-T VNS, GREISENS	4.5.1.5
6 -7.5	PYROPE	ISOM. A	3.58	NSI	MG-AL GARNET. XLS ARE RARE. ALTERS TO KELYPHITE. ULTRABASIC IG RKS	W486 1-97
6 -7.5	ANDRADITE	ISOM. A	3.86	NSI	CA-FE,TI GARNET. F-MAG GLBL. CONTACT META CALCAREOUS RKS, ALK IGS	W489 1-89
6 -7.5	SPESSARTINE	ISOM. A	4.19	NSI	MN-AL GARNET. EQUANT CRYSTALS. MN-RICH META RKS + VEINS, PEGS	W486 1-99
6 -7.5	GARNET	ISOM.	3.58-4.32	NSI	GROUP NAME. DOMINANT MOLECULE DETERMINES NAME. DISTINGUISHED BY S.G. + NONMEGASCOPIC MEANS. EQUANT XLS, MSV, VITR-DULL, DODEC PARTING. SLIGHTLY- TO IN-SOL IN HF	W483 1-77
6 -7.5	ALMANDINE	ISOM. A	4.32	NSI	FE-AL GARNET. F-MAG GLBL. META-ARGILLACEOUS RKS, RARE-IG RKS, PEGS	W486 1-85
6.5	PETALITE	MNCL.	2.41-2.42	TSI	FOL, MSV. 1 PERF + 1 GOOD-CLVGS AT 38.5°. LI-BEARING PEGMATITES	M260 4-271
6.5	AGATE	HEXA.	2.57-2.64	SIL	BANDED OR VARIEGATED. CHALCEDONY(MICROXLINE QTZ). NODULES IN BASALT	D-111 210
6.5	CHALCEDONY	HEXA.	2.57-2.64	SIL	MICROXLINE + MICROFIBR QTZ. MAMMILLARY. LO-T VEINS +OTHER CAVITIES	D-111 195
6.5	SARD	HEXA.	2.57-2.64	SIL	BRWNISH ORANGE + RED CHALCEDONY (MICROCRYSTALLINE QUARTZ)	D-111 206
6.5	CARNELIAN	HEXA.	2.57-2.64	SIL	RED TO BRWNISH RED CHALCEDONY (MICROCRYSTALLINE QUARTZ)	D-111 206
6.5	XONOTLITE	MNCL.	2.71	ISI	EXPOS-COLOR MAY FADE. ELONGATE XLS. CLVG // LENGTH. CNTCT META LS	W455 N
6.5	CHONDRODITE	MNCL.	3.16-3.26	NSI	ISOLATED TBLR GRNS. VITR-RESIN. GELAT. LIKE OTHER HUMITES-SKARNS	W513 1-48
6.5	KORNERUPINE	ORTH.	3.27-3.34	NSI	PRISM MASSES. PRISM CLVG. WITH CORDIERITE + SAPPHIRINE IN SCHISTS	W523 N
6.5	GENTHELVITE	ISOM.	3.44-3.70	TSI	HELVITE(SEE) GRP PROPS + OCCURRENCES. STALACTITIC, INCRUSTATIONS	W351 4-303
6.5	THALENITE	MNCL.	4.23-4.45	SSI	PINK-FLESH RED. TBLR, PRISM. RARE IN PEGMATITES	W477 N
6.5	BADDELEYITE	MNCL.	5.4 -6.02	OXD	FLAT PRISM. VITR. BASAL + 2 POOR CLVGS. DCMP-HOT H2SO4. PLACERS, ASSOCIATED WITH ULTRAMAFIC IGNEOUS ROCKS	5.1.1.
6.5-7	BLOODSTONE	HEXA.	2.57-2.65	SIL	SUBTRANSLUCENT. GREEN-RED SPOTS. CHALCEDONY (MICROCRYSTALLINE QTZ)	D-111 219

NONMETALLIC RED

HARDNESS	NAME	XL. SYS.	SPECIFIC GRAVITY	CHEM. CLASS	REMARKS	REFERENCES
6.5-7	JASPER	HEXA.	2.57-2.65	SIL	SUBTRANSLUCENT. SOME IS VARIEGATED OR BANDED. MICROGRNLR. CAVITIES	D-111 224
6.5-7	DIASPORE	ORTH.	3.3 -3.5	OXD	PLTY, FOL-MSV. VITR, PEARLY. WITH CORUNDUM IN EMERY ROCK	7.1.2.1
6.5-7.5	ANDALUSITE	ORTH.	3.13-3.16	NSI	BLUNT, NEARLY SQ PRISMS. COMMONLY ALTERED. CONTACT +REGIONAL META	W521 1-129
7	ELPIDITE	ORTH.	2.58	SSI	FIBR-MSV. PRISM CLVG. IN ALKALIC IGNEOUS ROCKS(E.G., OSLO EKERITE)	W454 N
7	QUARTZ	HEXA.	2.65	SIL	DOUBLY TERMINATED HEX PRISMS STRIATED ⊥ LENGTH(LO-T), BIPYRAMIDS (HI-T),MSV. VITR. CONCH. SOL-HF. WDSPRD IN MOST RK TYPES + VEINS	D-111 9
7	AMETHYST	HEXA.	2.65	SIL	PURPLE(SOME REDDISH OR BLUISH). XLINE. LO-T, IN CAVITIES IN BASALT	D-111 171
7	ROSE QUARTZ	HEXA.	2.65	SIL	PALE PINK, PALER-EXPOS. TURBID. GRSY. MSV. BLEACHES~CA.575C. PEGS	D-111 186
7	ELBAITE	HEXA.	3.03-3.10	CSI	NA- LI-TOURMALINE. INFUSIBLE. TYPICALLY PINK AND/OR GREEN. PEGS	W465 1-300
7	TOURMALINE	HEXA.	3.03-3.25	CSI	GROUP NAME. VARIETIES COMMONLY DISTINGUISHABLE MEGASCOPICALLY ON BASIS OF COLOR(S). LENGTHWISE STRIATED PRISMATIC XLS WITH CROSS-SECTIONS THAT RESEMBLE SPHERICAL TRIANGLES. SOME XLS EXHIBIT BOTH LENGTHWISE + CONCENTRIC COLOR ZONING. VITR-RESIN. BRTL. ELECTRICALLY CHARGED ON HEATING + COOLING. MG-RICH VARIETIES FLUO-YEL (ULTRAVIOLET - SWL). PEGS, HI-T VNS, IG AND META ROCKS	W465 1-300
7	DUMORTIERITE	ORTH.	3.3	NSI	FIBR-COMMONLY RADIATING. VITR. 1 GOOD,1 POOR CLVG. PEGS + META RKS	W259 N
7.5	PHENAKITE	HEXA.	2.98	NSI	RHOMBOHEDRAL(FLAT)-PRISMATIC. VITREOUS. PEGMATITES	W496 N
7.5	STAUROLITE	ORTH.	3.74-3.83	NSI	PRISM-CRUCIFORM TWINS COMMON. FIRES TO MAG POWDER. SCHIST + GNEISS	W522 1-151
7.5	ZIRCON	TETR.	4.6 -4.7	NSI	TERMINATED PRISMS, ADMN. SOME FLUO-YEL-ORANGE. ACCESS-IG, SANDS	W494 1-59
7.5-8	BERYL	HEXA.	2.66-2.83	CSI	GROOVED PRISM. POOR BASAL CLVG.SOME FLUO WEAKLY-YEL. PEGMATITES	W463 1-256
7.5-8	SPINEL	ISOM.	3.55	OXD	GROUP NAME. SOME HAVE DISTINCTIVE COLORS OR OTHER HAND-SPECIMEN PROPS. OCTAH. GRNLR, GLASSY. CONCH. SOME FLUO-RED TO YEL-GREEN. IG,META(ESP CALC RKS),PEGS	7.2.1.1
8.5	CHRYSOBERYL	ORTH.	3.65-3.85	OXD	TBLR-STRIATED. 3 CLVGS. OPALESCENT, ASTERATED. PEGS, PLACERS, META	7.2.3.
9	CORUNDUM	HEXA.	4.0 -4.1	OXD	STEEP PYRAMIDAL, PRISM. BASAL PARTING.WDSPRD IN SI-DEFICIENT RKS	4.4.1.1
10	DIAMOND	ISOM.	3.50-3.53	ELE	OCTAHEDRAL. ADMN-GRSY. BRTL. OCTAH CLVG. ULTRABASIC IGS + PLACERS	1.2.4.1

NONMETALLIC ORANGE

HARDNESS	NAME	XL. SYS.	SPECIFIC GRAVITY	CHEM. CLASS	REMARKS	REFERENCES
1 -2	MONTMORILLONITE (CLAY)	MNCL.	2.2 -2.7	PSI	IRREG+HEX-SHAPED PLTS. "SWELLING CLAY," ALTER-HI MG +CA,LO K RKS (ALKALINE CONDITIONS). SOILS, BENTONITES, ETC.	W398 3-226
1 -2.5	NONTRONITE (CLAY)	MNCL.	2.5 -3.0	PSI	SUBTRANSLUCENT, LATHS, MSV, GELAT. "SWELLING CLAY," ALTER-HI MG + CA,LO K RKS(ALK CONDITIONS). WITH OPAL +/OR QTZ IN VNS, BENTONITE	W398 3-226
1 -3	CLAY	MNCL.	1.85-3.0	PSI	NAME GIVEN SEVERAL HYDROUS AL-SILICATES WITH ALKALIES OR ALKALINE EARTHS ± MG OR FE. NAMES UNSETTLED, MOST DISTINGUISHED BY NONMEGASCOPIC MEANS. FINE-GRAINED AGGS(COMPACT,MEALY,ETC.). EARTHY-PEARLY,UNCTUOUS,PLASTIC WHEN WET,DEHYDRATED ON HEATING. SOME ARE ABSORBENT. MXING PRODS,HYDROTHERMAL ALTER, DETRITAL, DIAGENIC	W398 3-191
1.5-2	REALGAR	MNCL.	3.48-3.56	SLD	RED, OR-YEL. RESIN-GRSY. DCMP-HNO3. WITH AS-MINS IN PB,AG,AU-VEINS	2.6.10.
1.5-2	SMITHITE	MNCL.	4.88	SST	EQUANT, TBLR. ADMN. 1 PERF CLVG. RARE-WITH SULFIDES(VALAIS,SWITZER- LAND)	3.5.7.
1.5-2.5	SIDERONATRITE	ORTH.	2.15-2.35	SUF	FIBR CRUSTS,NODULAR. 1 CLVG. DCMP-BOILING H2O. WITH OTHER SULFATES IN ARID AREAS	31.5.3.
1.5-2.5	ERYTHRITE	MNCL.	C 3.18	ARS	RADIAL-STELLATE GROUPS,GLBL, ADMN-DULL, CLVGS. SOL-ACIDS. OXIDA- TION PRODUCT OF COBALT AND NICKEL ARSENIDES	40.2.15.2
2 -2.5	BOTRYOGEN	MNCL.	2.14	SUF	AGGS-RADIATING,WITH XLINE SURF. VITR. CLVGS, CONCH. SOL-HCL. SCNDRY-SULFATE CAPPINGS ATOP PYRITE DEPS, ESP IN ARID AREAS	31.6.7.
2 -3	PASCOITE	TRCL.	1.87	VAS	STRK-YEL. GRNLR CRUSTS OF LATHLK XLS. VITR.1 CLVG. CONCH. SOL-H2O. EFFLORESCENCES – VANADIUM DEPOSITS, WITH CARNOTITE IN SANDSTONES	47.1.13.
2 -3	PENNANTITE (CHLORITE)	MNCL.	3.06	PSI	TINY FLAKES. IN MANGANESE ORES + VEINS TRANSECTING THE ORES	W389 3-146
2 -3	PENNINITE (CHLORITE)	MNCL.	2.6 -3.1	PSI	PSEUDORHOMB XLS. WITH MAGNETITE + CHONDRODITE(SERP RKS), SCHISTS	W383 3-137
2 -3	CHLORITE	MNCL.	2.6 -3.3	PSI	GROUP NAME, MEMBER NAMES UNSETTLED. ONLY FEW RARE VARIETIES MEGASCOPICALLY DIS- TINGUISHABLE, DISSEM FLKS, PEARLY, 1 PERF CLVG TO FLEX,INELAST FOLIA. DCMP- IN BOILING H2SO4. SCNDRY-HYDROTHERMAL ALTER, LO-GRADE META, DETRITAL, AUTHIGENIC	W381 3-131
2 -3	XANTHOCONITE	MNCL.	5.40-5.68	SST	STRK-ORANGE-YEL. TBLR. LATHLK. ADMN. 1 CLVG. WITH RUBY SILVERS	3.2.2.3
2 -3	URANOSPHAERITE	ORTH.	6.36	OXD	YEL STRK. GLBL FMS. GRSY. 1 CLVG. OXIDATION PROD OF PITCHBLENDE	5.3.3.
2.5	AMARANTITE	TRCL.	2.19-2.29	SUF	STRK-YEL. ELONGATE,RADIATED AGGS. VITR. SOL-HCL,DCMP-H2O. SCNDRY- WITH COPIAPITE, ETC.	31.6.4.
2.5	METAVOLTINE	HEXA.	2.5	SUF	TBLR, GRNLR-SCALY AGGS. RESIN. 1 CLVG. PARTLY SOL-H2O,DCMP-DILUTE ACIDS(SLOWLY). WITH OTHER HYDROUS SULFATES IN DIVERSE ENVIRONMENTS	31.6.9.
2.5	BUTLERITE	MNCL.	2.55	SUF	TBLR. VITR. 1 CLVG. SCNDRY WITH OTHER SULFATES.	31.6.2.
2.5	PARABUTLERITE	ORTH.	2.55	SUF	PRISM-STRIATED, VITR. PODR PRISM CLVG. CONCH. SOL-DILUTE ACIDS. SCNDRY – ALTER PROD AFTER COPIAPITE, OXIDIZED ZONES OF PYRITIC VNS	31.6.3.
2.5	LOPEZITE	TRCL.	2.69	CHR	SHOT-LK AGGS. SOL-H2O.(ORANGE SOLUTION). VUGS IN MSV NITRATE ROCK	35.2.1.
2.5	EGLESTONITE	ISOM.	8.33-8.45	HAL	EXPOS-DK BRWN, BLK. YEL STRK. MSV, CRUSTS, DDDEC. RESIN. CONCH. DCMP-ACIDS(CALOMEL RESIDUE). MERCURY DEPOSITS	10.1.1.
2.5-3	COPIAPITE	TRCL.	2.08-2.17	SUF	CRUSTS-GRNLR OR SCALY AGGS. PEARLY CLVGS. SOL-H2O. SCNDRY-WITH OTHER SULFATES AS OXIDATION PRODS OF SULFIDES (E.G., PYRITE)	31.6.11.1

NONMETALLIC ORANGE

HARDNESS	NAME	XL. SYS.	SPECIFIC GRAVITY	CHEM. CLASS	REMARKS	REFERENCES
2.5-3	MAGNESIOCOPIAPITE	TRCL.	2.08-2.17	SUF	INDISTINGUISHABLE MEGASCOPICALLY FROM COPIAPITE, WHICH SEE	31.6.11.2
2.5-3	CUPROCOPIAPITE	TRCL.	2.08-2.17	SUF	INDISTINGUISHABLE MEGASCOPICALLY FROM COPIAPITE, WHICH SEE	31.6.11.3
2.5-3	CROCOITE	MNCL.	5.96-6.02	CHR	PRISM,MSV,ETC. ADMN-VITR. CLVGS. CONCH. SECTILE. SCNDRY-IN GOSSANS	35.3.1.
2.5-3	VANADINITE	HEXA.	6.88	VAN	PRISM,HAIRLK. RESIN. CONCH. SOME SHOW CONCENTRIC ZONING. SOL-HNO3 (YEL), HCL(GREEN), OXIDIZED ZONES OF LEAD DEPOSITS	41.7.2.3
2.5-3	WULFENITE	TETR.	6.5 -7.0	MBS	TBLR,DIPYRAMIDAL,MSV,GRNLR. RESIN-ADMN. CLVGS.UNEVN. DCMP-HCL, HNO3, SOL-H2SO4, ALKS. SCNDRY-OXIDIZED ZONES OF PB- + MO-BEARING RKS	48.1.4.1
2.5-5	GUMMITE		3.9 -6.4	OXD	FIELD TERM FOR HYDROUS OXIDES OF URANIUM OF UNKNOWN IDENTITY	5.2.1.
3	ZIPPEITE	ORTH.		SUF	CRUSTS. DULL-SILKY. FLUO-GREEN. SCNDRY ON ALTERING URANINITE	31.4.4.
3	HOHMANNITE	TRCL.	2.2	SUF	GRNLR, PRISM. VITR. CLVGS. EXPOS-DEHYDRATES TO METAHOHMANNITE. SOL- HCL,DCMP-HOT H2O. SECONDARY - COMMONLY WITH COPIAPITE	31-6.5.
3	CALCITE	HEXA.	2.71	CBT	VARIOUS TRIGONAL XL HABITS, MSV. VITR-PEARLY. RHOMB CLVG. SOME IS FLUO.,PHOSPHO,THERMOLUM. EFF-DILUTE HCL. WDSPRD-VNS, DIVERSE ROCKS	14.1.1.1
3 -3.5	GREENOCKITE	HEXA.	4.9	SLD	HEMIMORPH PYRAMIDAL XLS. ADMN-RESIN. SOL-HCL. COATINGS ON ZN-MINS	2.6.4.2
3 -3.5	DESCLOIZITE	ORTH.	6.1 -6.3	VAN	VARIABLE-XLS TO COMPACT MSV. GRSY.COMMONLY COLOR ZONED. SOL-ACIDS. OXIDIZED ZONES OF VANADIUM-BEARING ORE DEPOSITS	41.5.2.1
3.5	FAIRFIELDITE	TRCL.	3.08	PHO	FOLIATED AGGS, PRISM. PEARLY-GRSY. CLVGS. SOL-ACIDS. PEGMATITES	40.2.3.1
3.5	HUREAULITE	MNCL.	3.15-3.20	PHO	TBLR,MSV,SCALY,FIBR. VITR. 1 CLVG. SOL-ACIDS. WITH ALTERED LITHI-OPHILITE AND/OR TRIPHYLITE	39.1.3.
3.5-4	MIMETITE	HEXA.	7.24	ARS	PRISM,ACICULAR,GLBL,ETC. RESIN. RHOMB CLVG. UNEVN. SOL-HNO3. OXI-DIZED ZONES OF LEAD DEPOSITS	41.7.2.2
3.5-4	KLEINITE	HEXA.	8.0	HAL	REDDISH-DAYLIGHT, YEL TO ORANGE-OTHERWISE. PRISM. ADMN-GRSY. BASAL + RHOMB CLVGS. SOL-HCL,HNO3,NH4BR. RARE-WITH OTHER HG MINS (TEXAS)	10.2.10.
4	ZINCITE	HEXA.	5.64-5.68	OXD	HEMIMORPH PYRAMIDAL,MSV,GRNLR. SOME FLUO. SOL-ACIDS. RARE-FRNKLN	4.2.2.1
4 -4.5	ANCYLITE	ORTH.	3.95	CBT	PSEUDO-OCTAH,CRUSTS,VITR. SPLINTERY. SOL-ACIDS. DRUSES IN ALK PEGS	16.2.10.
4 -5	LITHIOPHILITE	ORTH.	3.50-3.58	PHO	MSV. ANHEDRAL. VITR. CLVGS. UNEVN. SOL-ACIDS. PEGMATITES	38.1.1.2
4 -5	CURITE	ORTH.	7.26	OXD	MSV. SACCH. STRIATED PRISMS. ADMN. 1 CLVG. SOL-ACIDS. ALTER AFTER URANINITE	5.3.2.
4 -5.5	LIMONITE		2.7 -4.3	OXD	FIELD TERM FOR HYDROUS IRON OXIDES OF UNKNOWN IDENTITES.	7.1.3.
4.5-5	BERZELIITE	ISOM.	4.08	ARS	MSV. ROUNDED GRAINS. UNEVN. SOL-HCL,HNO3. LS SKARNS	38.2.1.1
4.5-5	MANGANBERZELIITE	ISOM.	4.46	ARS	DISTINGUISHED FROM BERZELIITE BY CHEMICAL ANALYSIS	38.2.1.2
4.5-5	THORITE	TETR.	5.2 -5.4	NSI	TERMINATED PRISMS.PRISM CLVG. MOST IS ALTERED. RARE ACCESS-SYENITE	W496 N
4.5-5	SCHEELITE	TETR.	6.08-6.12	WOS	DIPYRAMIDAL,MSV,GRNLR,VITR.CLVGS.UNEVN.DCMP-HCL,HNO3.FLUO-BLUE(X-,SW ULTRAVI-OLET-, + CATHODE-RAYS),THERMOLUM. HI-T VNS,PEGS,GREISSENS,CNTCT META RKS,ETC.	48.1.3.1
5	DURANGITE	MNCL.	3.94-4.07	ARS	PYRAMIDAL. VITR. PRISM CLVG. SOL-H2SO4.WITH CASSITERITE(E.G.,PEGS)	41.5.7.

NONMETALLIC ORANGE

	Mineral		Crystal System	S.G.		Description	Ref
5 -6	MARIALITE (SCAPOLITE)		TETR.	2.50-2.62	TSI	ME-0-20. PURE MARIALITE UNKNOWN IN NATURE	W352 4-321
5 -6	DIPYRE (SCAPOLITE)		TETR.	2.57-2.69	TSI	ME-20-50.	W352 4-321
5 -6	MIZZONITE (SCAPOLITE)		TETR.	2.67-2.74	TSI	ME-50-80.	W352 4-322
5 -6	SCAPOLITE		TETR.	2.50-2.78	TSI	GROUP NAME. MOST TYPES INDISTINGUISHABLE MEGASCOPICALLY BUT S.G. MAY BE IN-DICATIVE. LONG,STRIATED PRISMS + AGGS OF COARSE XLS. SUBCONCH. POOR CLVG GIVING IRREGULAR, STRIATED-APPEARING SURFACES. MARIALITE IS INSOL; MEIONITE IS DCMP IN HCL. MOST FLUO—YEL-RED(ULTRAVIOLET,LW). CNTCT META CALC RKS,PEGS	W352 4-321
5 -6	(WERNERITE) (SCAPOLITE)		TETR.	2.50-2.78	TSI	GROUP NAME NO LONGER USED	W352 4-322
5 -6	MEIONITE (SCAPOLITE)		TETR.	2.73-2.78	TSI	ME-80-100. PURE MEIONITE UNKNOWN IN NATURE	W352 4-321
5 -6	ROSENBUSCHITE		TRCL.	3.3	NSI	PRISMATIC, ACICULAR. 1 PERF CLVG. SOL-HCL. NEPHELINE SYENITE-RARE	W518 N
5.5-6.5	OPAL		TETR.	1.8 -2.25	SIL	SUBMICROXLINE QTZ + H2O.RESIN-PEARLY. MSV.MAMILLARY CRUSTS,CONCH.SOL-HF +CAUS-TIC ALKS.SOME FLUO—YEL-GREEN. INCRUSTATIONS, IN CAVITIES(ESP IN VOLCANIC RKS)	D-111 287
6	HUMITE		ORTH.	3.20-3.32	NSI	GROUP NAME, ALSO APPLIED TO SPECIES OF GROUP. IRREGULAR-SHAPED GRAINS, VITR-RESIN, GELAT. SIMILAR TO OLIVINE. CONTACT METAMORPHOSED DOLOMITIC LIMESTONES	W514 1-49
6 -6.5	MICROCLINE (K-FELDSPAR)		TRCL.	2.56-2.63	TSI	DISTINGUISHED FROM ORTHOCLASE OPTICALLY. PLUTONIC IG RKS, PEGS	W308 4-7
6 -6.5	ORTHOCLASE (K-FELDSPAR)		TRCL.	2.55-2.63	TSI	DISTINGUISHED FROM MICROCLINE OPTICALLY. PLUTONIC IG RKS, PEGS	W303 4-7
6 -6.5	PERTHITE (FELDSPARS)		TRCL.	2.56-2.65	TSI	MEGASCOPICALLY, MICROSCOPICALLY,OR SUBMICROSCOPICALLY INTERDIGITA-TED MICROCLINE OR ORTHOCLASE + PLAGIOCLASE (TYPICALLY ALBITE)	W299 4-7
6 -6.5	OLIGOCLASE (PLAGIOCLASE)		TRCL.	2.62-2.67	TSI	AN10-30. SEE FELDSPAR. PLUTONIC IG RKS, FEW PEGS. SUNSTONE IS VAR.	W312 4-95
6 -6.5	FELDSPAR			2.55-3.39	TSI	INCLUDES PLAGIOCLASE(NA-CA)SERIES, ALKALI(K-NA) AND BA FELDSPARS. MOST EASI-LY DISTINGUISHED BY NONMEGASCOPIC MEANS. MONOCLINIC ONES HAVE 2 CLVGS AT 90° AND SIMPLE(IF ANY) TWINNING; TRICLINIC ONES HAVE 2 CLVGS UP TO 4° OFF 90° AND TYPICALLY HAVE POLYSYNTHETIC TWINNING (GIVING STRIATED APPEARANCE ON SOME CLVG SURFACES OF PLAGIOCLASE). PEGMATITES; IG, META, AND SOME SED ROCKS	W261 4-1
6 -7.5	SPESSARTINE		ISOM. A	4.19	NSI	MN-AL GARNET. EQUANT CRYSTALS. MN-RICH META RKS + VEINS, PEGS	W486 1-99
6 -7.5	GARNET		ISOM.	3.58-4.32	NSI	GROUP NAME. DOMINANT MOLECULE DETERMINES NAME. DISTINGUISHED BY S.G. + NONMEGA-SCOPIC MEANS.EQUANT XLS.MSV.VITR-DULL.DODEC PARTING. SLIGHTLY- TO IN-SOL IN HF	W483 1-77
6.5	SARD		HEXA.	2.57-2.64	SIL	BRWNISH ORANGE + RED CHALCEDONY (MICROCRYSTALLINE QUARTZ)	D-111 206
7	QUARTZ		HEXA.	2.65	SIL	DOUBLY TERMINATED HEX PRISMS STRIATED ⊥ LENGTH(LO-T), BIPYRAMIDS (HI-T).MSV.VITR.CONCH.SOL-HF.WDSPRD IN MOST RK TYPES + VEINS	D-111 9
7.5	ZIRCON		TETR.	4.6 -4.7	NSI	TERMINATED PRISMS. ADMN. SOME FLUO-YEL-ORANGE. ACCESS-IG, SANDS	W494 1-59
10	DIAMOND		ISOM.	3.50-3.53	ELE	OCTAHEDRAL. ADMN-GRSY. BRTL. OCTAH CLVG. ULTRABASIC IGS + PLACERS	1,2,4,1

NONMETALLIC YELLOW

HARDNESS	NAME	XL. SYS.	SPECIFIC GRAVITY	CHEM. CLASS	REMARKS	REFERENCES
1	SASSOLITE	TRCL.	1.46-1.50	OXD	SCALES. PEARLY. BASAL CLVG(FLEX). BITTER,SALTY TASTE. HOT SPRINGS	6.2.1.
1	HANNAYITE	TRCL.	1.89	PHO	STRIATED,THIN XLS. CLVGS. DATA LACKING. GUANO DEPOSITS	39.1.2.
1	GAMMA SULFUR	MNCL.	2.05	ELE	LIGHT YELLOW-COLORLESS. MINUTE EQUANT XLS. ADAMANTINE. FUMAROLES	1.2.3.3
1	MOLYSITE	HEXA. A	2.90	HAL	MSV. BASAL CLVG. DLQSNT. VOL SUBLIMATE	9.3.1.
1	SILLENITE	ISOM.	8.80	OXD	GRNLR. WAXY-DULL. RARE-SCNDRY AFTER BISMUTITE(DURANGO,MEXICO)	4.4.8.
1-1.5	NATRON	MNCL.	1.48	CBT	GRNLR CRUSTS. VITR. CLVGS. ALK TASTE. IN SOLUTIONS, EFFLORESCENCES	15.1.6.
1-1.5	THERMONATRITE	ORTH. A	2.26	CBT	CRUST-EFFLOR. VITR. 1 CLVG. ALK TASTE. RARE-WITH TRONA-ARID SOILS	15.1.1.
1-1.5	EPISTOLITE	MNCL.	2.89	NSI	PLATY XLS. 1 PERF, 1 GOOD CLVG. RARE IN PEG-JULIANEHAAB,GREENLAND	W508 N
1-2	KOENENITE	HEXA.	1.98	HAL	CRUSTS. PEARLY. BASAL CLVG(FLEX). DCMP-HOT H2O. K-DEPS(N. GERMANY)	10.2.8.
1-2	VERMICULITE	MNCL.	2.2-2.4	PSI	BRONZY. MICACEOUS. EXFOLIATES WITH SWELLING ON HEATING. ATTKD-ACID (SILICA RESID). ALTER PROD OF BIOTITE(ETC.)-HYDROTHERMAL OR WXING	W396 3-246
1-2	MONTMORILLONITE (CLAY)	MNCL.	2.2-2.7	PSI	IRREG+HEX-SHAPED PLTS. "SWELLING CLAY." ALTER-HI MG +CA,LO K RKS (ALKALINE CONDITIONS). SOILS, BENTONITES, ETC.	W398 3-226
1-2	PYROPHYLLITE	MNCL.	2.65-2.90	PSI	LAMELLAR. CLVG-INELAST PLTS. PARTLY DCMP-H2SO4. VNS, FEW SCHISTS	W361 3-115
1-2	BEIDELLITE (CLAY)	MNCL.	2.	PSI	"SWELLING CLAY." ALTER-HI MG +CA,LO K RKS(ALK CONDITIONS). RARE-GOUGE (COLORADO). BENTONITES	W398 3-226
1-2.5	NONTRONITE (CLAY)	MNCL.	2.5-3.0	PSI	SUBTRANSLUCENT. LATHS, MSV. GELAT. "SWELLING CLAY." ALTER-HI MG + CA,LO K RKS(ALK CONDITIONS). WITH OPAL +/OR QTZ IN VNS, BENTONITE	W398 3-226
1-3	GARNIERITE	MNCL.	2.27-2.87	PSI	NI-SERP. MAGNETIC AFTER HEATING. GELAT. SERPENTINE-RICH ROCKS	W379 3-175
1-3	CLAY (CHLORITE)	MNCL.	1.85-3.0	PSI	NAME GIVEN SEVERAL HYDROUS AL-SILICATES WITH ALKALIES OR ALKALINE EARTHS + MG OR FE. NAMES UNSETTLED. MOST DISTINGUISHED BY NONMEGASCOPIC MEANS. FINE-GRAINED AGGS(COMPACT,MEALY,ETC.). EARTHY-PEARLY,UNCTUOUS,PLASTIC WHEN WET,DEHYDRATED ON HEATING. SOME ARE ABSORBENT. WXING PRODS,HYDROTHERMAL ALTER, DETRITAL, DIAGENIC	W398 3-191
1.5	TESCHEMACHERITE	ORTH. A	1.57	CBT	COMPACT MASSES. CLVG. BRTL. SOL-H2O. EFF-ACIDS. GUANO DEPS	13.1.3.
1.5	APJOHNITE	MNCL.	1.78	SUF	FIBR. SILKY. SOL-H2O. EFFLORESCENCES IN SHELTERED PLACES	29.7.3.3
1.5	PICKERINGITE	MNCL.	1.73-1.79	SUF	FIBR. VITR. ASTRINGENT. EFFLORESCENCES IN SHELTERED PLACES	29.7.3.1
1.5	HALOTRICHITE	MNCL.	1.89	SUF	FIBR. VITR. ASTRINGENT. EFFLORESCENCES IN SHELTERED PLACES	29.7.3.2
1.5	FELSOBANYAITE	ORTH.	2.33	SUF	RADIAL AGGS LAMELLAR XLS. PEARLY. 2 CLVGS. WITH MARCASITE,STIBNITE AND BARITE (FELSOBANY, HUNGARY)	31.2.3.
1.5	ARSENOLITE	ISOM.	3.86-3.88	OXD	WHITE-COLOR TINTED. OCTAH, CRUSTS. VITR-SILKY. OCTAH CLVG. ASTRIN-GENT-SWEETISH TASTE. SCNDRY AFTER ARSENOPYRITE + OTHER AS-MINS	4.4.2.1
1.5	IODARGYRITE	HEXA.	5.69	HAL	EXPOS-YEL. MSV. PRISM-HEMIMORPHIC. RESIN. BASAL CLVG(FLEX). DCMP-H2SO4, HNO3, KI SOLUTION. OXIDIZED ZONES OF AG DEPOSITS	9.1.4.
1.5	CALOMEL	TETR.	7.15	HAL	EXPOS-DARKENS. TBLR. MSV. ADMN. PRISM CLVG. PLASTIC.FLUO-RED. SOL-AQ-REG. SCNDRY AFTER SEVERAL MERCURY MINERALS	9.1.5.

54

NONMETALLIC YELLOW

Hardness	Mineral	Crystal System	S.G.	Streak/Group	Description	Ref
1.5-2	SALAMMONIAC	ISOM.	1.53	HAL	SKELETAL, DENDRITIC. VITR.OCTAH CLVG. SALTY TASTE. VOL SUBLIMATE	9.1.2.
1.5-2	ALUNOGEN	TRCL.	1.77	SUF	FIBR MASSES. VITR-SILKY. SHARP, ACID TASTE. EFFLOR IN FES2-BEARING ROCKS	29.8.6.
1.5-2	HUMBOLDTINE	ORTH.	2.28	CCC	CAPILLARY,BOTRYOIDAL FIBR CRUSTS,GRNLR,DENSE. DULL-RESIN. CLVGS. SOL-ACIDS, ASSOCIATED WITH LIGNITE	50.1.3.
1.5-2	SODA-NITER	HEXA.	2.24-2.29	NIT	INCRUST, RHOMB. VITR. RHOMB + OTHER CLVGS. COOLING TASTE. IN ARID SOILS IN PROTECTED LOCATIONS	18.1.1.
1.5-2	REALGAR	MNCL.	3.48-3.56	SLD	RED, OR-YEL. RESIN-GRSY. DCMP-HNO3. WITH AS-MINS IN PB,AG,AU-VEINS	2.6.10.
1.5-2	COVELLITE	HEXA.	4.6 -4.76	SLD	SOME IRID. GRAY STRK. 1 PERF CLVG-FLEX PLTS. ALTER OF CU-SULFIDES	2.6.8.1
1.5-2.5	BILINITE	MNCL.	1.88	SUF	FIBR-RADIAL. SOL-H2O. RARE-ALTER OF FES2 IN LIGNITE(SCHWAZ,BOHEMIA)	29.7.3.5
1.5-2.5	SULFUR	ORTH.	2.07	ELE	CONCH-UNEVN. BRTL. BURNS-BLU FL + ACRID ODOR. SOL-CS2. VOLS + SED	1.2.3.1
1.5-2.5	SIDERONATRITE	ORTH.	2.15-2.35	SUF	FIBR CRUSTS,NODULAR. 1 CLVG. DCMP-BOILING H2O. WITH OTHER SULFATES IN ARID AREAS	31.5.3.
1.5-2.5	TYUYAMUNITE	ORTH.	3.3 -4.35	VAS	SCALES-RADIAL AGGS,MSV,POWDERY. ADMN,PEARLY. 3 CLVGS. SOL-HNO3, HCL,H2SO4. VNLTS IN LS + WITH CARNOTITE	47.1.2.
1.5-2.5	HELIOPHYLLITE	ORTH.	6.89	ASI	PSEUDOTETRAGONAL,PYRAMIDAL-STRIATED,MSV-FOL,GRNLR. VITR.CLVG. 1 PERF CLVG. SOL-HNO3, WM HCL, ALKS. WITH ECDEMITE	46.1.2.
1.5-4.5	ARSENIOSIDERITE	TETR.	3.60	ARS	STRK-YEL. SUBTRANSLUCENT. RADIAL FIBR AGGS. SUBMET-SILKY. 1 CLVG. SOL-HOT ACIDS. SECONDARY - DIVERSE OCCURRENCES	42.6.4.
2	TAYLORITE	UNKN.		SUF	COMPACT LUMPS. PUNGENT + BITTER TASTE. SOL-H2O,ACIDS. GUANO DEPS	28.2.1.3
2	STERCORITE	TRCL.	1.62	PHO	XLINE MASSES, NODULES. VITR. SOL-H2O. GUANO DEPOSITS	39.1.1.
2	TACHYHYDRITE	HEXA.	1.67	HAL	ROUNDED MASSES,VITR. RHOMB CLVG. DLQSNT. SHAPP + BITTER.SED-K-RICH SALINE DEPOSITS	11.1.3.
2	STRUVITE	ORTH.	1.71	PHO	DIVERSE XLS,MOST HEMIMORPHIC.VITR.2 CLVGS. SOL-ACIDS, H2O(SLIGHT-LY). GUANO DEPOSITS, URINARY SEDIMENT, ETC.	40.1.1.
2	BOUSSINGAULTITE	MNCL. A	1.72	SUF	CRUSTS, STALAC, PRISM. CLVG SOL-H2O. FUMAROLIC, GEYSER DEPOSITS	29.3-7.3
2	DIETRICHITE	MNCL.	1.8	SUF	FIBR. SILKY. SOL-H2O. RARE-EFFLORESCENCES IN SHELTERED PLACES	29.7.3.4
2	KIROVITE	MNCL.	1.76-1.87	SUF	DISTINGUISHED FROM MELANTERITE BY NONMEGASCOPIC MEANS. SAME OCCUR	29.6.8.3
2	MELANTERITE	MNCL.	1.84-1.90	SUF	STALAC, CRUSTS,ETC. VITR. CLVGS. SWEET,ASTRINGENT,METALLIC TASTE. EXPOS(DRY AIR)--YEL-WHT + OPQ. EFFLORESCENCE-BY OXIDATION OF SUFS	29.6.8.1
2	PISANITE	MNCL.	1.95	SUF	DISTINGUISHED FROM MELANTERITE BY NONMEGASCOPIC MEANS. SAME OCCUR	29.6.8.2
2	SYLVITE	ISOM.	1.98-2.00	HAL	CUBIC, GRNLR. VITR. CUBIC CLVG. BITTER TASTE. SED EVAPORITE DEPS	9.1.1.2
2	BRUGNATELLITE	HEXA.	2.14	OXD	FLKS, FOL-MSV. PEARLY. BASAL CLVG. EFF-HCL. INCRUST-CRACKS IN SER-PENTINE	6.1.7.
2	HALITE	ISOM.	2.17	HAL	CUBIC, GRNLR. VITR. CUBIC CLVG. SALTY TASTE. SED EVAPORITE DEPS	9.1.1.1
2	RHOMBOCLASE	ORTH.	2.23	SUF	THIN TBLR, STALACTITIC. SUBVITR. BASAL(FLEX),PRISM CLVGS. CONCH-FIBR. SOL-ACIDS,H2O(SLOWLY). RARE-ALTER PROD OF SULFIDES	29.1.1.
2	GYPSUM	MNCL.	2.31-2.32	SUF	TBLR, MSV. SUBVITR-PEARLY. 1 PERF CLVG(FLEX,INELAST). SOL-HCL. SED-BEDDED DEPOSITS, FUMAROLIC, EFFLORESCENCES, ETC.	29.6.3.
2	GLAUCONITE	MNCL.	2.4 -2.95	PSI	PLTS IN ROUND GRAINS. ATTKD-HCL. MARINE DEPOSITS - GREENSANDS	W377 3-35

55

NONMETALLIC YELLOW

HARDNESS	NAME	XL. SYS.	SPECIFIC GRAVITY	CHEM. CLASS	REMARKS	REFERENCES
2	TELLURITE	ORTH.	5.88-5.92	OXD	ACICULAR, LATHS-STRIATED. 1 PERF CLVG. SOL-HCL,HNO3,ALKALIES. OXIDATION PRODUCT OF NATIVE TELLURIUM AND TELLURIDES	4.5.4.
2	MASSICOT	ORTH.	A 9.56	OXD	MSV, SCALY, GRSY-DULL. 2 CLVGS.SOL-HCL,HNO3,ALKALIES. OXIDIZED PB-DEPOSITS	4.2.7.
2 -2.5	MELLITE	TETR.	1.64	CCC	MSV,GRNLR,CRUSTS,NODULAR, RESIN,PRISM CLVG, CONCH. SOL-HNO3. FLUO-BLUE(SW). SECONDARY IN OPEN SPACES IN LIGNITE	50.2.1.
2 -2.5	MASCAGNITE	ORTH.	1.77	SUF	CRUSTS, STALACTITIC. VITR-DULL. 1 CLVG. SHARP,BITTER. FUMAROLIC	28.2.1.1
2 -2.5	SEPIOLITE (CLAY)	ORTH.	2.08	PSI	POROUS(FLOATS ON H2O WHEN DRY)-COMPACT, SMOOTH FEEL, GELAT, WITH SERP OR OPAL, DESERT SOILS, CLAYEY SEDS WITH SALINES, LAKE SEDS. MEERSCHAUM IS A VARIETY	W444 N
2 -2.5	DICKITE (CLAY)	MNCL.	2.62	PSI	ALTER|ACID CONDITIONS|OF FELDSPARS, ETC. HYDROTHERMAL WITH QUARTZ, SULFIDES, ETC.	W363 3-194
2 -2.5	PHLOGOPITE	MNCL.	2.76-2.90	PSI	MG-MICA. ATTKD-HCL. CNTCT META-LS. ULTRABASIC IGNEOUS ROCKS	W373 3-42
2 -2.5	BASSETITE	MNCL.	3.10	PHO	FANLK + CHESSBOARDLK GROUPS OF PLTS. 2 CLVGS. WITH OTHER SCNDRY URANIUM MINERALS	42.8.15.
2 -2.5	AUTUNITE	TETR.	3.1-3.2	PHO	TBLR, SUBPARALLEL,FOL,SCALY AGGS, SERRATE CRUSTS, VITR,PEARLY. 2 CLVGS. SOL-ACIDS. FLUO-YEL-GREEN. ALTER PROD OF U MINS(PEGS,VNS)	42.8.13.2
2 -2.5	URANOCIRCITE	TETR.	3.53	PHO	HABIT LK AUTUNITE. PEARLY. 3 CLVGS. WITH OTHER SCNDRY URANIUM MINS	42.8.13.3
2 -2.5	HYDROZINCITE	MNCL.	3.5-4.0	CBT	MSV,INCRUST. SILKY-DULL.1 CLVG. FLUO-BLUE,LILAC. SOL-ACIDS.SCNDRY WITH OTHER ZINC MINERALS	16.1.3.
2 -2.5	LANARKITE	MNCL.	6.92	SUF	MSV. ELONGATE. ADMN,PEARLY. CLVGS(THIN LAMINAE-FLEX), FLUO-YELIX-RAYS + ULTRAVIOLET RADIATION). SOL-WARM HNO3. RARE-LEAD DEPOSITS	30.2.1.
2 -2.5	SCHWARTZEMBERGITE	TETR.	7.39	IOD	ROUNDED FLAT PYRAMIDAL XLS, CRUSTS. ADMN. 1 CLVG. SOL-HCL. PB DEPS	22.1.2.
2 -3	PHOSPHURANYLITE	TETR.		PHO	EARTHY,SCALY-CRUSTS, PLTY. 1 CLVG. SOL-ACIDS. MXED U-BEARING PEGS	41.6.11.
2 -3	PASCOITE	TRCL.	1.87	VAS	STRK-YEL. GRNLR CRUSTS OF LATHLK XLS. VITR.1 CLVG. CONCH. SOL-H2O. EFFLORESCENCES - VANADIUM DEPOSITS. WITH CARNOTITE IN SANDSTONES	47.1.13.
2 -3	VASHEGYITE	UNKN.	1.96	PHO	SUBTRANSLUCENT. MSV,MICROXLINE-FIBR,DULL,SOL-ACIDS. WITH VARISCITE	42.9.3.
2 -3	BETA SULFUR	MNCL.	1.96-1.98	ELE	LIGHT YEL-COLORLESS PURE, BROWN IMPURE. THICK TABULAR. FUMAROLES	1.2.3.2
2 -3	RICHELLITE	UNKN.	2.	PHO	MSV COMPACT-FOL,RADIAL FIBR GLBLS. GRSY-HORNLK. SOL-ACIDS. RARE-WITH HALLOYSITE, ETC. (RICHELLE, BELGIUM)	42.6.7.
2 -3	BIANCHITE	MNCL.	2.07	SUF	CRUSTS. VITR. SOL-H2O. RARE-EFFLOR ON MINE WALLS(IN JULIAN ALPS)	29.6.6.2
2 -3	ROSSITE	TRCL.	2.45	VAS	GLASSLK LUMPS, VITR.1 CLVG. BRTL. SOL-H2O(SLOW). VNS IN CARNOTITE-BEARING SANDSTONE (BULL PEN CANYON, COLORADO)	47.1.11.
2 -3	PITTICITE	UNKN.	2.2-2.5	ARS	MSV,BOTRYOIDAL,OPALINE CRUSTS,ETC. DULL-GRSY. SOL-ACIDS. OXIDATION ZONES OF AS-BEARING DEPOSITS, DEPOSITED FROM MINE + SPRING WATERS	43.2.6.
2 -3	FIBROFERRITE	ORTH.	1.84-2.52	SUF	FIBR CRUSTS. SILKY. 1 CLVG. DCMP-H2O. SCNDRY-BY OXIDATION OF FES2	31.6.6.
2 -3	COOKEITE (CLAY)	MNCL.	2.69	PSI	LI-BEARING CHLOR. WITH TOURMALINE IN PEGS, PROBABLY AS ALTER PROD	W378 3-147
2 -3	SHERIDANITE (CLAY)	MNCL.	2.65-2.80	PSI	INTERMEDIATE BETWEEN CLINOCHLORE + CORUNDOPHILITE. RARE-META ROCKS	WN 3-137
2 -3	TINTICITE (CLAY)	ORTH.	2.7-2.9	PHO	DENSE EARTHY-PORCELANEOUS MASSES.SCNDRY-WITH JAROSITE,LIMONITE,ETC	42.8.4.
2 -3	XANTHOXENITE	UNKN.	2.8-2.97	PHO	MASSES + CRUSTS-PLTY XLS. DULL-WAXY. 1 CLVG. PHOSPHATE-RICH PEGS	42.8.10.
2 -3	CHAMOSITE (CLAY)	MNCL.	2.9-3.1	PSI	FINE-GRAINED AGGS. IN LATERITES, AS OOLITES, IN MATRIX,ETC. OF IRONSTONES	WN 3-164

NONMETALLIC YELLOW

H	Mineral	Crystal	G	Habit	Cleavage/Characteristics	Class	Reference
2 -3	SALEEITE	TETR.	3.27	PHO	TBLR. CLVGS. WITH OTHER SCNDRY URANIUM MINERALS		42.8.13.4
2 -3	SHARPITE	ORTH.	3.3	CBT	CRUSTS OF RADIATING FIBERS. EFF-ACIDS. RARE-URANIUM DEP(KATANGA)		16.1.15.
2 -3	TROEGERITE	TETR.	3.3	ARS	TBLR.,SUBPARALLEL AGGS. PEARLY. CLVGS. SOL-ACIDS. SCNDRY-VEIN MINS		42.7.6.
2 -3	CHLORITE	MNCL.	2.6-3.3	PSI	GROUP NAME, MEMBER NAMES UNSETTLED. ONLY FEW RARE VARIETIES MEGASCOPICALLY DISTINGUISHABLE. DISSEM FLKS, PEARLY, 1 PERF CLVG TO FLEX,INELAST FOLIA. DCMP- IN BOILING H2SO4. SCNDRY-HYDROTHERMAL ALTER, LO-GRADE META, DETRITAL, AUTHIGENIC		W381 3-131
2 -3	FERGHANITE	ORTH.	3.31	VAS	SCALES. WAXY. 2 CLVGS. WXING PROD OF TYUYAMUNITE(?)-WITH U MINS		47.1.4.
2 -3	URANOSPINITE	TETR.	3.45	ARS	TBLR. PEARLY. 2 CLVGS. WITH OTHER SCNDRY URANIUM MINERALS		42.8.13.6
2 -3	GREENALITE	MNCL.	2.5-3.5	PSI	FINE GRAINED. RARE-IRON FORMATIONS (MESABI RANGE OF MINNESOTA)		W380 3-164
2 -3	LAMPROPHYLLITE	ORTH.	3.45-3.54	NSI	ELONGATE PLTS. 1 GOOD, 1 POOR CLVG. NEPHELINE SYENITES		W527 N
2 -3	URANOPHANE	MNCL.	3.8-4.	NSI	MSV. FIBR. RADIATING, TBLR. 1 CLVG. GELAT. IN GRANITE + PEGMATITES		W529 N
2 -3	CARNOTITE	UNKN.	4.1	VAS	POWDER,CRUSTS,MICRO XLS. DULL-SILKY.1 CLVG. SOL-ACIDS. SCNDRY-E.G.,SANDSTONE MATRICES		47.1.1.
2 -3	SCHOEPITE	ORTH.	4.8	OXD	TBLR. ADMN. BASAL CLVG. WITH OTHER SECONDARY URANIUM MINERALS		5.2.4.
2 -3	BECQUERELITE	ORTH.	5.2	OXD	YEL STRK. TBLR-STRIATED, MSV. ADMN. BASAL CLVG. RDACTV. OXIDATION PRODUCT OF URANINITE		5.2.3.
2 -3	XANTHOCONITE	MNCL.	5.40-5.68	SST	STRK-ORANGE-YEL. TBLR, LATHLK. ADMN. 1 CLVG. WITH RUBY SILVERS		3.2.2.3
2 -3	PHOSGENITE	TETR.	6.13	CBT	PRISM, MSV. ADMN. CLVGS(PERCUSSION FIGURE ON BASAL PLANE) FLUO-YEL (LW ULTRAVIOLET,CATHODE + X-RAYS). EFF-HNO3. SCNDRY AFTER PB MINS		16.1.7.
2 -3	URANOSPHAERITE	ORTH.	6.36	OXD	YEL STRK. GLBL FMS. GRSY. 1 CLVG. OXIDATION PROD OF PITCHBLENDE		5.3.3.
2 -3	BEYERITE	TETR.	6.50-6.56	CBT	PLTY, MSV. VITR(XLS). CONCH. EFF-ACIDS. SCNDRY-WITH BISMUTITE		16.2.5.
2 -3	TRIGONITE	MNCL.	6.1-7.1	ASI	DOMATIC. VITR-ADMN. CLVGS. UNEVN. SOL-DILUTE ACIDS. RARE-WITH NATIVE LEAD + DIXENITE IN CREVICES IN DOLO-HAUSMANNITE ORE(LANGBAN)		45.1.2.
2 -3	SAHLINITE	MNCL.	7.95	ARS	AGGS OF SCALES.1 CLVG. RARE-IN HAUSMANNITE-DOLO RK(LANGBAN,SWEDEN)		41.1.1.
2 -4	MICA	MNCL.	2.4-3.4	PSI	GROUP NAME. SOME SPECIES DISTINGUISHABLE ON BASIS OF COLOR. SERIES ARE NOT COMPLETE. HEX-SHAPED XLS,DISSEM. PERF CLVG TO FLEX,ELAST FLKS. PERCUSSION + PRESSURE FIGURES MAY BE IMPOSED ON FLKS. CHIEF OCCURRENCES ARE NOTED UNDER SPECIES		W365 3-1
2 -4	BAUXITE		2.	3.5 OXD	FIELD TERM FOR MATERIALS RICH IN HYDROUS ALUMINUM OXIDES.		6.2.3.
2.5	BISMOCLITE	TETR.	7.12	HAL	IN SERIES WITH DAUBREEITE. MSV, SCALY, GRSY-DULL. 1 CLVG. PLASTIC. SOL-ACIDS. SCNDRY-ALTER PROD OF BISMUTHINITE OR NATIVE BISMUTH		10.1.6.2
2.5	DAUBREEITE	TETR. C	7.56	HAL	IN SERIES WITH BISMOCLITE,WHICH SEE-SCNDRY AFTER BISMUTHINITE + BI		10.1.6.3
2.5	MITRIDATITE	UNKN.		PHO	MSV,NODULES,CRUSTS. DULL-RESIN. FRIABLE-DENSE. SOL-HOT ACIDS. OOLITIC SEDIMENTARY IRON ORES - RARE (SOUTHERN RUSSIA)		42.6.6.
2.5	OXAMMITE	ORTH.	1.4-1.6	CCC	LAMELLAR,POWDERY. 1 CLVG. WITH MASCAGNITE IN GUANO DEPOSITS(PERU)		50.1.4.
2.5	CARNALLITE	ORTH.	1.60	HAL	GRNLR, PSEUDOHEX PRISM. GRSY. CONCH. DLQSNT. BITTER. SED SALT DEPS		11.1.2.
2.5	PHOSPHORROESSLERITE	MNCL.	1.72-1.73	PHO	EQUANT,SKELETAL. VITR,EXPOS-DULL. VITR. 2 CLVGS. UNSTABLE ABOVE CIRCA 25°C IN COLD (30°C) WET MUD, MINE WORKINGS (SALZBG, AUSTRIA)		39.2.5.2
2.5	TRUDELLITE	HEXA.	1.93	HAL	MSV. VITR. RHOMB CLVG.ASTRINGENT. DLQSNT. RARE-TARAPACA PROV,CHILE		12.1.3.
2.5	PICROMERITE	MNCL. A	2.03	SUF	PRISM, CRUSTS. VITR. BITTER. FUMAROLIC, MARINE SALINE DEPS-AS BEDS		29.3.7.1
2.5	COQUIMBITE	HEXA.	2.10-2.12	SUF	PRISM, GRNLR. VITR. CLVGS. ASTRINGENT. SOL-ACIDS. WITH OTHER SO4'S		29.8.3.

NONMETALLIC YELLOW

HARDNESS	NAME	XL. SYS.	SPECIFIC GRAVITY	CHEM. CLASS	REMARKS	REFERENCES
2.5	PYROAURITE	HEXA.	2.10-2.14	OXD	TBLR-STRIATED. WAXY,PEARLY. BASAL CLVG(FLEX). EFF-HCL. LO-T VNS	6.1.5.3
2.5	SJOGRENITE	HEXA.	2.08-2.14	OXD	PLTS-STRIATED. WAXY,PEARLY. BASAL CLVG(FLEX).EFF-HCL. RARE-LANGBAN	6.1.6.3
2.5	CHRYSOTILE (SERPENTINE)	MNCL.	2.2	PSI	FIBROUS, SOME IS ASBESTIFORM. SECONDARY, COMMONLY IN VEINS	W379 3-170
2.5	ALUMOHYDROCALCITE	MNCL.	2.23	CBT	CHALKY MASSES. 2 CLVGS. SOL-ACIDS,DCMP-BOILING H2O. WITH ALLOPHANE	16.2.4.
2.5	UNGEMACHITE	HEXA.	2.29	SUF	TBLR. VITR. 1 CLVG. SOL-DILUTE HCL. VN-FILLINGS IN JAROSITE(CHU-QUICAMATA, CHILE)	31.4.2.
2.5	CLINO-UNGEMACHITE	MNCL.	2.29	SUF	MEGASCOPICALLY INDISTINGUISHABLE FROM UNGEMACHITE WITH WHICH IT OCCURS	31.4.3.
2.5	CHLORMANGANOKALITE	HEXA.	2.31	HAL	RHOMB. VITR. CONCH. DLQSNT. RARE-WITH HALITE,ETC.-CAVITIES IN VOLS	11.4.4.
2.5	BRUSHITE	MNCL.	2.33	PHO	NEEDLEK,ETC. VITR,PEARLY. 2 CLVGS. SOL-HCL. PHOSPHATE DEPOSITS	39.2.1.1
2.5	METASIDERONATRITE	ORTH.	2.46	SUF	XLINE AGGS. SILKY. 3 CLVGS. FIBR. SOL-DILUTE ACIDS, BOILING H2O. SCNDRY (E.G., WITH ALUMS, ETC. -- CHUQUICAMATA, CHILE)	31.5.2.
2.5	ERIOCHALCITE	ORTH. A	2.47	HAL	LICHENLK AGGS. VITR. PRISM,BASAL CLVGS. SOL-HCL, NH4OH. SCNDRY, FU-MAROLE SUBLIMATE	9.2.7.
2.5	METAVOLTINE	HEXA.	2.5	SUF	TBLR, GRNLR-SCALY AGGS. RESIN. 1 CLVG. PARTLY SOL-H2O,DCMP-DILUTE ACIDS(SLOWLY). WITH OTHER HYDROUS SULFATES IN DIVERSE ENVIRONMENTS	31.6.9.
2.5	SCHROECKINGERITE	HEXA.	2.51	CBT	GLBLR,SCALY. VITR. BASAL CLVG(FLEX).FLUO-YEL GREEN. SOL-H2O, ACIDS. ALTER PRODUCT OF URANINITE, EFFLORESCENCE	15.2.4.
2.5	CALCIOFERRITE	UNKN.	2.53	PHO	FOL NODULAR MASSES. PEARLY. CLVGS. RARE-NODULES IN CLAY(BAVARIA)	42.8.9.
2.5	SYNGENITE	MNCL.	2.58-2.60	SUF	CRUSTS, LAMELLAR AGGS. PRISM. VITR. CLVGS. SOL-H2O(IN PART WITH FM OF GYPSUM). RARE-MARINE SALT DEPS. PRODUCT OF VOLCANIC ACTION	29.3.1.
2.5	PHARMACOSIDERITE	TETR.	2.80	ARS	STRIATED CUBES;TETRA,GRNLR. ADMN.1 CLVG. UNEVN. SECTILE. OXIDATION PRODUCT OF OTHER ARSENIC MINERALS	42.9.1.
2.5	KRAUSITE	MNCL.	2.84	SUF	PRISM. VITR. 2 CLVGS. SOL-HCL, DCMP-H2O(BOTH SLOWLY). WITH OTHER SULFATES	29.5.1.
2.5	PARAGONITE	MNCL.	2.85	PSI	NA-MICA. DISTINGUISHED FROM MUSCOVITE BY CHEM OR X-RAY ANAL. WIDE-SPREAD IN SCHISTS AND PHYLLITES. ALSO IN VEINS	W370 3-31
2.5	SZOMOLNOKITE	MNCL.	3.03-3.07	SUF	BIPYRAMIDAL,STALAC.VITR. CONCH.SOL-H2O(SLOW). EFFLOR-ACID SOLUTION	29.6.2.2
2.5	SENGIERITE	ORTH.	4.	VAS	SIX-SIDED PLTS;CRUSTS. VITR. 1 CLVG. BRTL. SOL-ACIDS. SCNDRY	47.1.3.
2.5	TUNGSTITE	ORTH.	5.5	OXD	MSV. PLTY. RESIN. BASAL + PRISM CLVG. SOL-ALKALIES. OXIDATION PROD OF TUNGSTEN MINERALS	4.6.5.
2.5	CHLORARGYRITE	ISOM.	5.56	HAL	CRUSTS,CUBIC. RESIN. DUCTILE. EXPOS-VIOLET-BRWN. SCNDRY IN AG DEPS	9.1.1.4
2.5	MIERSITE	ISOM.	5.64	HAL	CRUSTS,AGGS. ADMN. DODEC CLVG. WITH IODYRITE(BROKEN HILL,AUSTRALIA)	9.1.3.2
2.5	MARSHITE	ISOM.	5.68	HAL	YEL STRK. TETRA-STRIATED. ADMN. FLUO-RED. RARE-CU DEPS	9.1.3.3
2.5	COTUNNITE	ORTH. A	5.80	HAL	GRNLR, LATHLK. ADMN-SILKY. 1 CLVG. SOL-H2O,ACIDS. VOL SUBLIMATE, ALTERATION PRODUCT OF GALENA	9.2.6.
2.5	BROMARGYRITE	ISOM.	6.50	HAL	CRUSTS,CUBIC. RESIN. DUCTILE. SECONDARY IN SILVER DEPOSITS	9.1.1.5
2.5	MENDIPITE	ORTH.	7.24	HAL	FIBR. RADIATED. SILKY,RESIN. PRISM + OTHER CLVGS. SOL-HNO3. SCNDRY WITH CERUSSITE, ETC.	10.1.4.
2.5	EGLESTONITE	ISOM.	8.33-8.45	HAL	EXPOS-DK BRWN, BLK. YEL STRK. MSV, CRUSTS, DODEC. RESIN. CONCH. DCMP-ACIDS(CALOMEL RESIDUE). MERCURY DEPOSITS	10.1.1.

NONMETALLIC YELLOW

Hardness	Mineral	Sp.Gr.	Crystal Sys.	Cleav.	Habit, Properties	Ref.
2.5	TERLINGUAITE	8.73	MNCL.	HAL	EXPOS-OLIVE GREEN. PRISM, POWDERY. ADMN. 1 CLVG. SOL-ACIDS. RARE- MERCURY DEPOSIT (TERLINGUA, TEXAS)	10.1.2.
2.5	MINIUM	8.9-9.2	UNKN.	OXD	MSV. EARTHY-GRSY-DULL. SOL-HCL(CL?),DCMP-HNO3.ALTER PROD OF GALENA	4.3.1.
2.5-3	GAYLUSSITE	1.99	MNCL.	CBT	TBLR,STRIATED. VITR. PRISM CLVG. EFF-ACIDS. WITH TRONA,ETC.	15.2.3.
2.5-3	TRONA	2.14	MNCL.	CBT	FIBR,ELONGATE. VITR. 1 PERF,OTHER CLVGS. ALK TASTE.SOL-H2O. EFF-ACIDS. SALINE LAKE DEPS, EFFLORESCENT ON SOILS IN ARID REGIONS	13.1.4.
2.5-3	KAINITE	2.15	MNCL.	SUF	GRNLR, TBLR. VITR. 1 CLVG. SALTY,BITTER. IN MARINE POTASH DEPS	31.4.1.
2.5-3	COPIAPITE	2.08-2.17	TRCL.	SUF	CRUSTS-GRNLR OR SCALY AGGS. PEARLY. CLVGS. SOL-H2O. SCNDRY-WITH OTHER SULFATES AS OXIDATION PRODUCTS OF SULFIDES(E.G., PYRITE)	31.6.11.1
2.5-3	MAGNESIOCOPIAPITE	2.08-2.17	TRCL.	SUF	INDISTINGUISHABLE MEGASCOPICALLY FROM COPIAPITE, WHICH SEE	31.6.11.2
2.5-3	CUPROCOPIAPITE	2.08-2.17	TRCL.	SUF	INDISTINGUISHABLE MEGASCOPICALLY FROM COPIAPITE, WHICH SEE	31.6.11.3
2.5-3	AMARILLITE	2.19	MNCL.	SUF	EQUANT, // GROUPS. VITR-ADMN. PRISM CLVG-ASTRINGENT. VNS IN COQUIM-BITE	29.5.3.2
2.5-3	LEONITE	2.20	MNCL.	SUF	ANHEDRAL, TBLR. WAXY. CONCH. BITTER. SCNDRY IN MARINE SALINE DEPS	29.3.5.2
2.5-3	WHEWELLITE	2.23	MNCL.	CCC	HEART-SHAPED XLS. VITR,PEARLY. CLVGS. CONCH. SOL-ACIDS. ORGANIC ASSOCIATION, VEINS	50.1.1.
2.5-3	LIEBIGITE	2.41	ORTH.	CBT	EQANT,GRNLR,SCALY. VITR-PEARLY. FLUO-GREEN. EFF-ACIDS. MINE WALLS	15.2.9.
2.5-3	LOEWEITE	2.37-2.42	UNKN.	SUF	ANHEDRAL GRAINS. VITR. CONCH. BITTER. DISSEM IN MARINE SALINE DEPS	29.3.4.
2.5-3	THENARDITE	2.66	ORTH.	SUF	DIPYRAMIDAL, CRUSTS. VITR. CLVGS. FAINTLY SALTY. SALINE LAKE DEPS, EFFLORESCENCES, FUMAROLIC	28.2.4.
2.5-3	LANTHANITE	2.61-2.74	ORTH.	CBT	PLTY-TBLR, GRNLR, PEARLY. MICACEOUS CLVG. SOL-ACIDS. COATING ON CERITE, ALLANITE, ETC.	15.2.10.
2.5-3	GLAUBERITE	2.75-2.85	MNCL.	SUF	TBLR, PRISM, DIPYRAMIDAL-STRIATED. VITR,PEARLY. CLVGS. CONCH. BRTL. SALTY TASTE. SOL-HCL. SALINE DEPS,DISSEM IN CLASTIC SEDS, FUMAROLIC	28.4.2.
2.5-3	PARSONSITE	5.37	MNCL.	PHO	CRUSTS,POWDER,LATHLK. SUBADMN. SOL-ACIDS. SCNDRY IN U-BEARING DEPS	41.8.4.
2.5-3	VALENTINITE	5.76	ORTH.	OXD	PRISM, TBLR, MSV. ADMN. PRISM CLVG. SOL-HCL. SCNDRY AFTER SB-MINS	4.4.4.
2.5-3	CROCOITE	5.96-6.02	MNCL.	CHR	PRISM,MSV,ETC. ADMN-VITR. CLVGS. CONCH. SECTILE. SCNDRY-IN GOSSANS	35.3.1.
2.5-3	ANGLESITE	6.37-6.39	ORTH.	SUF	TBLR,PRISM,MSV,ETC. ADMN. CLVGS. CONCH. BRTL. SOME FLUO-YEL. SOL-HNO3(DIFFICULTLY). SECONDARY - BY OXIDATION OF GALENA	28.3.1.3
2.5-3	LEADHILLITE	6.55	MNCL.	CBT	PSEUDOHEX, GRNLR. RESIN, PEARLY. 1 CLVG. FLUO-YEL. EFF-HNO3. SCNDRY IN OXIDIZED ZONES OF LEAD DEPOSITS	17.1.3.
2.5-3	VANADINITE	6.88	HEXA.	VAN	PRISM,HAIRLK. RESIN. CONCH. SOME SHOW CONCENTRIC ZONING. SOL-HNO3 (YEL), HCL(GREEN). OXIDIZED ZONES OF LEAD DEPOSITS	41.7.2.3
2.5-3	WULFENITE	6.5 -7.0	TETR.	MBS	TBLR,DIPYRAMIDAL,MSV,GRNLR. RESIN-ADMN. CLVGS.UNEVN. DCMP-HCL, HNO3, SOL-H2SO4,ALKS. SCNDRY-OXIDIZED ZONES OF PB- + MO-BEARING RKS	48.1.4.1
2.5-3	MATLOCKITE	7.12	TETR.	HAL	TBLR, PLTY AGGS, ETC. ADMN. 1 CLVG. SOL-HNO3, DCMP-H2SO4. PB-DEPS	10.1.6.1
2.5-3	ECDEMITE	7.14	TETR.	ASI	CRUSTS-TBLR XLS,FOL MSV. VITR. 1 CLVG. SOL-HNO3,HCL(WARM),ALKS. RARE - E.G.,WITH HELIOPHYLLITE (LANGBAN, SWEDEN)	46.1.1.
2.5-3	LORETTOITE	7.39-7.95	TETR.	HAL	YEL-STRK, MSV- BLDD, FIBR. ADMN. 1 CLVG. SOL-ACIDS. RARE-PB DEPS	10.1.3.

NONMETALLIC YELLOW

HARDNESS	NAME	XL. SYS.	SPECIFIC GRAVITY	CHEM. CLASS	REMARKS	REFERENCES
2.5-3	STOLZITE	TETR.	7.9-8.34	WOS	DIPYRAMIDAL-STRIATED. RESIN. CLVGS. CONCH. SOL-HCL(YEL). WITH LIMONITE, ETC. - OXIDIZED DEPOSITS WITH TUNGSTEN MINERALS	48.1.4.2
2.5-3	RASPITE	MNCL.	8.46	WOS	TBLR-STRIATED. ADMN.1 CLVG. DCMP-HCL(YEL). WITH STOLZITE(MXED VNS)	48.1.5.
2.5-3.5	GIBBSITE	MNCL.	2.3-2.42	OXD	TBLR, MAMMILLARY, PEARLY-VITR. 1 CLVG. CLAYLK ODOR WHEN DAMP. MX-ING PRODUCT, BAUXITE DEPOSITS, LOW TEMPERATURE VEINS	6.2.2.
2.5-3.5	SERPENTINE		2.5-2.6	PSI	GROUP NAME, NOMENCLATURE FOR MEMBERS IS UNSETTLED. MSV-ASBESTIFM. WAXY-GRSY. GELAT. SOME FLUO-CREAM YEL. ALTER PROD AFTER MG-SILICATES(E.G.,OLIVINE). VNS	W381 3-170
2.5-3.5	ANTIGORITE (SERPENTINE)	MNCL.	2.65	PSI	COLORS COMMONLY VARIEGATED OR MOTTLED. PLATY. SECONDARY	W381 3-170
2.5-3.5	AMMONIOJAROSITE	HEXA.	C 3.11	SUF	NODULAR, TBLR. WAXY-DULL. SOL-HCL. RARE WITH JAROSITE, ETC.	30.2.4.4
2.5-3.5	JAROSITE	HEXA.	2.91-3.26	SUF	CRUSTS,GRNLR,ETC. RESIN.1 CLVG. SOL-HCL. ON FE-BEARING ORES + RKS	30.2.4.3
2.5-3.5	BISMUTITE	TETR.	6.1-7.7	CBT	GRAY STRK. MSV, RADIAL FIBR CRUSTS. VITR-DULL. EFF-ACIDS. SCNDRY AFTER BISMUTH MINERALS	16.1.8.
2.5-4	ZINNWALDITE	MNCL.	2.90-3.02	PSI	LI-,FE-MICA. WITH LI-BEARING MINS-PEGS, CASSITERITE-BEARING VEINS	W371 3-92
2.5-5	GUMMITE		3.9-6.4	OXD	FIELD TERM FOR HYDROUS OXIDES OF URANIUM OF UNKNOWN IDENTITY	5.2.1.
3	ZIPPEITE	ORTH.		SUF	CRUSTS. DULL-SILKY. FLUO-GREEN. SCNDRY ON ALTERING URANINITE	31.4.4.
3	FLUELLITE	ORTH.	2.14-2.17	HAL	DIPYRAMIDAL. VITR. 1 + POOR PYRAMIDAL CLVGS. VNS, PEGS	11.5.9.
3	RINNEITE	HEXA.	2.35	HAL	EXPOS-BRWN. GRNLR. SILKY. CONCH. ASTRINGENT. SCNDRY-SED SALT DEPS	11.4.3.
3	SHORTITE	ORTH.	2.60	CBT	WEDGE-SHAPED XLS. VITR. 1 CLVG. FLUO-AMBER. DCMP-H2O. WITH TRONA	14.3.2.
3	CALCITE	HEXA.	2.71	CBT	VARIOUS TRIGONAL XL HABITS. MSV. VITR-PEARLY. RHOMB CLVG. SOME IS FLUO,PHOSPHO,THERMOLUM. EFF-DILUTE HCL. WDSPRD-VNS,DIVERSE RKS	14.1.1.1
3	NATROJAROSITE	HEXA.	3.18	SUF	CRUSTS, SCALY. VITR. 1 CLVG. CONCH. SOL-HCL(SLOW). WITH OTHER SUL-FATES FORMED BY OXIDATION OF PYRITE, ETC.	30.2.4.5
3	ASTROPHYLLITE	TRCL.	3.3-3.4	SSI	SUBMET-PEARLY. BRONZY-GLDN YEL. CLVG-BRTL FLKS. DCMP-HCL. ALK RKS	W480 N
3	ARGENTOJAROSITE	HEXA.	3.66	SUF	SCALY. BRILLANT. 1 CLVG. RARE-SCNDRY WITH ANGLESITE,ETC.(TINTIC)	30.2.4.6
3	OLIVENITE	ORTH.	3.9-4.46	ARS	ELONGATE,GLBL FIBR,MSV,ETC. ADMN-SILKY. CLVGS. CONCH. SOL-ACIDS + NH4OH. SCNDRY - OXIDIZED ZONES OF ORE DEPOSITS	41.6.5.1
3	PILBARITE (MIXTURE ?)	UNKN.	4.6	NSI	CANARY YEL. MAY BE A VARIETY OF THORITE. RARE (E.G.,ORE DEPS)	W496 N
3	ROEMERITE	TRCL.	2.17	SUF	VIOLET OR YEL TINGE. CRUSTS, CUBICLK. OILY. 2 CLVGS. SALTY,ASTRIN-GENT. WITH COPIAPITE - BY OXIDATION OF PYRITE	29.7.2.
3 -3.5	LAUMONTITE (ZEOLITE)	MNCL.	2.2 -2.3	TSI	PERF CLVGS-3 DIRECTIONS IN 1 ZONE. GELAT. CAVITIES IN SEVERAL RKS	W342 4-401
3 -3.5	HANKSITE	HEXA.	2.56	SUF	STRIATED PRISMS. VITR-DULL. 1 CLVG. SALTY. FLUO-YEL(LW). SOL-H2O, EFF-DILUTE ACIDS. SALINE LAKE DEPOSITS	32.1.1.
3 -3.5	SZAIBELYITE	ORTH.	2.62	BOR	IN SERIES WITH SUSSEXITE(SEE). VNS, SALT DEPS, SKARN ZONES	26.1.5.2
3 -3.5	HISINGERITE	AMOR.	3.	PSI	RESINOUS. CONCH. 1 POOR CLVG. DCMP-HCL. SCNDRY(E.G.,AFTER OLIVINE)	W400 N
3 -3.5	LANDESITE	UNKN.	3.03	PHO	OCTAHLK XLS (PSEUDOMORPHOUS). 2 CLVGS(AT 90°). ALTER PROD OF RED-DINGITE - IN PEGMATITE	40.2.6.
3 -3.5	HOPEITE	ORTH.	3.00-3.10	PHO	TBLR,PRISM,TUFTED AGGS,CRUSTS. VITR,PEARLY. CLVGS. SOL-DILUTE HCL. WITH HEMIMORPHITE, ETC.	40.2.11.

NONMETALLIC YELLOW

Hardness	Name	Crystal System	Sp.Gr.	Group	Description	Dana #
3 -3.5	REDDINGITE	ORTH.	3.0 -3.2	PHO	MSV,GRNLR,FIBR,TBLR. VITR. 1 CLVG. UNEVN. SOL-ACIDS. PEGMATITES	40.2.5.1
3 -3.5	SUSSEXITE	ORTH.	3.30	BOR	IN SERIES WITH SZAIBELYITE. FIBR(INFLEX),DENSE-CHALKY AGGS. SILKY-DULL. SOL-ACIDS(SLOWLY). VNS(E.G., FRANKLIN,N.J.)	26.1.5.1
3 -3.5	WITHERITE	ORTH.	4.29	CBT	PSEUDOHEX, GLBLR,ETC. VITR. FLUO(ETC.)-LK ARAGONITE.SOL-HCL. VNS	14.1.3.2
3 -3.5	BARITE	ORTH.	4.50	SUF	TBLR, MSV,ETC. VITR-PEARLY. CLVGS. SOME FLUO, PHOSPHO, THERMOLUM. SOME FETID WHEN RUBBED. COMMON-VNS,SED RKS,CAVITIES IN IG RKS	28.3.1.1
3 -3.5	GREENOCKITE	HEXA.	4.9	SLD	HEMIMORPH PYRAMIDAL XLS. ADMN-RESIN. SOL-HCL. COATINGS ON ZN-MINS	2.6.4.2
3 -3.5	PHOENICOCHROITE	ORTH.	5.75	CHR	TBLR,MSV,COATINGS. RESIN. 1 CLVG. SOL-HCL. GOSSANS WITH CROCOITE	35.3.2.
3 -4	FAUJASITE (ZEOLITE)	ISOM.	1.91-1.93	TSI	OCTAHEDRAL XLS + CLVG. DCMP-ACID. RARE-LIMBURGITE(BADEN)	W333 4-400
3 -4	MORDENITE (ZEOLITE)	ORTH.	2.12-2.15	TSI	STRIATED PRISMS. ACICULAR, COTTON-LK AGGS. ZEOL-ASSOC, AUTHIGENIC	W339 4-401
3 -4	EVANSITE	UNKN.	1.8 -2.2	PHO	MSV,COLLOFM COATINGS. VITR. CONCH. BRTL. SOL-ACIDS. WITH LIMONITE	42.2.6.
3 -4	GISMONDINE (ZEOLITE)	ORTH.	2.1 -2.3	TSI	PSEUDOTETRAG BIPYRAMIDS. 1 CLVG. GELAT. LEUCITE RKS, ALTER OF PLAG	W339 4-401
3 -4	WAVELLITE	ORTH.	2.36	PHO	RADIAL FIBR,STELLATE,CRUSTS. VITR-PEARLY.CLVGS. SOL-ACIDS. SCNDRY-WDSPRD-OPENINGS IN DIVERSE ROCKS,LIMONITE,PHOSPHATE ROCKS, ETC.	42.7.4.
3 -4	CACOXENITE	HEXA.	2.2 -2.4	PHO	ACICULAR,TUFTS,RADIAL AGGS,FIBR CRUSTS. SILKY. SOL-ACIDS. SCNDRY-DIVERSE OCCURRENCES (WITH DUFRENITE, WAVELLITE, ETC.)	42.9.2.
3 -4	DIADOCHITE	TRCL.	2.0 -2.4	PHO	NODULAR,PLTY,GEL-LK,COLLOFM. DULL-WAXY. UNEVN-CONCH. SOL-ACIDS. IN GOSSANS AND AS EFFLORESCENCES	43.2.4.
3 -4	PARSETTENSITE	MNCL.	2.59-2.85	PSI	MN-RICH STILPNOMELANE, WHICH SEE.	W390 3-103
3 -4	STILPNOMELANE	MNCL.	2.59-2.96	PSI	MICACEOUS,BUT 2ND CLVG ⊥ 1ST). SOL-HF+H2SO4(1/1). LO-GRD META RKS	W390 3-103
3 -4	STILPNOCHLORANE	UNKN.	2.5 -3.0	PSI	YELLOWISH TO BRONZY RED. SCALES.	WN N
3 -4	SODDYITE	ORTH.	4.63	NSI	STRIATED PRISM + PYRAMIDAL XLS. GELAT. RDACTV. RARE-WITH CURITE	W530 N
3 -4	FOURMARIERITE	ORTH.	6.05	OXD	TBLR-STRIATED. ADMN. BASAL CLVG. SOL-ACIDS. SCNDRY URANIUM MINERAL	5.3.1.
3 -5.5	ZEOLITE		1.9 -2.45	TSI	GROUP OF MIN FAMILIES WITH SOMEWHAT RELATED APPEARANCES, OCCURRENCES, AND COMPOSITIONS. MANY ARE INDISTINGUISHABLE MEGASCOPICALLY. DIVERSE HABITS(E.G., BUNDLES, FIBROUS), PEARLY-GLASSY, SUBCONCH. SOME FLUO-ORANGE OR YEL-GREEN. CONTINUOUS,IN PART REVERSIBLE, DEHYDRATION(I.E.,WATER CAN BE EXPELLED WITHOUT DESTROYING XL STRUCTURE). SCNDRY IN OPEN SPACES; COMMONLY TWO OR MORE OCCUR TOGETHER IN CAVITIES IN BASALTIC RKS - THIS OCCURRENCE IS REFERRED TO HEREIN AS THE ZEOLITE ASSOCIATION (ABBREVIATED "ZEOL-ASSOC").	W330 4-351
3.5	MOSESITE	ISOM.		HAL	EXPOS-OLIVE GREEN. OCTAH. ADMN. OCTAH CLVG.WHITENS IN HCL. HG-DEPS	10.2.11.
3.5	CALCIOVOLBORTHITE	ORTH.		VAN	SUBTRANSLUCENT. SCALY AGGS,FIBR-DENSE. VITR-PEARLY. 1 CLVG. SOL-ACIDS. SCNDRY-IN SS + OXIDIZED ZONES OF VANADIUM-BEARING DEPS	41.5.2.4
3.5	PINNOITE	TETR.	2.27	BOR	GRNLR,FIBR,NODULES. VITR. UNEVN. SOL-DILUTE ACIDS. SALT DEPISTASSFURT, GERMANY)	25.1.1.
3.5	SULFOHALITE	ISOM.	2.50	SUF	DDDEC, OCTAH. VITR. CONCH. SALTY. SOL-H2O(SLOWLY). SALINE DEPOSITS	30.1.6.
3.5	KIESERITE	MNCL.	2.57	SUF	MSV,GRNLR. VITR. CLVGS. FRIABLE. SOL-H2O(SLOW). MARINE SALINE DEPS	29.6.2.1
3.5	MONETITE	TRCL. A	2.93	PHO	CRUSTS,AGGS. VITR. 3 CLVGS. EXPOS-HYDRATES TO BRUSHITE. GUANO DEPS	37.1.1.
3.5	FAIRFIELDITE	TRCL.	3.08	PHO	FOLIATED AGGS, PRISM. PEARLY-GRSY. CLVGS. SOL-ACIDS. PEGMATITES	40.2.3.1
3.5	AKROCHORDITE	MNCL.	3.19	ARS	WARTLK AGGS. 2 CLVGS. SOL-H2SO4(PURPLE). RARE-LANGBAN, SWEDEN	42.3.3.

61

NONMETALLIC YELLOW

HARDNESS	NAME	XL. SYS.	SPECIFIC GRAVITY	CHEM. CLASS	REMARKS	REFERENCES
3.5	HUREAULITE	MNCL.	3.15-3.20	PHO	TBLR,MSV,SCALY,FIBR, VITR. 1 CLVG. SOL-ACIDS. WITH ALTERED LITHIOPHILITE AND/OR TRIPHYLITE	39.1.3.
3.5	DIETZEITE	MNCL.	3.62-3.70	IOD	TBLR,FIBR CRUSTS. 1 POOR CLVG. SOL-H2O. NITRATE DEPOSITS	23.1.1.
3.5	CHALCOCYANITE	ORTH.	3.60-3.70	SUF	TBLR. HYGROSCOPIC TO BLUE PENTAHYDRATE OR CHALCANTHITE. FUMAROLIC	28.3.3.
3.5	STRONTIANITE	ORTH.	3.64-3.78	CBT	SPEAR-LK, MSV. VITR. CLVGS. FLUO(ETC.)-LK ARAGONITE. SOL-HCL. VNS	14.1.3.3
3.5	VOLBORTHITE	MNCL.	3.5 -3.8	VAN	SUBTRANSLUCENT. SCALY,SPONGY,FIBR CRUSTS. VITR-PEARLY.1 CLVG. SOL-ACIDS. SECONDARY – DIVERSE OCCURRENCES	41.5.3.
3.5	RHABDOPHANE	HEXA.	3.94-4.01	PHO	RADIATED FIBR MAMMILLARY CRUSTS. GRSY. UNEVN. SOL-HCL. WITH LIMONITE	40.3.5.
3.5	ADAMITE	ORTH.	4.32-4.48	ARS	ELONGATE,CRUSTS,RADIAL AGGS. VITR. CLVGS. UNEVN. SOL-DILUTE ACIDS. SOME FLUO YEL. OXIDIZED ZONES OF DEPS WITH ZN-+ AS-RICH MINERALS	41.6.5.3
3.5	WALPURGITE	TRCL	C	PHO	LATHLK,SUBPARALLEL,RADIAL AGGS. ADMN-GRSY. 1 CLVG. SCNDRY IN U-VNS	41.4.1.
3.5	GEORGIADESITE	MNCL.	7.1	ARS	PSEUDOHEX PLTS,STRIATED. RESIN. SOL-HNO3. RARE-1 SPECIMEN(LAURIUM)	41.3.3.
3.5	RUSSELLITE	TETR.	7.33-7.37	OXD	GRNLR. RARE-WITH WOLFRAMITE,ETC., PROBABLY AS ALTER PROD OF BI (CASTLE-AN-DINAS MINE, CORNWALL)	4.6.4.
3.5	HEULANDITE (ZEOLITE)	MNCL.	2.1 -2.2	TSI	PLTY GRP. TBLR-BROAD CENTRAL PORTIONS(COFFIN-LIKE). ZEOL-ASSOC	M347 4-377
3.5	STILBITE (ZEOLITE)	MNCL.	2.1 -2.2	TSI	PLTY GRP. SHEAF-LIKE AGGS. 1 GOOD CLVG. DCMP-HCL. ZEOL-ASSOC	M345 4-377
3.5	NORTHUPITE	ISOM.	2.38	CBT	OCTAH. VITR. CONCH. EFF-ACIDS,DCMP-HOT H2O. SALINE CLAY DEPS	16.2.2.
3.5	BREWSTERITE (ZEOLITE)	MNCL.	2.45	TSI	PLTY GRP. ELONGATE,STRIATED PRISMS. 1 CLVG. DCMP-HCL. ZEOL-ASSOC	M347 4-383
3.5	LANGBEINITE	ISOM.	2.83	SUF	NODULAR, DISSEM. VITR.CONCH. SOL-H2O(SLOWLY). MARINE SALTS	28.4.3.1
3.5-4	DOLOMITE	HEXA.	2.84-2.86	CBT	RHOMB-CURVED, GRNLR. VITR-PEARLY. RHOMB CLVG. SOME FLUO. EFF-WARM HCL IF POWDERED. SED RKS, XLS IN CAVITIES, ETC.	14.2.1.1
3.5-4	ALUNITE	HEXA.	2.6 -2.9	SUF	GRNLR, MSV. VITR,PEARLY. CLVGS. SOL-H2SO4. "ALUNITIZED" ROCKS	30.2.4.1
3.5-4	NATROALUNITE	HEXA.	2.6 -2.9	SUF	GRNLR, MSV. VITR,PEARLY. CLVGS. SOL-H2SO4. RARE,BUT VARIED OCCUR.	30.2.4.2
3.5-4	ARAGONITE	ORTH.	2.94-2.95	CBT	ACICULAR, RADIAL FIBR, STALACTITIC.VITR. CLVGS. FLUO(ULTRA-VIOLET, X-RAYS,ELECTRON BEAMS),THERMOLUM. EFF-ACIDS. LO-T DEPOSITION	14.1.3.1
3.5-4	ANKERITE	HEXA	3.01-3.03	CBT	DISTINGUISHED FROM DOLOMITE(SEE) BY NONMEGASCOPIC METHODS. VNS	14.2.1.2
3.5-4	SCORODITE	ORTH.	3.28	ARS	PYRAMIDAL,ETC.-AGGS,CRUSTS,ETC. VITR. CLVGS. SOL-ACIDS. SCNDRY-AFTER ARSENOPYRITE AND OTHER AS-BEARING MINERALS	40.3.1.3
3.5-4	DICKINSONITE	MNCL.	3.38-3.41	PHO	TBLR,SCALY. VITR,PEARLY.1 CLVG. SOL-ACIDS. PEGS-WITH LITHIOPHILITE ETC.	40.2.1.
3.5-4	TARBUTTITE	TRCL	4.12	PHO	CRUSTS,SHEAFLK AGGS OF ROUNDED,STRIATED,EQUANT XLS. VITR,PEARLY. 1 CLVG. OXIDIZED ZONES OF ZINC DEPOSITS	41.6.7.
3.5-4	POWELLITE	TETR.	4.21-4.25	MBS	PYRAMIDAL,TBLR,MSV-FOL,POWDER,CRUSTS,SUBADMN-PEARLY. CLVGS. UNEVN. FLUO-YEL. DCMP-HCL,HNO3. ALTER PROD OF MOLYBDENITE	48.1.3.2
3.5-4	LAUTARITE	MNCL.	4.59	IOD	RADIAL AGGS-STRIATED PRISMS.CLVGS.SOL-HCL. GYPSUM BANDS-NITER DEPS	21.1.1.
3.5-4	NADORITE	ORTH.	7.02	ATI	TBLR,PRISM,SUBPARALLEL-DIVERGENT GROUPS. RESIN-ADMN. CLVG. DIVERSE	46.1.4.

62

NONMETALLIC YELLOW

H	Mineral	Crystal	S.G.	Luster	Description	Ref
3.5-4	PYROMORPHITE	HEXA.	7.00-7.08	PHO	PRISM,GLBL,ETC. RESIN. RHOMB CLVG. UNEVN. SOL-HNO3. OXIDIZED ZONES OF LEAD DEPOSITS	41.7.2.1
3.5-4	MIMETITE	HEXA.	7.24	ARS	PRISM,ACICULAR,GLBL,ETC. RESIN. RHOMB CLVG. UNEVN. SOL-HNO3. OXI-DIZED ZONES OF LEAD DEPOSITS	41.7.2.2
3.5-4	KLEINITE	HEXA.	8.0	HAL	REDDISH-DAYLIGHT, YEL TO ORANGE-OTHERWISE. PRISM. ADMN-GRSY. BASAL, RHOMB CLVGS. SOL-HCL,HNO3,NH4BR. RARE-WITH OTHER HG MINS (TEXAS).	10.2.10.
3.5-4.5	MAGNESITE	HEXA.	2.98-3.02	CBT	GRNLR, RHOMB. VITR. RHOMB CLVG. FLUO-GREEN,BLUE. SOL-WARM HCL. AL-TERATION OF MG-RICH RKS (E.G., SERPENTINE)	14.1.1.2
3.5-4.5	FERRI-SICKLERITE	ORTH.	3.2 -3.4	PHO	SUBTRANSLUCENT. MSV. 1 CLVG. SOL-ACIDS. WXING PROD OF TRIPHLITE, LITHIOPHILITE (PEGMATITES)	38.1.3.1
3.5-4.5	SICKLERITE	ORTH.	3.2 -3.45	PHO	SUBTRANSLUCENT. DISTINGUISHED BY CHEM ANAL FROM FERRI-SICKERLITE	38.1.3.2
3.5-4.5	CHENEVIXITE	UNKN.	3.93	ARS	MSV EARTHY-OPALINE,ACICULAR. SUBCONCHOIDAL.SOL-ACIDS. ORE VEINS	41.5.9.
3.5-4.5	SIDERITE	HEXA.	3.95-3.97	CBT	TARN-IRID. MSV. RHOMB,ETC. VITR.RHOMB CLVG. SOL-HOT ACID. SED, VNS	14.1.1.3
3.5-4.5	CORKITE	HEXA.	4.30	PHO	RHOMB,PSEUDOCUBIC. VITR. 1 CLVG. SOL-WARM HCL. SCNDRY-DIVERSE	43.1.1.2
3.5-6	MARGARITE (BRITTLE MICA)	MNCL. 3	3.1	PSI	AGGS OF LAMELLAE. ATTKD-H2SO4. WITH DIASPORE + CORUNDUM-EMERY DEPS	W392 3-95
3.5-6	CLINTONITE (BRITTLE MICA)	MNCL. 3	3.1	PSI	WITH TALC-CHLORITE SCHIST, WITH SPINEL IN METASOMATIZED CALC RKS. DISTINGUISHED FROM XANTHOPHYLLITE BY OPTICAL OR X-RAY ANALYSES	M391 3-99
3.5-6	XANTHOPHYLLITE (BRITTLE MICA)	MNCL. 3	3.1	PSI	WITH TALC-CHLORITE SCHIST, WITH HUMITES IN METASOMATIZED CALC RKS. DISTINGUISHED FROM CLINTONITE BY OPTICAL AN/OR X-RAY ANALYSES	W391 3-99
3.5-7	BRITTLE MICA	MNCL. 2.6 -3.2		PSI	GROUP OF MORE OR LESS RELATED MINS THAT RESEMBLE MICAS BUT CLEAVE TO BRTL FLKS AND HAVE HARDNESSES OF 3.5-4 VS. 2-3(MICA) ON CLVG FLKS AND 6 VS. 4⊥ TO FLKS. WITH CORUNDUM + DIASPORE IN EMERY DEPS, CHLORITE + MICA SCHISTS, SKARN	W392 3-95
4	RENARDITE	ORTH.		PHO	DRUSY CRUSTS.1 CLVG.SOL-ACIDS. DATA LACKING. RARE-SCNDRY(KATANGA)	42.3.5.
4	WEDDELLITE	TETR.	1.93-1.95	CCC	PYRAMIDAL,AGGS. SUBCONCH. ORGANIC ASSOC. BOTTOM MUDS(WEDDELL SEA)	50.1.2.
4	EPISTILBITE (ZEOLITE)	MNCL.	2.1 -2.3	TSI	PLTY GRP. SHEAF-LK + SPHERICAL AGGS. DCMP-HCL.PIEZOELECTRIC. ZEOL-ASSOCIATION	W346 4-377
4	SALMONSITE	ORTH.	2.88	PHO	FIBR MASSES. 2 CLVGS(AT 90°).ALTER PROD OF HUREAULITE IN PEGMATITE	40.2.8.
4	LEUCOPHANE	ORTH.	2.96	SSI	TBLR. 2 PERF CLVGS. TWINNING COMMON(SOME PENTRATION). RARE-PEGS	W476 N
4	SEAMANITE	ORTH.	3.08	BOR	ACICULAR.1 CLVG. BRTL.SOL-DILUTE ACIDS. RARE-MN DEP(IRON CO. MICH.	27.1.4.
4	FLUORITE	ISOM.	3.18	HAL	CUBIC, MSV. VITR. OCTAH CLVG. FLUO-VIOLET. DCMP-H2SO4. WDSPRD-VNS, CAVITIES, DISSEMINATED IN IGNEOUS ROCKS, ETC.	9.2.1.
4	BARYTOCALCITE	MNCL.	3.66-3.71	CBT	PRISM-STRIATED. MSV. VITR. CLVG. WEAKLY FLUO. SOL-HCL. LO-T DEPS	14.2.3.
4	HOLDENITE	ORTH.	4.11	ARS	EQUANT. VITR. 1 POOR CLVG. SUBCONCH. VN CUTTING ORE(FRANKLIN,N.J.)	41.1.2.
4	CARYINITE	ORTH.	4.29	ARS	SUBTRANSLUCENT. MSV, GRNLR. GRSY. CLVGS. SOL-HNO3. VNS IN SKARN-LANGBAN, SWEDEN	38.2.2.
4	AMPANGABEITE	ORTH.	3.36-4.64	OXD	PRISM, RADIAL AGGS. HORNLK. CONCH. DCMP-ACIDS. K-RICH PEGMATITES	8.4.3.
4	ZINCITE	HEXA.	5.64-5.68	OXD	HEMIMORPHIC PYRAMIDAL,MSV,GRNLR. SOME FLUO. SOL-ACIDS. RARE-FRNKLN	4.2.2.1
4	PUCHERITE	ORTH. C	6.57	VAS	TBLR,ACICULAR,MSV,CRUSTS. VITR-ADMN,1 CLVG. SUBCONCH.SOL-HCL(RED). SECONDARY, WITH BISMUTITE AND BEYERITE	47.1.8.
4	HUEBNERITE	MNCL.	7.12	WDS	SUBMET. TARN-IRID. PRISM-STRIATED, -RADIATING XL GRPS. 1 CLVG. DCMP-AQ-REG.H2SO4 OR HCL(SLOWLY). DIVERSE-VNS,CNTCT META-PLACERS	48.1.1.1
-4.5	PHILLIPSITE (ZEOLITE)	MNCL.	2.2	TSI	PENETRATION TWINS, SPHERULITES. 2 GOOD CLVGS. GELAT. DISTINGUISHED FROM HARMOTOME BY OPTICAL METHODS. ZEOL-ASSOC, DEEP-SEA RED CLAYS	W343 4-386

63

NONMETALLIC YELLOW

HARDNESS	NAME	XL. SYS.	SPECIFIC GRAVITY	CHEM. CLASS	REMARKS	REFERENCES
4 -4.5	CARPHOSIDERITE	HEXA.	2.50-2.91	SUF	SCALY.RESIN-DULL.SOL-HCL.INCRUSTATION WHERE PYRITE IS WEATHERED	30.2.4.7
4 -4.5	ALSTONITE	ORTH.	3.67-3.71	CBT	PSEUDOHEX.STRIATED.VITR.FLUO-YEL(LW).SOL-HCL.LO-T DEPS	14.2.2.
4 -4.5	VOLTZITE	UNKN.	3.7 -3.8	SLD	SECONDARY INCRUSTS.VITR-GRSY.GIVES H2S IN HCL.WITH SULFIDES	2.6.4.3
4 -4.5	ANCYLITE	ORTH.	3.95	CBT	PSEUDO-OCTAH,CRUSTS.VITR.SPLINTERY.SOL-ACIDS.DRUSES IN ALK PEGS	16.2.10.
4 -4.5	AUSTINITE	ORTH.	4.13	ARS	BLADED,FIBR CRUSTS + NODULES.SUBADMN,SILKY.1 CLVG.SCNDRY-RARE	41.5.1.3
4 -4.5	SMITHSONITE	HEXA.	4.40-4.45	CBT	BOTRYOIDAL.MSV.VITR-PEARLY.RHOMB CLVG.SOME FLUO-GREEN,BLUE.EFF-ACIDS.SCNDRY-OXIDIZED ZONES OF ZINC DEPOSITS	14.1.1.6
4 -4.5	THOROGUMMITE	TETR.	4.5	NSI	SUBTRANSLUCENT.DULL.COMMONLY SECONDARY(E.G.,AFTER THORITE) PEGS	WN N
4 -4.5	BASTNASITE	HEXA.	4.9 -5.2	CBT	TBLR-GROOVED,GRNLR.VITR,PEARLY.CLVGS.SOL-HOT ACIDS.CNTCT META	16.2.9.
4 -4.5	BINDHEIMITE	ISOM.	4.6 -5.6	ANT	DENSE-EARTHY CRUSTS,NODULAR-CONCENTRIC.SOME IS OPALINE.RESIN-DULL.DCMP-HNO3,HCL.OXIDATION ZONES OF ANTIMONY-BEARING LEAD DEPOSITS	44.1.1.1
4 -4.5	LITHIOPHILITE	ORTH.	3.50-3.58	PHO	MSV,ANHEDRAL.VITR.CLVGS.UNEVN.SOL-ACIDS.PEGMATITES	38.1.1.2
4 -5	SVABITE	HEXA.	3.5 -3.8	ARS	SHORT PRISM,MSV.VITR-CLVG.SOL-DILUTE ACIDS.RARE-E.G.,FRNKLN,N.J.	41.7.3.1
4 -5	SARKINITE	MNCL.	4.08-4.18	ARS	TBLR,SPHERICAL GROUPS.GRNLR.GRSY.1 CLVG.UNEVN.SOL-DILUTE ACIDS.DIVERSE (E.G.,VEINS)	41.6.3.3
4 -5	XENOTIME	TETR.	4.4 -5.1	PHO	PRISM + PYRAMIDAL.VITR.1 CLVG.ACID + ALK IG RKS + PEGS.DETRITAL	38.4.1.
4 -5	FLUOCERITE	HEXA.	5.93-6.14	HAL	FRESH-YEL.MSV,TBLR.RESIN,PEARLY.BASAL CLVG.SOL-H2SO4.PEGS	9.3.2.
4 -5	KASOLITE	MNCL.	5.96-6.46	NSI	ELONGATE XLS.1 PERF,2 POOR CLVGS.GELAT.RDACTV.RARE-WITH CURITE	W530 N
4 -5	CERVANTITE	ORTH.	A 6.64	OXD	MSV,FIBR,ACICULAR.GRSY-PEARLY.OXIDATION PROD OF SB-MINS	4.5.6.
4 -5.5	LIMONITE		2.7 -4.3	OXD	FIELD TERM FOR HYDROUS IRON OXIDES OF UNKNOWN IDENTIES.	7.1.3.
4 -5.5	BETAFITE	ISOM.	3.7 -5.	OXD	SUBMET.OCTAH.WAXY.CONCH.METAMICT.DCMP-ACIDS.GRANITIC PEGS	8.4.1.
4 -5.5	STIBICONITE	ISOM.	5.58	OXD	MSV,POWDER,CRUSTS.EARTHY-GLASSY.OXIDATION PROD OF SB-MINS	4.5.7.
4.5	CHABAZITE (ZEOLITE)	HEXA.	2.05-2.10	TSI	RHOMBS(NEARLY CUBES).RHOMB CLVG.DCMP-HCL.ZEOL-ASSOC.RK CAVITIES	W334 4-387
4.5	GMELINITE (ZEOLITE)	HEXA.	2.0 -2.2	TSI	PRISM CLVG.DCMP-HCL.ZEOL-ASSOC	W335 4-387
4.5	LEVYNITE (ZEOLITE)	ORTH.	2.0 -2.2	TSI	CHEM OR X-RAY ANAL DISTINGUISH CHABAZITE +LEVYNITE.RARE-ZEOL-ASSOC	W335 4-387
4.5	COLEMANITE	MNCL.	2.42-2.43	BOR	EQUANT,GRNLR.VITR.2 CLVGS.SOL-HOT HCL.EVAPORITE BORATE DEPS	25.1.9.
4.5	HARMOTOME (ZEOLITE)	MNCL.	2.41-2.47	TSI	HABIT LK PHILLIPSITE(ABOVE).+ CRUCIFM TWINS.DCMP-HCL.ZEOL-ASSOC.ALSO WITH AL-WXING PRODS,IN GNEISSES,+ WITH MN-MINERALIZATION	W344 4-386
4.5	RALSTONITE	ISOM.	2.56-2.62	HAL	OCTAH.VITR.OCTAH CLVG.DCMP-HF.RARE AT SOME CRYOLITE LOCALITIES	11.5.10.
4.5	LACROIXITE	MNCL.	3.13	PHO	XLS.VITR-RESIN.CLVGS.SOL-HCL,H2SO4.RARE-DRUSES IN GRANITE(SAX-ONY,GERMANY)	41.2.1.
4.5	FILLOWITE	MNCL.	3.43	PHO	GRNLR,PSEUDORHOMB.GRSY.1 CLVG.SOL-ACIDS.PEG-WITH TRIPLOIDITE.	40.2.2.

NONMETALLIC YELLOW

Hardness	Mineral	Crystal System	S.G.	Habit	Description	Ref
4.5	SYNCHISITE	HEXA.	3.90	CBT	ACUTE RHOMB, TBLR-HEMIMORPHIC. GRSY. 1 CLVG. SOL-ACIDS. ALK PEGS	16.2.8.
4.5	LINDGRENITE	MNCL.	4.26	MBS	PLTY,TBLR,MSV. CLVGS. SOL-HCL,HNO3.SCNDRY-RARE(CHUQUICAMATA,CHILE)	49.1.4.
4.5	CORDYLITE	HEXA.	4.31	CBT	PRISM, DIPYRAMIDAL. GRSY,PEARLY. 1 CLVG. SOL-ACIDS. ALK PEGS	16.2.7.
4.5	CONICHALCITE	ORTH.	4.1 -4.33	ARS	SUBTRANSLUCENT, EQUANT,BOTRYOIDAL,RADIAL FIBR. VITR. UNEVN. SOL-HCL,HNO3. SECONDARY-OXIDIZED ZONE OF COPPER DEPOSITS	41.5.1.2
4.5	PARISITE	HEXA.	4.33-4.39	CBT	STRIATED BIPYRAMIDS. VITR,PEARLY. SOL-HOT ACIDS. VNS, ALK PEGS	16.2.6.
4.5	LESSINGITE	HEXA.	4.69	NSI	REDDISH OR GREENISH YEL. VITR. RARE IN PLACERS WITH OTHER CE-MINS	W512 N
4.5	BRANNERITE	UNKN.	4.5 -5.43	OXD	BLK, EXPOS-BRWNSH YEL. PRISM. CONCH. RDACTV.DCMP-H2SO4.RARE-PLAC-ER (KELLY GULCH, IDAHO)	8.2.4.
4.5	BAYLDONITE	MNCL.	5.5	ARS	SUBTRANSLUCENT. FIBR MAMMILLARY CONCRETIONS CRUSTS POWDER. RESIN. SOL-HCL(DIFFICULTY). OXIDIZED ZONES OF COPPER DEPOSITS	42.4.1.
4.5	HEDYPHANE	HEXA.	5.82	ARS	PRISM,TBLR,MSV. GRSY. RHOMB CLVG. SOL-HNO3. RARE-E.G.,FRNKLN, N.J.	41.7.3.2
4.5	EULYTINE	ISOM.	6.6	NSI	MINUTE TETRAHEDRA. POOR DODEC CLVG. GELAT. WITH BISMUTH + QTZ	W494 N
4.5	BISMITE	MNCL.	8.64-9.22	OXD	GRAY-YEL STRK. MSV, GRNLR. VITR-DULL. OXIDATION PROD OF BI-MINS	4.4.7.
4.5-5	APOPHYLLITE	TETR.	2.33-2.37	PSI	VITR SQ PRISMS, PEARLY BASES. BASAL + PRISM CLVG. DCMP-HCL(SILICA RESIDUE). SECONDARY IN CAVITIES, COMMONLY WITH ZEOLITES, ETC.	W394 3-258
4.5-5	AUGELITE	MNCL.	2.70	PHO	TBLR,ACICULAR,MSV. VITR,PEARLY. CLVGS. SOL-HCL(SLOW). VNS,ETC. RARE	41.6.8.
4.5-5	PARAWOLLASTONITE	MNCL.	2.92	ISI	DISTINGUISHED FROM WOLLASTONITE BY X-RAY ANAL. CONTACT META LS-	W455 2-167
4.5-5	GOYAZITE	HEXA.	3.26	PHO	RHOMB-STRIATED,ROUNDED GRAINS. GRSY,PEARLY.1 CLVG.SOL-ACIDS(SLOW). DIAMOND SANDS, CREVICES	41.5.8.3
4.5-5	NATROPHILITE	ORTH.	3.41	PHO	GRNLR, MSV. RESIN-PEARLY. CLVGS SOL-ACIDS. RARE-PEG(BRANCHVILLE, CONNECTICUT)	38.1.2.
4.5-5	TRIPLOIDITE	MNCL.	3.66	PHO	//-DIVERGENT FIBR-COLUMNAR AGGS,STRIATED PRISM. VITR-ADMN. CLVGS. UNEVEN-SUBCONCH. SOL-ACIDS. PEGMATITES - PROBABLY SECONDARY	41.6.3.1
4.5-5	PLUMBOGUMMITE	HEXA.	4.01	PHO	CONCENTRIC BOTRYOIDAL. GUM-LIKE LUSTER. UNEVN. SOL-HOT ACIDS. SECONDARY-IN LEAD DEPOSITS WITH PYROMORPHITE	41.5.8.1
4.5-5	BERZELIITE	ISOM.	4.08	ARS	MSV, ROUNDED GRAINS. UNEVN. SOL-HCL,HNO3. LS SKARNS	38.2.1.1
4.5-5	THORITE	TETR.	5.2 -5.4	NSI	TERMINATED PRISMS.PRISM CLVG. MOST IS ALTERED. RARE ACCESS-SYENITE	W496 N
4.5-5	TRIPUHYITE	UNKN.	5.82	ANT	MICROXLINE AGGS. DATA LACKING. VNS IN DACITE, PLACERS.	44.1.3.
4.5-5	SCHEELITE	TETR.	6.08-6.12	WOS	DIPYRAMIDAL,MSV,GRNLR.VITR.CLVGS.UNEVN.DCMP-HCL,HNO3.FLUO-BLUE(X-,SW ULTRAVI-OLET-, + CATHODE-RAYS).THERMOLUM. HI-T VNS,PEGS,GREISSENS,CNTCT META RKS,ETC.	48.1.3.1
4.5-5	ATELESTITE	MNCL.	6.82	ARS	MINUTE TBLR XLS,MAMMILLARY. RESIN-ADMN. 1 POOR CLVG. SOL-HCL,HNO3 (DIFFICULTY). RARE WITH BISMUTITE	41.3.4.
4.5-5.5	WILKEITE	HEXA.	3.1	PHO	ROUNDED XLS,GRNLR. SUBRESIN. 1 CLVG. GELAT. CNTCT META LS	41.7.7.
4.5-5.5	LEGRANDITE	MNCL.	3.96-4.06	ARS	RADIATING AGGS OF PRISMS. 1 CLVG. RARE-1 SPECIMEN(WITH SPHALERITE)	42.7.1.
4.5-5.5	EMMONSITE	MNCL.	4.52-4.53	TLI	FIBR CRUSTS,GLBL AGGS. VITR. CLVGS. SOL-ACIDS. ALTER OF TELLURIDES	34.2.1.
4.5-6	MONIMOLITE	ISOM.	5.9 -7.3	ANT	OCTAH,CUBIC. GRSY. OCTAH CLVG. CONCH-SPLINTERY. RARE-E.G.,LANGBAN	44.1.2.

NONMETALLIC YELLOW

HARDNESS	NAME	XL. SYS.	SPECIFIC GRAVITY	CHEM. CLASS	REMARKS	REFERENCES
5 ●	NATROLITE (ZEOLITE)	ORTH.	2.20-2.26	TSI	FIBR GRP. SQ NEEDLES. PRISM CLVG. GELAT. PYROELECTRIC. ZEOL-ASSOC	W340 4-358
5	MESOLITE (ZEOLITE)	MNCL.	2.25-2.27	TSI	FIBR GRP. TUFTS OF HAIR-LIKE XLS. GELAT. PYROELECTRIC. ZEOL-ASSOC	W341 4-358
5 ●	SCOLECITE (ZEOLITE)	MNCL.	2.25-2.29	TSI	FIBR GRP. THIN STRIATED PRISMS. PRISM CLVG. GELAT. ZEOL-ASSOC. AL-SO HYDROTHERMAL + IN METAMORPHOSED CALCAREOUS ROCKS	W341 4-358
5	OKENITE	TRCL.	2.28-2.33	ISI	FIBR, TBLR, MSV. 1 PERF CLVG-ELAST. GELAT. IN AMYGDULES IN BASALT	W358 N
5	STERRETTITE	ORTH.	2.44	PHO	PRISM-SOME STRIATED. VITR. CLVGS. IN ALTERED VARISCITE NODULES	42.7.5.
5	CRANDALLITE	HEXA.	2.78-2.92	PHO	CONCENTRIC NODULAR AGGS,ETC. VITR-DULL. 1 CLVG. SOL-ACIDS(DIFFI-CULTLY). SECONDARY - DIVERSE OCCURRENCES	41.5.8.4
5	DELTAITE (MIXTURE ?)	HEXA.	2.90-3.00	PHO	FIBR CRUSTS,PRISM. 1 CLVG. RARE-WITH CRANDALLITE(FAIRFIELD, UTAH)	41.5.8.5
5	WHITLOCKITE	HEXA.	3.12	PHO	RHOMB, GRNLR. VITR. UNEVN. SOL-DILUTE ACIDS. PEGS, PHOSPHATE ROCKS	38.3.1.
5	APATITE	HEXA.	2.9 -3.2	PHO	GROUP NAME. MOST TYPES INDISTINGUISHABLE MEGASCOPICALLY. HEX PRISMS-COMMONLY ROUNDED,MSV,GRNLR. VITR. SOL-ACIDS. SOME FLUO--YEL-ORANGE. IG RKS, VNS, PEGS	41.7.1
5	SVANBERGITE	HEXA.	3.22	PHO	RHOMB,PSEUDOCUBIC,GRNLR,VITR-ADMN.1 CLVG. DIVERSE META RKS,PLACERS	43.1.1.4
5	CHILDRENITE	ORTH.	3.22-3.28	PHO	EQUANT-PLTY. VITR. 1 CLVG. UNEVN. SOL-ACIDS. DIVERSE-E.G.,VNS,PEGS	42.5.2.1
5	RINKITE	MNCL.	3.46	NSI	PRISM, TBLR. 1 CLVG. ATTACKED BY H2SO4. ALKALIC SYENITES-RARE	W517 N
5	TITANITE(=SPHENE)	MNCL.	3.45-3.55	NSI	DOUBLE WEDGE-SHAPED XLS, MSV-GRNLR. ADAMANTINE. ALTERS TO LEUCOX-	W525 1-69
5 ●	SPHENE	MNCL.	3.45-3.55	NSI	ENE. DCMP-H2SO4. COMMON ACCESSARY CONSTITUENT OF IG + META ROCKS	W525 1-69
5	HUHNERKOBELITE	ORTH.	3.5 -3.6	PHO	MSV, GRNLR. VITR. 2 CLVGS. PEGS. DISTINGUISHED FROM VARULITE BY CHEMICAL ANALYSIS PEGS + HYDROTHERMAL ALTER OF TRIPHYLITE	38.1.1.3
5	VARULITE	ORTH.	3.5 -3.6	PHO	MSV, GRNLR. VITR. 2 CLVGS. PEGS. DISTINGUISHED FROM HUHNERKOBELITE BY CHEMICAL ANALYSIS PEGS + HYDROTHERMAL ALTER OF TRIPHYLITE	38.1.1.4
5	ADELITE	ORTH.	3.70-3.76	ARS	MSV,ELONGATE. RESIN. CONCH. SOL-DILUTE ACIDS. MN DEPOSITS	41.5.1.1
5	BECKELITE	HEXA.	4.15	NSI	CUBIC OR DODECAHEDRAL. CUBIC CLVG. SOL-ACIDS. SYENITE (RARE)	W520 N
5	BISMUTOTANTALITE	ORTH.	8.26	OXD	SUBMET. EXPOS-PINKISH YEL. PRISM. RARE-PEGMATITE(SW UGANDA)	8.1.9.
5	CARPHOLITE	ORTH.	2.94	ISI	RADIATING FIBERS. 1 PERF CLVG. VEINS, SOME OF WHICH ARE SN-BEARING	W402 N
5 -5.5	HERDERITE	MNCL.	2.95	PHO	PRISM,TBLR,BOTRYOIDAL-RADIAL FIBR. VITR. CLVG. SOL-ACIDS. PEGS-	41.5.4.1
5 -5.5	MELINOPHANE	TETR.	3.0	SSI	LOW PYRAMIDS. BASAL CLVG. RARE-LANGESUND FJORD, NORWAY	WN N
5 -5.5 ●	DATOLITE	MNCL.	2.96-3.00	NSI	TINTED WHT-COLORLESS. GLASSY. GELAT. SCNDRY WITH ZEOLITES-CAVITIES	W355 1-171
5 -5.5	HYDROXYL-HERDERITE	MNCL.	3.01	PHO	PRISM,TBLR,BOTRYOIDAL-RADIAL FIBR. VITR. CLVG. SOL-ACIDS. PEGS-LATE STAGE	41.5.4.2
5 -5.5	MAGNERITE	MNCL.	3.15	PHO	NEARLY OPQ-ALTERED. PRISM STRIATED,MSV. VITR. CLVGS. UNEVN-SPLIN-TERY. SOL-ACIDS. DIVERSE-E.G., VEINS AND VOLCANIC ROCKS)	41.6.1.
5 -5.5	ALLUAUDITE	UNKN.	3.4 -3.5	PHO	SUBTRANSLUCENT. MSV, GRNLR, FIBR-GLBL AGGS. 2 CLVGS. SOL-ACIDS. ALTERATION PRODUCT OF VARULITE-HUHNERKOBELITE (PEGMATITES)	38.1.4.1
5 -5.5	MANGAN-ALLUAUDITE	UNKN.	3.4 -3.5	PHO	SUBTRANSLUCENT. DISTINGUISHED FROM ALLUAUDITE BY CHEMICAL ANALYSIS	38.1.4.2
5 -5.5	KAINOSITE	ORTH.	3.34-3.61	CSI	SHORT PRISMS. 1 CLVG. EFF-HCL. ORE VEINS-RARE(E.G.,N. BURGESS,ONT.	W463 N

NONMETALLIC YELLOW

Hardness	Mineral	Crystal	S.G.	Type	Description	Ref
5 -5.5	GOETHITE	ORTH.	3.3 -4.29	OXD	SUBMET. YELLOWISH STRK. MSV, FIBR, ETC. SILKY-ADMN. SOL-HCL. MXING 7.1.2.2 PRODUCT	
5 -5.5	PYROCHLORE	ISOM.	4.45	OXD	OCTAH. VITR. SUBCONCH. DCMP-H2SO4(SLOW). ALK PEGS + IG RKS	8.1.1.1
5 -5.5	MONAZITE	MNCL.	4.6 -5.4	PHO	TBLR-STRIATED, WAXY-ADMN. CLVGS. UNEVN. DCMP-ACIDS(SLOW). ACCESS IN GRANITIC IG + GNEISSIC RKS, PEGS, AND DETRITAL SED ROCKS	38.4.2.
5 -5.5	MICROLITE	ISOM.	6.38-6.46	OXD	OCTAH. VITR. SUBCONCH. DCMP-H2SO4(SLOW).GRANITIC PEGS WITH TA MINS	8.1.1.2
5 -6	VISHNEVITE	HEXA.	2.32-2.42	TSI	SULFATE-RICH END MEMBER OF CANCRINITE FAMILY. OCCURRENCE LIKE CANCRINITE.	WN 4-310
5 -6	CANCRINITE	HEXA.	2.42-2.51	TSI	CARBONATE-RICH MEMBERS EFFERVESC AS WELL AS GELATINIZE WITH HCL MSV (RARE PRISM XLS). PRISM CLVG. EFF-HCL. NEPHELINE-BEARING RKS	M354 4-310
5 -6	MARIALITE (SCAPOLITE)	TETR.	2.50-2.62	TSI	ME-0-20. PURE MARIALITE UNKNOWN IN NATURE	M352 4-321
5 -6	DIPYRE (SCAPOLITE)	TETR.	2.57-2.69	TSI	ME-20-50.	M352 4-321
5 -6	MIZZONITE (SCAPOLITE)	TETR.	2.67-2.74	TSI	ME-50-80.	M352 4-321
5 -6	SCAPOLITE	TETR.	2.50-2.78	TSI	GROUP NAME. MOST TYPES INDISTINGUISHABLE MEGASCOPICALLY BUT S.G. MAY BE IN-DICATIVE. LONG,STRIATED PRISMS + AGGS OF COARSE XLS. SUBCONCH. POOR CLVG GIVING IRREGULAR, STRIATED-APPEARING SURFACES. MARIALITE IS INSOL; MEIONITE IS DCMP IN HCL. MOST FLUO-YEL-RED(ULTRAVIOLET,LW). CNTCT META CALC RKS,PEGS	
5 -6	MEIONITE (SCAPOLITE)	TETR.	2.73-2.78	TSI	ME-80-100. PURE MEIONITE UNKNOWN IN NATURE	M352 4-321
5 -6	MELILITE	TETR.	2.95-3.05	SSI	SHORT PRISMS. VITR-RESIN. BASAL CLVG. GELAT. BASIC,CA-RICH ALK IG	W473 1-236
5 -6	ENSTATITE (ORTHOPYROXENE)	ORTH.	3.21	ISI	MG PYROXENE. WITH OLIVINE IN BASIC + ULTRABASIC IG RKS, GRANULITES	W405 2-8
5 -6	RICHTERITE (AMPHIBOLE)	MNCL.	2.97-3.45	ISI	THERMALLY META LS + SKARNS, HYDROTHERMAL VNS IN ALK IGNEOUS ROCKS	M435 2-352
5 -6	FLORENCITE	HEXA.	3.59-3.71	PHO	RHOMB-PSEUDOCUBIC. GRSY-RESIN. CLVGS. SPLINTERY. SOL-HCL(PARTLY). IN DIAMOND SANDS	41.5.8.6
5 -6	PRIORITE	ORTH.	4.85-5.05	OXD	SUBMET. IN SERIES WITH ESCHYNITE(SEE). RED-YEL STRK. GRANITIC PEGS	8.3.5.2
5 -6	ESCHYNITE	ORTH.	5.14-5.24	OXD	SUBMET. IN SERIES WITH PRIORITE. BLK-BRWN STRK. PRISM, MSV, WAXY, EXPOS-DULL.POOR CLVG. METAMICT. ALK(E.G.,NEPH SYENITE) PEGMATITES	8.3.5.1
5 -6	"SAMARSKITE"	UNKN.	5.67	OXD	SUBMET. DISTINGUISHED FROM SAMARSKITE ONLY BY X-RAY ANALYSIS	8.3.6.2
5 -6	SAMARSKITE	ORTH.	5.69	OXD	SUBMET. RED-BRWN TO BLK STRK. PRISM, MSV, VITR. RDACTV. PEGS	8.3.6.1
5 -7	PYROXENE		2.96-3.96	ISI	GROUP NAME. SEVERAL END AND OTHER MEMBERS WITH NAMES UNSETTLED FOR SOME. MANY MEGASCOPICALLY INDISTINGUISHABLE; OCCURRENCE MAY BE INDICATIVE. ORTH + MNCL SERIES.SHORT PRISMS.PRISM CLVG AT 87° + 92°. ALTERS TO AMPHIBOLE.IG + META RKS	W402 2-1
5 -7	OLIVINE	ORTH.	3.22-4.39	NSI	GRP NAME FOR MINS OF FAYALITE-FORSTERITE + FAYALITE-TEPHROITE SERIES, MONTI-CELLITE + GLAUCOCHROITE. GRNLR MASSES, DISSEM. CONCH. GELAT. COMMONLY ALTERED TO SERPENTINE. BASIC + ULTRABASIC IG RKS, CNTCT META DOLOMITIC LS, METEORITES	W498 1-2
5.5	BRAZILIANITE	MNCL.	2.98-2.99	PHO	EQUANT STRIATED,GLBL-RADIAL FIBR. VITR. 1 CLVG. CONCH. DCMP-HF,HOT H2SO4. PEGMATITES	41.5.10.
5.5	BITYITE	MNCL.	3.	PSI	SIMILAR TO COMMON MICAS, DISTINGUISHED OPTICALLY. LI-BEARING PEGS	M391 N
5.5	GUARINITE	TRCL.	3.2 -3.5	NSI	TBLR. 2 CLVGS AT 90 DEGREES. GELAT. MEGASCOPICALLY LIKE MOEHLERITE	W518 N
5.5	JOAQUINITE	ORTH.	3.89	SSI	TYPICALLY HONEY-YEL. TBLR-EQUANT XLS. WITH BENITOITE IN VEIN	M463 N
5.5	PEROVSKITE	MNCL.	3.97-4.05	OXD	SUBMET. CUBIC, OCTAH, GRNLR, ADMN. 1 CLVG. DCMP-HF, H2SO4.BASIC IG	7.4.2.1

NONMETALLIC YELLOW

HARDNESS	NAME	XL. SYS.	SPECIFIC GRAVITY	CHEM. CLASS	REMARKS	REFERENCES
5.5	WILLEMITE	HEXA.	3.9 -4.1	NSI	PRISMS, MSV. MAY FLUO-GRN, SOME PHOS + TRIBOLUM. GELAT. ZINC DEPS	W497 N
5.5	TRITOMITE	HEXA.	4.2	CSI	ACUTE TRIGONAL PYRAMIDS. 1 POOR CLVG. GELAT. RARE-LANGESUND, NORWAY	W509 N
5.5	RUTHERFORDINE	UNKN.	4.82	CBT	MATTED FIBERS. SOL-ACIDS. RARE-PSEUDOMORPHS AFTER URANINITE(PEGS)	16.1.14.
5.5	STIBIOCOLUMBITE	ORTH.	5.68	OXD	IN SERIES WITH STIBIOTANTALITE, WHICH SEE. RARE IN PEGMATITES	8.1.8.2
5.5	DJALMAITE	ISOM.	5.75-5.88	OXD	MODIFIED OCTAH. SUBCONCH. (=TA-BETAFITE?). RARE-GRANITIC PEGMATITE	8.4.2.
5.5	STIBIOTANTALITE	ORTH.	7.34	OXD	IN SERIES WITH STIBIOCOLUMBITE. YEL-BRWN STRK.PRISM-STRIATED. RESIN-ADMN.1 GOOD,1 POOR-CLVGS. SOL-HF.PYROELECT. PEGS,PLACERS	8.1.8.1
5.5	SODALITE	ISOM.	2.27-2.33	TSI	DODEC, MSV. POOR DODEC CLVG. GELAT. SOME FLUO-YEL-ORANGE. IN NEPH-ELINE-BEARING RKS. DISSOLVE-HNO3,EVAPORATE SLOWLY, HALITE REMAINS	M348 4-289
5.5-6	BERYLLONITE	MNCL.	2.81	PHO	TBLR, COMPLEX. VITR,PEARLY. CLVGS. CONCH. SOL-ACIDS(SLOW). PEGS	38.1.6.
5.5-6	MONTEBRASITE	TRCL.	2.98	PHO	AMBLYGONITE SERIES(SEE FOR PROPERTIES + OCCURRENCES.)	41.5.5.2
5.5-6	NATROMONTEBRASITE	TRCL.	3.04-3.1	PHO	AMBLYGONITE SERIES(SEE FOR PROPERTIES + OCCURRENCES.)	41.5.5.3
5.5-6	AMBLYGONITE	TRCL.	3.11	PHO	EQUANT, LARGE MASSES. VITR,PEARLY. CLVGS. SOL-ACIDS(DIFFICULTY). LI-, PO4-RICH PEGS	41.5.5.1
5.5-6	WOEHLERITE	MNCL.	3.42	NSI	PRISMATIC, TBLR. 1 GOOD CLVG. SOL-HCL. PEGS + SOME SILICIC IG RKS	W516 N
5.5-6	ANTHOPHYLLITE (AMPHIBOLE)	ORTH.	2.85-3.57	ISI	WITH CORDIERITE IN GNEISSES, WITH TALC IN META ULTRABASIC IG RKS	M425 2-211
5.5-6	GEDRITE (AMPHIBOLE)	ORTH.	2.85-3.57	ISI	RELATED IN COMPOSITION + OCCURRENCE TO ANTHOPHYLLITE	M425 2-211
5.5-6	ANATASE	TETR.	3.90	OXD	SUBMET. PALE YEL STRK. ACUTE PYRAMIDAL. ADMN. BASAL + PYRAMIDAL CLVG. CONVERTS TO RUTILE ON HEATING. VNS,ACCESS-IG RKS,DETRITAL	4.5.2.
5.5-6	BROOKITE	ORTH.	4.08-4.20	OXD	GRAY TO YEL STRK.TBLR XLS. ADMN. POOR CLVGS. VNS, DISSEM, DETRITAL	4.5.3.
5.5-6	NORDENSKIOLDINE	HEXA.	4.20	BOR	TBLR, LENSLK XLS. VITR,PEARLY. BASAL CLVG. DCMP-HCL. ALK PEG	24.1.8.
5.5-6.5	OPAL		1.8 -2.25	SIL	SUBMICROXLINE QTZ + H2O.RESIN-PEARLY. MSV,MAMILLARY CRUSTS,CONCH.SOL-HF +CAUS-TIC ALKS.SOME FLUO-YEL-GREEN. INCRUSTATIONS, IN CAVITIES(ESP IN VOLCANIC RKS)	D-111 287
5.5-6.5	FERGUSONITE	TETR. C	5.38	OXD	SUBMET. IN SERIES WITH FORMANITE. PRISM. COMMONLY HEMIHEDRAL, IR-REGULAR MASSES. VITR. METAMICT.DCMP-HF. GRANITIC PEGMATITES	8.1.3.1
5.5-6.5	ROMEITE	ISOM.	4.7 -5.4	ANT	OCTAH,MSV. GRSY. OCTAH CLVG. DIVERSE(E.G.,MN DEPS)	44.1.1.2
5.5-6.5	FORMANITE	TETR. C	7.03	OXD	SUBMET. IN SERIES WITH FERGUSONITE,WHICH SEE. PLACERS(W. AUSTRALIA)	8.1.3.2
5.5-7.5	NARSARSUKITE	TETR.	2.75	PSI	TBLR. PRISM CLVG. SOL-HF. IN PEGS ASSOC WITH + CAVITIES IN ALK IGS	W358 N
6	BERTRANDITE	ORTH.	2.6	ISI	PLATY. 1 PERF, OTHER CLVGS. PEGMATITES, TYPICALLY WITH BERYL	W479 N
6	CATAPLEIITE	HEXA.	2.75	CSI	THIN PLATES. 2 PERF CLVGS. RARE-ALKALIC RKS, INCLUDING PEGMATITES	W454 N
6	CLINOENSTATITE (CLINOPYROXENE)	MNCL.	3.19	ISI	HI-T FORM. RARE IN IGNEOUS ROCKS, IN METEORITES.	M408 N
6	HUMITE	ORTH.	3.20-3.32	NSI	GROUP NAME, ALSO APPLIED TO SPECIES OF GROUP. IRREGULAR-SHAPED GRAINS. VITR-RESIN. GELAT. SIMILAR TO OLIVINE. CONTACT METAMORPHOSED DOLOMITIC LIMESTONES	W514 1-49
6	CLINOHUMITE	MNCL.	3.21-3.35	NSI	COMPLEX XLS. GELAT. SKARNS, TALC SCHISTS, + SERPENTINITES	W515 1-49

NONMETALLIC YELLOW

H	Mineral	Crystal	SG	Type	Notes	Ref
6	DANALITE	ISOM.	3.28-3.44	TSI	HELVITE(SEE) GRP PROPS. FLESH-RED TO GRAY. OCTAH. GRANITES, SKARNS	W351 4-303
6	HELVITE	ISOM.	3.20-3.44	TSI	TETRA,MSV. POOR OCTAH CLVG.GELAT(H2SF),PYROELECTRIC.BOIL IN HCL + AS2O3,DECANT, WASH,HELVITE IS STAINED CANARY YELLOW (AS2S3). RARE-GNEISS,GRANITE,CNTCT ZONES	M351 4-303
6	EPIDOTE	MNCL.	3.38-3.49	SSI	GROUP NAME. NO GENERALLY ACCEPTED NOMENCLATURE FOR GROUP MEMBERS. CLOSELY RE-LATED TO ZOISITE + ALLANITE. PISTACHIO-BLACKISH GREEN-BROWNISH. LONG THIN GROOVED XLS, GRNLR MASSES. VNS, OTHER CAVITIES, REGIONAL META + IG ROCKS	W448 1-193
6	PISTACITE	MNCL.	3.3 -3.49	SSI	NAME USED BY SOME FOR INTERMEDIATE COPOSITION MEMBERS-EPIDOTE GRP	W448 1-184
6	LAVENITE	MNCL.	3.5	NSI	PRISMATIC,TBLR. 1 GOOD CLVG. SOL-HCL(SLOWLY). ALKALIC IGNEOUS RKS	W517 N
6	AEGIRINE-AUGITE (CLINOPYROXENE)	MNCL.	3.40-3.55	ISI	TYPICAL PYROXENE OF ALKALIC ROCKS - ESPECIALLY SYENITIC ONES	W411 2-79
6	ARDENNITE	ORTH.	3.6 -3.65	SSI	PRISM XLS. 1 PERF, 1 GOOD-CLVGS. 1 PARTING. RARE-PEGMATITES + VNS	W529 N
6 -6.5	MICROCLINE (K-FELDSPAR)	TRCL.	2.56-2.63	TSI	DISTINGUISHED FROM ORTHOCLASE OPTICALLY. PLUTONIC IG RKS, PEGS	W308 4-7
6 -6.5	ORTHOCLASE (K-FELDSPAR)	TRCL.	2.55-2.63	TSI	DISTINGUISHED FROM MICROCLINE OPTICALLY. PLUTONIC IG RKS, PEGS	W303 4-7
6 -6.5	PERTHITE (FELDSPARS)	TRCL.	2.56-2.65	TSI	MEGASCOPICALLY, MICROSCOPICALLY OR SUBMICROSCOPICALLY INTERDIGITA-TED MICROCLINE OR ORTHOCLASE + PLAGIOCLASE (TYPICALLY ALBITE)	W299 4-7
6 -6.5	HYALOPHANE (BA-FELDSPAR)	MNCL.	2.58-2.82	TSI	BA EQUIVALENT OF ADULARIA. PRISM, MSV. MN-RICH VEINS + ROCKS	W304 4-166
6 -6.5	PREHNITE	ORTH.	2.90-2.95	SSI	ROSETTES-TBLR XLS. 1 CLVG. SOL-HCL(SLOW). CAVITIES(WITH ZEOLITES)	W359 3-263
6 -6.5	PARACELSIAN (BA-FELDSPAR)	MNCL.	3.31-3.32	TSI	PSEUDO-ORTHORHOMBIC. TOPAZ-LK XLS. MN-RICH VEINS AND ROCKS	W307 4-167
6 -6.5	FELDSPAR		2.55-3.39	TSI	INCLUDES PLAGIOCLASE(NA-CA)SERIES, ALKALI(K-NA) AND BA FELDSPARS. MOST EASI-LY DISTINGUISHED BY NONMEGASCOPIC MEANS. MONOCLINIC ONES HAVE 2 CLVGS AT 90° AND SIMPLE(IF ANY) TWINNING; TRICLINIC ONES HAVE 2 CLVGS UP TO 4° OFF OF 90° AND TYPICALLY HAVE POLYSYNTHETIC TWINNING (GIVING STRIATED APPEARANCE ON SOME CLVG SURFACES OF PLAGIOCLASE). PEGMATITES IG, META, AND SOME SED ROCKS	W261 4-1
6 -6.5	CELSIAN (BA-FELDSPAR)	MNCL.	3.10-3.39	TSI	SEE FELDSPAR. TYPICALLY CLEAVABLE MASSES. MN-RICH VEINS + ROCKS	W307 4-166
6 -6.5	RUTILE	TETR.	4.21-4.25	OXD	SUBMET. PRISM-STRIATED.ADMN. POOR CLVGS. VNS, ACCESS-META + IG RKS	4.5.1.1
6 -7	VESUVIANITE	TETR.	3.33-3.43	SSI	PRISM, MSV. 1 POOR CLVG. SUBCONCH. ATTKD-HCL. META-ESP CNTCT LS	W508 1-113
6 -7	CASSITERITE	TETR.	6.99	OXD	SUBMET. RADIAL CONCRETIONARY MASSES. ADMN-DULL. HI-T VNS, GREISENS	4.5.1.5
6 -7.5	GROSSULAR	ISOM. A	3.59	NSI	CA-AL GARNET. EQUANT XLS. CONTACT + REGIONAL META CALCAREOUS RKS	W489 1-93
6 -7.5	ANDRADITE	ISOM. A	3.86	NSI	CA-FE,TI GARNET. F-MAG GLBL. CONTACT META CALCAREOUS RKS, ALK IGS	W489 1-89
6 -7.5	GARNET	ISOM.	3.58-4.32	NSI	GROUP NAME. DOMINANT MOLECULE DETERMINES NAME. DISTINGUISHED BY S.G. + NONMEGA-SCOPIC MEANS.EQUANT XLS,MSV.VITR-DULL.DODEC PARTING. SLIGHTLY- TO IN-SOL IN HF	W483 1-77

NONMETALLIC YELLOW

HARDNESS	NAME	XL. SYS.	SPECIFIC GRAVITY	CHEM. CLASS	REMARKS	REFERENCES
6.5	CHRYSOPRASE	HEXA.	2.57-2.64	SIL	APPLE-GREEN CHALCEDONY(MICROFIBR QTZ). SECONDARY-VNS IN SERPENTINE	D-111 218
6.5	CHALCEDONY	HEXA.	2.57-2.64	SIL	MICROXLINE + MICROFIBR QTZ. MAMMILLARY. LO-T VEINS +OTHER CAVITIES	D-111 195
6.5	CHONDRODITE	MNCL.	3.16-3.26	NSI	ISOLATED TBLR GRNS. VITR-RESIN. GELAT. LIKE OTHER HUMITES-SKARNS	W513 1-48
6.5	JEREMEJEVITE	HEXA.	3.28	BOR	PRISMS. VITR. CONCH. RARE IN RESIDUUM ABOVE GRANITE(E.SIBERIA)	24.1.7.
6.5	KORNERUPINE	ORTH.	3.27-3.34	NSI	PRISM MASSES. PRISM CLVG. WITH CORDIERITE + SAPPHIRINE IN SCHISTS	W523 N
6.5	CLINOZOISITE	MNCL.	3.21-3.38	SSI	CALCIUM-ALUMINUM END-MEMBER OF EPIDOTE GROUP. SAME OCCUR	W448 1-193
6.5	BADDELEYITE	MNCL.	5.4 -6.02	OXD	FLAT PRISM. VITR. BASAL + 2 POOR CLVGS. DCMP-HOT H2SO4. PLACERS, ASSOCIATED WITH ULTRAMAFIC IGNEOUS ROCKS	5.1.1
6.5	FAYALITE	ORTH.	4.2 -4.39	NSI	FAYALITE-FORSTERITE (FE-MG)OLIVINE SERIES CONSISTS OF FORSTERITE (FA0-10),CHRYSOLITE(FA10-30),HYALOSIDERITE(FA30-50),HORTONOLITE	W498 1-2
6.5-7	FERROHORTONOLITE	ORTH.	4.0 -4.3	NSI	(FA50-70),FERROHORTONOLITE(FA70-90) + FAYALITE(FA90-100). PROPS	W499 1-22
6.5-7	HORTONOLITE	ORTH.	3.7 -4.1	NSI	RANGE WITH COMPOSITION. MG-RICH ONES IN BASIC + ULTRABASIC IG RKS	W499 1-22
6.5-7	HYALOSIDERITE	ORTH.	3.5 -3.75	NSI	CONSTITUTING NEARLY 100 PER CENT OF SOME, + IN CONTACT METAMOR-	W498 1-22
6.5-7	CHRYSOLITE	ORTH.	3.3 -3.6	NSI	PHOSED DOLOMITES. FE-RICH ONES IN ACIDIC + ALKALIC IGNEOLS ROCKS	W498 1-22
7	FORSTERITE	ORTH.	3.22	NSI		W498 1-22
6.5-7	JASPER	HEXA.	2.57-2.65	SIL	SUBTRANSLUCENT. SOME IS VARIEGATED OR BANDED. MICROGRNLR. CAVITIES	D-111 224
6.5-7	SPODUMENE	MNCL.	3.03-3.22	ISI	COMMONLY FLUO(+PHOSPHORESCES)-ORANGE. THERMOLUM. LI-BEARING PEGS	W418 2-92
6.5-7	AXINITE	TRCL.	3.26-3.36	CSI	STRIATED WEDGE-SHAPED XLS. 1 GOOD CLVG. SOL-HCL(SLOW). CNTCT META	W505 1-320
6.5-7	DIASPORE	ORTH.	3.3 -3.5	OXD	PLTY, FOL-MSV. VITR, PEARLY. WITH CORUNDUM IN EMERY ROCK	7.1.2.1
6.5-7,5	SILLIMANITE	ORTH.	3.23-3.27	NSI	PRISMATIC XLS, FIBR MASSES. VITR. 1 CLVG. HI-GRADE META ROCKS	W520 1-121
7	ELPIDITE	ORTH.	2.58	SSI	FIBR-MSV. PRISM CLVG. IN ALKALIC IGNEOUS ROCKS(E.G., OSLO EKERITE)	W454 N
7	QUARTZ	HEXA.	2.65	SIL	DOUBLY TERMINATED HEX PRISMS STRIATED ⊥ LENGTH(LO-T), BIPYRAMIDS (HI-T).MSV. VITR,CONCH.SOL-HF. WDSPRD IN MOST RK TYPES + VEINS	D-111 9
7	CITRINE	HEXA.	2.65	SIL	YELLOW. XLINE. RARE--TYPICALLY WITH AMETHYST IN CAVITIES IN BASALT	D-111 181
7	DANBURITE	ORTH.	3.0	TSI	PRISM. 1 POOR CLVG. VITR. GELAT(AFTER IGN). PEGS, WITH FELDSPAR IN DOLOMITE	W258 N
7	ELBAITE	HEXA.	3.03-3.10	CSI	NA- LI-TOURMALINE. INFUSIBLE. TYPICALLY PINK AND/OR GREEN. PEGS	M465 1-300
7	TOURMALINE	HEXA.	3.03-3.25	CSI	GROUP NAME. VARIETIES COMMONLY DISTINGUISHABLE MEGASCOPICALLY ON BASIS OF COLOR(S), LENGTHWISE STRIATED PRISMATIC XLS WITH CROSS-SECTIONS THAT RESEM- BLE SPHERICAL TRIANGLES. SOME XLS EXHIBIT BOTH LENGTHWISE + CONCENTRIC COLOR ZONING. VITR-RESIN, BRTL. ELECTRICALLY CHARGED ON HEATING + COOLING, MG-RICH VARIETIES FLUO-YEL (ULTRAVIOLET - SW). PEGS, HI-T VNS, IG AND META ROCKS	W465 1-300

70

NONMETALLIC YELLOW

Hardness	Mineral	Crystal	SG	Class	Description	Ref
7 -7.5	BORACITE	ORTH.	2.91-2.97	BOR	CUBIC,ETC.,GRNLR,FIBR. VITR. SOL-HCL(SLOWLY). SALT-GYPSUM BEDS	26.1.7.
7.5	HAMBERGITE	ORTH.	2.36	BOR	PRISM-STRIATED. VITR. 2 CLVGS. SOL-HF. ALKALIC PEGMATITES	26.1.2.
7.5	PHENAKITE	HEXA.	2.98	NSI	RHOMBOHEDRAL(FLAT)-PRISMATIC. VITREOUS. PEGMATITES	W496 N
7.5	STAUROLITE	ORTH.	3.74-3.83	NSI	PRISM-CRUCIFORM TWINS COMMON. FIRES TO MAG POWDER. SCHIST + GNEISS	W522 1-151
7.5	ZIRCON	TETR.	4.6 -4.7	NSI	TERMINATED PRISMS. ADMN. SOME FLUO-YEL-ORANGE. ACCESS-IG. SANDS	W494 1-59
7.5-8	BERYL	HEXA.	2.66-2.83	CSI	GROOVED PRISM. POOR BASAL CLVG.SOME FLUO WEAKLY-YEL. PEGMATITES	W463 1-256
7.5-8	SPINEL	ISOM.	3.55	OXD	GROUP NAME. SOME HAVE DISTINCTIVE COLORS OR OTHER HAND-SPECIMEN PROPS. OCTAH. GRNLR, GLASSY, CONCH. SOME FLUO—RED TO YEL-GREEN. IG,META(ESP CALC RKS),PEGS	7.2.1.1
						7.2.1.3
7.5-8	GAHNITE	ISOM.	4.62	OXD	SPINEL(SEE). OCTAM. SCHISTS, PEGS, CNTCT META-LS	44.2.2.
7.5-8.5	SWEDENBORGITE	HEXA.	4.29	ANT	PRISM. 1 CLVG. SUBCONCH. RARE-SKARN(LANGBAN, SWEDEN)	
8	RHODIZITE	ISOM.	3.31-3.38	BOR	DODEC. VITR. OCTAH CLVG. WITH RED TOURMALINE IN PEGMATITES-RARE	24.1.6.
8	TOPAZ	ORTH.	3.49-3.57	NSI	STRIATED PRISMS. PERF BASAL CLVG. CNTCT ZONES, GREISENS, PEGS	W509 1-145
8.5	CHRYSOBERYL	ORTH.	3.65-3.85	OXD	TBLR-STRIATED. 3 CLVGS. OPALESCENT, ASTERATED. PEGS, PLACERS, META	7.2.3.
9	CORUNDUM	HEXA.	4.0 -4.1	OXD	STEEP PYRAMIDAL, PRISM. BASAL PARTING.WDSPRD IN SI-DEFICIENT RKS	4.4.1.1
10	DIAMOND	ISOM.	3.50-3.53	ELE	OCTAHEDRAL. ADMN-GRSY. BRTL. OCTAH CLVG. ULTRABASIC IGS + PLACERS	1.2.4.1

71

NONMETALLIC GREEN

HARDNESS	NAME	XL. SYS.	SPECIFIC GRAVITY	CHEM. CLASS	REMARKS	REFERENCES
1	GLAUCOCERINITE	UNKN.	2.75	SUF	WARTY MASSES, RADIAL + CONCENTRIC. WAXY. WITH SMITHSONITE,ETC.(LAURIUM SPECIMEN)	LAU-31,1,2.
1	TALC	MNCL.	2.58-2.83	PSI	FOL MSV. BASAL CLVG. SECTILE. SOAPY FEEL. PEARLY. SCHIST, META IG	W364 3-121
1	MOLYSITE	HEXA. A	2.90	HAL	MSV. BASAL CLVG. DLQSNT. VOL SUBLIMATE	9.3.1.
1	LAWRENCITE	HEXA. A	3.16	HAL	MSV. BASAL CLVG. SOL-H2O. DLQSNT. METEORIC, VOL SUBLIMATE	9.2.3.1
1	PARALAURIONITE	MNCL.	6.15	HAL	LATHLK. SUBADMN. 1 CLVG. SOL-HNO3. XLS BENDIGLIDING). SCNDRY AFTER LEAD MINERALS	10.1.8.
1	SILLENITE	ISOM.	8.80	OXD	GRNLR. WAXY-DULL. RARE-SCNDRY AFTER BISMUTITE(DURANGO,MEXICO)	4.4.8.
1-2	VERMICULITE	MNCL.	2.2 -2.4	PSI	BRONZY. MICACEOUS. EXFOLIATES WITH SWELLING ON HEATING. ATTKD-ACID (SILICA RESIDUE). ALTER PROD OF BIOTITE(ETC.) - HYDROTHERMAL,WXING	W396 3-246
1-2	MONTMORILLONITE (CLAY)	MNCL.	2.2 -2.7	PSI	IRREG+HEX-SHAPED PLTS. SWELLING CLAY. ALTER-HI MG +CA,LO K RKS (ALKALINE CONDITIONS). SOILS, BENTONITES, ETC.	W398 3-226
1-2	CELADONITE	MNCL.	2.6 -2.9	PSI	PHYSICALLY + CHEMICALLY LIKE(=?) GLAUCONITE. CAVITIES IN BASALTS	W378 3-217
1-2	PYROPHYLLITE	MNCL.	2.65-2.90	PSI	LAMELLAR. CLVG-INELAST PLTS. PARTLY DCMP-H2SO4. VNS, FEW SCHISTS	W361 3-115
1-2	BEIDELLITE (CLAY)	MNCL.	2.	PSI	"SWELLING CLAY." ALTER-HI MG +CA,LO K RKS(ALK CONDITIONS). RARE-GOUGE (COLORADO), BENTONITES	W398 3-226
1-2	HYDROBIOTITE	MNCL.	2.6 -3.3	PSI	FMED UNDER ALK CONDITIONS. CLASTIC + DIAGENIC-SEDS. HYDROTHERMAL	W396 3-213
1-2	AURICHALCITE	ORTH.	3.64	CBT	ACICULAR,LATHLK,GRNLR. SILKY. 1 CLVG. SOL-ACIDS. SCNDRY CU,ZN-DEPS	16.1.4.
1-2.5	NONTRONITE (CLAY)	MNCL.	2.5 -3.0	PSI	SUBTRANSLUCENT, LATHS, MSV. GELAT. SWELLING CLAY. ALTER-HI MG + CA,LO K RKS(ALK CONDITIONS), WITH OPAL +/OR QTZ IN VNS, BENTONITE	W398 3-226
1-3	GARNIERITE (CHLORITE)	MNCL.	2.27-2.87	PSI	NI-SERP. MAGNETIC AFTER HEATING. GELAT. SERPENTINE-RICH ROCKS	W379 3-175
1-3	CLAY	MNCL.	1.85-3.0	PSI	NAME GIVEN SEVERAL HYDROUS AL-SILICATES WITH ALKALINE EARTHS + MG OR FE. NAMES UNSETTLED. MOST DISTINGUISHED BY NONMEGASCOPIC MEANS. FINE-GRAINED AGGS(COMPACT,MEALY,ETC.). EARTHY-PEARLY,UNCTUOUS,PLASTIC WHEN WET,DEHYDRATED ON HEATING. SOME ARE ABSORBENT. WXING PRODS,HYDROTHERMAL ALTER, DETRITAL, DIAGENIC	W398 3-191
1.5	WATER	HEXA.	.92	OXD	LIQUID ABOVE 0°C.(32°F)+ BELOW 100°C.(212°F). S.G. FOR ICE	4.1.3.
1.5	APJOHNITE	MNCL.	1.78	SUF	FIBR. SILKY. SOL-H2O. EFFLORESCENCES IN SHELTERED PLACES	29.7.3.3
1.5	MALOTRICHITE	MNCL.	1.89	SUF	FIBR. VITR. ASTRINGENT. EFFLORESCENCES IN SHELTERED PLACES	29.7.3.2
1.5	IODARGYRITE	HEXA.	5.69	HAL	EXPOS-YEL. MSV. PRISM-HEMIMORPHIC. RESIN. BASAL CLVG(FLEX). DCMP-H2SO4, HNO3, KI SOLUTION. OXIDIZED ZONES OF SILVER DEPOSITS	9.1.4.
1.5-2	VIVIANITE	MNCL.	2.67-2.69	PHO	PRISM. ROUNDED, GLBL. VITR-DULL. CLVGS. FIBR-FLEX. DARKENS ON EXPOSURE. SOL-ACIDS. SCNDRY OF DIVERSE OCCURRENCES	40.2.15.1
1.5-2.5	SULFUR	ORTH.	2.07	ELE	CONCH-UNEVN. BRTL. BURNS-BLU FL + ACRID ODOR. SOL-CS2. VOLS + SED	1.2.3.1
1.5-2.5	CHRYSOCOLLA	UNKN.	2.3 -2.5	CSI	MSV. VITR-EARTHY. CONCH. DCMP-HCL. OXIDE ZONES CU-DEPS IN SIL RKS	W420 N
1.5-2.5	ANNABERGITE	MNCL. C	3.23	ARS	XLINE COATINGS,EARTHY. ADMN-DULL. CLVGS. SOL-ACIDS. OXIDATION PROD OF COBALT AND NICKEL ARSENIDES	40.2.15.3
1.5-2.5	TYROLITE	ORTH.	3.0 -3.25	ARS	FAN-SHAPED AGGS,CRUSTS,ETC. VITR.PEARLY, 1 CLVG(FLEX). SOL-ACIDS. SECONDARY - WITH OXIDATION PRODUCTS OF COPPER DEPOSITS	42.3.2.

NONMETALLIC GREEN

1.5-2.5	TYUYAMUNITE	ORTH.	3.3 -4.35	VAS	SCALES-RADIAL AGGS,MSV,POWDERY, ADMN,PEARLY. 3 CLVGS. SOL-HNO3, HCL,H2SO4. VNLTS IN LS + WITH CARNOTITE	47.1.2.
1.5-2.5	HELIOPHYLLITE	ORTH.	6.89	ASI	PSEUDOTETRAGONAL,PYRAMIDAL-STRIATED,MSV-FOL,GRNLR. VITR.CLVG. 1 PERF CLVG. SOL-HNO3, WM HCL, ALKS. WITH ECDEMITE	46.1.2.
2	KIROVITE	MNCL.	1.76-1.87	SUF	DISTINGUISHED FROM MELANTERITE BY NONMEGASCOPIC MEANS. SAME OCCUR	29.6.8.3
2	MELANTERITE	MNCL.	1.84-1.90	SUF	STALAC, CRUSTS,ETC. VITR. CLVGS. SWEET,ASTRINGENT,METALLIC TASTE. EXPOS(DRY AIR)-YEL-WHT + DPQ. EFFLORESCENCE - BY OXIDATION OF SULS	29.6.8.1
2	RHOMBOCLASE	ORTH.	2.23	SUF	THIN TBLR, STALACTITIC, SUBVITR. BASAL(FLEX),PRISM CLVGS. CONCH-FIBR. SOL-ACIDS(H2O(SLOWLY). RARE-ALTER PROD OF SULFIDES	29.1.1.
2	CHALCOPHYLLITE	HEXA.	2.67	ARS	TBLR,FOL MSV,ROSETTES. VITR,PEARLY. BASAL, RHOMB CLVGS. SOL-ACIDS. NH4OH OXIDIZED ZONES OF COPPER DEPOSITS	43.2.1.
2	GLAUCONITE	MNCL.	2.4 -2.95	PSI	PLTS IN ROUND GRAINS. ATTKD-HCL. MARINE DEPOSITS - GREENSANDS	W377 3-35
2	GERHARDTITE	ORTH.	3.40-3.43	NIT	STRIATED PRISM, FLEX XLS. 2 CLVGS. SOL-DILUTE ACIDS. WITH CUPRITE	19.1.1.
2 -2.5	EPSOMITE	ORTH.	A 1.68	SUF	CRUSTS. VITR-EARTHY. CLVGS. BITTER, SALTY, METALLIC. COMMON EFFLOR	29.6.9.1
2 -2.5	BORAX	MNCL.	1.71-1.72	BOR	HABIT-LK PYROXENE. VITR-EARTHY. CLVGS. SWEETISH. SOL-H2O(TO ALK SOLUTION). BORAX DEPOSITS (SALINE LAKE DEPOSITS, ETC.)	25.1.4.
2 -2.5	MORENOSITE	ORTH.	1.95	SUF	CRUSTS. VITR,1 CLVG. CONCH. METALLIC,ASTRINGENT. EFFLOR-AT NI-SUL-FIDE DEPOSITS	29.6.9.3
2 -2.5	GOSLARITE	ORTH.	1.98	SUF	CRUSTS. VITR-SILKY. 1 CLVG. ASTRINGENT,METALLIC,NAUCEOUS. EFFLOR-TYPICALLY BY ALTERATION OF SPHALERITE	29.6.9.2
2 -2.5	SEPIOLITE (CLAY)	ORTH.	2.08	PSI	POROUS(FLOATS ON H2O WHEN DRY)-COMPACT. SMOOTH FEEL. GELAT. WITH SERP OR OPAL; DESERT SOILS, CLAYEY SEDS WITH SALINES, LAKE SEDS. MEERSCHAUM IS A VARIETY	W444 N
2 -2.5	LINDACKERITE	MNCL.	2.0 -2.5	ARS	LATHS IN ROSETTES,CRUSTS. VITR. 1 CLVG. WITH OTHER SCNDRY MINERALS	43.1.2.
2 -2.5	PHLOGOPITE	MNCL.	2.76-2.90	PSI	MG-MICA. ATTKD-HCL. CNTCT META-LS. ULTRABASIC IGNEOUS ROCKS	W373 3-42
2 -2.5	LIROCONITE	MNCL.	2.9 -3.01	ARS	THIN,STRIATED XLS. VITR. CLVGS. CONCH. SOL-ACIDS.OXIDIZED ZONES OF COPPER DEPOSITS	42.2.5.
2 -2.5	AUTUNITE	TETR.	3.1 -3.2	PHO	TBLR, SUBPARALLEL,FOL,SCALY AGGS, SERRATE CRUSTS. VITR,PEARLY. 2 CLVGS. SOL-ACIDS. FLUO-YEL-GREEN.ALTER PROD OF U MINS(VNS + PEGS)	42.8.13.2
2 -2.5	ZEUNERITE	TETR.	3.2	ARS	NOT KNOWN TO OCCUR IN NATURE.	42.8.13.5
2 -2.5	TORBERNITE	TETR.	3.22	PHO	TBLR, SUBPARALLEL,FOL,SCALY AGGS. VITR,PEARLY. 2 CLVGS. SOL-ACIDS OXIDATION PRODUCT OF URANINITE, ESPECIALLY IN GOSSANS	42.8.13.1
2 -2.5	JOHANNITE	TRCL.	3.32	SUF	PRISM,DRUSY AGGS OF LATHLK FIBERS. VITR. 1 CLVG. BITTER. SOL-ACIDS SECONDARY - FORMED BY ALTERATION OF URANINITE	31.5.4.
2 -2.5	URANOCIRCITE	TETR.	3.53	PHO	HABIT LK AUTUNITE. PEARLY. 3 CLVGS. WITH OTHER SCNDRY URANIUM MINS	42.8.13.3
2 -2.5	METAZEUNERITE	TETR.	3.64	ARS	TBLR,ACUTE PYRAMIDAL-STRIATED. VITR,PEARLY. FLUO-YEL GREEN. SOL-ACIDS. WITH OTHER SECONDARY URANIUM MINERALS	42.8.14.2
2 -2.5	LANARKITE	MNCL.	6.92	SUF	MSV, ELONGATE. ADMN,PEARLY. CLVGS(THIN LAMINAE-FLEX). FLUO-YEL(X-RAYS + ULTRAVIOLET RADIATION).SOL-WARM HNO3. RARE - LEAD DEPOSITS	30.2.1.
2 -3	ALLOPHANE (CHLORITE)	AMOR.	1.85-1.89	PSI	INCRUST. "BOOKS." GELAT. ALTER(ACID CONDITIONS) OF FELDSPARS, ETC. IN COAL, ORE DEPOSITS, AND VEINS	W531 3-194
2 -3	VASHEGYITE	UNKN.	1.96	PHO	SUBTRANSLUCENT. MSV,MICROXLINE-FIBR.DULL.SOL-ACIDS. WITH VARISCITE	42.9.3.
2 -3	FIBROFERRITE	ORTH.	1.84-2.52	SUF	FIBR CRUSTS. SILKY. 1 CLVG. DCMP-H2O. SCNDRY-BY OXIDATION OF FES2	31.6.6.
2 -3	COOKEITE (CHLORITE)	MNCL.	2.69	PSI	LI-BEARING CHLOR. WITH TOURMALINE IN PEGS, PROBABLY AS ALTER PROD	W378 3-147
2 -3	SHERIDANITE (CHLORITE)	MNCL.	2.65-2.80	PSI	INTERMEDIATE BETWEEN CLINOCHLORE + CORUNDOPHILITE. RARE-META ROCKS	WN 3-137

NONMETALLIC GREEN

HARDNESS	NAME	XL. SYS.	SPECIFIC GRAVITY	CHEM. CLASS	REMARKS	REFERENCES
2 -3	AMESITE (SERPENTINE?)	HEXA.	2.75-2.85	PSI	FINE GRAINED. HEX PLTS. WITH DIASPORE,CORUNDOPHILITE,ETC.-EMERY RK	W383 3-164
2 -3	TINTICITE	ORTH.	2.7 -2.9	PHO	DENSE EARTHY-PORCELANEOUS MASSES.SCNDRY-WITH JAROSITE,LIMONITE,ETC	42.8.4.
2 -3	CORUNDOPHILITE (CHLORITE)	MNCL.	2.7 -2.9	PSI	A THURINGITIC SUBFAMILY CHLOR. COUNTRY RKS OF EMERY DEP AT CHESTER, MASSACHUSETTS	W383 3-137
2 -3	DIABANTITE (CHLORITE)	MNCL.	2.7 -2.9	PSI	IN CAVITIES + VEINS IN BASALTIC IGNEOUS ROCKS	W383 3-137
2 -3	DELESSITE (CHLORITE)	MNCL.	2.7 -2.9	PSI	COMMONLY AS REPLACEMENT OF BIOTITE + OTHER FERROMAGNESIAN MINERALS	W383 3-137
2 -3	CLINOCHLORE (CHLORITE)	MNCL.	2.6 -3.0	PSI	HEX-OUTLINE (COMMONLY SLENDER)XLS,MSV. VERY COMMON,E.G.IN TALC RKS	W383 3-137
2 -3	SYMPLESITE	TRCL.	3.01	ARS	SPHERICAL AGGS-FIBR. VITR,PEARLY. CLVGS. SOL-ACIDS. SCNDRY AS-DEPS	40.2.15.5
2 -3	RHIPIDOLITE (CHLORITE)	MNCL.	2.88-3.08	PSI	INTERMEDIATE BETWEEN SHERIDANITE + DAPHNITE. FELDSPAR SCHISTS, VNS	W383 3-137
2 -3	PENNINITE (CHLORITE)	MNCL.	2.6 -3.1	PSI	PSEUDORHOMB XLS. WITH MAGNETITE + CHONDRODITE(SERP RKS), SCHISTS	W383 3-137
2 -3	BRUNSVIGITE (CHLORITE)	MNCL.	2.8 -3.1	PSI	CHAMOSITIC SUBFAMILY CHLOR. MSV,RADIATING AGGS. CAVITIES IN GABBRO	W383 3-137
2 -3	DAPHNITE (CHLORITE)	MNCL.	2.75-3.1	PSI	FE END MEMBER. REPLACEMENT MIN AFTER BIOTITE + OTHER FERROMAG MINS	W383 3-137
2 -3	CHAMOSITE	MNCL.	2.9 -3.1	PSI	FINE-GRAINED AGGS. IN LATERITES, AS OOLITES, IN MATRIX,ETC. OF IRONSTONES	WN 3-164
2 -3	SPANGOLITE	HEXA.	3.14	SUF	TBLR, SOME IS HEMIMORPHIC. VITR. BASAL, RHOMB CLVGS. CONCH. SOL-ACIDS, SCNDRY WITH CUPRITE AND OTHER COPPER MINERALS	31.1.5.
2 -3	SHARPITE	ORTH.	3.3	CBT	CRUSTS OF RADIATING FIBERS. EFF-ACIDS. RARE-URANIUM DEP(KATANGA)	16.1.15.
2 -3	CHLORITE	MNCL.	2.6 -3.3	PSI	GROUP NAME. MEMBER NAMES UNSETTLED. ONLY FEW RARE VARIETIES MEGASCOPICALLY DISTINGUISHABLE. DISSEM FLKS. PEARLY. 1 PERF CLVG TO FLEX,INELAST FOLIA. DOMP-IN BOILING H2SO4. SCNDRY-HYDROTHERMAL ALTER, LO-GRADE META, DETRITAL, AUTHIGENIC	W381 3-131
2 -3	THURINGITE (CHLORITE)	MNCL.	2.8 -3.3	PSI	SCALES, MSV. IRON FORMATIONS, SCNDRY AFTER FERROMAG MINS, GNEISSES	W383 3-137
2 -3	URANOSPINITE	TETR.	3.45	ARS	TBLR. PEARLY. 2 CLVGS. WITH OTHER SCNDRY URANIUM MINERALS	42.8.13.6
2 -3	GREENALITE (SERPENTINE)	MNCL.	2.5 -3.5	PSI	FINE GRAINED. RARE-IRON FORMATIONS (MESABI RANGE OF MINNESOTA)	W380 3-164
2 -3	URANOPHANE	MNCL.	3.8 -4.	NSI	MSV, FIBR, RADIATING, TBLR. 1 CLVG. GELAT. IN GRANITE + PEGMATITES	W529 N
2 -3	CARNOTITE	UNKN.	4.1	VAS	POWDER,CRUSTS,MICRO XLS. DULL-SILKY,1 CLVG. SOL-ACIDS. SCNDRY-E.G. SANDSTONE MATRICES	47.1.1.
2 -3	PHOSGENITE	TETR.	6.13	CBT	PRISM. MSV. ADMN. CLVGS(PERCUSSION FIGURE ON BASAL PLANE) FLUO-YEL (LW ULTRAVIOLET,CATHODE + X-RAYS). EFF-HNO3. SCNDRY AFTER PB MINS	16.1.7.
2 -3	BEYERITE	TETR.	6.50-6.56	CBT	PLTY, MSV. VITR(XLS). CONCH. EFF-ACIDS. SCNDRY-WITH BISMUTITE	16.2.5.
2 -4	MICA	MNCL.	2.4 -3.4	PSI	GROUP NAME. SOME SPECIES DISTINGUISHABLE ON BASIS OF COLOR. SERIES ARE NOT COMPLETE. HEX-SHAPED XLS,DISSEM. PERF CLVG TO FLEX,ELAST FLKS. PERCUSSION + PRESSURE FIGURES MAY BE IMPOSED ON FLKS. CHIEF OCCURRENCES ARE NOTED UNDER SPECIES	W365 3-1
2 -4	BAUXITE		2	3.5	OXD	FIELD TERM FOR MATERIALS RICH IN HYDROUS ALUMINUM OXIDES.	6.2.3.
2.5	MITRIDATITE	UNKN.		PHO	MSV,NODULES,CRUSTS. DULL-RESIN. FRIABLE-DENSE. SOL-HOT ACIDS. OOLITIC SEDIMENTARY IRON ORES - RARE (SOUTHERN RUSSIA)	42.6.6.
2.5	RETGERSITE	TETR.	2.04	SUF	FIBR CRUSTS, TWISTED TUFTS. VITR. BASAL, PRISM CLVGS. BITTER, METALLIC TASTE. EFFLORESCENCE-SCNDRY BY OXIDATION OF NICKEL MINS.	29.6.7.

NONMETALLIC GREEN

	Mineral	Crystal	SG		Description	Ref
2.5	● COQUIMBITE	HEXA.	2.10-2.12	SUF	PRISM, GRNLR. VITR. CLVGS. ASTRINGENT. SOL-ACIDS. WITH OTHER SO4'S	29.8.3.
2.5	PYROAURITE	HEXA.	2.10-2.14	OXD	TBLR-STRIATED. WAXY,PEARLY. BASAL CLVG(FLEX). EFF-HCL. LO-T VNS	6.1.5.3
2.5	● CHRYSOTILE (SERPENTINE)	MNCL.	2.2	PSI	FIBROUS, SOME IS ASBESTIFORM. SECONDARY, COMMONLY IN VEINS	W379 3-170
2.5	● CHALCANTHITE	TRCL. A	2.29	SUF	PRISM,STALAC,GRNLR. VITR. CLVGS. METALLIC,NAUSEOUS TASTE. SCNDRY-IN OXIDE ZONES OF COPPER SULFIDE DEPOSITS	29.6.5.1
2.5	CHALCOALUMITE	TRCL.	2.29	SUF	BOTRYOIDAL,LATHLK.DULL-VITR. CLVGS. SOL-DILUTE ACIDS(SLOW). CRUSTS IN OXIDE ZONES OF COPPER SULFIDE DEPOSITS	31.1.9.
2.5	● BRUCITE	HEXA.	2.38-2.40	OXD	WHT STRK, TBLR, FOL MSV. WAXY-PEARLY. BASAL CLVG(FLEX). SOL-ACIDS. LOW TEMPERATURE VEINS	6.1.1.1
2.5	MITSCHERLICHITE	TETR.	2.42	HAL	PYRAMIDAL, PRISM. VITR. RARE-STALACTITIC ON FLOOR OF VESUVIUS	11.3.2.
2.5	ERIOCHALCITE	ORTH. A	2.47	HAL	LICHENLK AGGS. VITR. PRISM,BASAL CLVGS. SOL-HCL,NH4OH. SCNDRY, FU-MAROLE SUBLIMATE	9.2.7.
2.5	METAVOLTINE	HEXA.	2.5	SUF	TBLR, GRNLR-SCALY AGGS. RESIN. 1 CLVG. PARTLY SOL-H2O,DCMP-DILUTE ACIDS(SLOWLY).WITH OTHER HYDROUS SULFATES IN DIVERSE ENVIRONMENTS	31.6.9.
2.5	SCHROECKINGERITE	HEXA.	2.51	CBT	GLBLR,SCALY. VITR. BASAL CLVG(FLEX).FLUO-YEL GREEN. SOL-H2O, ACIDS. ALTER PRODUCTION OF URANINITE, EFFLORESCENCE	15.2.4.
2.5	CALCIOFERRITE	UNKN.	2.53	PHO	FOL NODULAR MASSES. PEARLY. CLVGS. RARE-NODULES IN CLAY(BAVARIA)	42.8.9.
2.5	LIZARDITE (SERPENTINE)	MNCL.	2.5-2.6	PSI	SCALES, MASSIVE. SECONDARY (MASSIVE MATRIX MATERIALS)	WN 3-170
2.5	FERRINATRITE	HEXA.	2.55-2.61	SUF	FIBR AGGS, PRISM, MSV. VITR. CLVGS. SPLINTERY. SCNDRY WITH OTHER IRON SULFATES IN ARID REGIONS	29.4.1.
2.5	VOLKONSKOITE (CLAY)	MNCL.	2.2 -2.7	PSI	CHROMIUM-NONTRONITE(?).LAMELLAR. UNSTABLE IN H2O. GELAT. "SWELL-ING CLAY," ALTER-HI MG +CA +O K RXSIALK CONDITIONS). CLAY DEP	W399 N
2.5	PHARMACOSIDERITE	TETR.	2.80	ARS	STRIATED CUBES,TETRA,GRNLR. ADMN.1 CLVG. UNEVN. SECTILE. OXIDATION PRODUCT OF OTHER ARSENIC MINERALS	42.9.1.
2.5	BANDYLITE	TETR.	2.81	BOR	TBLR, LICHENLK. VITR,PEARLY. 1 CLVG(FLEX). DCMP-H2O, SOL-NH4OH,HCL,HNO3. RARE-SCNDRY WITH ATACAMITE, ETC.(CALAMA, CHILE)	26.1.4.
2.5	ROSCOELITE	MNCL.	2.97	PSI	V-MICA. RARE-E.G., ASSOCIATED WITH GOLD ORES	W369 3-14
2.5	DEVILLINE	MNCL.	3.13	SUF	PLTY-LATHLK CRUSTS. VITR,PEARLY. CLVGS. SOL-HNO3. SCNDRY-CU DEPS	31.3.3.
2.5	PYROCHROITE	HEXA.	3.23-3.27	OXD	TBLR, RHOMB, FOL MSV. PEARLY-DULL. BASAL CLVG(FLEX). SOL-HCL. WITH OTHER MN MINERALS IN LOW TEMPERATURE HYDROTHERMAL VEINS	6.1.1.2
2.5	METATORBERNITE	TETR.	3.5 -3.70	PHO	TBLR,ROSETTES + SHEAFLK AGGS OF CURVED + COMPOSITE XLS.VITR,PEARLY 1 CLVG. SOL-ACIDS. LO-T HYDROTHERMAL DEPOSITION	42.8.14.1
2.5	TRICHALCITE	ORTH.	3.5 -3.86	ARS	PSEUDOHEX AGGS, RADIATING GROUPS,ETC. VITR-SILKY. 1 CLVG. SOL-HCL. WITH COPPER-ARSENIC MINERALS	40.2.13.
2.5	SENGIERITE	ORTH.	4.	VAS	SIX-SIDED PLTS,CRUSTS. VITR. 1 CLVG. BRTL. SOL-ACIDS. SCNDRY	47.1.3.
2.5	NANTOKITE	ISOM.	4.14	HAL	MSV, GRNLR. ADMN. DODEC CLVG. DCMP-H2O,SOL-HCL,HNO3,NH4OH. CU-DEPS	9.1.3.1
2.5	TUNGSTITE	ORTH.	5.5	OXD	MSV, PLTY. RESIN. BASAL + PRISM CLVG. SOL-ALKALIES. OXIDATION PROD OF TUNGSTEN MINERALS	4.6.5.
2.5	COTUNNITE	ORTH. A	5.80	HAL	GRNLR, LATHLK. ADMN-SILKY. 1 CLVG. SOL-H2O,ACIDS. VOL SUBLIMATE, ALTERATION PRODUCT OF GALENA	9.2.6.
2.5	BROMARGYRITE	ISOM.	6.50	HAL	CRUSTS,CUBIC. RESIN. DUCTILE. SECONDARY IN SILVER DEPOSITS	9.1.1.5
2.5	CHLOROXIPHITE	MNCL.	6.76-6.93	HAL	BLDS-STRIATED. RESIN. CLVGS. BRTL. SOL-HNO3. RARE-SOMERSET,ENGLAND	10.2.6.
2.5	TERLINGUAITE	MNCL.	8.73	HAL	EXPOS-OLIVE GREEN, PRISM, PDWDERY, ADMN. 1 CLVG. SOL-ACIDS. RARE-MERCURY DEPOSIT (TERLINGUA, TEXAS)	10.1.2.

75

NONMETALLIC GREEN

HARDNESS	NAME	XL. SYS.	SPECIFIC GRAVITY	CHEM. CLASS	REMARKS	REFERENCES
2.5-3	COPIAPITE	TRCL.	2.08-2.17	SUF	CRUSTS-GRNLR OR SCALY AGGS. PEARLY, CLVGS. SOL-H2O. SCNDRY-WITH OTHER SULFATES AS OXIDATION PRODUCTS OF SULFIDES(E.G., PYRITE)	31.6.11.1
2.5-3	MAGNESIOCOPIAPITE	TRCL.	2.08-2.17	SUF	INDISTINGUISHABLE MEGASCOPICALLY FROM COPIAPITE, WHICH SEE	31.6.11.2
2.5-3	CUPROCOPIAPITE	TRCL.	2.08-2.17	SUF	INDISTINGUISHABLE MEGASCOPICALLY FROM COPIAPITE, WHICH SEE	31.6.11.3
2.5-3	AMARILLITE	MNCL.	2.19	SUF	EQUANT,// GROUPS. VITR-ADMN. PRISM CLVG.ASTRINGENT. VNS IN COQUIMBITE	29.5.3.2
2.5-3	BLOEDITE	MNCL.	2.22-2.28	SUF	PRISM, GRNLR. VITR. CONCH. FAINTLY SALTY + BITTER. EVAPORITE DEPS	29.3.5.1
2.5-3	LIEBIGITE	ORTH.	2.41	CBT	EQANT,GRNLR,SCALY. VITR-PEARLY. FLUO-GREEN. EFF-ACIDS. MINE WALLS	15.2.9.
2.5-3	SERICITE	MNCL.	2.77-2.88	PSI	FINE-GRAINED WHITE (NA OR K) MICA. DEUTERIC-IG RKS, META ROCKS	W369 3-14
2.5-3	KROEHNKITE	MNCL.	2.90	SUF	FIBR AGGS,MSV,GRNLR. VITR. 1 PERF CLVG. SOL-H2O TO ACID SOLUTION. BECOMES GREEN + OPQ ON EXPOSURE.EFFLORESCENCE ON CU-SULFIDES,ETC.	29.3.3.
2.5-3	BIOTITE	MNCL.	2.7 -3.3	PSI	MG-,FE-MICA. BLEACHED + ATTKD-H2SO4.COMMON-E.G.,ACID IG,PEGS, META	W373 3-55
2.5-3	SIDEROPHYLLITE	MNCL.	2.8 -3.4	PSI	FE-MICA. RARE IN PEGMATITES	W373 3-65
2.5-3	LANGITE	ORTH.	3.48-3.50	SUF	FIBR-LAMELLAR CRUSTS, SILKY. 2 CLVGS. SOL-ACIDS,NH4OH. SCNDRY-OXIDATION OF COPPER SULFIDES	31.2.2.
2.5-3	CLINOCLASE	MNCL.	4.38	ARS	ELONGATE,ROSETTES,CRUSTS,FIBR. VITR,PEARLY. 1 CLVG. SOL-ACIDS + NH4OH. WITH OLIVENITE + OTHER SECONDARY COPPER MINERALS	41.3.1.
2.5-3	CALEDONITE	ORTH.	5.6 -5.77	SUF	COATINGS, DIVERGENT GROUPS. RESIN. CLVGS. EFF-HNO3. OXIDE ZONES OF COPPER-LEAD DEPOSITS	32.1.2.
2.5-3	VAUQUELINITE	MNCL.	6.02	CHR	WEDGE-SHAPED,MAMMILLARY FIBR. ADMN-RESIN. UNEVN. SOL-HNO3(PARTLY). SECONDARY - WITH CROCOITE, ETC.	36.1.1.
2.5-3	ANGLESITE	ORTH.	6.37-6.39	SUF	TBLR,PRISM,MSV,ETC. ADMN. CLVGS. CONCH. BRTL. SOME FLUO-YEL. SOL-HNO3(DIFFICULTLY). SCNDRY - BY OXIDATION OF GALENA	28.3.1.3
2.5-3	LEADHILLITE	MNCL.	6.55	CBT	PSEUDOHEX, GRNLR. RESIN, PEARLY. 1 CLVG. FLUO-YEL. EFF-HNO3. SCNDRY IN OXIDIZED ZONES OF LEAD DEPOSITS	17.1.3.
2.5-3	WULFENITE	TETR.	6.5 -7.0	MBS	TBLR,DIPYRAMIDAL,MSV,GRNLR. RESIN-ADMN. CLVGS.UNEVN. DCMP-HCL. HNO3, SOL-H2SO4,ALKS. SCNDRY-OXIDIZED ZONES OF PB- + MO-DEPOSITS	48.1.4.1
2.5-3	MATLOCKITE	TETR.	7.12	HAL	TBLR, PLTY AGGS, ETC. ADMN. 1 CLVG. SOL-HNO3, DCMP-H2SO4. PB-DEPS	10.1.6.1
2.5-3	ECDEMITE	TETR.	7.14	ASI	CRUSTS-TBLR XLS,FOL MSV. VITR. 1 CLVG. SOL-HNO3,HCL(WARM),ALKS. RARE-E.G., WITH HELIOPHYLLITE (LANGBAN, SWEDEN)	46.1.1.
2.5-3	STOLZITE.	TETR.	7.9 -8.34	MOS	DIPYRAMIDAL-STRIATED. RESIN. CLVGS. CONCH. SOL-HCL(YEL). WITH LIMONITE,ETC.-OXIDIZED DEPOSITS WITH TUNGSTEN MINERALS	48.1.4.2
2.5-3.5	GIBBSITE	MNCL.	2.3 -2.42	OXD	TBLR, MAMMILLARY. PEARLY-VITR. 1 CLVG. CLAYLK ODOR WHEN DAMP. WXING PRODUCT, BAUXITE DEPOSITS, LOW TEMPERATURE VEINS	6.2.2.
2.5-3.5	SERPENTINE	MNCL.	2.5 -2.6	PSI	GROUP NAME. NOMENCLATURE FOR MEMBERS IS UNSETTLED. MSV-ASBESTIFM. WAXY-GRSY. GELAT, SOME FLUO-CREAM YEL. ALTER PROD AFTER MG-SILICATES(E.G.,OLIVINE), VNS	W381 3-170
2.5-3.5	ANTIGORITE (SERPENTINE)	MNCL.	2.65	PSI	COLORS COMMONLY VARIEGATED OR MOTTLED. PLATY. SECONDARY	W381 3-170
2.5-3.5	BISMUTITE	TETR.	6.1 -7.7	CBT	GRAY STRK. MSV. RADIAL FIBR CRUSTS. VITR-DULL. EFF-ACIDS. SCNDRY AFTER BISMUTH MINERALS	16.1.8.
2.5-3	MUSCOVITE	MNCL.	2.77-2.88	PSI	K-MICA. COMMON RK-FORMING MIN-E.G., ACID IG, PEGS, GREISENS, CNTCT METASOMATIZED(FLUORINE) ZONES, REGIONAL META-ARGILLACEOUS ROCKS	W367 3-11
3	HYDROCALUMITE	MNCL.	2.15	OXD	MSV. VITR, PEARLY. BASAL + POOR CLVGS. SOL-HCL. RARE-CNTCT META-LS	6.2.4.
3	METAVAUXITE	MNCL.	2.35	PHO	PRISM,ACICULAR,//-RADIAL AGGS. VITR-SILKY. WX PROD(I.E.,SUPERGENE)	42.8.5.

NONMETALLIC GREEN

H	Mineral	Cryst	SG	Opt	Description	Ref
3	PARAVAUXITE	TRCL.	2.36	PHO	PRISM,SUBPARALLEL-RADIAL AGGS. VITR,PEARLY. 1 CLVG. CONCH. WITH VAUXITE, ETC. IN TIN VEINS	42.8.6.
3	CALCITE	HEXA.	2.71	CBT	VARIOUS TRIGONAL XL HABITS. VITR-PEARLY. RHOMB CLVG. SOME IS FLUO,PHOSPHO,THERMOLUM. EFF-DILUTE HCL. WDSPRD-VNS,DIVERSE ROCKS	14.1.1.1
3	APHTHITALITE	HEXA.	2.65-2.71	SUF	TRIGONAL-TBLR, MAMMILLARY,MSV. VITR. CLVGS. CONCH. SOL-H2O, ACIDS. FUMAROLIC, OCEANIC + LACUSTRINE SALINE DEPOSITS	28.2.2.
3	ZEOPHYLLITE	HEXA.	2.76	PSI	SPHERES, RADIAL STRUCTURE. BASAL CLVG. WITH ZEOLITES IN AMYGDULES	M359 N
3	VOLTAITE	ISOM.	2.6 -2.8	SUF	SUBTRANSLUCENT. CUBIC, OCTAH, MSV. GRNLR, RESIN, CONCH. SOL-ACIDS. DCMP-H2O TO ACID SOLUTION. SOLFATARAS, FUMAROLES, EFFLORESCENCES	29.5.2.
3	LEIGHTONITE	TRCL.	2.95	SUF	LATHLK.VITR. RARE-VNS + CAVITIES WITH ATACAMITE,ETC.(CHUQUICAMATA)	29.4.3.
3	TAGILITE	ORTH.	3.4 -3.6	PHO	EARTHY CRUSTS,FIBR CONCRETIONS. DATA LACKING. SCNDRY-CU-DEPOSITS	42.4.3.
3	PARATACAMITE	HEXA.	3.74	HAL	GRNLR,CRUSTS.VITR.RHOMB CLVG.SOL-ACIDS. SCNDRY AFTER TENORITE,ETC.	10.1.12.
3	KAMAREZITE	ORTH.	3.98	SUF	STRIATED ELONGATE XLS.1 CLVG. SOL-ACIDS,NH4OH. NEEDS VERIFICATION.	31.3.1.
3	OLIVENITE	ORTH.	3.9 -4.46	ARS	ELONGATE,GLBL FIBR,MSV,ETC. ADMN-SILKY. CLVGS. CONCH. SOL-ACIDS + NH4OH. SCNDRY-OXIDIZED ZONES OF ORE DEPOSITS	41.6.5.1
3	SALESITE	ORTH.	4.72-4.82	IOD	PRISM-PYRAMIDS. VITR. PRISM CLVG. SOL-HNO3. OXIDIZED ORE(CHUQUICA-MATA, CHILE)	22.1.1.
3	DUFTITE	ORTH.	C 6.98	ARS	SUBTRANSLUCENT. AGGS CURVED XLS. VITR-DULL. WITH AZURITE(TSUMEB)	41.5.1.4
3 -3.5	PHOSPHOPHYLLITE	MNCL.	3.08-3.13	PHO	THICK TBLR. VITR. CLVGS. FLUO-VIOLET(SW). SOL-ACIDS. PEG-WITH FE, MN PHOSPHATES AND SPHALERITE	40.2.12.
3 -3.5	PHOSPHOFERRITE	ORTH.	3.0 -3.2	PHO	MSV,GRNLR,FIBR,TBLR. VITR. 1 CLVG. UNEVN. SOL-ACIDS. PEGMATITES	40.2.5.2
3 -3.5	ATACAMITE	ORTH.	3.75-3.78	HAL	APPLE GREEN STRK. PRISM-STRIATED, GRNLR, FIBR. VITR-ADMN. CLVGS. SOL-ACIDS. SCNDRY AFTER CU MINS, ESP UNDER SALINE-ARID CONDITIONS	10.1.11.1
3 -3.5	CELESTITE	ORTH.	3.96-3.98	SUF	TYPICALLY COLORLESS-BLUISH. TBLR,ETC. VITR-PEARLY. CLVGS. SOME FLUO,THERMOLUM.SOL-H2O(SLIGHTLY).SED RKS(ESP GYPSUM + LS), VEINS	28.3.1.2
3 -3.5	WITHERITE	ORTH.	4.29	CBT	PSEUDOHEX, GLBLR,ETC. VITR. FLUO(ETC.)-LK ARAGONITE.SOL-HCL. VNS	14.1.3.2
3 -3.5	BARITE	ORTH.	4.50	SUF	TBLR, MSV,ETC. VITR-PEARLY. CLVGS. SOME FLUO, PHOSPHO, THERMOLUM. SOME FETID WHEN RUBBED. COMMON-VNS,SED RKS, CAVITIES IN IG RKS	28.3.1.1
3 -3.5	MOTTRAMITE	ORTH.	5.8 -6.0	VAN	VARIABLE-XLS TO COMPACT MSV. GRSY.COMMONLY COLOR ZONED. SOL-ACIDS. OXIDIZED ZONES OF VANADIUM-BEARING ORE DEPOSITS	41.5.2.2
3 -3.5	CERUSSITE	ORTH.	6.53-6.57	CBT	TWINNED CLUSTERS, MSV. ADMN-RESIN. CLVGS. FLUO-YELIX-RAYS +(LW)UL-TRA VIOLET). OXIDIZED ZONES OF LEAD DEPOSITS	14.1.3.4
3 -4	EVANSITE	UNKN.	1.8 -2.2	PHO	MSV,COLLOFM COATINGS. VITR. CONCH. BRTL. SOL-ACIDS. WITH LIMONITE	42.2.6.
3 -4	WAVELLITE	ORTH.	2.36	PHO	RADIAL FIBR,STELLATE,CRUSTS. VITR-PEARLY,CLVGS. SOL-ACIDS. SCNDRY-WDSPRD-OPENINGS IN DIVERSE ROCKS,LIMONITE,PHOSPHATE RKS, ETC.	42.7.4.
3 -4	CACOXENITE	HEXA.	2.2 -2.4	PHO	ACICULAR,TUFTS,RADIAL AGGS,FIBR CRUSTS. SILKY. SOL-ACIDS. SCNDRY-DIVERSE OCCURRENCES (WITH DUFRENITE,WAVELLITE, ETC.)	42.9.2.
3 -4	DIADOCHITE	TRCL.	2.0 -2.4	PHO	NODULAR,PLTY,GEL-LK,COLLOFM. DULL-WAXY. UNEVN-CONCH. SOL-ACIDS. IN GOSSANS AND AS EFFLORESCENCES	43.2.4.
3 -4	METAVARISCITE	MNCL.	2.54	PHO	LATHLK, GRNLR. VITR. 1 CLVG. SOL-ALKALIES. WITH VARISCITE	40.3.2.1
3 -4	STILPNOMELANE	MNCL.	2.59-2.96	PSI	MICACEOUS(BUT 2ND CLVG ⊥ 1ST). SOL-HF+H2SO4(1/1). LO-GRD META RKS	M390 3-103
3 -4	MIXITE	HEXA.	3.79	ARS	HAIRLK-STRIATED,CONCENTRIC FIBR MASSES. DULL. WITH OTHER BI-MINS	42.5.7.

NONMETALLIC GREEN

HARDNESS	NAME	XL. SYS.	SPECIFIC GRAVITY	CHEM. CLASS	REMARKS	REFERENCES
3 -4	MOLYBDOPHYLLITE	HEXA.	4.72	SSI	FOLIATED MASSES. 1 PERF CLVG(BASAL). WITH HAUSMANNITE AT LANGBAN	W479 N
3 -5.5	ZEOLITE		1.9 -2.5	TSI	GROUP OF MIN FAMILIES WITH SOMEWHAT RELATED APPEARANCES, OCCURRENCES, AND COMPOSITIONS. MANY ARE INDISTINGUISHABLE MEGASCOPICALLY. DIVERSE HABITS(E.G., BUNDLES, FIBROUS). PEARLY-GLASSY. SUBCONCH. SOME FLUO-ORANGE OR YEL-GREEN. CONTINUOUS, IN PART REVERSIBLE, DEHYDRATION(I.E.,WATER CAN BE EXPELLED WITHOUT DESTROYING XL STRUCTURE). SCNDRY IN OPEN SPACES; COMMONLY TWO OR MORE OCCUR TOGETHER IN CAVITIES IN BASALTIC RKS — THIS OCCURRENCE IS REFERRED TO HEREIN AS THE ZEOLITE ASSOCIATION (ABBREVIATED "ZEOL-ASSOC").	W330 4-351
3.5	MOSESITE	ISOM.		HAL	EXPOS-OLIVE GREEN. OCTAH. ADMN. OCTAH CLVG.WHITENS IN HCL. MG-DEPS	10.2.11.
3.5	CALCIOVOLBORTHITE	ORTH.		VAN	SUBTRANSLUCENT. SCALY AGGS,FIBR-DENSE. VITR-PEARLY. 1 CLVG. SOL-ACIDS. SCNDRY-IN SS + OXIDIZED ZONES OF VANADIUM-BEARING DEPOSITS	41.5.2.4
3.5	GORDONITE	TRCL.	2.23	PHO	SHEAFLK A-GS. VITR,PEARLY. CLVGS. CONCH. SOL-ACIDS. IN VARISCITE NODULES	42.8.8.
3.5	PINNOITE	TETR.	2.27	BOR	GRNLR,FIBR,NODULES. VITR. UNEVN. SOL-DILUTE ACIDS. SALT DEPS(STASS-FURT, GERMANY)	25.1.1.
3.5	VAUXITE	TRCL.	2.39	PHO	TBLR,SUBPARALLEL-RADIAL AGGS + NODULES.VITR. WITH WAVELLITE,SN-VNS	42.8.7.
3.5	SULFOHALITE	ISOM.	2.50	SUF	DODEC, OCTAH. VITR. CONCH. SALTY. SOL-H2O(SLOWLY). SALINE DEPOSITS	30.1.6.
3.5	ZARATITE	ISOM.	2.57-2.69	CBT	INCRUST, MSV. VITR-GRSY.CONCH. EFF-WARM HCL. SCNDRY ON NI-SULFIDE	16.1.2.
3.5	ANAPAITE	TRCL.	2.81	PHO	TBLR,ROSETTELK AGGS, CRUSTS. VITR. 2 CLVGS. SOL-HCL,HNO3. DIVERSE	40.2.9.
3.5	KEMPITE	ORTH.	2.94	HAL	PRISM. SOL-DILUTE ACIDS. RARE-WITH OTHER MN MINS(ALUM RK PK,CALIF)	10.1.11.2
3.5	FAIRFIELDITE	TRCL.	3.08	PHO	FOLIATED AGGS, PRISM. PEARLY-GRSY. CLVGS. SOL-ACIDS. PEGMATITES	40.2.3.1
3.5	LUDLAMITE	MNCL.	3.12-3.19	PHO	TBLR-// AGGS,GRNLR. VITR. 2 CLVGS. SOL-ACIDS. SCNDRY-DIVERSE	42.6.3.
3.5	VERNADSKITE	UNKN.	3.3	SUF	XL AGGS. SOL-HCL. DATA LACKING. ALTER OF DOLEROPHANITE(VESUVIUS)	31.5.5.
3.5	CRONSTEDTITE (SERPENTINE)	HEXA.	3.45	PSI	FINE GRAINED. MORE COMMON THAN GREENALITE, IN IRON FORMATIONS	W389 3-164
3.5	CHLOROPHOENICITE	MNCL.	3.46	ARS	LONG PRISM, STRIATED. VITR,PEARLY. 1 CLVG. SOL-ACIDS. FRNKLN, N.J.	41.1.4.1
3.5	CHALCOCYANITE	ORTH.	3.60-3.70	SUF	TBLR. HYGROSCOPIC TO BLUE PENTAHYDRATE OR CHALCANTHITE. FUMAROLIC	28.3.3.
3.5	DUSSERTITE	HEXA.	3.75	ARS	PLTY CRUSTS,ROSETTES. RARE	41.5.8.7
3.5	STRONTIANITE	ORTH.	3.64-3.78	CBT	SPEAR-LK. MSV. VITR. CLVGS. FLUO(ETC.)-LK ARAGONITE. SOL-HCL. VNS	14.1.3.3
3.5	VOLBORTHITE	MNCL.	3.5 -3.8	VAN	SUBTRANSLUCENT. SCALY,SPONGY,FIBR CRUSTS. VITR-PEARLY.1 CLVG. SOL-ACIDS. SECONDARY - DIVERSE OCCURRENCES	41.5.3.
3.5	ANTLERITE	ORTH.	3.88	SUF	TBLR, FIBR, GRNLR. VITR. 2 CLVGS. SOL-H2SO4. OXIDE ZONES-CU DEPS (ESPECIALLY IN ARID AREAS)	30.1.2.
3.5	ADAMITE	ORTH.	4.32-4.48	ARS	ELONGATE,CRUSTS,RADIAL AGGS. VITR. CLVGS. UNEVN. SOL-DILUTE ACIDS. SOME FLUO YEL. OXIDIZED ZONES OF DEPS WITH ZN- + AS-RICH MINERALS	41.6.5.3
3.5	TSUMEBITE	MNCL.	6.13	PHO	CRUSTS,TBLR,TWINNED. VITR. UNEVN. SOL-HCL,HNO3(SLOW). SCNDRY-WITH SMITHSONITE, CERUSSITE, AND AZURITE	42.2.2.
3.5	HYDROCERUSSITE	HEXA.	6.80	CBT	SCALY,TBLR,ADMN,PEARLY,BASAL CLVG. EFF-ACIDS. SCNDRY AFTER PB MINS	16.1.12.
3.5	RUSSELLITE	TETR.	7.33-7.37	OXD	GRNLR. RARE-WITH WOLFRAMITE,ETC., PROBABLY AS ALTER PROD OF BI. (CASTLE-AN-DINAS MINE, CORNWALL	4.6.4.

NONMETALLIC GREEN

H	Mineral	Sys	SG	Code	Description	Ref
3.5-4	OVERITE	ORTH.	2.53	PHO	PLTY,SUBPARALLEL PLTY AGGS. VITR.2 CLVGS. SOL-HOT HNO3. IN ALTERED VARISCITE NODULES	42.8.12.
3.5-4	METASTRENGITE	MNCL.	2.76	PHO	TBLR,BOTRYOIDAL,RADIAL FIBR CRUSTS. VITR. 2 CLVGS. SOL-HCL. WITH STRENGITE	40.3.2.2
3.5-4	LANGBEINITE	ISOM.	2.83	SUF	NODULAR, DISSEM. VITR.CONCH. SOL-H2O(SLOWLY). MARINE SALTS	28.4.3.1
3.5-4	DOLOMITE	HEXA.	2.84-2.86	CBT	RHOMB-CURVED,GRNLR. VITR-PEARLY. RHOMB CLVG. SOME FLUO. EFF-WARM HCL IF POWDERED. SED RKS, XLS IN CAVITIES, ETC.	14.2.1.1
3.5-4	ARAGONITE	ORTH.	2.94-2.95	CBT	ACICULAR, RADIAL FIBR, STALACTITIC.VITR. CLVGS. FLUO(ULTRA-VIOLET, X-RAYS, ELECTRON BEAMS),THERMOLUM EFF-ACIDS. LO-T DEPOSITION	14.1.3.1
3.5-4	BERAUNITE	MNCL.	2.8 -3.08	PHO	STRK-YEL,OLIVE. RADIATED FOL GLBLS,CRUSTS,ETC. VITR. 1 CLVG. SOL-ACIDS. SCNDRY—WITH OTHER PHOSPHATES + LIMONITE—FE DEPS + PEGS	42.7.2.
3.5-4	SCORODITE	ORTH.	3.28	ARS	PYRAMIDAL,ETC.-AGGS,CRUSTS,ETC. VITR. CLVGS. SOL-ACIDS. SCNDRY-AFTER ARSENOPYRITE AND OTHER AS-BEARING MINERALS	40.3.1.3
3.5-4	LAUBMANNITE	ORTH.	3.33	PHO	SUBTRANSLUCENT. CRUSTS. VITR-SILKY. CLVG. RARE WITH LIMONITE	41.4.6.
3.5-4	DICKINSONITE	MNCL.	3.38-3.41	PHO	TBLR,SCALY. VITR,PEARLY.1 CLVG. SOL-ACIDS. PEGS—WITH LITHIOPHILITE, ETC.	40.2.1.
3.5-4	EUCHROITE	ORTH.	3.44	ARS	PRISM,TBLR,EQUANT,STRIATED. VITR. CLVGS. SOL-ACIDS. RARE—WITH OLI-VENITE (NEUSOHL, HUNGARY)	42.4.6.
3.5-4	VESZELYITE	MNCL.	3.3 -3.5	PHO	PRISM,TBLR,GRNLR. VITR. CLVGS. SOL-ACIDS. WITH SCNDRY CU MINERALS	42.2.1.
3.5-4	BROCHANTITE	MNCL.	3.97	SUF	ACICULAR,CRUSTS. VITR. 1 CLVG. SOL-ACIDS. OXIDE ZONES OF CU DEPS	30.1.1.
3.5-4	MALACHITE	MNCL.	3.6 -4.05	CBT	INCRUST,MAMMILLARY. VITR-DULL. CLVGS. SOL-ACIDS. OXIDIZED ZONES-CU DEPOSITS	16.1.6.
3.5-4	POWELLITE	TETR.	4.21-4.25	MBS	PYRAMIDAL,TBLR,MSV-FOL,POWDER,CRUSTS,SUBADMN-PEARLY. CLVGS. UNEVN. FLUO-YEL. DCMP-HCL,HNO3. ALTER PRODUCT OF MOLYBDENITE	48.1.3.2
3.5-4	PYROMORPHITE	HEXA.	7.00-7.08	PHO	PRISM,GLBL,ETC. RESIN. RHOMB CLVG. UNEVN. SOL-HNO3. OXIDIZED ZONES OF LEAD DEPOSITS	41.7.2.1
3.5-4.5	SAMPLEITE	ORTH.	3.20	PHO	LATHLK. PEARLY. CLVGS. RARE—IN OXIDIZED ORE(CHUQUICAMATA,CHILE)	42.6.1.
3.5-4.5	DUFRENITE	MNCL.	3.1 -3.34	PHO	BOTRYOIDAL—RADIAL,CRUSTS. VITR-SILKY. CLVGS. SOL-DILUTE ACIDS. WITH LIMONITE IN GOSSANS	41.6.9.
3.5-4.5	CHENEVIXITE	UNKN.	3.93	ARS	MSV EARTHY-OPALINE,ACICULAR. SUBCONCHOIDAL.SOL-ACIDS. ORE VEINS	41.5.9.
3.5-4.5	SIDERITE	HEXA.	3.95-3.97	CBT	TARN-IRID. MSV. RHOMB,ETC. VITR.RHOMB CLVG. SOL-HOT ACID. SED, VNS	14.1.1.3
3.5-4.5	BEUDANTITE	HEXA.	4.	ARS	RHOMB,PSEUDOCUBIC. VITR. 1 CLVG. SOL-HCL. SCNDRY-DIVERSE	43.1.1.1
3.5-4.5	CORKITE	HEXA.	4.30	PHO	RHOMB,PSEUDOCUBIC. VITR. 1 CLVG. SOL-WARM HCL. SCNDRY-DIVERSE	43.1.1.2
3.5-6	MARGARITE (BRITTLE MICA)	MNCL.	3	PSI	AGGS OF LAMELLAE. ATTKD-H2SO4. WITH DIASPORE + CORUNDUM—EMERY DEPS	W392 3-95
3.5-6	CLINTONITE (BRITTLE MICA)	MNCL.	3	PSI	WITH TALC-CHLORITE SCHIST, WITH SPINEL IN METASOMATIZED CALC RKS. DISTINGUISHED FROM XANTHOPHYLLITE BY OPTICAL OR X-RAY ANALYSES	W391 3-99
3.5-6	XANTHOPHYLLITE (BRITTLE MICA)	MNCL.	3	PSI	WITH TALC-CHLORITE SCHIST, WITH HUMITES IN METASOMATIZED CALC RKS. DISTINGUISHED FROM CLINTONITE BY OPTICAL+/OR X-RAY ANALYSES	W391 3-99
3.5-7	BRITTLE MICA	MNCL.	2.6 -3.2	PSI	GROUP OF MORE OR LESS RELATED MINS THAT RESEMBLE MICAS BUT CLEAVE TO BRTL FLKS AND HAVE HARDNESSES OF 3.5-4 VS. 2-3(MICA) ON CLVG FLKS AND 6 VS. 4 ⊥ TO FLKS. WITH CORUNDUM + DIASPORE IN EMERY DEPS, CHLORITE + MICA SCHISTS, SKARN	W392 3-95
4	MONTGOMERYITE	MNCL.	2.48-2.58	PHO	LATHS,SUBPARALLEL PLTY AGGS. VITR. 2 CLVGS. IN ALTERED VARISCITE	42.8.11.
4	LEUCOPHANE	ORTH.	2.96	SSI	TBLR. 2 PERF CLVGS. TWINNING COMMON(SOME PENTRATION). RARE—PEGS	W476 N
4	FLUORITE	ISOM.	3.18	HAL	CUBIC, MSV. VITR. OCTAH CLVG. FLUO-VIOLET. DCMP-H2SO4. SOL-HCL. ALK IG RKS CAVITIES, DISSEMINATED IN IGNEOUS ROCKS, ETC.	9.2.1.
4	JOHNSTRUPITE	MNCL.	3.3	NSI	BRWNISH GREEN. STRIATED PRISM PLATES. 1 CLVG. SOL-HCL. ALK IG RKS	W516 N

NONMETALLIC GREEN

HARDNESS	NAME	XL. SYS.	SPECIFIC GRAVITY	CHEM. CLASS	REMARKS	REFERENCES
4	ANDREWSITE	ORTH.	3.48	PHO	RADIAL FIBR BOTRYOIDAL. SILKY. CLVG LENGTH FIBERS. WITH LIMONITE (CORNWALL, ENGLAND)	41.4.5.
4	BARYTOCALCITE	MNCL.	3.66-3.71	CBT	PRISM-STRIATED. MSV. VITR. CLVG. WEAKLY FLUO. SOL-HCL. LO-T DEPS	14.2.3.
4	SARCOPSIDE	MNCL.	3.64-3.73	PHO	RED TO BROWNISH,EXPOS-DK BROWN,BLUE,VIOLET,GREEN.FIBR MASSES.SILKY. 2 CLVGS (∥ + ⊥ FIBERS). SOL-ACIDS. RARE - PEGMATITES	41.6.4.
4	LIBETHENITE	ORTH.	3.97	PHO	STRIATED PRISM. VITR-GRSY. 2 CLVGS. CONCH. SOL-ACIDS + NH4OH. OXI-DIZED ZONES OF ORE DEPS-COMMONLY WITH MALACHITE, LIMONITE, ETC.	41.6.5.2
4	BELLINGERITE	TRCL.	4.88-4.90	IOD	STRIATED PRISMS. SUBCONCH. SOL-DILUTE HCL. RARE-VNS(CHUQUICAMATA, CHILE)	21.1.2.
4	VANDENBRANDITE	TRCL.	5.03	OXD	LATHLK, MSV. BASAL CLVG. SOL-ACIDS TO GREEN-YEL SOLUTION. SCNDRY	5.3.4.
4	WALTHERITE	UNKN.	5.32	CBT	YEL STRK. PRISM. VITR. CLVGS. EFF-ACIDS. RARE-SCNDRY(JOACHIMOV, BOHEMIA)	16.1.9.
4-4.5	VARISCITE	ORTH.	2.2 -2.57	PHO	NODULES,CRUSTS-SOME OPALINE. WAXY. 2 CLVGS. SPLINTERY-CONCH. SOL-ALKALIES. NEAR-SURFACE DEPOSITION IN CAVITIES (E.G., BRECCIAS)	40.3.1.1
4-4.5	SMITHSONITE	HEXA.	4.40-4.45	CBT	BOTRYOIDAL, MSV. VITR-PEARLY. RHOMB CLVG. SOME FLUO-GREEN. BLUE. EFF-ACIDS. SECONDARY - OXIDIZED ZONES OF ZINC DEPOSITS	14.1.1.6
4-4.5	BINDHEIMITE	ISOM.	4.6 -5.6	ANT	DENSE-EARTHY CRUSTS,NODULAR-CONCENTRIC,SOME IS OPALINE. RESIN-DULL. DCMP-HNO3, HCL. OXIDATION ZONES OF ANTIMONY-BEARING LEAD DEPOSITS	44.1.1.1
4-5	PYROSMALITE	HEXA.	3.06-3.19	PSI	HEX PLTS. PERF BASAL CLVG. GELAT. RARE MN-,FE-DEPS (FRNKLN, N.J.)	W359 N
4-5	TRIPHYLITE	ORTH.	3.34-3.50	PHO	MSV. ANHEDRAL. VITR. CLVGS. UNEVN. SOL-ACIDS. PEGMATITES	38.1.1.1
4-5	SVABITE	HEXA.	3.5 -3.8	ARS	SHORT PRISM,MSV. VITR.CLVG. SOL-DILUTE ACIDS.RARE-E.G.,FRNKLN,N.J.	41.7.3.1
4-5	CORNETITE	ORTH.	4.10	PHO	CRUSTS,PRISM. VITR. SOL-HCL. RARE WITH OTHER SECONDARY CU MINERALS	41.3.2.
4-5	ROSASITE	MNCL.	4.0 -4.2	CBT	MAMMILLARY CRUSTS, FIBR. 2 CLVGS AT 90°. SOL-ACIDS. SCNDRY CU-ZN, PB DEPOSITS	16.1.5.
4-5	XENOTIME	TETR.	4.4 -5.1	PHO	PRISM + PYRAMIDAL. VITR. 1 CLVG. ACID + ALK IG RKS + PEGS,DETRITAL	38.4.1.
4-5.5	BETAFITE	ISOM.	3.7 -5.	OXD	SUBMET. OCTAH. WAXY. CONCH. METAMICT. DCMP-ACIDS. GRANITIC PEGS	8.4.1.
4.5	CUPROTUNGSTITE	UNKN.		MOS	FRIABLE-COMPACT,MICROXLINE MASSES. VITR-EARTHY. SOL-ACIDS. DATA LACKING. ALTERATION PRODUCT OF SCHEELITE	49.1.1.
4.5	CHABAZITE (ZEOLITE)	HEXA.	2.05-2.10	TSI	RHOMBS(NEARLY CUBES). RHOMB CLVG. DCMP-HCL.ZEOL-ASSOC. RK CAVITIES	W334 4-387
4.5	GMELINITE (ZEOLITE)	HEXA.	2.0 -2.2	TSI	PRISM CLVG. DCMP-HCL. ZEOL-ASSOC	W335 4-387
4.5	LEVYNITE (ZEOLITE)	ORTH.	2.0 -2.2	TSI	CHEM OR X-RAY ANAL DISTINGUISH CHABAZITE +LEVYNITE.RARE-ZEOL-ASSOC	W335 4-387
4.5	PROSOPITE	MNCL.	2.88-2.90	HAL	TBLR, MSV, GRNLR. VITR-DULL. PYRAMIDAL CLVG. DCMP-H2SO4. VNS, GRE1-11.5.7. SENS, PEGMATITES -- RARE	
4.5	ROSCHERITE	MNCL.	2.92	PHO	6 OR 8 SIDED PRISMS.PLTY VERMICULAR AGGS. 2 CLVGS. SOL-ACIDS. PEG DRUSES IN GRANITE	42.8.2.
4.5	LACROIXITE	MNCL.	3.13	PHO	XLS. VITR-RESIN. CLVGS. SOL-HCL,H2SO4. RARE-DRUSES IN GRANITE(SAX-ONY, GERMANY)	41.2.1.
4.5	CHALCOSIDERITE	TRCL.	3.22	PHO	CRUSTS,SHEAFLK AGGS OF XLS. VITR. 2CLVGS. SOL-HCL(DIFFICULTLY). SECONDARY-ESPECIALLY IN FERRUGINOUS GOSSANS	42.6.2.2
4.5	FRONDELITE	ORTH.	3.3 -3.49	PHO	SUBTRANSLUCENT. RADIAL FIBR BOTRYOIDAL(SOME HAVE CONCENTRIC COLOR BANDING). VITR-DULL.CLVGS.UNEVN.SOL-HCL.SCNDRY-PO4 PEG MINERALS	41.6.6.1
4.5	ROCKBRIDGEITE	ORTH.	3.3 -3.49	PHO	IN SERIES WITH FRONDELITE(SEE FOR PROPS + OCCUR.) WITH LIMONITE	41.6.6.2

80

NONMETALLIC GREEN

Hardness	Mineral	Crystal System	Density	Cleavage/Fracture	Description	Ref
4.5	NATROCHALCITE	MNCL.	3.47-3.51	SUF	PYRAMIDAL. VITR. 1 CLVG. SOL-ACIDS,H2OISLOW). SCNDRY(CHUQUICAMATA)	31.5.1.
4.5	HINSDALITE	HEXA.	3.65	PHO	TBLR,PSEUDOCUBIC,MSV,GRNLR. VITR.1 CLVG. RARE-VN(HINSDALE CO,COLO)	43.1.1.3
4.5	FLINKITE	ORTH.	3.87	ARS	FEATHERLK AGGS. VITR. SOL-ACL. RARE-VNS IN ORE(PERSBERG, SWEDEN)	41.3.5.
4.5	ERINITE	ORTH.	4.04	ARS	SUBTRANSLUCENT. CONCENTRIC,FIBR,MAMMILLARY CRUSTS. DULL-RESIN. 1 CLVG. SOL-HNO3.SECONDARY-BY OXIDATION OF ARSENICAL COPPER ORES	41.4.2.
4.5	CORNWALLITE	UNKN.	4.17	ARS	RADIAL FIBR BOTRYOIDAL CRUSTS.DATA LACKING.WITH OLIVENITE + TENOR-ITE	42.3.1.
4.5	LINDGRENITE	MNCL.	4.26	MBS	PLTY,TBLR,MSV. CLVGS. SOL-HCL,HNO3.SCNDRY-RARE(CHUQUICAMATA,CHILE)	49.1.4.
4.5	CONICHALCITE	ORTH.	4.1 -4.33	ARS	SUBTRANSLUCENT. EQUANT,BOTRYOIDAL,RADIAL FIBR. VITR. UNEVN. SOL-HCL, HNO3. SECONDARY-OXIDIZED ZONES OF COPPER DEPOSITS	41.5.1.2
4.5	LESSINGITE	HEXA.	4.69	NSI	REDDISH OR GREENISH YEL. VITR. RARE IN PLACERS WITH OTHER CE-MINS	W512 N
4.5	MACKAYITE	TETR.	4.86	TLI	PRISM, PYRAMIDAL. VITR. SUBCONCH. OXIDE ZONE OF TE + PYRITE VEIN	34.2.2.
4.5	CARACOLITE	UNKN.	5.0 -5.2	SUF	CRUSTS OF PSEUDOHEX XLS. VITR. RARE-WITH BINDHEIMITE,ETC.(CHILE)	30.1.3.
4.5	BAYLDONITE	MNCL.	5.5	ARS	SUBTRANSLUCENT. FIBR MAMMILLARY CONCRETIONS CRUSTS POWDER. RESIN. SOL-HCL(DIFFICULTLY). OXIDIZED ZONES OF COPPER DEPOSITS	42.4.1.
4.5	BISMITE	MNCL.	8.64-9.22	OXD	GRAY-YEL STRK. MSV, GRNLR. VITR-DULL. OXIDATION PROD OF BI-MINS	4.4.7.
4.5-5	APOPHYLLITE	TETR.	2.33-2.37	PSI	VITR SQ PRISMS, PEARLY BASES. BASAL + PRISM CLVG. DCMP-HCLISILICA RESIDUE), SECONDARY IN CAVITIES, COMMONLY WITH ZEOLITES, ETC.	W394 3-258
4.5-5	WOLLASTONITE	TRCL.	2.87-3.09	ISI	TBLR, MSV. CLVGS-84° 96.DCMP-HCL. FLUO-YEL-ORANGE. CNTCT META-LS	W456 2-167
4.5-5	WOLFEITE	MNCL.	3.83	PHO	IN SERIES WITH TRIPLOIDITE(SEE FOR PROPERTIES + OCCURRENCE.)	41.6.3.2
4.5-5	PLUMBOGUMMITE	HEXA.	4.01	PHO	CONCENTRIC BOTRYOIDAL. GUM-LIKE LUSTER. UNEVN. SOL-HOT ACIDS. SECONDARY - IN LEAD DEPOSITS WITH PYROMORPHITE	41.5.8.1
4.5-5	PSEUDOMALACHITE	MNCL.	4.00-4.40	PHO	CONCENTRIC,RADIAL FIBR BOTRYOIDAL. VITR. 1 CLVG. SOL-ACIDS. SCNDRY IN OXIDIZED ZONES OF COPPER DEPOSITS	41.4.3.
4.5-5	TRIPUHYITE	UNKN.	5.82	ANT	MICROXLINE AGGS. DATA LACKING. VNS IN DACITE, PLACERS.	44.1.3.
4.5-5	SCHEELITE	TETR.	6.08-6.12	MOS	DIPYRAMIDAL,MSV,GRNLR,VITR,CLVGS,UNEVN,DCMP-HCL,HNO3,FLUO-BLUE(X-,SW ULTRAVI- OLET-, + CATH(ODE-RAYS),THERMOLUM. HI-T VNS,PEGS,GREISSENS,CNTCT META RKS,ETC.	48.1.3.1
4.5-5	ATELESTITE	MNCL.	6.82	ARS	MINUTE TBLR XLS,MAMMILLARY. RESIN-ADMN. 1 POOR CLVG. SOL-HCL,HNO3 (DIFFICULTLY). RARE WITH BISMUTITE	41.3.4.
4.5-5.5	LEWISTONITE	HEXA.	3.08	PHO	FIBR CRUSTS,PRISMS-SUB // SHEAFLK GROUPS.CLVGS. WITH DEHRNITE(FAIR-FIELD, UTAH)	41.7.5.
4.5-5.5	DEHRNITE	HEXA.	3.04-3.09	PHO	BOTRYOIDAL CRUSTS,PRISMS-SUB // SHEAFLK GROUPS. VITR. 1 CLVG. SOL-ACIDS. WITH OTHER PHOSPHATE MINERALS	41.7.4.
4.5-5.5	EMMONSITE	MNCL.	4.52-4.53	TLI	FIBR CRUSTS,GLBL AGGS. VITR. CLVGS. SOL-ACIDS. ALTER OF TELLURIDES	34.2.1.
4.5-6	MONIMOLITE	ISOM.	5.9 -7.3	ANT	OCTAH,CUBIC. GRSY. OCTAH CLVG. CONCH-SPLINTERY. RARE-E.G.,LANGBAN	44.1.2.
5	TURANITE	ORTH.		VAN	CRUSTS,RADIAL FIBR CONCRETIONS. WITH VA + U MINS IN CAVITIES IN LS (E.G.,TURKESTAN)	41.5.2.5
5	MARDITE	TETR.	2.81-2.87	PHO	PYRAMIDAL-STRIATED,CRUSTS-RADIAL FIBR,ETC. VITR. 1 CLVG. SOL-ACIDS (DIFFICULTLY). WITH MILLSITE IN VARISCITE, AFTER AMBLYGONITE-PEGS	42.5.4.
5	HYDROXYLAPATITE	HEXA.	2.9 -3.1	PHO	APATITE GROUP(SEE FOR PROPS.) RARE-TALC SCHIST	41.7.1.3

NONMETALLIC GREEN

HARDNESS	NAME	XL. SYS.	SPECIFIC GRAVITY	CHEM. CLASS	REMARKS	REFERENCES
5	CARBONATE-APATITE	HEXA.	2.9 –3.1	PHO	APATITE GROUP(SEE FOR PROPS.)RARE-NODULES(PHOSPHATE RK),GUANO,ETC.	41.7.1.4
5	CHLORAPATITE	HEXA. A	3.17	PHO	APATITE GROUP(SEE FOR PROPS.) RARE-VNS IN GABBRO, METEORITES	41.7.1.2
5	FLUORAPATITE	HEXA. A	3.18	PHO	APATITE GROUP(SEE FOR PROPS.) COMMON-IGS, METAS, MARINE SEDS, ETC.	41.7.1.1
5	APATITE	HEXA.	2.9 –3.2	PHO	GROUP NAME. MOST TYPES INDISTINGUISHABLE MEGASCOPICALLY. HEX PRISMS-COMMONLY ROUNDED,MSV,GRNLR. VITR. SOL-ACIDS. SOME FLUO-ORANGE. IG RKS,VNS,PEGS	41.7.1.
5	HEMIMORPHITE	ORTH.	3.45	SSI	HEMIMORPHIC XLS, SHEAFLK. MAMMILLARY, ETC. 1 PERF + 1 POOR CLVG. GELAT. PYROELECTRIC. SOME FLUO-PALE ORANGE(LONG WAVE). ZINC DEPS	W481 N
5	CALAMINE	ORTH.	3.45	SSI	(=HEMIMORPHITE)	W481 N
5	DIOPTASE	HEXA.	3.28–3.5	CSI	PRISMATIC. RHOMB CLVG. GELAT. OXIDIZED ZONES OF COPPER DEPOSITS	W453 N
5	ARROJADITE	MNCL.	3.55	PHO	MSV. VITR.CLVGS. UNEVN. SOL-DILUTE ACIDS. PEGS(E.G.,SOUTH DAKOTA)	38.1.7.
5	TITANITE(=SPHENE)	MNCL.	3.45–3.55	NSI	DOUBLE WEDGE-SHAPED XLS, MSV-GRNLR. ADAMANTINE. ALTERS TO LEUCOX-	W525 1-69
5	SPHENE	MNCL.	3.45–3.55	NSI	ENE. DCMP-H2SO4. COMMON ACCESSORY CONSTITUENT OF IG + META ROCKS	W525 1-69
5	LUDWIGITE	ORTH.	3.6	BOR	SUBTRANSLUCENT. IN SERIES WITH PAIGEITE. FIBR MASSES. SILKY. SOL-ACIDS. HIGH TEMPERATURE CONTACT METAMORPHIC ROCKS	24.1.1.1
5	HUHNERKOBELITE	ORTH.	3.5 –3.6	PHO	MSV. GRNLR. VITR. 2 CLVGS. DISTINGUISHED FROM VARULITE BY CHEM ANAL. PEGS & HYDROTHERMAL ALTER OF TRIPHYLITE	38.1.1.3
5	VARULITE	ORTH.	3.5 –3.6	PHO	MSV. GRNLR. VITR. 2 CLVGS. DISTINGUISHED FROM HUHNERKOBELITE BY CHEM ANAL. PEGS & HYDROTHERMAL ALTER OF TRIPHYLITE	38.1.1.4
5	ADELITE	ORTH.	3.70–3.76	ARS	MSV=ELONGATE. RESIN. CONCH. SOL-DILUTE ACIDS. MN DEPOSITS	41.5.1.1
5	TILASITE	MNCL.	3.75–3.79	ARS	ELONG,MSV,RESIN. CLVGS + PARTINGS. SOL-HCL,HNO3.WITH OTHER MN-MINS	41.5.6.
5	DURANGITE	MNCL.	3.94–4.07	ARS	PYRAMIDAL. VITR. PRISM CLVG. SOL-H2SO4.WITH CASSITERITE(E.G.,PEGS)	41.5.7.
5	PAIGEITE	ORTH.	4.7	BOR	SUBTRANSLUCENT. IN SERIES WITH LUDWIGITE(SEE). HI-T CNTCT META	24.1.1.2
5 –5.5	HERDERITE	MNCL.	2.95	PHO	PRISM,TBLR,BOTRYOIDAL-RADIAL FIBR. VITR. CLVG. SOL-ACIDS. PEGS-LATE STAGE	41.5.4.1
5 –5.5	DATOLITE	MNCL.	2.96–3.00	NSI	TINTED WHT-COLORLESS. GLASSY. GELAT. SCNDRY WITH ZEOLITES-CAVITIES	W355 1-171
5 –5.5	HYDROXYL-HERDERITE	MNCL.	3.01	PHO	PRISM,TBLR,BOTRYOIDAL-RADIAL FIBR. VITR. CLVG. SOL-ACIDS. PEGS-LATE STAGE	41.5.4.2
5 –5.5	WAGNERITE	MNCL.	3.15	PHO	NEARLY OPQ-ALTERED. PRISM STRIATED,MSV. VITR. CLVGS. UNEVN-SPLIN-TERY. SOL-ACIDS. DIVERSE (E.G., VEINS AND VOLCANIC ROCKS)	41.6.1.
5 –5.5	ALLUAUDITE	UNKN.	3.4 –3.5	PHO	SUBTRANSLUCENT. MSV, GRNLR, FIBR-GLBL AGGS. 2 CLVGS. SOL-ACIDS. ALTERATION PRODUCT OF VARULITE-HUHNERKOBELITE (PEGMATITES)	38.1.4.1
5 –5.5	MANGAN-ALLUAUDITE	UNKN.	3.4 –3.5	PHO	SUBTRANSLUCENT. DISTINGUISHED FROM ALLUAUDITE BY CHEMICAL ANALYSIS	38.1.4.2
5 –5.5	MONAZITE	MNCL.	4.6 –5.4	PHO	TBLR-STRIATED. WAXY-ADMN. CLVGS. UNEVN. DCMP-ACIDS(SLOW). ACCESS IN GRANITIC IG + GNEISSIC RKS,PEGS, AND DETRITAL SEDIMENTARY RKS	38.4.2.
5 –5.5	MICROLITE	ISOM.	6.38–6.46	OXD	OCTAH. VITR. SUBCONCH. DCMP-H2SO4(SLOW).GRANITIC PEGS WITH TA MINS	8.1.1.2
5 –6	MARIALITE (SCAPOLITE)	TETR.	2.50–2.62	TSI	ME-0-20. PURE MARIALITE UNKNOWN IN NATURE	W352 4-321
5 –6	DIPYRE (SCAPOLITE)	TETR.	2.57–2.69	TSI	ME-20-50.	W352 4-321

82

NONMETALLIC GREEN

5 -6	MIZZONITE (SCAPOLITE)	TETR.	2.67-2.74	TSI	ME-50-80.		M352 4-322
5 -6	SCAPOLITE	TETR.	2.50-2.78	TSI	GROUP NAME. MOST TYPES INDISTINGUISHABLE MEGASCOPICALLY BUT S.G. MAY BE INDICATIVE. LONG,STRIATED PRISMS + AGGS OF COARSE XLS. SUBCONCH. POOR CLVG GIVING IRREGULAR, STRIATED-APPEARING SURFACES. MARIALITE IS INSOL; MEIONITE IS DCMP IN HCL. MOST FLUO—YEL-RED(ULTRAVIOLET,LW). CNTCT META CALC RKS,PEGS		M352 4-321
5 -6	(WERNERITE) (SCAPOLITE)	TETR.	2.50-2.78	TSI	GROUP NAME NO LONGER USED		M352 4-322
5 -6	MEIONITE (SCAPOLITE)	TETR.	2.73-2.78	TSI	ME-80-100. PURE MEIONITE UNKNOWN IN NATURE		M352 4-321
5 -6	TURQUOIS	TRCL.	2.6 -2.84	PHO	XLS-RARE,CRUSTS,STALAC. WAXY, 2 CLVGS, SOL-HCL(DIFFICULTLY). WXED ALUMINOUS ROCKS (ESPECIALLY IN ARID AREAS)		42.6.2.1
5 -6	AKERMANITE	TETR.	2.94	SSI	ISOMORPHOUS WITH GEHLENITE + MELILITE. RARE IN NATURE, IN SLAGS		M473 1-236
5 -6	EKERMANNITE (AMPHIBOLE)	MNCL.	3.00	ISI	PLUTONIC ALK IG RKS + ASSOC PEGS. (LESS COMMON THAN ARFVEDSONITE)		M439 2-364
5 -6	GEHLENITE	TETR.	3.04	SSI	SHORT SQ PRISMS. VITR-RESIN,CONCH,GELAT. CNTCT META-IMPURE CO3-RKS		M473 1-236
5 -6	MELILITE	TETR.	2.95-3.05	SSI	SHORT PRISMS. VITR-RESIN. BASAL CLVG. GELAT. BASIC,CA-RICH ALK IG		M473 1-236
5 -6	ENSTATITE (ORTHOPYROXENE)	ORTH.	3.21	ISI	MG PYROXENE. WITH OLIVINE IN BASIC + ULTRABASIC IG RKS, GRANULITES		M405 2-8
5 -6	OMPHACITE (CLINOPYROXENE)	MNCL.	3.29-3.37	ISI	ALTERS TO DK GREEN AMPHIB. WITH PYROPE-ALMANDINE IN ECLOGITES		WN 2-154
5 -6	HARDYSTONITE	TETR. A	3.40	SSI	GEHLENITE GROUP. GRNLR. SOME FLUO-PURPLE(SHORT-WAVE). RARE(FRNKLN)		M473 N
5 -6	ACTINOLITE (AMPHIBOLE)	MNCL.	3.02-3.44	ISI	REGIONAL METAIGREENSCHIST + AMPHIBOLITES) OF SI-BEARING LS + DOLO		M434 2-249
5 -6	FERROACTINOLITE (AMPHIBOLE)	MNCL.	3.02-3.44	ISI	REGIONAL + CONTACT METAMORPHISM, ESP OF SI-BEARING LS + DOLOMITE		M434 2-249
5 -6	BYSSOLITE (AMPHIBOLE)	MNCL.	3.02-3.44	ISI	STIFF FIBR HABIT. COMMONLY AS TUFTS IN CAVITIES (E.G.,WITH ZEOLITE)		WN N
5 -6	BRONZITE (ORTHOPYROXENE)	ORTH.	3.3 -3.45	ISI	SUBMET.(BRONZY). FS12-30. IN BASIC + ULTRABSIC IG RKS, SOME METAS		M405 2-9
5 -6	RICHTERITE (AMPHIBOLE)	MNCL.	2.97-3.45	ISI	THERMALLY META LS + SKARNS, HYDROTHERMAL VNS IN ALK IGNEOUS ROCKS		M435 2-352
5 -6	TSCHERMAKITE (AMPHIBOLE)	MNCL.	3.02-3.45	ISI	KYANITE AMPHIBOLITES DERIVED FROM CALC OR DOLO SHALES OR ECLOGITE		M434 2-263
5 -6	FERROTSCHERMAKITE (AMPHIBOLE)	MNCL.	3.02-3.45	ISI	MORE OR LESS END-MEMBER COMPOSITION IN HORNBLENDE(SEE) SERIES		M435 2-263
5 -6	EDENITE (AMPHIBOLE)	MNCL.	3.02-3.45	ISI	MORE OR LESS END-MEMBER COMPOSITION IN HORNBLENDE(SEE) SERIES		M434 2-263
5 -6	FERROEDENITE (AMPHIBOLE)	MNCL.	3.02-3.45	ISI	MORE OR LESS END-MEMBER COMPOSITION IN HORNBLENDE(SEE) SERIES		M435 2-263
5 -6	COMMON HORNBLENDE (AMPHIBOLE)	MNCL.	3.02-3.45	ISI	IN MANY IGNEOUS RKS + METAMORPHIC RKS OF LOW TO HIGH GRADE		M430 2-263
5 -6	URALITE (AMPHIBOLE)	MNCL.	3.02-3.45	ISI	SECONDARY AFTER PYROXENE, ESPECIALLY IN IGNEOUS ROCKS		M436 N
5 -6	HASTINGSITE (AMPHIBOLE)	MNCL.	3.0 -3.5	ISI	INTERMEDIATE IGNEOUS ROCKS AND LOCALLY IN CONTACT ZONES		M434 2-266
5 -6	ARFVEDSONITE (AMPHIBOLE)	MNCL.	3.50	ISI	PLUTONIC ALKALIC IGNEOUS MASSES AND ASSOCIATED PEGMATITE MASSES		M439 2-364
5 -6	FERROHASTINGSITE (AMPHIBOLE)	MNCL.	3.50	ISI	ALKALIC GRANITES AND SYENITES AND ASSOCIATED PEGMATITES		M435 2-264
5 -6	AMPHIBOLE		3.0 -3.57	ISI	GROUP NAME. SEVERAL END AND OTHER MEMBERS WITH NAMES UNSETTLED FOR SOME. MANY MEGASCOPICALLY INDISTINGUISHABLE BUT SOME COLORS AND OCCURRENCES ARE INDICATIVE. ORTH AND MNCL SERIES. LONG SLENDER XLS, PRISM CLVG AT 56° + 124°. SOL-HF(SLOWLY). ALTERS TO BIOTITE +/OR CHLORITE. IGNEOUS AND METAMORPHIC RKS, VNS		M422 2-203

NONMETALLIC GREEN

HARDNESS	NAME	XL. SYS.	SPECIFIC GRAVITY	CHEM. CLASS	REMARKS	REFERENCES
5 -6	HYPERSTHENE (ORTHOPYROXENE)	ORTH.	3.43-3.60	ISI	F530-50. BASIC(E.G.,NORITE) + ULTRABASIC IG RKS, SOME META ROCKS	W405 2-9
5 -6	CUMMINGTONITE (AMPHIBOLE)	MNCL.	3.10-3.60	ISI	CNTCT META ZONES, ORE DEPS, + LESS COMMONLY REGIONAL META ROCKS	W427 2-234
5 -6	GRUNERITE (AMPHIBOLE)	MNCL.	3.10-3.60	ISI	TYPICALLY IN METAMORPHOSED IRON-RICH SEDIMENTARY ROCKS	W427 2-234
5 -6	FERROHYPERSTHENE (ORTHOPYROXENE)	ORTH.	3.59-3.75	ISI	F550-70. BASIC + ULTRABASIC IGNEOUS ROCKS, SOME METAMORPHIC ROCKS	WN 2-9
5 -6	EULITE (ORTHOPYROXENE)	ORTH.	3.73-3.86	ISI	F570-88. RARE IN GRANITE, IN FE-RICH DOLERITES + META-SED ROCKS	WN 2-9
5 -6	ORTHOFERROSILITE (ORTHOPYROXENE)	ORTH.	3.96	ISI	FE PYROXENE. RARE, IN SOME FE-RICH METAMORPHIC ROCKS	W405 2-8
5 -7	PYROXENE		2.96-3.96	ISI	GROUP NAME. SEVERAL END AND OTHER MEMBERS WITH NAMES UNSETTLED FOR SOME, MANY MEGASCOPICALLY INDISTINGUISHABLE; OCCURRENCE MAY BE INDICATIVE. ORTH + MNCL SERIES.SHORT PRISMS.PRISM CLVG AT 87° + 92°. ALTERS TO AMPHIBOLE.IG + META RKS	W402 2-1
5 -7	OLIVINE	ORTH.	3.22-4.39	NSI	GRP NAME FOR MINS OF FAYALITE-FORSTERITE + FAYALITE-TEPHROITE SERIES, MONTI-CELLITE + GLAUCOCHROITE. GRNLR MASSES, DISSEM. CONCH. GELAT. COMMONLY ALTERED TO SERPENTINE. BASIC + ULTRABASIC IG RKS, CNTCT META DOLOMITIC LS, METEORITES	W498 1-2
5.5	NOSEAN	ISOM.	2.30-2.40	TSI	SODALITE(SEE) GROUP PROPS. IN PHONOLITE. (DISSOLVE-HNO3. EVAPORATE SLOWLY.NO XLS REMAIN, ADD CACL2, HALITE + GYPSUM FORM)	W350 4-289
5.5	HILLEBRANDITE	MNCL.	2.69	NSI	RADIATING FIBR, PRISM CLVG. SOL-HCL. CNTCT META CALCAREOUS ROCKS	W506 N
5.5	BRAZILIANITE	MNCL.	2.98-2.99	PHO	EQUANT STRIATED,GLBL-RADIAL FIBR. VITR. 1 CLVG. CONCH. DCMP-HF+HOT H2SO4. PEGMATITES	41.5.10.
5.5	AUGITE (CLINOPYROXENE)	MNCL.	3.23-3.52	ISI	THE COMMON PYROXENE OF SUB-ALKALIC IGNEOUS RKS(E.G., GABBROS)	W411 2-109
5.5	WILLEMITE	HEXA.	3.9 -4.1	NSI	PRISMS, MSV. MAY FLUO-GRN, SOME PHOS + TRIBOLUM. GELAT. ZINC DEPS	W497 N
5.5	MANGANOSITE	ISOM.	5.36	OXD	GREEN, EXPOS-BLK. STRK-BRWN. OCTAH, GRNLR. CUBIC CLVG. FIBR BREAK. LANGBAN, SWEDEN AND FRANKLIN, NEW JERSEY	4.2.1.3
5.5	STIBIOCOLUMBITE	ORTH.	5.68	OXD	IN SERIES WITH STIBIOTANTALITE, WHICH SEE. RARE IN PEGMATITES	8.1.8.2
5.5	DJALMAITE	ISOM.	5.75-5.88	OXD	MODIFIED OCTAH. SUBCONCH. (=TA-BETAFITE?). RARE-GRANITIC PEGMATITE	8.4.2.
5.5	BUNSENITE	ISOM.	6.90	OXD	STRK-BROWNISH BLK. OCTAH. VITR. RARE-WITH BI,ETC.(OXIDIZED NI-U VN AT JOHANNGEORGENSTADT, SAXONY)	4.2.1.2
5.5	STIBIOTANTALITE	ORTH.	7.34	OXD	IN SERIES WITH STIBIOCOLUMBITE. YEL, YEL-BRWN STRK.PRISM-STRIATED. RESIN-ADMN. 1 GOOD,1 POOR-CLVGS. SOL-HF.PYROELECT. PEGS + PLACERS	8.1.8.1
5.5-6	SODALITE	ISOM.	2.27-2.33	TSI	DODEC, MSV. POOR DODEC CLVG. GELAT. SOME FLUO-YEL-ORANGE. IN NEPH-ELINE-BEARING RKS.DISSOLVE-HNO3.EVAPORATE SLOWLY, HALITE REMAINS	W348 4-289
5.5-6	HAUYNE	ISOM.	2.44-2.50	TSI	SODALITE(SEE) GROUP PROPS. NEPH-BEARING IG RKS. DISSOLVE-HNO3. EVAPORATE SLOWLY, GYPSUM CRYSTALS REMAIN.	W350 4-289
5.5-6	MILARITE	HEXA.	2.57	CSI	PRISMATIC. GLASSY. RARE-PEGMATITE NEAR GLETSCH, SWITZERLAND	W257 N
5.5-6	MONTEBRASITE	TRCL.	2.98	PHO	AMBLYGONITE SERIES(SEE FOR PROPERTIES + OCCURRENCES.)	41.5.5.2
5.5-6	LAZULITE	MNCL.	3.08	PHO	SUBTRANSLUCENT. ACUTE PYRAMIDAL.MSV.GRNLR. VITR. CLVGS. SOL-HOT ACIDS(SLOWLY). PEGS,QTZ VNS,SI-RICH ALUMINOUS META(HI-GRADE) RKS	41.8.1.1
5.5-6	SOUZALITE	MNCL.	3.09	PHO	FIBR MASSES. 2 CLVGS. SOL-HCL(DIFFICULTY). PEG-ALTER AFTER SCOR-ZALITE	41.8.2.
5.5-6	NATROMONTEBRASITE	TRCL.	3.04-3.1	PHO	AMBLYGONITE SERIES(SEE FOR PROPERTIES + OCCURRENCES.)	41.5.5.3
5.5-6	AMBLYGONITE	TRCL.	3.11	PHO	EQUANT, LARGE MASSES. VITR,PEARLY. CLVGS. SOL-ACIDS(DIFFICULTY). LI-, PO4-RICH PEGS	41.5.5.1
5.5-6	BABINGTONITE	TRCL.	3.36	ISI	SMALL NEARLY EQUANT STRIATED XLS, BRILLIANT,GLASSY. 1 GOOD, 1 POOR CLVGS. FUSES TO BLK MAG GLBL. IN CAVITIES, COMMONLY WITH ZEOLITES	W462 N

NONMETALLIC GREEN

H	Mineral	Cryst	Density	Type	Description	Ref
5.5-6	SCORZALITE	MNCL.	3.38	PHO	IN SERIES WITH LAZULITE(SEE FOR PROPS). PEGMATITE	41.8.1.2
5.5-6	ANTHOPHYLLITE (AMPHIBOLE)	ORTH.	2.85-3.57	ISI	WITH CORDIERITE IN GNEISSES, WITH TALC IN META ULTRABASIC IG RKS	W425 2-211
5.5-6	GEDRITE (AMPHIBOLE)	ORTH.	2.85-3.57	ISI	RELATED IN COMPOSITION + OCCURRENCE TO ANTHOPHYLLITE	W425 2-211
5.5-6	TITANAUGITE (CLINOPYROXENE)	MNCL.	3.3 -3.6	ISI	AUGITE WITH NOTABLE TI-CONTENT. ALKALIC BASIC IGNEOUS ROCKS	W411 2-113
5.5-6	ANATASE	TETR.	3.90	OXD	SUBMET. PALE YEL STRK. ACUTE PYRAMIDAL. ADMN. BASAL + PYRAMIDAL CLVG. CONVERTS TO RUTILE ON HEATING. VNS,ACCESS-IG RKS,DETRITAL	4.5.2.
5.5-6.5	OPAL	TETR.	1.8 -2.25	SIL	SUBMICROXLINE QTZ + H2O.RESIN-PEARLY. MSV,MAMILLARY CRUSTS,CONCH,SOL-HF +CAUS-TIC ALKS,SOME FLUO--YEL-GREEN. INCRUSTATIONS, IN CAVITIES(ESP IN VOLCANIC RKS)	D-111 287
5.5-6.5	DIOPSIDE (CLINOPYROXENE)	MNCL.	3.22-3.38	ISI	SOME LIGHT ONES FLUO-BLUE. WHT-PALE GREEN. META CALC,SIL RKS	W411 2-42
5.5-6.5	SALITE (CLINOPYROXENE)	MNCL.	3.2 -3.6	ISI	INTERMEDIATE BETWEEN DIOPSIDE + HEDENBERGITE. META CA-RICH SED + IGNEOUS ROCKS	W411 2-42
5.5-6.5	FERROSALITE (CLINOPYROXENE)	MNCL.	3.2 -3.6	ISI	INTERMEDIATE BETWEEN DIOPSIDE + HEDENBERGITE. OCCURS LIKE SALITE	W411 2-42
5.5-6.5	DIALLAGE (CLINOPYROXENE)	MNCL.	3.2 -3.6	ISI	DIOPSIDE + AUGITE WITH GOOD PARTING OFTEN OCCUPIED BY MAGNETITE OR ILMENITE	W411 2-43
5.5-6.5	YTTRIALITE	MNCL.	4.3 -4.6	SSI	PRISMATIC, MASSIVE. RARE - WITH GADOLINITE, ETC. IN PEGMATITES	W477 N
5.5-6.5	EUXENITE	ORTH.	4.9 -5.9	OXD	SUBMET. IN SERIES WITH POLYCRASE. STRK YEL,GRAY,RED-BRWN. PRISM, MSV,GRSY,CONCH,DCMP-HOT HCL,HF,H2SO4. GRANITIC PEGMATITES	8.3-3.3.1
5.5-6.5	POLYCRASE	ORTH.	4.9 -5.9	OXD	SUBMET. IN SERIES WITH EUXENITE, WHICH SEE. PEGMATITES	8.3.3.2
5.5-7	KYANITE	TRCL.	3.53-3.65	NSI	HARDNESS-5 ⊥ TO + 7 // TO LENGTH OF BLADE-LIKE XLS. GNEISS +SCHIST	W527 1-137
6	MERWINITE	MNCL.	3.15	NSI	MULTIPLE TWINNING COMMON. 1 PERF CLVG. GELAT. CONTACT META ZONES	W505 N
6	CLINOENSTATITE (CLINOPYROXENE)	MNCL.	3.19	ISI	HI-T FORM. RARE IN IGNEOUS ROCKS, IN METEORITES.	W408 N
6	PUMPELLYITE	MNCL.	3.18-3.23	SSI	FIBR-PLTY. 1 GOOD + 1 POOR CLVG. CAVITIES + LOW-GRADE META ROCKS	W519 1-227
6	FASSAITE (CLINOPYROXENE)	MNCL.	2.96-3.34	ISI	NAME GIVEN AL-RICH,NA-POOR PYROXENES IN META LIMESTONES +DOLOMITES	W503 2-161
6	ZOISITE	ORTH.	3.15-3.37	SSI	DISTINGUISHED FROM CLINOZOISITE BY OPTICAL OR X-RAY ANAL.	W446 1-185
6	JADEITE (CLINOPYROXENE)	MNCL.	3.24-3.43	ISI	RELATIVELY UNCOMMON, TYPICALLY WITH ALBITE, HI-T + ALSO LO-GRADE METAMORPHIC ROCKS	W411 2-99
6	HELVITE	ISOM.	3.20-3.44	TSI	TETRA,MSV. POOR OCTAH CLVG,GELAT(H2S),PYROELECTRIC,BOIL IN HCL + AS2O3,DECANT, WASH,HELVITE IS STAINED CANARY YELLOW (AS2S3). RARE-GNEISS,GRANITE,CNTCT ZONES	W351 4-303
6	PIGEONITE (CLINOPYROXENE)	MNCL.	3.30-3.46	ISI	COMMON PYROXENE OF MANY DOLERITES(INCLUDING DIABASES) + BASALTS	W408 2-143
6	GLAUCOCHROITE	ORTH.	3.48	NSI	BLUISH GREEN, CA-MN OLIVINE. RARE (FRANKLIN FURNACE, NEW JERSEY)	W504 N
6	EPIDOTE	MNCL.	3.38-3.49	SSI	GROUP NAME. NO GENERALLY ACCEPTED NOMENCLATURE FOR GROUP MEMBERS. CLOSELY RE-LATED TO ZOISITE + ALLANITE. PISTACHIO-BLACKISH GREEN-BROWNISH. LONG THIN GROOVED XLS, GRNLR MASSES. VNS, OTHER CAVITIES, REGIONAL META + IG ROCKS	W448 1-193
6	PISTACITE	MNCL.	3.3 -3.49	SSI	NAME USED BY SOME FOR INTERMEDIATE COMPOSITION MEMBERS-EPIDOTE GRP	W448 1-184
6	CLINOHYPERSTHENE (CLINOPYROXENE)	MNCL.	3.2 -3.55	ISI	HI-T FORM. ONLY IN METEORITES.	W408 N
6	JOHANNSENITE (CLINOPYROXENE)	MNCL.	3.44-3.55	ISI	MN-ANALOG OF DIOPSIDE-HEDENBERGITE. SKARNS, CNTCT ZONES. VEINS	W411 2-75

85

NONMETALLIC GREEN

HARDNESS	NAME	XL. SYS.	SPECIFIC GRAVITY	CHEM. CLASS	REMARKS	REFERENCES
6	AEGIRINE-AUGITE (CLINOPYROXENE)	MNCL.	3.40-3.55	ISI	TYPICAL PYROXENE OF ALKALIC ROCKS - ESPECIALLY SYENITIC ONES	W411 2-79
6	HEDENBERGITE (CLINOPYROXENE)	MNCL.	3.50-3.56	ISI	BRWNISH OR DK GREEN TO BLK. SKARNS, META FE-RICH SEDS, ORE DEPS	W411 2-42
6	AEGIRINE (CLINOPYROXENE)	MNCL.	3.55-3.60	ISI	TYPICAL PYROXENE OF ALKALIC ROCKS - ESPECIALLY GRANITIC ONES	W411 2-79
6	CLINOFERROSILITE (CLINOPYROXENE)	MNCL.	3.5 -3.7	ISI	HI-T FORM. NOT DEFINITELY IDENTIFIED IN NATURAL MATERIALS	W408 N
6	CAPPELENITE	HEXA.	4.4	CSI	GREENISH BRWN. PRISMATIC. RARE-VNS IN SYENITE(LANGESUND, NORWAY)	W509 N
6 -6.5	MICROCLINE (K-FELDSPAR)	TRCL.	2.56-2.63	TSI	DISTINGUISHED FROM ORTHOCLASE OPTICALLY. PLUTONIC IG RKS, PEGS	W308 4-7
6 -6.5	ORTHOCLASE (K-FELDSPAR)	TRCL.	2.55-2.63	TSI	DISTINGUISHED FROM MICROCLINE OPTICALLY. PLUTONIC IG RKS, PEGS	W303 4-7
6 -6.5	AMAZONITE (K-FELDSPAR)	TRCL.	2.56-2.63	TSI	GREEN MICROCLINE. PEGMATITES. (AMAZONITE IS COMMONLY PERTHITIC.)	WN 4-34
6 -6.5	PERTHITE (FELDSPARS)	TRCL.	2.56-2.65	TSI	MEGASCOPICALLY, MICROSCOPICALLY,OR SUBMICROSCOPICALLY INTERDIGITATED MICROCLINE OR ORTHOCLASE + PLAGIOCLASE(TYPICALLY ALBITE)	W299 4-7
6 -6.5	ANDESINE (PLAGIOCLASE)	TRCL.	2.64-2.69	TSI	AN30-50. SEE FELDSPAR. INTERMEDIATE SI-CONTENT IGNEOUS ROCKS	W312 4-95
6 -6.5	PREHNITE	ORTH.	2.90-2.95	SSI	ROSETTES-TBLR XLS. 1 CLVG. SOL-HCL(SLOW). CAVITIES(WITH ZEOLITES)	W359 3-263
6 -6.5	FELDSPAR		2.55-3.39	TSI	INCLUDES PLAGIOCLASE(NA-CA)SERIES, ALKALI(K-NA) AND BA FELDSPARS. MOST EASILY DISTINGUISHED BY NONMEGASCOPIC MEANS. MONOCLINIC ONES HAVE 2 CLVGS AT 90° AND SIMPLE(IF ANY) TWINNING; TRICLINIC ONES HAVE 2 CLVGS UP TO 4° OFF 90° AND TYPICALLY HAVE POLYSYNTHETIC TWINNING (GIVING STRIATED APPEARANCE ON SOME CLVG SURFACES OF PLAGIOCLASE). PEGMATITES; IG, META, AND SOME SED ROCKS	W261 4-1
6 -6.5	RUTILE	TETR.	4.21-4.25	OXD	SUBMET. PRISM-STRIATED,ADMN.POOR CLVGS. VNS, ACCESS-META + IG RKS	4,5,1,1
6 -7	VESUVIANITE	TETR.	3.33-3.43	SSI	PRISM, MSV. 1 POOR CLVG. SUBCONCH. ATTKD-HCL. META-ESP CNTCT LS	W508 1-113
6 -7	THORTVEITITE	MNCL.	3.58	SSI	GRAYISH GREEN. TAPERED, PRISM. PRISM CLVG. RARE-GRANITIC PEGS	W477 N
6 -7.5	GROSSULAR	ISOM. A	3.59	NSI	CA-AL GARNET. EQUANT XLS. CONTACT + REGIONAL META CALCAREOUS RKS	W469 1-93
6 -7.5	HYDROGROSSULAR	ORTH.	3.13-3.59	NSI	GROSSULAR +H2O. SLOWLY SOL-HCL + HNO3. META CALCAREOUS ROCKS	W493 1-104
6 -7.5	UVAROVITE	ISOM.	3.90	NSI	CA-CR GARNET. DODEC XLS. WITH CHROMITE IN SERPENTINITES, SKARNS	W489 1-101
6 -7.5	GARNET	ISOM.	3.58-4.32	NSI	GROUP NAME. DOMINANT MOLECULE DETERMINES NAME. DISTINGUISHED BY S.G. + NONMEGASCOPIC MEANS.EQUANT XLS,MSV,VITR-DULL,DODEC PARTING. SLIGHTLY- TO IN-SOL IN HF	W483 1-77
6.5	PETALITE	MNCL.	2.41-2.42	TSI	FOL, MSV. 1 PERF + 1 GOOD-CLVGS AT 38.5°. LI-BEARING PEGMATITES	M260 4-271
6.5	CHRYSOPRASE	HEXA.	2.57-2.64	SIL	APPLE-GREEN CHALCEDONY(MICROFIBR QTZ). SECONDARY-VNS IN SERPENTINE	D-111 218
6.5	CHALCEDONY	HEXA.	2.57-2.64	SIL	MICROXLINE + MICROFIBR QTZ. MAMMILLARY. LO-T VEINS +OTHER CAVITIES	D-111 195
6.5	KORNERUPINE	ORTH.	3.27-3.34	NSI	PRISM MASSES. PRISM CLVG. WITH CORDIERITE + SAPPHIRINE IN SCHISTS	W523 N
6.5	CLINOZOISITE	MNCL.	3.21-3.38	SSI	CALCIUM-ALUMINUM END-MEMBER OF EPIDOTE GROUP. SAME OCCUR	W448 1-193
6.5	CHLORITOID	MNCL.	3.51-3.80	PSI	LK BRTL-MICAS IN FM. ADDITIONAL PRISM CLVG. DCMP-H2SO4. META + VNS	M393 1-161

NONMETALLIC GREEN

Hardness	Mineral	Crystal System	Density	Type	Description	Ref
6.5	BADDELEYITE	MNCL. 5.4-6.02	OXD	FLAT PRISM. VITR. BASAL + 2 POOR CLVGS. DCMP-HOT H_2SO_4. PLACERS, ASSOCIATED WITH ULTRAMAFIC IGNEOUS ROCKS	5.1.1
6.5	FAYALITE	ORTH. 4.2-4.39	NSI	FAYALITE-FORSTERITE (FE-MG)OLIVINE SERIES CONSISTS OF FORSTERITE (FA0-10),CHRYSOLITE(FA10-30),HYALOSIDERITE(FA30-50),HORTONOLITE	W498 1-2
6.5-7	FERROHORTONOLITE	ORTH. 4.0-4.3	NSI	(FA50-70),FERROHORTONOLITE(FA70-90) + FAYALITE(FA90-100). PROPS	W499 1-22
6.5-7	HORTONOLITE	ORTH. 3.7-4.1	NSI	RANGE WITH COMPOSITION. MG-RICH ONES IN BASIC + ULTRABASIC IG RKS	W499 1-22
6.5-7	HYALOSIDERITE	ORTH. 3.5-3.75	NSI	CONSTITUTING NEARLY 100 PER CENT OF SOME, + IN CONTACT METAMOR-	W498 1-22
6.5-7	CHRYSOLITE	ORTH. 3.3-3.6	NSI	PHOSED DOLOMITES. FE-RICH ONES IN ACIDIC + ALKALIC IGNEOUS ROCKS	W498 1-22
7	FORSTERITE	ORTH. 3.22	NSI		W498 1-22
6.5-7	BLOODSTONE	HEXA. 2.57-2.65	SIL	SUBTRANSLUCENT. GREEN-RED SPOTS. CHALCEDONY (MICROCRYSTALLINE QTZ)	D-111 219
6.5-7	HELIOTROPE	HEXA. 2.57-2.65	SIL	SUBTRANSLUCENT. GREEN-RED SPOTS. CHALCEDONY(MICROCRYSTALLINE QTZ)	D-111 219
6.5-7	PRASE	HEXA. 2.57-2.65	SIL	TRANSLUCENT. DULL LEEK GREEN CHALCEDONY (MICROFIBROUS QUARTZ)	D-111 219
6.5-7	PLASMA	HEXA. 2.57-2.65	SIL	SUBTRANSLUCENT. VARIOUS GREENS. CHALCEDONY (MICROCRYSTALLINE QTZ)	D-111 218
6.5-7	JASPER	HEXA. 2.57-2.65	SIL	SUBTRANSLUCENT. SOME IS VARIEGATED OR BANDED. MICROGRNLR. CAVITIES	D-111 224
6.5-7	SPODUMENE	MNCL. 3.03-3.22	ISI	COMMONLY FLUO(+PHOSPHORESCES)-ORANGE. THERMOLUM. LI-BEARING PEGS	W418 2-92
6.5-7	AXINITE	TRCL. 3.26-3.36	CSI	STRIATED WEDGE-SHAPED XLS. 1 GOOD CLVG. SOL-HCL(SLOW). CNTCT META	W505 1-320
6.5-7	DIASPORE	ORTH. 3.3-3.5	OXD	PLTY, FOL-MSV. VITR, PEARLY. WITH CORUNDUM IN EMERY ROCK	7.1.2.1
6.5-7	GADOLINITE	MNCL. 4.0-4.6	NSI	PRISMATIC. GELAT. COMMONLY METAMICT. WITH FLUORITE IN SOME PEGS	W356 N
7	QUARTZ	HEXA. 2.65	SIL	DOUBLY TERMINATED HEX PRISMS STRIATED LENGTH(LO-T), BIPYRAMIDS (HI-T),MSV. VITR.CONCH.SOL-HF. WIDESPREAD-MOST RK TYPES + VEINS	D-111 9
7	ELBAITE	HEXA. 3.03-3.10	CSI	NA- LI-TOURMALINE. INFUSIBLE. TYPICALLY PINK AND/OR GREEN. PEGS	W465 1-300
7	TOURMALINE	HEXA. 3.03-3.25	CSI	GROUP NAME. VARIETIES COMMONLY DISTINGUISHABLE MEGASCOPICALLY ON BASIS OF COLOR(S). LENGTHWISE STRIATED PRISMATIC XLS WITH CROSS-SECTIONS THAT RESEMBLE SPHERICAL TRIANGLES. SOME XLS EXHIBIT BOTH LENGTHWISE + CONCENTRIC COLOR ZONING. VITR-RESIN. BRTL. ELECTRICALLY CHARGED ON HEATING + COOLING. MG-RICH VARIETIES FLUO-YEL (ULTRAVIOLET - SW). PEGS, HI-T VNS, IG AND META ROCKS	W465 1-300
7	DUMORTIERITE	ORTH. 3.3	NSI	FIBR-COMMONLY RADIATING. VITR. 1 GOOD,1 POOR CLVG. PEGS + META RKS	W259 N
7-7.5	BORACITE	ORTH. 2.91-2.97	BOR	CUBIC,ETC.,GRNLR,FIBR. VITR. SOL-HCL(SLOWLY). SALT-GYPSUM BEDS	26.1.7.
7.5	GRANDIDIERITE	ORTH. 3.0	NSI	GREENISH BLUE. XLS UNKNOWN. 2 GOOD CLVGS. PEGS + APLITES-RARE	W497 N
7.5	EUCLASE	MNCL. 3.0-3.1	PSI	PRISM XLS. 1 GOOD, 2 POOR CLVGS. PEGS, PLACERS, GRANITES, SCHISTS	W357 N
7.5	SAPPHIRINE	MNCL. 3.40-3.58	OXD	TBLR, DISSEM. VITR. HI-T META RKS WITH CORUNDUM, SPINEL, ETC.	7.3.1.2
7.5	ZIRCON	TETR. 4.6-4.7	NSI	TERMINATED PRISMS. ADMN. SOME FLUO-YEL-ORANGE. ACCESS-IG, SANDS	W494 1-59

87

NONMETALLIC GREEN

HARDNESS	NAME	XL. SYS.	SPECIFIC GRAVITY	CHEM. CLASS	REMARKS	REFERENCES
7.5-8	BERYL	HEXA.	2.66-2.83	CSI	GROOVED PRISM. POOR BASAL CLVG. SOME FLUO WEAKLY-YEL. PEGMATITES	W463 1-256
7.5-8	SPINEL	ISOM.	3.55	OXD	GROUP NAME. SOME HAVE DISTINCTIVE COLORS OR OTHER HAND-SPECIMEN PROPS. OCTAH. GRNLR. GLASSY. CONCH. SOME FLUO-RED TO YEL-GREEN. IG,META(ESP CALC RKS),PEGS	7.2.1.1
7.5-8	GAHNITE	ISOM.	4.62	OXD	SPINEL(SEE). OCTAH. SCHISTS, PEGS, CNTCT META-LS	7.2.1.3
8.5	CHRYSOBERYL	ORTH.	3.65-3.85	OXD	TBLR-STRIATED. 3 CLVGS. OPALESCENT, ASTERATED. PEGS, PLACERS, META	7.2.3.
9	CORUNDUM	HEXA.	4.0 -4.1	OXD	STEEP PYRAMIDAL, PRISM. BASAL PARTING. WDSPRD IN SI-DEFICIENT RKS	4.4.1.1
9.5	MOISSANITE	HEXA.	3.1 -3.21	ELE	SUBMET. TABULAR. CONCHOIDAL FRACT. METEOR. (ARTIF=CARBORUNDUM)	1.1.7.4
10	DIAMOND	ISOM.	3.50-3.53	ELE	OCTAHEDRAL, ADMN-GRSY. BRTL. OCTAH CLVG. ULTRABASIC IGS + PLACERS	1.2.4.1

NONMETALLIC BLUE

HARDNESS	NAME	XL. SYS.	SPECIFIC GRAVITY	CHEM. CLASS	REMARKS	REFERENCES
1	GLAUCOCERINITE	UNKN.	2.75	SUF	WARTY MASSES, RADIAL + CONCENTRIC. WAXY. WITH SMITHSONITE,ETC.(LAU-31.1.2. RIUM SPECIMEN)	
1 -2	PYROPHYLLITE	MNCL.	2.65-2.90	PSI	LAMELLAR. CLVG-INELAST PLTS. PARTLY DCMP-H2SO4. VNS, FEW SCHISTS	W361 3-115
1 -2	SAPONITE (CLAY)	MNCL.	2.-3.	PSI	IRREG PLTS, FIBR, MSV. SWELLING CLAY. ALTER-HI MG +CA,LO K RKS	W399 3-226
1 -2	AURICHALCITE	ORTH.	3.64	CBT	ACICULAR,LATHLK,GRNLR. SILKY. 1 CLVG. SOL-ACIDS. SCNDRY CU,ZN-DEPS	16.1.4.
1 -3	CLAY	MNCL.	1.85-3.0	PSI	NAME GIVEN SEVERAL HYDROUS AL-SILICATES WITH ALKALIES OR ALKALINE EARTHS ± MG OR FE. NAMES UNSETTLED; MOST DISTINGUISHED BY NONMEGASCOPIC MEANS. FINE-GRAINED AGGS(COMPACT,MEALY,ETC.). EARTHY-PEARLY,UNCTUOUS,PLASTIC WHEN WET.DEHYDRATED ON HEATING. SOME ARE ABSORBENT. WXING PRODS,HYDROTHERMAL ALTER, DETRITAL, DIAGENIC	W398 3-191
1.5	WATER	HEXA.	.92	OXD	LIQUID ABOVE 0°C.(32°F)+ BELOW 100°C.(212°F). S.G. FOR ICE	4.1.3.
1.5	ARSENOLITE	ISOM.	3.86-3.88	OXD	WHITE-COLOR TINTED. OCTAH, CRUSTS. VITR-SILKY. OCTAH CLVG. ASTRIN-GENT-SWEETISH TASTE. SCNDRY AFTER ARSENOPYRITE AND OTHER AS-MINS	4.4.2.1
1.5-2	VIVIANITE	MNCL.	2.67-2.69	PHO	PRISM, ROUNDED, GLBL. VITR-DULL. CLVGS. FIBR-FLEX. DARKENS ON EX-POSURE. SOL-ACIDS. SECONDARY OF DIVERSE OCCURRENCES	40.2.15.1
1.5-2	COVELLITE	HEXA.	4.6-4.76	SLD	SOME IRID. GRAY STRK. 1 PERF CLVG-FLEX PLTS. ALTER OF CU-SULFIDES	2.6.8.1
1.5-2.5	CHRYSOCOLLA	UNKN.	2.3 -2.5	CSI	MSV. VITR-EARTHY. CONCH. DCMP-HCL. OXIDE ZONES CU-DEPS IN SIL RKS	W420 N
1.5-2.5	TYROLITE	ORTH.	3.0 -3.25	ARS	FAN-SHAPED AGGS,CRUSTS,ETC. VITR,PEARLY. 1 CLVG(FLEX). SOL-ACIDS. SECONDARY - WITH OXIDATION PRODUCTS OF COPPER DEPOSITS	42.3.2.
2	MELANTERITE	MNCL.	1.84-1.90	SUF	STALAC, CRUSTS,ETC. VITR. CLVGS. SWEET,ASTRINGENT,METALLIC TASTE. EXPOS(DRY AIR) --YEL-WHT + OPQ. EFFLORESCENCE-BY OXIDATION OF SUFS	29.6.8.1
2	PISANITE	MNCL.	1.95	SUF	DISTINGUISHED FROM MELANTERITE BY NONMEGASCOPIC MEANS. SAME OCCUR	29.6.8.2
2	SYLVITE	ISOM.	1.98-2.00	HAL	CUBIC, GRNLR. VITR. CUBIC CLVG. BITTER TASTE. SED EVAPORITE DEPS	9.1.1.2
2	MANASSEITE	HEXA.	2.00-2.10	OXD	FOL-MSV. GRSY. BASAL CLVG(FLEX). SOL-HCL. WITH HYDROTALCITE + SER-PENTINE	6.1.6.1
2	HALITE	ISOM.	2.17	HAL	CUBIC, GRNLR. VITR. CUBIC CLVG. SALTY TASTE. SED EVAPORITE DEPS	9.1.1.1
2	RHOMBOCLASE	ORTH.	2.23	SUF	THIN TBLR, STALACTITIC. SUBVITR. BASAL(FLEX),PRISM CLVGS. CONCH-FIBR. SOL-ACIDS,H2O(SLOWLY). RARE-ALTER PROD OF SULFIDES	29.1.1.
2	CHALCOPHYLLITE	HEXA.	2.67	ARS	TBLR,FOL MSV,ROSETTES. VITR,PEARLY. BASAL, RHOMB CLVGS. SOL-ACIDS, NH4OH. OXIDIZED ZONES OF COPPER DEPOSITS	43.2.1.
2	GLAUCONITE	MNCL.	2.4 -2.95	PSI	PLTS IN ROUND GRAINS. ATTKD-HCL. MARINE DEPOSITS - GREENSANDS	W377 3-35
2	BORAX	MNCL.	1.71-1.72	BOR	HABIT-LK PYROXENE. VITR-EARTHY. CLVGS. SWEETISH. SOL-H2O(TO ALK SOLUTION). BORAX DEPOSITS (SALINE LAKE DEPOSITS, ETC.)	25.1.4.
2 -2.5	GOSLARITE	ORTH.	1.98	SUF	CRUSTS. VITR-SILKY. 1 CLVG. ASTRINGENT,METALLIC,NAUCEOUS. EFFLOR-TYPICALLY BY ALTERATION OF SPHALERITE	29.6.9.2
2 -2.5	BOOTHITE	MNCL.	2.0 -2.2	SUF	MSV, FIBR. VITR-PEARLY. 1 CLVG. SOL-H2O. EFFLOR WITH MELANTERITE	29.6.8.4
2 -2.5	KAOLINITE (CLAY)	TRCL.	2.60-2.68	PSI	ALTER(ACID CONDITIONS) OF FELDSPARS, ETC. HYDROTHERMAL + WXING PRODUCT	W362 3-194
2 -2.5	LIROCONITE	MNCL.	2.9 -3.01	ARS	THIN,STRIATED XLS. VITR. CLVGS. CONCH. SOL-ACIDS. OXIDIZED ZONES OF COPPER DEPOSITS	42.2.5.
2 -2.5	CHALCOMENITE	ORTH.	3.35	SLI	ACICULAR-STRIATED. VITR.SOL-ACIDS. ALTER PROD OF CU + PB SELENIDE	34.1.1.
2 -2.5	SCHWARTZEMBERGITE	TETR.	7.39	IOD	ROUNDED FLAT PYRAMIDAL XLS. CRUSTS. ADMN. 1 CLVG. SOL-HCL. PB DEPS	22.1.2.
2 -3	ALLOPHANE (CLAY)	AMOR.	1.85-1.89	PSI	INCRUST, "BOOKS." GELAT, ALTER(ACID CONDITIONS) OF FELDSPARS, ETC. IN COAL, ORE DEPOSITS, AND VEINS	W531 3-194

NONMETALLIC BLUE

HARDNESS	NAME	XL. SYS.	SPECIFIC GRAVITY	CHEM. CLASS	REMARKS	REFERENCES
2-3	AMESITE (SERPENTINE?)	HEXA.	2.75-2.85	PSI	FINE GRAINED. HEX PLTS. WITH DIASPORE,CORUNDOPHILITE,ETC.--EMERY RK W383	3-164
2-3	SYMPLESITE	TRCL.	3.01	ARS	SPHERICAL AGGS-FIBR. VITR,PEARLY. CLVGS. SOL-ACIDS. SCNDRY AS-DEPS	40.2.15.5
2-3	DAPHNITE (CHLORITE)	MNCL.	2.75-3.1	PSI	FE END MEMBER. REPLACEMENT MIN AFTER BIOTITE + OTHER FERROMAG MINS W383	3-137
2-3	SPANGOLITE	HEXA.	3.14	SUF	TBLR, SOME IS HEMIMORPHIC. VITR. BASAL, RHOMB CLVGS. CONCH. SOL-ACIDS. SCNDRY WITH CUPRITE AND OTHER COPPER MINERALS	31.1.5.
2-3	CHLORITE	MNCL.	2.6-3.3	PSI	GROUP NAME. MEMBER NAMES UNSETTLED. ONLY FEW RARE VARIETIES MEGASCOPICALLY DISTINGUISHABLE, DISSEM FLKS. PEARLY. 1 PERF CLVG TO FLEX,INELAST FOLIA. DCMP-IN BOILING H2SO4. SCNDRY-HYDROTHERMAL ALTER, LO-GRADE META, DETRITAL, AUTHIGENIC	W381 3-131
2-4	MICA	MNCL.	2.4-3.4	PSI	GROUP NAME. SOME SPECIES DISTINGUISHABLE ON BASIS OF COLOR. SERIES ARE NOT COMPLETE, HEX-SHAPED XLS,DISSEM. PERF CLVG TO FLEX,ELAST FLKS. PERCUSSION + PRESSURE FIGURES MAY BE IMPOSED ON FLKS. CHIEF OCCURRENCES ARE NOTED UNDER SPECIES	W365 3-1
2 -6.5	WAD		2.8-4.4	OXD	FIELD TERM FOR HYDROUS MN-OXIDES OF UNKNOWN IDENTITY. (MN-CEPS)	4.5.1.3
2.5	BISMOCLITE	TETR.	7.12	HAL	IN SERIES WITH DAUBREEITE. MSV, SCALY, GRSY-DULL. 1 CLVG. PLASTIC. SOL-ACIDS. SCNDRY-ALTER PROD OF BISMUTHINITE OR NATIVE BISMUTH	10.1.6.2
2.5	DAUBREEITE	TETR. C	7.56	HAL	IN SERIES WITH BISMOCLITE,WHICH SEE. SCNDRY AFTER BISMUTHINITE + BI	10.1.6.3
2.5	PERCYLITE	ISOM.		HAL	MSV, CUBIC, DODEC. VITR. SOL-BOILING HNO3. SCNDRY-ORE DEPS	10.2.4.
2.5	CARNALLITE	ORTH.	1.60	HAL	GRNLR, PSEUDOHEX PRISM. GRSY. CONCH. DLQSNT. BITTER. SED SALT DEPS	11.1.2.
2.5	RETGERSITE	TETR.	2.04	SUF	FIBR CRUSTS, TWISTED TUFTS. VITR. BASAL, PRISM CLVGS. BITTER, METALLIC TASTE. EFFLORESCENCE-SCNDRY BY OXIDATION OF NICKEL MINS	29.6.7.
2.5	ALUMOHYDROCALCITE	MNCL.	2.23	CBT	CHALKY MASSES. 2 CLVGS. SOL-ACIDS,DCMP-BOILING H2O. WITH ALLOPHANE	16.2.4.
2.5	CHALCANTHITE	TRCL. A	2.29	SUF	PRISM,STALAC,GRNLR. VITR. CLVGS. METALLIC,NAUSEOUS TASTE. SCNDRY-IN OXIDE ZONES OF CU-SULFIDE DEPOSITS	29.6.5.1
2.5	CHALCOALUMITE	TRCL.	2.29	SUF	BOTRYOIDAL,LATHLK,DULL-VITR. CLVGS. SOL-DILUTE ACIDS(SLOW). CRUSTS ON LIMONITE STALACTITES (BISBEE, ARIZONA)	31.1.9.
2.5	BRUCITE	HEXA.	2.38-2.40	OXD	WHT STRK. TBLR, FOL MSV. WAXY-PEARLY. BASAL CLVG(FLEX). SOL-ACIDS. LOW TEMPERATURE VEINS	6.1.1.1
2.5	MITSCHERLICHITE	TETR.	2.42	HAL	PYRAMIDAL, PRISM. VITR. RARE-STALACTITIC ON FLOOR OF VESUVIUS	11.3.2.
2.5	ERIOCHALCITE	ORTH. A	2.47	HAL	LICHENLK AGGS. VITR. PRISM,BASAL CLVGS. SOL-HCL, NH4OH. SCNDRY, FUMAROLE SUBLIMATE	9.2.7.
2.5	FERRINATRITE	HEXA.	2.55-2.61	SUF	FIBR AGGS, PRISM. MSV. VITR. CLVGS. SPLINTERY. SCNDRY WITH OTHER IRON SULFATES IN ARID REGIONS	29.4.1.
2.5	RANSOMITE	ORTH.	2.63	SUF	CRUSTS. VITR-PEARLY. 1 CLVG. WITH OTHER HYDRATED SULFATES-RARE	29.7.1.
2.5	CHLOROTHIONITE	ORTH.	2.67	SUF	CRUSTS. FUMAROLIC(VESUVIUS-1872)	30.1.4.
2.5	VOLCHONSKOITE (CLAY)	MNCL.	2.2-2.7	PSI	CHROMIUM-NONTRONITE(?).LAMELLAR. UNSTABLE IN H2O. GELAT. "SWELLING CLAY" ALTER--HI MG +CA;LO K RKS(ALK CONDITIONS).RARE-CLAY DEP	W399 N
2.5	BANDYLITE	TETR.	2.81	BOR	TBLR, LICHENLK. VITR,PEARLY. 1 CLVG(FLEX). DCMP-H2O, SOL-NH4OH, HCL, HNO3. RARE-SCNDRY WITH ATACAMITE, ETC. (CALAMA, CHILE)	26.1.4.
2.5	SZOMOLNOKITE	MNCL.	3.03-3.07	SUF	BIPYRAMIDAL,STALAC.VITR. CONCH.SOL-H2O(SLOW). EFFLOR-ACID SOLUTION	29.6.2.2
2.5	DEVILLINE	MNCL.	3.13	SUF	PLTY-LATHLK CRUSTS. VITR,PEARLY. CLVGS. SOL-HNO3. SCNDRY-CU DEPS	31.3.3.
2.5	PYROCHROITE	HEXA.	3.23-3.27	OXD	TBLR, RHOMB, FOL MSV. PEARLY-DULL. BASAL CLVG(FLEX). SOL-HCL. WITH OTHER MN MINERALS IN LO-T HYDROTHERMAL VEINS	6.1.1.2
2.5	TEINEITE	ORTH.	3.80	TEL	CRUSTS, AGGS, PRISM. 3 CLVGS. SOL-HCL(TO BLUE),HNO3(TO YEL). OXIDATION PRODUCT OF TETRAHEDRITE AND NATIVE TELLURIUM	33.1.2.

NONMETALLIC BLUE

Hardness	Mineral	Crystal Sys.	S.G.	Luster	Description	Ref.
2.5	TRICHALCITE	ORTH.	3.5 -3.86	ARS	PSEUDOHEX AGGS, RADIATING GROUPS,ETC. VITR-SILKY. 1 CLVG. SOL-HCL. WITH COPPER-ARSENIC MINERALS	40.2.13.
2.5	CUMENGITE	TETR.	4.67	HAL	OCTAH. VITR. CLVGS. SOL-HNO3. RARE-WITH BOLEITE(BAJA, MEXICO)	10.2.2.
2.5	PSEUDOBOLEITE	TETR.	4.85	HAL	//GROWTH WITH BOLEITE. PEARLY. CLVGS. SOL-HNO3. RARE-BAJA, MEXICO	10.2.3.
2.5	LINARITE	MNCL.	5.35	SUF	ELONGATE, CRUSTS. VITR. 2 CLVGS. SOL-DILUTE HNO3. OXIDE ZONES CU AND LEAD DEPOSITS	30.2.3.
2.5	DIABOLEITE	TETR.	5.41-5.43	HAL	TBLR, PLTY. 1 CLVG. CONCH. SOL-HNO3. SCNDRY WITH CERUSSITE, ETC.	10.2.5.
2.5	MENDIPITE	ORTH.	7.24	HAL	FIBR, RADIATED. SILKY,RESIN. PRISM + OTHER CLVGS. SOL-HNO3. SCNDRY WITH CERUSSITE, ETC.	10.1.4.
2.5-3	KAINITE	MNCL.	2.15	SUF	GRNLR, TBLR. VITR. 1 CLVG. SALTY,BITTER. IN MARINE POTASH DEPS	31.4.1.
2.5-3	ZINCALUMINITE	HEXA.	2.26	SUF	CRUSTS, PLTY. SOL-ACIDS + ALKALIES. SCNDRY AT ZINC DEPS(LAURIUM)	31.1.7.
2.5-3	BLOEDITE	MNCL.	2.22-2.28	SUF	PRISM, GRNLR. VITR. CONCH. FAINTLY SALTY + BITTER. EVAPORITE DEPS	29.3.5.1
2.5-3	CORVUSITE	UNKN.	2.82	OXD	OPQ. MSV. SOL-ACIDS. WDSPRD IN UTAH-COLO U-ORE AREA-IMPREGNATING SS	4.6.2.
2.5-3	KROEHNKITE	MNCL.	2.90	SUF	FIBR AGGS,MSV,GRNLR. VITR. 1 PERF CLVG. SOL-H2O TO ACID SOLUTION. BECOMES GREEN + OPQ ON EXPOSURE. EFFLORESCENCE ON CU-SULFIDES,ETC.	29.3.3.
2.5-3	LANGITE	ORTH.	3.48-3.50	SUF	FIBR-LAMELLAR CRUSTS. SILKY. 2 CLVGS. SOL-ACIDS,NH4OH. SCNDRY-OXI-DATION OF COPPER SULFIDES	31.2.2.
2.5-3	CLINOCLASE	MNCL.	4.38	ARS	ELONGAT,ROSETTES,CRUSTS,FIBR. VITR,PEARLY. 1 CLVG. SOL-ACIDS + NH4OH. WITH OLIVENITE + OTHER SECONDARY COPPER MINERALS	41.3.1.
2.5-3	CALEDONITE	ORTH.	5.6 -5.77	SUF	COATINGS, DIVERGENT GROUPS. RESIN. CLVGS. EFF-HNO3. OXIDE ZONES OF COPPER-LEAD DEPOSITS	32.1.2.
2.5-3	ANGLESITE	ORTH.	6.37-6.39	SUF	TBLR,PRISM,MSV,ETC. ADMN. CLVGS. CONCH. BRTL. SOME FLUO-YEL. SOL-HNO3(DIFFICULTLY). SECONDARY-BY OXIDATION OF GALENA	28.3.1.3
2.5-3	LEADHILLITE	MNCL.	6.55	CBT	PSEUDOHEX, GRNLR. RESIN,PEARLY. 1 CLVG. FLUO-YEL. EFF-HNO3. SCNDRY IN OXIDIZED ZONES OF LEAD DEPOSITS	17.1.3.
2.5-3.5	SERPENTINE		2.5 -2.6	PSI	GROUP NAME. NOMENCLATURE FOR MEMBERS IS UNSETTLED. MSV-ASBESTIFM. WAXY-GRSY. GELAT. SOME FLUO-CREAM YEL. ALTER PROD AFTER MG-SILICATES(E.G.,OLIVINE). VNS	W381 3-170
2.5-3.5	ANTIGORITE (SERPENTINE)	MNCL.	2.65	PSI	COLORS COMMONLY VARIEGATED OR MOTTLED. PLATY. SECONDARY	W381 3-170
2.5-3.5	BISMUTITE	TETR.	6.1 -7.7	CBT	GRAY STRK, MSV, RADIAL FIBR CRUSTS. VITR-DULL. EFF-ACIDS. SCNDRY AFTER BISMUTH MINERALS	16.1.8.
3	TORREYITE	UNKN.	2.67	SUF	MSV,GRNLR,FOL. VITR-PEARLY. 1 CLVG. SOL-ACIDS. VNS(STERLING HILL, NEW JERSEY)	31.1.4.
3	CALCITE	HEXA.	2.71	CBT	VARIOUS TRIGONAL XL HABITS. MSV. VITR-PEARLY. RHOMB CLVG. SOME IS FLUO,PHOSPHO,THERMOLUM. EFF-DILUTE HCL. WDSPRD-VNS,DIVERSE ROCKS	14.1.1.1
3	APHTHITALITE	HEXA.	2.65-2.71	SUF	TRIGONAL-TBLR, MAMMILLARY,MSV. VITR. CLVGS. CONCH. SOL-H2O, ACIDS. FUMAROLIC, OCEANIC + LACUSTRINE SALINE DEPOSITS	28.2.2.
3	LEIGHTONITE	TRCL.	2.95	SUF	LATHLK.VITR. RARE-VNS + CAVITIES WITH ATACAMITE,ETC.(CHUQUICAMATA)	29.4.3.
3	LOSEYITE	MNCL.	3.27	CBT	LATHS-SUBPARALLEL TO RADIAL AGGS. IN CALCITE VNLETS AT FRNKLN,N.J.	16.1.1.
3	BUTTGENBACHITE	HEXA.	3.33	SUF	INDISTINGUISHABLE MEGASCOPICALLY FROM CONNELLITE. RARE-1 LOCALITY	31.1.1.2
3	CONNELLITE	HEXA.	3.36	SUF	ACICULAR,STRIATED,FELTED AGGS. VITR. SOL-ACIDS, NH4OH. WITH SCNDRY COPPER MINERALS	31.1.1.1
3	SALESITE	ORTH.	4.72-4.82	IOD	PRISM-PYRAMIDS. VITR. PRISM CLVG. SOL-HNO3. OXIDIZED ORE(CHUQUICA-MATA, CHILE)	22.1.1.

NONMETALLIC BLUE

HARDNESS	NAME	XL. SYS.	SPECIFIC GRAVITY	CHEM. CLASS	REMARKS	REFERENCES
3 -3.5	PHOSPHOPHYLLITE	MNCL.	3.08-3.13	PHO	THICK TBLR. VITR. CLVGS. FLUO-VIOLET(SW). SOL-ACIDS. PEG-WITH FE, MN PHOSPHATES AND SPHALERITE	40.2.12.
3 -3.5	CELESTITE	ORTH.	3.96-3.98	SUF	TYPICALLY COLORLESS-BLUISH. TBLR,ETC. VITR-PEARLY. CLVGS. SOME FLUO,THERMOLUM. SOL-H2O(SLIGHTLY). SED RKS(ESP GYPSUM + LS),VNS	28.3.1.2
3 -3.5	BARITE	ORTH.	4.50	SUF	TBLR, MSV,ETC. VITR-PEARLY. CLVGS. SOME FLUO, PHOSPHO, THERMOLUM. SOME FETID WHEN RUBBED. COMMON-VNS, SED RKS, CAVITIES IN IG RKS	28.3.1.1
3 -3.5	BOLEITE	TETR.	5.05	HAL	GREEN-BLUE STRK. PSEUDOCUBIC. VITR-PEARLY. CLVGS. SOL-HNO3. SCNDRY	10.2.1.
3 -3.5	CERUSSITE	ORTH.	6.53-6.57	CBT	TWINNED CLUSTERS, MSV. ADMN-RESIN. CLVGS. FLUO-YELX-RAYS +(LW,UL-TRA VIOLET). OXIDIZED ZONES OF LEAD DEPOSITS	14.1.3.4
3 -4	EVANSITE	UNKN.	1.8 -2.2	PHO	MSV,COLLOFM COATINGS. VITR. CONCH. BRTL. SOL-ACIDS. WITH LIMONITE	42.2.6.
3 -4	WAVELLITE	ORTH.	2.36	PHO	RADIAL FIBR, STELLATE,CRUSTS. VITR-PEARLY.CLVGS. SOL-ACIDS. SCNDRY-WDSPRD-OPENINGS IN DIVERSE ROCKS,LIMONITE-PHOSPHATE ROCKS, ETC.	42.7.4.
3 -4	MIXITE	HEXA.	3.79	ARS	HAIRLK-STRIATED,CONCENTRIC FIBR MASSES. DULL. WITH OTHER BI-MINS	42.5.7.
3.5	VAUXITE	TRCL.	2.39	PHO	TBLR,SUBPARALLEL-RADIAL AGGS + NODULES.VITR. WITH WAVELLITE,SN-VNS	42.8.7.
3.5	ANHYDRITE	ORTH.	2.98	SUF	MSV, GRNLR, FIBR. VITR-PEARLY. 2 CLVGS. SOL-ACIDS. SED ROCKS	28.3.2.
3.5	CHALCOCYANITE	ORTH.	3.60-3.70	SUF	TBLR. HYGROSCOPIC TO BLUE PENTAHYDRATE OR CHALCANTHITE. FUMAROLIC	28.3.3.
3.5	VRBAITE	ORTH.	5.27-5.33	SST	SUBMET.DK BLUISH GRAY,DK RED WITH TRANSMITTED LIGHT. STRK-RED WITH YEL TINGE. TBLR,PYRAMIDAL. 1 CLVG. SOL-HNO3,H2SO4. WITH REALGAR	3.8.13.
3.5	KOLBECKITE	MNCL.	2.39	PHO	PRISM. VITR-PEARLY. 1 CLVG. BRTL. DCMP-ACIDS. IN QTZ-WOLFRAMITE VN	43.2.7.
3.5-4	ALUNITE	HEXA.	2.6 -2.9	SUF	GRNLR, MSV, VITR,PEARLY. CLVGS. SOL-H2SO4. "ALUNITIZED" ROCKS	30.2.4.1
3.5-4	NATROALUNITE	HEXA.	2.6 -2.9	SUF	GRNLR, MSV. VITR,PEARLY. CLVGS. SOL-H2SO4. RARE,BUT VARIED OCCUR.	30.2.4.2
3.5-4	ARAGONITE	ORTH.	2.94-2.95	CBT	ACICULAR, RADIAL FIBR, STALACTITIC.VITR. CLVGS. FLUO(ULTRA-VIOLET, X-RAYS;ELECTRON BEAMS),THERMOLUM. EFF-ACIDS. LO-T DEPOSITION	14.1.3.1
3.5-4	SCORODITE	ORTH.	3.28	ARS	PYRAMIDAL,ETC.,AGGS,CRUSTS,ETC. VITR. CLVGS. SOL-ACIDS. SCNDRY-AFTER ARSENOPYRITE AND OTHER AS-BEARING MINERALS	40.3.1.3
3.5-4	VESZELYITE	MNCL.	3.3 -3.5	PHO	PRISM,TBLR,GRNLR. VITR. CLVGS. SOL-ACIDS. WITH SCNDRY CU MINERALS	42.2.1.
3.5-4	AZURITE	MNCL.	3.77	CBT	TBLR-AGGS. VITR. CLVGS. SOL-ACIDS,ETC. SCNDRY ZONES OF CU DEPS	16.1.11.
3.5-4	POWELLITE	TETR.	4.21-4.25	MBS	PYRAMIDAL,TBLR,MSV-FOL POWDER,CRUSTS.SUBADMN-PEARLY. CLVGS. UNEVN. FLUO-YEL. DCMP-HCL,HNO3. ALTER PROD OF MOLYBDENITE	48.1.3.2
3.5-4.5	SAMPLEITE	ORTH.	3.20	PHO	LATHLK. PEARLY. CLVGS. RARE-IN OXIDIZED ORE(CHUQUICAMATA,CHILE)	42.6.1.
4	FLUORITE	ISOM.	3.18	HAL	CUBIC, MSV. VITR. OCTAH CLVG. FLUO-VIOLET, DCMP-H2SO4. WESPRD-VNS, CAVITIES, DISSEMINATED IN IGNEOUS ROCKS, ETC.	9.2.1.
4	ANDREWSITE	ORTH.	3.48	PHO	RADIAL FIBR BOTRYOIDAL. SILKY. CLVG // LENGTH FIBERS. WITH LIMONITE (CORNWALL, ENGLAND)	41.4.5.
4	SARCOPSIDE	MNCL.	3.64-3.73	PHO	RED TO BROWNISH-EXPOS-DK BROWN.BLUE VIOLET,GREEN,FIBR MASSES.SILKY. 2 CLVGS⊥ + // FIBERS). SOL-ACIDS. RARE-PEGMATITES	41.6.4.
4	YEATMANITE	TRCL.	5.0	NSI	PSEUDO-ORTHORHOMBIC. COMMONLY TWINNED. RARE- DEPS AT FRANKLIN,N.J.	M529 N
4 -4.5	VARISCITE	ORTH.	2.2 -2.57	PHO	NODULES,CRUSTS-SOME OPALINE. WAXY. 2 CLVGS. SPLINTERY-CONCH. SOL-ALKALIES. NEAR-SURFACE DEPOSITION IN CAVITIES(E.G., BRECCIAS)	40.3.1.1
4 -4.5	VOLTZITE	UNKN.	3.7 -3.8	SLD	SECONDARY INCRUSTS. VITR-GRSY. GIVES H2S IN HCL. WITH SULFIDES	2.6.4.3

NONMETALLIC BLUE

4 -4.5	SMITHSONITE	HEXA.	4.40-4.45	CBT	BOTRYOIDAL, MSV, VITR-PEARLY. RHOMB CLVG. SOME FLUO-GREEN, BLUE. EFF-ACIDS. SECONDARY-OXIDIZED ZONES OF ZINC DEPOSITS	14.1.1.6	
4 -4.5	THOROGUMMITE	TETR.	4.5	NSI	SUBTRANSLUCENT. DULL. COMMONLY SECONDARY(E.G.,AFTER THORITE) PEGS	WN	N
4 -5	TRIPHYLITE	ORTH.	3.34-3.50	PHO	MSV, ANHEDRAL. VITR. CLVGS. UNEVN. SOL-ACIDS. PEGMATITES	38.1.1.1	
4 -5	CORNETITE	ORTH.	4.10	PHO	CRUSTS,PRISM. VITR. SOL-HCL. RARE WITH OTHER SECONDARY CU MINERALS	41.3.2.	
4 -5	ROSASITE	MNCL.	4.0 -4.2	CBT	MAMMILLARY CRUSTS, FIBR. 2 CLVGS AT 90°. SOL-ACIDS. SCNDRY CU-ZN, PB DEPOSITS	16.1.5.	
4.5	WOODHOUSEITE	HEXA.	3.01	PHO	PSEUDOCUBIC,TBLR -CURVED,STRIATED. VITR,PEARLY. 1 CLVG. VUGS,ETC.	43.1.1.5	
4.5	HEDYPHANE	HEXA.	5.82	ARS	PRISM,TBLR,MSV. GRSY. RHOMB CLVG. SOL-HNO3. RARE-E.G.,FRNKLN, N.J.	41.7.3.2	
4.5-5	PLUMBOGUMMITE	HEXA.	4.01	PHO	CONCENTRIC BOTRYOIDAL, GUM-LIKE LUSTER. UNEVN. SOL-HOT ACIDS. SECONDARY - IN LEAD DEPOSITS WITH PYROMORPHITE	41.5.8.1	
4.5-5	PSEUDOMALACHITE	MNCL.	4.00-4.40	PHO	CONCENTRIC,RADIAL FIBR BOTRYOIDAL. VITR. 1 CLVG. SOL-ACIDS. SCNDRY IN OXIDIZED ZONES OF COPPER DEPOSITS	41.4.3.	
5	OKENITE	TRCL.	2.28-2.33	ISI	FIBR, TBLR, MSV. 1 PERF CLVG-ELAST. GELAT. IN AMYGDULES IN BASALT	W358	N
5	COERULEOLACTITE	UNKN.	2.57-2.69	PHO	MICROXLINE FIBR CRUSTS + BOTRYOIDAL AGGS. CONCH. SOL-ACIDS + ALKS. RARE - WITH LIMONITE AND/OR WAVELLITE	42.7.3.	
5	WARDITE	TETR.	2.81-2.87	PHO	PYRAMIDAL-STRIATED,CRUSTS-RADIAL FIBR,ETC. VITR.1 CLVG. SOL-ACIDS (DIFFICULTY).WITH MILLISITE IN VARISCITE, AFTER AMBLYGONITE--PEGS	42.5.4.	
5	DELTAITE (MIXTURE ?)	HEXA.	2.90-3.00	PHO	FIBR CRUSTS,PRISM. 1 CLVG. RARE-WITH CRANDALLITE(FAIRFIELD, UTAH)	41.5.8.5	
5	HYDROXYLAPATITE	HEXA.	2.9 -3.1	PHO	APATITE GROUP(SEE FOR PROPS.) RARE-TALC SCHIST	41.7.1.3	
5	CARBONATE-APATITE	HEXA.	2.9 -3.1	PHO	APATITE GROUP(SEE FOR PROPS.)RARE-NODULES(PHOSPHATE RK),GUANO,ETC.	41.7.1.4	
5	CHLORAPATITE	HEXA.	A 3.17	PHO	APATITE GROUP(SEE FOR PROPS.) RARE-VNS IN GABBRO, METEORITES	41.7.1.2	
5	FLUORAPATITE	HEXA.	A 3.18	PHO	APATITE GROUP(SEE FOR PROPS). COMMON-IGS, METAS, MARINE SEDS, ETC.	41.7.1.1	
5	APATITE	HEXA.	2.9 -3.2	PHO	GROUP NAME. MOST TYPES INDISTINGUISHABLE MEGASCOPICALLY. HEX PRISMS-COMMONLY ROUNDED,MSV,GRNLR. VITR. SOL-ACIDS. SOME FLUO--YEL-ORANGE. 16 RKS,VNS,PEGS	41.7.1.	
5	RIEBECKITE (AMPHIBOLE)	MNCL.	3.02-3.42	ISI	QTZ-BEARING SODIC IG RKS. MAGNESIO-RIEB.-AUTHIGENIC IN SEDS,IN VNS	W439	2-333
5	HEMIMORPHITE	ORTH.	3.45	SSI	HEMIMORPHIC XLS), SHEAFLK, MAMMILLARY, ETC. 1 PERF + 1 POOR CLVG. GELAT. PYROELECTRIC. SOME FLUO-PALE ORANGE(LONG WAVE). ZINC DEPS	W481	N
5	CALAMINE	ORTH.	3.45	SSI	(=HEMIMORPHITE)	W481	N
5	KATOPHORITE (AMPHIBOLE)	MNCL.	3.20-3.50	ISI	RARE IN INTERMEDIATE TO BASIC ALKALIC IGNEOUS ROCKS	W440	2-359
5	ADELITE	ORTH.	3.70-3.76	ARS	MSV,ELONGATE. RESIN. CONCH. SOL-DILUTE ACIDS. MN DEPOSITS	41.5.1.1	
5 -6	VISHNEVITE	HEXA.	2.32-2.42	TSI	SULFATE-RICH END MEMBER OF CANCRINITE FAMILY. OCCURRENCE LIKE CANCRINITE. CARBONATE-RICH MEMBERS EFFERVESCE AS WELL AS GELATINIZE WITH HCL	WN	4-310
5 -6	LAZURITE	ISOM.	2.38-2.45	TSI	SODALITE(SEE) GROUP PROPS. CNTCT META LS. IN HNO3 EVOLVES H2S	W350	4-294
5 -6	CANCRINITE	HEXA.	2.42-2.51	TSI	MSV (RARE PRISM XLS). PRISM CLVG. EFF-HCL. NEPHELINE-BEARING RKS	W354	4-310
5 -6	MARIALITE (SCAPOLITE)	TETR.	2.50-2.62	TSI	ME-0-20. PURE MARIALITE UNKNOWN IN NATURE	W352	4-321

93

NONMETALLIC BLUE

HARDNESS	NAME	XL. SYS.	SPECIFIC GRAVITY	CHEM. CLASS	REMARKS	REFERENCES
5 -6	DIPYRE (SCAPOLITE)	TETR.	2.57-2.69	TSI	ME-20-50.	W352 4-321
5 -6	MIZZONITE (SCAPOLITE)	TETR.	2.67-2.74	TSI	ME-50-80.	W352 4-322
5 -6	SCAPOLITE	TETR.	2.50-2.78	TSI	GROUP NAME. MOST TYPES INDISTINGUISHABLE MEGASCOPICALLY BUT S.G. MAY BE INDICATIVE. LONG,STRIATED PRISMS + AGGS OF COARSE XLS. SUBCONCH. POOR CLVG GIVING IRREGULAR, STRIATED-APPEARING SURFACES. MARIALITE IS INSOL. MEIONITE IS DCMP IN HCL. MOST FLUO—YEL-RED(ULTRAVIOLET,LW). CNTCT META CALC RKS,PEGS	W352 4-321
5 -6	(WERNERITE) (SCAPOLITE)	TETR.	2.50-2.78	TSI	GROUP NAME NO LONGER USED	W352 4-322
5 -6	MEIONITE (SCAPOLITE)	TETR.	2.73-2.78	TSI	ME-80-100. PURE MEIONITE UNKNOWN IN NATURE	W352 4-321
5 -6	TURQUOIS	TRCL.	2.6 -2.84	PHO	XLS-RARE; CRUSTS, STALAC. WAXY. 2 CLVGS. SOL-HCL(DIFFICULTLY). MXED ALUMINOUS ROCKS(ESPECIALLY IN ARID AREAS)	42.6.2.1
5 -6	EKERMANNITE (AMPHIBOLE)	MNCL.	3.00	ISI	PLUTONIC ALK IG RKS + ASSOC PEGS. (LESS COMMON THAN ARFVEDSONITE)	W439 2-364
5 -6	HOLMQUISTITE (AMPHIBOLE)	ORTH.	3.06-3.13	ISI	CONTAINS NOTABLE LI. NEAR CNTCTS BETWEEN LI-RICH PEGS +COUNTRY RKS	W440 2-230
5 -6	TORENDRIKITE (AMPHIBOLE)	MNCL.	3.21	ISI	ALKALIC IGNEOUS ROCKS	W442 N
5 -6	CROSSITE (AMPHIBOLE)	MNCL.	3.1 -3.4	ISI	INTERMEDIATE BETWEEN RIEBECKITE + GLAUCOPHANE. SCHISTS, LK GLAUCOPHANE	W440 2-334
5 -6	CROCIDOLITE (AMPHIBOLE)	MNCL.	3.0 -3.4	ISI	FIBR RIEBECKITE. IN VEINS IN METAMORPHOSED BEDDED IRON-STONES	W440 2-334
5 -6	AMPHIBOLE		3.0 -3.57	ISI	GROUP NAME. SEVERAL END AND OTHER MEMBERS WITH NAMES UNSETTLED FOR SOME. MANY MEGASCOPICALLY INDISTINGUISHABLE BUT SOME COLORS AND OCCURRENCES ARE INDICATIVE. ORTH AND MNCL SERIES. LONG SLENDER XLS. PRISM CLVG AT 56° + 124°. SOL-HF(SLOWLY). ALTERS TO BIOTITE +/OR CHLORITE. IGNEOUS AND METAMORPHIC RKS, VNS	W422 2-203
5 -7	PYROXENE		2.96-3.96	ISI	GROUP NAME. SEVERAL END AND OTHER MEMBERS WITH NAMES UNSETTLED FOR SOME. MANY MEGASCOPICALLY INDISTINGUISHABLE; OCCURRENCE MAY BE INDICATIVE. ORTH + MNCL SERIES. SHORT PRISMS. PRISM CLVG AT 87° + 92°. ALTERS TO AMPHIBOLE. IG + META RKS	W402 2-)
5 -7	OLIVINE	ORTH.	3.22-4.39	NSI	GRP NAME FOR MINS OF FAYALITE-FORSTERITE + FAYALITE-TEPHROITE SERIES, MONTICELLITE + GLAUCOCHROITE. GRNLR MASSES, DISSEM. CONCH. GELAT. COMMONLY ALTERED TO SERPENTINE. BASIC + ULTRABASIC IG RKS, CNTCT META DOLOMITIC LS, METEORITES	W498 1-2
5.5	NOSEAN	ISOM.	2.30-2.40	TSI	SODALITE(SEE) GROUP PROPS. IN PHONOLITE. (DISSOLVE-HNO3, EVAPORATE SLOWLY, NO XLS REMAIN, ADD CACL2, HALITE + GYPSUM FORM)	W350 4-289
5.5	GUARINITE	TRCL.	3.2 -3.5	NSI	TBLR. 2 CLVGS AT 90 DEGREES. GELAT. MEGASCOPICALLY LIKE WOEHLERITE	W518 N
5.5-6	SODALITE	ISOM.	2.27-2.33	TSI	DODEC, MSV. POOR DODEC CLVG. GELAT. SOME FLUO-YEL-ORANGE. IN NEPHELINE-BEARING RKS. DISSOLVE-HNO3, EVAPORATE SLOWLY, HALITE REMAINS	W348 4-289
5.5-6	HAUYNE	ISOM.	2.44-2.50	TSI	SODALITE(SEE) GROUP PROPS. NEPH-BEARING IG RKS. DISSOLVE-HNO3, EVAPORATE SLOWLY, GYPSUM CRYSTALS REMAIN.	W350 4-289
5.5-6	MONTEBRASITE	TRCL.	2.98	PHO	AMBLYGONITE SERIES(SEE FOR PROPERTIES + OCCURRENCES.)	41.5.5.2
5.5-6	LAZULITE	MNCL.	3.08	PHO	SUBTRANSLUCENT. ACUTE PYRAMIDAL. MSV. GRNLR. VITR. CLVGS. SOL-HOT ACIDS(SLOWLY). PEGS, QTZ VNS, SI-RICH ALUMINOUS META(HI-GRADE)RKS	41.8.1.1
5.5-6	NATROMONTEBRASITE	TRCL.	3.04-3.1	PHO	AMBLYGONITE SERIES(SEE FOR PROPERTIES + OCCURRENCES.)	41.5.5.3
5.5-6	AMBLYGONITE	TRCL.	3.11	PHO	EQUANT, LARGE MASSES. VITR, PEARLY. CLVGS. SOL-ACIDS(DIFFICULTY). LI-, PO4-RICH PEGS	41.5.5.1
5.5-6	SCORZALITE	MNCL.	3.38	PHO	IN SERIES WITH LAZULITE(SEE FOR PROPS). PEGMATITE	41.8.1.2

NONMETALLIC BLUE

Hardness	Mineral	SG	Sys	Class	Notes	Ref
5.5-6	ANATASE	3.90	TETR.	OXD	SUBMET, PALE YEL STRK. ACUTE PYRAMIDAL, ADMN. BASAL + PYRAMIDAL CLVG. CONVERTS TO RUTILE ON HEATING. VNS,ACCESS-IG RKS,DETRITAL	4.5.2. D-111 287
5.5-6.5	OPAL	1.8-2.25	TETR.	SIL	SUBMICROXLINE QTZ + H2O. RESIN-PEARLY. MSV, MAMMILLARY CRUSTS.	
5.5-6.5	FRANKLINITE	5.07-5.22	ISOM.	OXD	SUBMET, MAGNETITE SERIES OF SPINEL GROUP. RED-BRWN STRK. OCTAH. OCTAH PARTING.SOL-HCL.WEAKLY MAGNETIC. RARE-FRANKLIN,N.J.	7.2.1.7
5.5-7	KYANITE	3.53-3.65	TRCL.	NSI	HARDNESS-5 ⊥ TO + 7 // TO LENGTH OF BLADE-LIKE XLS. GNEISS +SCHIST	W527 1-137
5.5-7.5	NARSARSUKITE	2.75	TETR.	PSI	TBLR. PRISM CLVG. SOL-HF. IN PEGS ASSOC WITH + CAVITIES IN ALK IGS	W358 N
6	CATAPLEITE	2.75	HEXA.	CSI	THIN PLATES. 2 PERF CLVGS. RARE-ALKALIC RKS, INCLUDING PEGMATITES	W454 N
6	LAWSONITE	3.05-3.10	ORTH.	SSI	PRISM, TBLR.CLVG-2 PERF, 1 POOR. LO-GRADETE.G.,GLAUCOPHANE)SCHISTS	W482 1-221
6	PUMPELLYITE	3.18-3.23	MNCL.	SSI	FIBR-PLTY. 1 GOOD + 1 POOR CLVG. CAVITIES + LOW-GRADE META ROCKS	W519 1-227
6	GLAUCOPHANE (AMPHIBOLE)	3.08-3.30	MNCL.	ISI	WITH LAWSONITE, ETC. IN SCHISTS + GNEISSES (COMMONLY NA-RICH ONES)	W439 2-333
6	JADEITE (CLINOPYROXENE)	3.24-3.43	MNCL.	ISI	RELATIVELY UNCOMMON, TYPICALLY WITH ALBITE. HI-T + ALSO LO-GRADE METAMORPHIC ROCKS	W411 2-99
6	GLAUCOCHROITE	3.48	ORTH.	NSI	BLUISH GREEN, CA-MN OLIVINE. RARE (FRANKLIN FURNACE, NEW JERSEY)	W504 N
6	TEPHROITE	3.78-4.1	ORTH.	NSI	MN-OLIVINE. OCCURS IN METAMORPHIC ROCKS AND FE-MN ORE DEPOSITS	W498 1-34
6 -6.5	ALBITE (PLAGIOCLASE)	2.57-2.63	TRCL.	TSI	AN0-10. SEE FELDSPAR. IG, META RKS, PEGS. MOONSTONE +CLEAVELANDITE ARE VARIETIES	W312 4-94
6 -6.5	LABRADORITE (PLAGIOCLASE)	2.68-2.72	TRCL.	TSI	AN50-70. BLUISH REFLECTIONS. SEE FELDSPAR. BASIC IGNEOUS ROCKS	W312 4-95
6 -6.5	FELDSPAR	2.55-3.39		TSI	INCLUDES PLAGIOCLASE(NA-CAISERIES), ALKALI(K-NA) AND BA FELDSPARS. MOST EASI-LY DISTINGUISHED BY NONMEGASCOPIC MEANS. MONOCLINIC ONES HAVE 2 CLVGS AT 90° AND SIMPLE(IF ANY) TWINNING. TRICLINIC ONES HAVE 2 CLVGS UP TO 4° OFF 90° AND TYPICALLY HAVE POLYSYNTHETIC TWINNING (GIVING STRIATED APPEARANCE ON SOME CLVG SURFACES OF PLAGIOCLASE). PEGMATITES; IG, META, AND SOME SED ROCKS	W261 4-1
6 -6.5	BENITOITE	3.65	HEXA.	CSI	UNEVENLY COLORED. TBLR. SOL-HF. FLUO-BLUE(SW).FROM 1 NATROLITE VN	W453 N
6 -6.5	RUTILE	4.21-4.25	TETR.	OXD	SUBMET. PRISM-STRIATED.ADMN.POOR CLVGS. VNS, ACCESS-META + IG RKS	4.5.1.1
6 -7	VESUVIANITE	3.33-3.43	TETR.	SSI	PRISM, MSV. 1 POOR CLVG. SUBCONCH. ATTKD-HCL. META-ESP CNTCT LS	W508 1-113
6.5	CHALCEDONY	2.57-2.64	HEXA.	SIL	MICROKLINE + MICROFIBR QTZ. MAMMILLARY. LO-T VEINS +OTHER CAVITIES	D-111 195
6.5-7	JASPER	2.57-2.65	HEXA.	SIL	SUBTRANSLUCENT. SOME IS VARIEGATED OR BANDED. MICROGRNLR. CAVITIES	D-111 224
6.5-7	SERENDIBITE	3.42	TRCL.	NSI	POLYSYNTHETIC TWINNING. NEARLY INSOL-ACIDS. CNTCT META ZONES-RARE	W527 N
6.5-7	DIASPORE	3.3 -3.5	ORTH.	OXD	PLTY, FOL-MSV. VITR, PEARLY. WITH CORUNDUM IN EMERY ROCK	7.1.2.1
7	QUARTZ	2.65	HEXA.	SIL	DOUBLY TERMINATED HEX PRISMS STRIATED ⊥ LENGTH(LO-T), BIPYRAMIDS (HI-T),MSV,VITR,CONCH.SOL-HF.WDSPRD IN MOST ROCK TYPES + VEINS	D-111 9
7	AMETHYST	2.65	HEXA.	SIL	PURPLE(SOME REDDISH OR BLUISH). XLINE, LO-T, IN CAVITIES IN BASALT	D-111 171
7	BLUE QUARTZ	2.65	HEXA.	SIL	GRSY. MSV. META + IG RKS, PEGMATITES-COMMONLY WITH TI-BEARING MINS	D-111 189

95

NONMETALLIC BLUE

HARDNESS	NAME	XL. SYS.	SPECIFIC GRAVITY	CHEM. CLASS	REMARKS	REFERENCES
● 7	CORDIERITE	ORTH.	2.53-2.78	CSI	COLOR DIFFERS WITH DIRECTION. DISSEM GRNS. 1 GOOD CLVG. META RKS	W470 1-268
7	ELBAITE	HEXA.	3.03-3.10	CSI	NA- LI-TOURMALINE. INFUSIBLE. TYPICALLY PINK AND/OR GREEN. PEGS	W465 1-300
● 7	TOURMALINE	HEXA.	3.03-3.25	CSI	GROUP NAME. VARIETIES COMMONLY DISTINGUISHABLE MEGASCOPICALLY ON BASIS OF COLOR(S). LENGTHWISE STRIATED PRISMATIC XLS WITH CROSS-SECTIONS THAT RESEMBLE SPHERICAL TRIANGLES. SOME XLS EXHIBIT BOTH LENGTHWISE + CONCENTRIC COLOR ZONING. VITR-RESIN. BRTL. ELECTRICALLY CHARGED ON HEATING + COOLING. MG-RICH VARIETIES FLUO-YEL (ULTRAVIOLET - SW). PEGS, HI-T VNS, IG AND META ROCKS	W465 1-300
● 7	DUMORTIERITE	ORTH.	3.3	NSI	FIBR-COMMONLY RADIATING. VITR. 1 GOOD,1 POOR CLVG. PEGS + META RKS	W259 N
7 -7.5	BORACITE	ORTH.	2.91-2.97	BOR	CUBIC,ETC.,GRNLR,FIBR. VITR. SOL-HCL(SLOWLY). SALT-GYPSUM BEDS	26.1.7.
7.5	GRANDIDIERITE	ORTH.	3.0	NSI	GREENISH BLUE. XLS UNKNOWN. 2 GOOD CLVGS. PEGS + APLITES-RARE	W497 N
7.5	EUCLASE	MNCL.	3.0 -3.1	PSI	PRISM XLS. 1 GOOD, 2 POOR CLVGS. PEGS, PLACERS, GRANITES, SCHISTS	W357 N
7.5	SAPPHIRINE	MNCL.	3.40-3.58	OXD	TBLR, DISSEM. VITR. HI-T META RKS WITH CORUNDUM, SPINEL, ETC.	7.3.1.2
● 7.5-8	BERYL	HEXA.	2.66-2.83	CSI	GROOVED PRISM. POOR BASAL CLVG.SOME FLUO WEAKLY-YEL. PEGMATITES	W463 1-256
● 7.5-8	SPINEL	ISOM.	3.55	OXD	GROUP NAME. SOME HAVE DISTINCTIVE COLORS OR OTHER HAND-SPECIMEN PROPS. OCTAH. GRNLR, GLASSY. CONCH. SOME FLUO—RED TO YEL-GREEN. IG,META(ESP CALC RKS),PEGS	7.2.1.1
7.5-8	GAHNITE	ISOM.	4.62	OXD	SPINEL(SEE). OCTAH. SCHISTS, PEGS, CNTCT META-LS	7.2.1.3
● 9	CORUNDUM	HEXA.	4.0 -4.1	OXD	STEEP PYRAMIDAL, PRISM. BASAL PARTING.WDSPRD IN SI-DEFICIENT RKS	4.4.1.1
9.5	MOISSANITE	HEXA.	3.1 -3.21	ELE	SUBMET. TABULAR. CONCHOIDAL FRACT. METEOR. (ARTIF=CARBORUNDUM)	1.1.7.4
● 10	DIAMOND	ISOM.	3.50-3.53	ELE	OCTAHEDRAL, ADMN-GRSY. BRTL. OCTAH CLVG. ULTRABASIC IGS + PLACERS	1.2.4.1

NONMETALLIC PURPLE

HARDNESS	NAME	XL. SYS.	SPECIFIC GRAVITY	CHEM. CLASS	REMARKS	REFERENCES
1	MOLYSITE	HEXA. A	2.90	HAL	MSV. BASAL CLVG. DLQSNT. VOL SUBLIMATE	9.3.1.
1	PARALAURIONITE	MNCL.	6.15	HAL	LATHLK. SUBADMN. 1 CLVG. SOL-HNO3. XLS BEND(GLIDING). SCNDRY AFTER LEAD MINERALS	10.1.8.
1.5-2	BARBERTONITE	HEXA.	2.05-2.15	OXD	PLTY.,FIBR-MSV.,WAXY,PEARLY.BASAL CLVG(FLEX). VNLETS IN SERPENTINE	6.1.6.2
1.5-2	STICHTITE	HEXA.	2.16	OXD	PLTY, FIBR-MSV. WAXY, BASAL CLVG(FLEX). EFF-HCL. VNS IN SERPENTINE	6.1.5.2
2	HALITE	ISOM.	2.17	HAL	CUBIC, GRNLR. VITR. CUBIC CLVG. SALTY TASTE. SED EVAPORITE DEPS	9.1.1.1
2 -2.5	HYDROZINCITE	MNCL.	3.5 -4.0	CBT	MSV,INCRUST, SILKY-DULL.1 CLVG. FLUO-BLUE,LILAC. SOL-ACIDS.SCNDRY WITH OTHER ZINC MINERALS	16.1.3.
2 -3	IANTHINITE	ORTH.		OXD	SUBMET. BRWN-VIOLET STRK. PLTY, TBLR. BASAL CLVG. ALTER AFTER URANINITE	5.3.5.
2 -3	KAEMMERERITE (CHLORITE)	MNCL.		PSI	SMALL HEX-SHAPED PYRAMIDAL XLS IN ULTRABASIC RKS + CHROMIUM DEPS	W386 3-146
2 -3	PENNINITE (CHLORITE)	MNCL.	2.6 -3.1	PSI	PSEUDORHOMB XLS. WITH MAGNETITE + CHONDRODITE(SERP RKS), SCHISTS	W383 3-137
2 -4	MICA	MNCL.	2.4 -3.4	PSI	GROUP NAME, MEMBER NAMES UNSETTLED. ONLY FEW RARE VARIETIES MEGASCOPICALLY DISTINGUISHABLE, DISSEM FLKS. PEARLY. 1 PERF CLVG TO FLEX,INELAST FOLIA. DCMP- IN BOILING H2SO4. SCNDRY-HYDROTHERMAL ALTER, LO-GRADE META, DETRITAL, AUTHIGENIC	W365 3-1
2.5	COQUIMBITE	HEXA.	2.10-2.12	SUF	GROUP NAME. SOME SPECIES DISTINGUISHABLE ON BASIS OF COLOR. SERIES ARE NOT COMPLETE. HEX-SHAPED XLS,DISSEM. PERF CLVG TO FLEX,ELAST FLKS. PERCUSSION + PRESSURE FIGURES MAY BE IMPOSED ON FLKS. CHIEF OCCURRENCES ARE NOTED UNDER SPECIES PRISM, GRNLR. VITR. CLVGS. ASTRINGENT. SOL-ACIDS. WITH OTHER SO4'S	29.8.3.
2.5	PARACOQUIMBITE	HEXA.	2.10-2.12	SUF	RHOMB,EQUANT,GRNLR. VITR. CLVGS. ASTRINGENT. WITH COQUIMBITE,ETC.	29.8.4.
2.5	QUENSTEDTITE	TRCL.	2.15	SUF	AGGS TBLR XLS. 2 CLVGS. SOL-H2O. RARE-WITH COQUIMBITE	29.8.5.
2.5	ALUMOHYDROCALCITE	MNCL.	2.23	CBT	CHALKY MASSES. 2 CLVGS. SOL-ACIDS,DCMP-BOILING H2O. WITH ALLOPHANE	16.2.4.
2.5	FERRINATRITE	HEXA.	2.55-2.61	SUF	FIBR AGGS, PRISM. MSV. VITR. CLVGS. SPLINTERY. SCNDRY WITH OTHER IRON SULFATES IN ARID REGIONS	29.4.1.
2.5	CHLORARGYRITE	ISOM.	5.56	HAL	CRUSTS,CUBIC. RESIN. DUCTILE. EXPOS-VIOLET-BRWN. SCNDRY IN AG DEPS	9.1.1.4
2.5-3	CHLOROCALCITE	ORTH.		HAL	PSEUDOCUBIC,TBLR. CUBELK CLVG. DLQSNT. BITTER TASTE. VOL SUBLIMATE	11.1.1.
2.5-4	LEPIDOLITE	MNCL.	2.80-2.90	PSI	LI-MICA. TABULAR XLS, AGGS OF SMALL FLAKES. DISTINGUISHED FROM ROSE-COLORED MUSCOVITE BY CHEM OR X-RAY ANAL.RARE, LI-BEARING PEGS	W370 3-85
2.5-4	ZINNWALDITE	MNCL.	2.90-3.02	PSI	LI-FE-MICA. WITH LI-BEARING MINS-PEGS, CASSITERITE-BEARING VEINS	W371 3-92
3	RINNEITE	HEXA.	2.35	HAL	EXPOS-BRWN. GRNLR. SILKY. CONCH. ASTRINGENT. SCNDRY-SED SALT DEPS	11.4.3.
3	CALCITE	HEXA.	2.71	CBT	VARIOUS TRIGONAL XL HABITS. MSV. VITR-PEARLY. RHOMB CLVG. SOME IS FLUO.,PHOSPHO,THERMOLUM. EFF-DILUTE HCL. WDSPRD-VNS+DIVERSE ROCKS	14.1.1.1
3 -3.5	ROEMERITE	TRCL.	2.17	SUF	VIOLET OR YEL TINGE. CRUSTS, CUBICLK. OILY. 2 CLVGS. SALTY,ASTRINGENT. WITH COPIAPITE – BY OXIDATION OF PYRITE	29.7.2.
3.5	ANHYDRITE	ORTH.	2.98	SUF	MSV, GRNLR, FIBR. VITR-PEARLY. 2 CLVGS. SOL-ACIDS. SED ROCKS	28.3.2.
3.5	HUREAULITE	MNCL.	3.15-3.20	PHO	TBLR,MSV,SCALY,FIBR. VITR. 1 CLVG. SOL-ACIDS. WITH ALTERED LITHIOPHILITE AND/OR TRIPHYLITE	39.1.3.

NONMETALLIC PURPLE

HARDNESS	NAME	XL. SYS.	SPECIFIC GRAVITY	CHEM. CLASS	REMARKS	REFERENCES
3.5	CHLOROPHOENICITE	MNCL.	3.46	ARS	LONG PRISM, STRIATED. VITR,PEARLY. 1 CLVG. SOL-ACIDS. FRNKLN, N.J.	41.1.4.1
3.5	ADAMITE	ORTH.	4.32-4.48	ARS	ELONGATE,CRUSTS,RADIAL AGGS. VITR. CLVGS. UNEVN. SOL-DILUTE ACIDS. SOME FLUO YEL.OXIDIZED ZONES OF DEPS WITH ZN-+AS-RICH MINERALS	41.6.5.3
3.5-4	CREEDITE	MNCL.	2.71-2.73	HAL	PRISM, RADIATING AGGS. VITR.1 CLVG. SOL-ACIDS(SLOWLY).FLUORITE VNS	12.1.1.
3.5-4	METASTRENGITE	MNCL.	2.76	PHO	TBLR,BOTRYOIDAL+RADIAL FIBR CRUSTS. VITR. 2 CLVGS. SOL-HCL. WITH STRENGITE	40.3.2.2
3.5-4	LANGBEINITE	ISOM.	2.83	SUF	NODULAR, DISSEM. VITR.CONCH. SOL-H2O(SLOWLY). MARINE SALTS	28.4.3.1
3.5-4	STRENGITE	ORTH.	2.5-2.87	PHO	TBLR,PSEUDO-OCTAH,BOTRYOIDAL AGGS,ETC.WAXY. 2 CLVGS. SOL-HCL. ALTERATION PRODUCT OF FE-BEARING PHOSPHATES	40.3.1.2
3.5-4	ARAGONITE	ORTH.	2.94-2.95	CBT	ACICULAR, RADIAL FIBR, STALACTITIC.VITR. CLVGS. FLUO(ULTRA-VIOLET, X-RAYS, ELECTRON BEAMS),THERMOLUM. EFF-ACIDS. LOW TEMPERATURE DEPS	14.1.3.1
3.5-4	SCORODITE	ORTH.	3.28	ARS	PYRAMIDAL,ETC.-AGGS,CRUSTS,ETC. VITR. CLVGS. SOL-ACIDS. SCNDRY-AFTER ARSENOPYRITE AND OTHER AS-BEARING MINERALS	40.3.1.3
4	FLUORITE	ISOM.	3.18	HAL	CUBIC, MSV. VITR. OCTAH CLVG. FLUO-VIOLET. DCMP-H2SO4. WDSPRD-VNS, CAVITIES, DISSEMINATED IN IGNEOUS ROCKS, ETC.	9.2.1.
4	SARCOPSIDE	MNCL.	3.64-3.73	PHO	RED TO BROWNISH-EXPOS-DK BROWN,BLUE,VIOLET,GREEN.FIBR MASSES,SILKY. 2 CLVGS(// + ⊥ FIBERS).SOL-ACIDS. RARE-PEGMATITES	41.6.4.
4-4.5	HETEROSITE	ORTH.	3.2-3.4	PHO	SUBTRANSLUCENT. EXPOS-BRWNISH. SATINY-DULL. 2 CLVGS(CURVED OR CRIN-KLED SURFACES).SOL-HCL.OXIDATION PROD OF TRIPHYLITE AND LITHIOPHILITE	38.1.5.1
4-4.5	PURPURITE	ORTH.	3.2-3.4	PHO	SUBTRANSLUCENT. DISTINGUISHED BY CHEM ANAL FROM HETEROSITE	38.1.5.2
4.5	ALLACTITE	MNCL.	3.83	ARS	ELONGATE,ROSETTEK AGGS. VITR. 1 CLVG. SOL-HCL. VNS(E.G.,LANGBAN)	41.2.4.
5	DELTAITE (MIXTURE?)	HEXA.	2.90-3.00	PHO	FIBR CRUSTS,PRISM. 1 CLVG. RARE-WITH CRANDALLITE(FAIRFIELD, UTAH)	41.5.8.5
5	HYDROXYLAPATITE	HEXA.	2.9-3.1	PHO	APATITE GROUP(SEE FOR PROPS.) RARE-TALC SCHIST	41.7.1.3
5	CARBONATE-APATITE	HEXA.	2.9-3.1	PHO	APATITE GROUP(SEE FOR PROPS.)RARE-NODULES(PHOSPHATE RK),GUANO,ETC.	41.7.1.4
5	CHLORAPATITE	HEXA. A	3.17	PHO	APATITE GROUP(SEE FOR PROPS.) RARE-VNS IN GABBRO, METEORITES	41.7.1.2
5	FLUORAPATITE	HEXA. A	3.18	PHO	APATITE GROUP(SEE FOR PROPS.) COMMON-IGS, METAS, MARINE SEDS, ETC.	41.7.1.1
5	APATITE	HEXA.	2.9-3.2	PHO	GROUP NAME. MOST TYPES INDISTINGUISHABLE MEGASCOPICALLY. HEX PRISMS-COMMONLY ROUNDED,MSV,GRNLR. VITR. SOL-ACIDS. SOME FLUO—YEL-ORANGE. IG RKS,VNS,PEGS	41.7.1.
5	TILASITE	MNCL.	3.75-3.79	ARS	ELONG,MSV,RESIN. CLVGS + PARTINGS. SOL-HCL,HNO3.WITH OTHER MN-MINS	41.5.6.
5-6	MARIALITE (SCAPOLITE)	TETR.	2.50-2.62	TSI	ME-0-20. PURE MARIALITE UNKNOWN IN NATURE	W352 4-321
5-6	DIPYRE (SCAPOLITE)	TETR.	2.57-2.69	TSI	ME-20-50.	W352 4-321
5-6	MIZZONITE (SCAPOLITE)	TETR.	2.67-2.74	TSI	ME-50-80.	W352 4-322
5-6	SCAPOLITE	TETR.	2.50-2.78	TSI	GROUP NAME. MOST TYPES INDISTINGUISHABLE MEGASCOPICALLY BUT S.G. MAY BE IN-DICATIVE. LONG,STRIATED PRISMS + AGGS OF COARSE XLS. SUBCONCH. POOR CLVG GIVING IRREGULAR, STRIATED-APPEARING SURFACES. MARIALITE IS INSOL; MEIONITE IS DCMP IN HCL. MOST FLUO—YEL-RED(ULTRAVIOLET,LW). CNTCT META CALC RKS,PEGS	W352 4-321
5-6	(WERNERITE) (SCAPOLITE)	TETR.	2.50-2.78	TSI	GROUP NAME NO LONGER USED	W352 4-322

98

NONMETALLIC PURPLE

H	Mineral	Xl Sys	SG	Streak	Comments	Ref
5 -6	MEIONITE (SCAPOLITE)	TETR.	2.73-2.78	TSI	ME-80-100. PURE MEIONITE UNKNOWN IN NATURE	W352 4-321
5 -6	HOLMQUISTITE (AMPHIBOLE)	ORTH.	3.06-3.13	ISI	CONTAINS NOTABLE LI. NEAR CNTCTS BETWEEN LI-RICH PEGS +COUNTRY RKS	W440 2-230
5 -6	TORENDRIKITE (AMPHIBOLE)	MNCL.	3.21	ISI	ALKALIC IGNEOUS ROCKS	W442 N
5 -6	CROSSITE (AMPHIBOLE)	MNCL.	3.1 -3.4	ISI	INTERMEDIATE BETWEEN RIEBECKITE + GLAUCOPHANE. SCHISTS, LK GLAUCOPHANE	W440 2-334
5 -6	AMPHIBOLE		3.0 -3.57	ISI	GROUP NAME. SEVERAL END AND OTHER MEMBERS WITH NAMES UNSETTLED FOR SOME. MANY MEGASCOPICALLY INDISTINGUISHABLE BUT SOME COLORS AND OCCURRENCES ARE INDICATIVE. ORTH AND MNCL SERIES. LONG SLENDER XLS. PRISM CLVG AT 56° + 124°. SOL-HF(SLOWLY). ALTERS TO BIOTITE +/OR CHLORITE. IGNEOUS AND METAMORPHIC RKS, VNS	W422 2-203
5 -7	PYROXENE		2.96-3.96	ISI	GROUP NAME. SEVERAL END AND OTHER MEMBERS WITH NAMES UNSETTLED FOR SOME. MANY MEGASCOPICALLY INDISTINGUISHABLE; OCCURRENCE MAY BE INDICATIVE. ORTH + MNCL SERIES.SHORT PRISMS,PRISM CLVG AT 87° + 92°. ALTERS TO AMPHIBOLE,IG + META RKS	M402 2-1
5.5	CLINOHEDRITE	MNCL.	3.33	SSI	CLINOHEDRAL. 1 PERF CLVG. WITH WILLEMITE, ETC. AT FRANKLIN, N.J.	W482 N
5.5	AUGITE (CLINOPYROXENE)	MNCL.	3.23-3.52	ISI	THE COMMON PYROXENE OF SUB-ALKALIC IGNEOUS RKS(E.G., GABBROS)	W411 2-109
5.5-6	TITANAUGITE (CLINOPYROXENE)	MNCL.	3.3 -3.6	ISI	AUGITE WITH NOTABLE TI-CONTENT. ALKALIC BASIC IGNEOUS ROCKS	W411 2-113
5.5-6	LEUCOPHENICITE	MNCL.	3.85	NSI	PURPLISH RED. STRIATED ELONGATE XLS. GELAT. RARE-FRNKLN, N.J.	W516 N
5.5-6	ANATASE	TETR.	3.90	OXD	SUBMET. PALE YEL STRK. ACUTE PYRAMIDAL. ADMN. BASAL + PYRAMIDAL CLVG. CONVERTS TO RUTILE ON HEATING. VNS,ACCESS-IG RKS,DETRITAL	4.5.2.
6	GLAUCOPHANE (AMPHIBOLE)	MNCL.	3.08-3.30	ISI	WITH LAWSONITE, ETC. IN SCHISTS + GNEISSES (COMMONLY NA-RICH ONES)	W439 2-333
6 -6.5	BENITOITE	HEXA.	3.65	CSI	UNEVENLY COLORED. TBLR. SOL-HF. FLUO-BLUE(SW).FROM 1 NATROLITE VN	W453 N
6 -6.5	RUTILE	TETR.	4.21-4.25	OXD	SUBMET. PRISM-STRIATED.ADMN.POOR CLVGS. VNS, ACCESS-META + IG RKS	4.5.1.1
6.5	GENTHELVITE	ISOM.	3.44-3.70	TSI	HELVITE(SEE) GRP PROPS + OCCURRENCES. STALACTITIC, INCRUSTATIONS	M351 4-303
6.5-7	SPODUMENE	MNCL.	3.03-3.22	ISI	COMMONLY FLUO(+PHOSPHORESCES)-ORANGE. THERMOLUM. LI-BEARING PEGS	W418 2-92
6.5-7	AXINITE	TRCL.	3.26-3.36	CSI	STRIATED WEDGE-SHAPED XLS. 1 GOOD CLVG. SOL-HCL(SLOW). CNTCT META	W505 1-320
6.5-7	DIASPORE	ORTH.	3.3 -3.5	OXD	PLTY, FOL-MSV. VITR, PEARLY. WITH CORUNDUM IN EMERY ROCK	7.1.2.1
6.5-7.5	ANDALUSITE	ORTH.	3.13-3.16	NSI	BLUNT, NEARLY SQ PRISMS. COMMONLY ALTERED. CONTACT +REGIONAL META	W521 1-129
7	QUARTZ	HEXA.	2.65	SIL	DOUBLY TERMINATED HEX PRISMS STRIATED ⊥ LENGTH(LO-T). BIPYRAMIDS (HI-T),MSV,VITR.CONCH.SOL-HF.WDSPRD IN MOST RK TYPES + VEINS	D-111 9
7	AMETHYST	HEXA.	2.65	SIL	PURPLE(SOME REDDISH OR BLUISH). XLINE. LO-T, IN CAVITIES IN BASALT	D-111 171
7	BLUE QUARTZ	HEXA.	2.65	SIL	GRSY. MSV. META + IG RKS, PEGMATITES-COMMONLY WITH TI-BEARING MINS	D-111 189
7					GROUP NAME. VARIETIES COMMONLY DISTINGUISHABLE MEGASCOPICALLY ON BASIS OF COLOR(S). LENGTHWISE STRIATED PRISMATIC XLS WITH CROSS-SECTIONS THAT RESEMBLE SPHERICAL TRIANGLES. SOME XLS EXHIBIT BOTH LENGTHWISE + CONCENTRIC COLOR ZONING. VITR-RESIN. BRTL. ELECTRICALLY CHARGED ON HEATING + COOLING. MG-RICH VARIETIES FLUO-YEL (ULTRAVIOLET - SW). PEGS, HI-T VNS, IG AND META ROCKS	
7	DUMORTIERITE	ORTH.	3.3	NSI	FIBR-COMMONLY RADIATING. VITR. 1 GOOD,1 POOR CLVG. PEGS + META RKS	W259 N
9	CORUNDUM	HEXA.	4.0 -4.1	OXD	STEEP PYRAMIDAL, PRISM. BASAL PARTING.WDSPRD IN SI-DEFICIENT RKS	4.4.1.1
10	DIAMOND	ISOM.	3.50-3.53	ELE	OCTAHEDRAL. ADMN-GRSY. BRTL. OCTAH CLVG. ULTRABASIC IGS + PLACERS	1.2.4.1

NONMETALLIC COLORLESS

HARDNESS	NAME	XL. SYS.	SPECIFIC GRAVITY	CHEM. CLASS	REMARKS	REFERENCES
1	GAMMA SULFUR	MNCL.	2.05	ELE	LIGHT YELLOW-COLORLESS. MINUTE EQUANT XLS. ADAMANTINE. FUMAROLES	1.2.3.3
1	TALC	MNCL.	2.58-2.83	PSI	FOL MSV. BASAL CLVG. SECTILE. SOAPY FEEL. PEARLY. SCHIST, META IG	W364 3-121
1	SCACCHITE	HEXA.	2.98	HAL	MSV. BASAL CLVG. SOL-H2O. DLQSNT. WITH OTHER SALTS-VOL SUBLIMATE	9.2.3.2
1	PARALAURIONITE	MNCL.	6.15	HAL	LATHLK. SUBADMN. 1 CLVG. SOL-HNO3. XLS BENDIGLIDING). SCNDRY AFTER LEAD MINERALS	10.1.8.
1 -1.5	NATRON	MNCL.	1.48	CBT	GRNLR CRUSTS. VITR. CLVGS. ALK TASTE. IN SOLUTIONS, EFFLORESCENCES	15.1.6.
1 -1.5	THERMONATRITE	ORTH. A	2.26	CBT	CRUST-EFFLOR. VITR. 1 CLVG. ALK TASTE. RARE-WITH TRONA-ARID SOILS	15.1.1.
1 -2	BISCHOFITE	MNCL.	1.60	HAL	GRNLR, PRISM. VITR-DULL. CONCH. DLQSNT. BITTER TASTE. SALINE DEPS	9.2.8.
1 -2	KOENENITE	HEXA.	1.98	HAL	CRUSTS, PEARLY. BASAL CLVG(FLEX). DCMP-HOT H2O. K-DEPS(N. GERMANY)	10.2.8.
1 -2	VERMICULITE	MNCL.	2.2 -2.4	PSI	BRONZY, MICACEOUS, EXFOLIATES WITH SWELLING ON HEATING, ATTKD-ACID (SILICA RESIDUE).ALTER PROD OF BIOTITE(ETC.)-HYDROTHERMAL OR WXING	W396 3-246
1 -3	CLAY	MNCL.	1.85-3.0	PSI	NAME GIVEN SEVERAL HYDROUS AL-SILICATES WITH ALKALIES OR ALKALINE EARTHS + MG OR FE. NAMES UNSETTLED, MOST DISTINGUISHED BY NONMEGASCOPIC MEANS. FINE-GRAINED AGGS(COMPACT,MEALY,ETC.). EARTHY-PEARLY,UNCTUOUS.PLASTIC WHEN WET.DEHYDRATED ON HEATING, SOME ARE ABSORBENT. WXING PRODS,HYDROTHERMAL ALTER, DETRITAL, DIAGENIC	W398 3-191
1.5	WATER	HEXA.	.92	OXD	LIQUID ABOVE 0°C.(32°F)+ BELOW 100°C.(212°F). S.G. FOR ICE	4.1.3.
1.5	TESCHEMACHERITE	ORTH. A	1.57	CBT	COMPACT MASSES. CLVG. BRTL. SOL-H2O. EFF-ACIDS. GUANO DEPS	13.1.3.
1.5	AMMONIA ALUM	ISOM.	1.65	SUF	FIBR, CRUSTS. VITR-SILKY. CONCH. SWEET,ASTRINGENT. IN COALY BEDS	29.5.5.3
1.5	APJOHNITE	MNCL.	1.78	SUF	FIBR. SILKY. SOL-H2O. EFFLORESCENCES IN SHELTERED PLACES	29.7.3.3
1.5	PICKERINGITE	MNCL.	1.73-1.79	SUF	FIBR. VITR. ASTRINGENT. EFFLORESCENCES IN SHELTERED PLACES	29.7.3.1
1.5	HALOTRICHITE	MNCL.	1.89	SUF	FIBR. VITR. ASTRINGENT. EFFLORESCENCES IN SHELTERED PLACES	29.7.3.2
1.5	ISOCLASITE	MNCL.	2.92	PHO	PRISM,FIBR. VITR-PEARLY. 1 CLVG. SOL-ACIDS. RARE-WITH CHALCEDONY + DOLOMITE (JOACHIMOV, BOHEMIA)	42.4.5.
1.5	IODARGYRITE	HEXA.	5.69	HAL	EXPOS-YEL. MSV. PRISM-HEMIMORPHIC. RESIN. BASAL CLVG(FLEX). DCMP-H2SO4,HNO3.KI SOLUTION. OXIDIZED ZONES OF SILVER DEPOSITS	9.1.4.
1.5	CALOMEL	TETR.	7.15	HAL	EXPOS-DARKENS. TBLR, MSV. ADMN. PRISM CLVG. PLASTIC.FLUO-RED. SOL-AQ-REG. SECONDARY AFTER SEVERAL MERCURY MINERALS	9.1.5.
1.5-2	MIRABILITE	MNCL.	1.49	SUF	PRISM, ACICULAR, MSV CRUSTS. VITR. CLVGS. COOL, THEN SALTY + BIT-TER. LOSES H2O IN DRY AIR TO POWDER. EVAPORITE DEPS, EFFLORESCENCES	29.2.2.
1.5-2	SALAMMONIAC	ISOM.	1.53	HAL	SKELETAL, DENDRITIC. VITR.OCTAH CLVG. SALTY TASTE. VOL SUBLIMATE	9.1.2.
1.5-2	ALUNOGEN	TRCL.	1.77	SUF	FIBR MASSES. VITR-SILKY. SHARP, ACID TASTE. EFFLOR IN FES2-BEARING ROCKS	29.8.6.
1.5-2	SODA-NITER	HEXA.	2.24-2.29	NIT	INCRUST, RHOMB. VITR. RHOMB + OTHER CLVGS. COOLING TASTE. IN ARID SOILS IN PROTECTED LOCATIONS	18.1.1.
1.5-2	VIVIANITE	MNCL.	2.67-2.69	PHO	PRISM, ROUNDED, GLBL. VITR-DULL. CLVGS. FIBR-FLEX. DARKENS ON EX-POSURE. SOL-ACIDS. SCNDRY OF DIVERSE OCCURRENCES	40.2.15.1
1.5-2.5	HYDROPARAGONITE	MNCL.	2.6 -2.85	PSI	(=BRAMMELLITE). RELATED TO PARAGONITE AS ILLITE IS TO MUSCOVITE. ALTER(ALK CONDITIONS) OF MICA, FELDSPAR, ETC.IN SHALE WITH COAL	W370 3-213
2	TACHYHYDRITE	HEXA.	1.67	HAL	ROUNDED MASSES.VITR. RHOMB CLVG. DLQSNT. SHARP + BITTER.SED-K-RICH SALINE DEPOSITS	11.1.3.

NONMETALLIC COLORLESS

H	Mineral	Cryst.	S.G.	Class	Description	Ref
2	STRUVITE	ORTH.	1.71	PHO	DIVERSE XLS, MOST HEMIMORPHIC. VITR. 2 CLVGS. SOL-ACIDS, H2O(SLIGHTLY). GUANO DEPOSITS, URINARY SEDIMENT, ETC.	40.1.1.
2	BOUSSINGAULTITE	MNCL. A	1.72	SUF	CRUSTS, STALAC, PRISM. CLVG. FUMAROLIC, GEYSER DEPOSITS	29.3.7.3
2	INYOITE	MNCL.	1.88	BOR	TBLR, GRNLR. VITR. 2 CLVGS. SOL-HOT H2O, ACIDS. EVAPORITE DEPOSITS	25.1.13.
2	SYLVITE	ISOM.	1.98-2.00	HAL	CUBIC, GRNLR. VITR. CUBIC CLVG. BITTER TASTE. SED EVAPORITE DEPS	9.1.1.2
2	NITER	ORTH.	2.10-2.11	NIT	CRUSTS, GRNLR. VITR. CLVGS. SALTY, COOLING TASTE. EFFLOR-CAVES, ETC.	18.1.2.
2	MEYERHOFFERITE	TRCL.	2.12	BOR	ELONGATE, FIBR. VITR, SILKY. CLVGS. SOL-ACIDS. COLEMANITE DEPOSITS	25.1.12.
2	HALITE	ISOM.	2.17	HAL	CUBIC, GRNLR. VITR. CUBIC CLVG. SALTY TASTE. SED EVAPORITE DEPS	9.1.1.1
2	PORTLANDITE	HEXA.	2.23	OXD	PLTS, PEARLY. BASAL CLVG(FLEX). SOL-H2O, HCL. CNTCT META-LS, FUMAROLES	6.1.1.3
2	RHOMBOCLASE	ORTH.	2.23	SUF	THIN TBLR, STALACTITIC. SUBVITR. BASAL(FLEX), PRISM CLVGS. CONCH-FIBR. SOL-ACIDS, H2O(SLOWLY). RARE-ALTER PROD OF SULFIDES	29.1.1.
2	GYPSUM	MNCL.	2.31-2.32	SUF	TBLR, MSV. SUBVITR-PEARLY. 1 PERF CLVG(FLEX, INELAST), SOL-HCL. SEDIMENTARY BEDDED DEPOSITS, FUMAROLIC EFFLORESCENCES, ETC.	29.6.3.
2	VEATCHITE	MNCL.	2.58-2.69	BOR	PLTY, FIBR. VITR, PEARLY. 2 CLVGS. RARE-BORATE DEP(LANG, CALIF.)	25.1.8.
2	GLAUCONITE	MNCL.	2.4-2.95	PSI	PLTS IN ROUND GRAINS. ATTKD-HCL. MARINE DEPOSITS - GREENSANDS	M377 3-35
2	THOMSENOLITE	MNCL.	2.98	HAL	PRISM, OPALINE CRUSTS. VITR. BASAL, PYRAMIDAL CLVGS. SOL-H2SO4. AT CRYOLITE LOCALITIES	11.5.4.
2	LECONTITE	ORTH.		SUF	PRISM. VITR. BITTER, SALINE TASTE. RARE-IN BAT GUANO(HONDURAS)	29.2.1.
2	EPSOMITE	ORTH. A	1.68	SUF	CRUSTS. VITR-EARTHY. CLVGS. BITTER, SALTY, METALLIC. COMMON EFFLOR	29.6.9.1
2 -2.5	BORAX	MNCL.	1.71-1.72	BOR	HABIT-LK PYROXENE. VITR-EARTHY. CLVGS. SWEETISH. SOL-H2O ITO ALK SOLUTION). BORAX DEPOSITS (SALINE LAKE DEPOSITS, ETC.)	25.1.4.
2 -2.5	POTASH ALUM	ISOM.	1.76	SUF	MSV, STALAC. VITR. OCTAH CLVG. CONCH. SWEET, ASTRINGENT. EFFLOR	29.5.5.1
2 -2.5	MASCAGNITE	ORTH.	1.77	SUF	CRUSTS, STALACTITIC. VITR-DULL. 1 CLVG. SHARP, BITTER. FUMAROLIC	28.2.1.1
2 -2.5	ETTRINGITE	HEXA.	1.77	SUF	PRISM. PRISMATIC CLVG. SOL-ACIDS, DCMP-H2O TO ALK SOLUTION. META LS	31.3.2.
2 -2.5	GOSLARITE	ORTH.	1.98	SUF	CRUSTS, VITR-SILKY. 1 CLVG. ASTRINGENT, METALLIC, NAUCEOUS. EFFLOR-TYPICALLY BY ALTERATION OF SPHALERITE	29.6.9.2
2 -2.5	BOBIERRITE	MNCL. A	2.20	PHO	ACICULAR OR FIBR XLS, GLBL, ETC. VITR. 1 CLVG. SOL-ACIDS. GUANO, ETC.	40.2.16.
2 -2.5	ANAUXITE (CLAY)	TRCL.	2.52	PSI	HEX-SHAPED PLTS. BASAL CLVG. ALTER(ACID CONDITIONS) OF FELDSPAR, ETC.	W400 3-194
2 -2.5	PHARMACOLITE	MNCL.	2.53-2.73	ARS	FIBERS. VITR, PEARLY. 1 CLVG(FLEX). SOL-ACIDS. OXIDIZED AS-DEPS	39.2.1.2
2 -2.5	VILLIAUMITE	ISOM.	2.79	HAL	MSV, GRNLR. CUBIC CLVG. SOL-H2O. CAVITIES IN ALK IG RKS	9.1.1.3
2 -2.5	HAIDINGERITE	ORTH.	2.85	ARS	BOTRYOIDAL, FIBR COATINGS. VITR, PEARLY. 1 CLVG(FLEX). SOL-ACIDS. OXIDIZED PARTS OF ARSENIC DEPOSITS	39.2.2.
2 -2.5	PHLOGOPITE	MNCL.	2.76-2.90	PSI	MG-MICA. ATTKD-HCL. CNTCT META-LS. ULTRABASIC IGNEOUS ROCKS	M373 3-42
2 -2.5	SENARMONTITE	ISOM.	5.50	OXD	OCTAH, GRNLR. RESIN-ADMN. OCTAH CLVG. SOL-HCL. SCNDRY AFTER SB-MINS	4.4.2.2

101

NONMETALLIC COLORLESS

HARDNESS	NAME	XL. SYS.	SPECIFIC GRAVITY	CHEM. CLASS	REMARKS	REFERENCES
2 -3	ROESSLERITE	MNCL. A	1.94	ARS	HAIRLK CRUSTS. VITR-DULL. CLVG. SOL-DILUTE HCL,H2O(SLIGHT). OXI-DIZED ZONES OF ARSENIC DEPOSITS	39.2.5.1
2 -3	BETA SULFUR	MNCL.	1.96-1.98	ELE	LIGHT YEL-COLORLESS PURE, BROWN IMPURE. THICK TABULAR. FUMAROLES	1.2.3.2
2 -3	HYDROBORACITE	MNCL.	2.17	BOR	LAMELLAR,FIBR. VITR,SILKY. 2 CLVGS. SOL-ACIDS. EVAPORITE DEPOSITS	25.1.10.
2 -3	MANANDONITE (CLAY)	MNCL.	2.69	PSI	RARE-PEGMATITE IN MADAGASCAR	W395 N
2 -3	SHERIDANITE (CLAY)	MNCL.	2.65-2.80	PSI	INTERMEDIATE BETWEEN CLINOCHLORE + CORUNDOPHILITE. RARE-META ROCKS WN	3-137
2 -3	CLINOCHLORE (CHLORITE)	MNCL.	2.6 -3.0	PSI	HEX-OUTLINE (COMMONLY SLENDER)XLS,MSV. VERY COMMON,E.G.IN TALC RKS W383	3-137
2 -3	CHLORITE	MNCL.	2.6 -3.3	PSI	GROUP NAME. MEMBER NAMES UNSETTLED. ONLY FEW RARE VARIETIES MEGASCOPICALLY DIS- W381 TINGUISHABLE. DISSEM FLKS. PEARLY, 1 PERF CLVG TO FLEX,INELAST FOLIA. DCMP-IN BOILING H2SO4. SCNDRY-HYDROTHERMAL ALTER, LO-GRADE META; DETRITAL, AUTHIGENIC	3-131
2 -3	PHOSGENITE	TETR.	6.13	CBT	PRISM, MSV. ADMN. CLVGS(PERCUSSION FIGURE ON BASAL PLANE) FLUO-YEL (LW ULTRAVIOLET, CATHODE + X-RAYS). EFF-HNO3. SCNDRY AFTER 3B MINS	16.1.7.
2 -4	MICA	MNCL.	2.4 -3.4	PSI	GROUP NAME. SOME SPECIES DISTINGUISHABLE ON BASIS OF COLOR. SERIES ARE NOT COM- W365 PLETE. HEX-SHAPED XLS,DISSEM. PERF CLVG TO FLEX,ELAST FLKS. PERCUSSION + PRES- SURE FIGURES MAY BE IMPOSED ON FLKS. CHIEF OCCURRENCES ARE NOTED UNDER SPECIES	3-1
2.5	CARNALLITE	ORTH.	1.60	HAL	GRNLR, PSEUDOHEX PRISM. GRSY. CONCH. DLQSNT. BITTER. SED SALT DEPS	11.1.2.
2.5	LANSFORDITE	MNCL.	1.69	CBT	STALACTITIC, PRISM. VITR, EXPOS-DULL. 2 CLVGS. EFF-ACIDS. EFFLOR	15.1.5.
2.5	PHOSPHORROESSLERITE	MNCL.	1.72-1.73	PHO	EQUANT,SKELETAL. VITR,EXPOS-DULL, CONCH. UNSTABLE ABOVE CIRCA 25° C, 39.2.5.2 COLD (10°C) WET MUD OF MINE WORKINGS (SALZBG, AUSTRIA)	
2.5	NESQUEHONITE	ORTH.	1.85	CBT	PRISM-STRIATED, RADIAL AGGS. VITR. PYRAMID, BASAL CLVGS. SOL-CO2 CHARGED H2O. LO-T DEPOSITION IN OPEN SPACES	15.1.2.
2.5	KERNITE	MNCL.	1.91	BOR	EQUANT-STRIATED,MSV. VITR,SATINY. CLVGS. SOL-HOT H2O,ACIDS. BORATE DEPOSITS	25.1.2.
2.5	ULEXITE	TRCL.	1.96	BOR	NODULAR, CRUSTS. SILKY-VITR. CLVGS. DCMP-HOT H2O. PLAYA SALT LAKES	25.1.7.
2.5	CRYPTOHALITE	ISOM.	2.00	HAL	MAMMILLARY,DENDRITIC. VITR. OCTAH CLVG. SALTY TASTE. VOL SUBLIMATE	11.4.1.2
2.5	PICROMERITE	MNCL. A	2.03	SUF	PRISM, CRUSTS. VITR. BITTER. FUMAROLIC. MARINE SALINE DEPS-AS BEDS	29.3.7.1
2.5	PYROAURITE	HEXA.	2.10-2.14	OXD	TBLR-STRIATED. WAXY,PEARLY. BASAL CLVG(FLEX). EFF-HCL. LO-T VNS	6.1.5.3
2.5	DARAPSKITE	MNCL.	2.20	NIT	PSEUDO-TETRAG, TBLR. 2 CLVGS. SOL-H2O. IN NITRATE DEPOSITS	20.1.1.
2.5	UNGEMACHITE	HEXA.	2.29	SUF	TBLR. VITR. 1 CLVG. SOL-DILUTE HCL. VN-FILLINGS IN JAROSITE(CHU- QUICAMATA, CHILE)	31.4.2.
2.5	CLINO-UNGEMACHITE	MNCL.	2.29	SUF	MEGASCOPICALLY INDISTINGUISHABLE FROM UNGEMACHITE WITH WHICH IT OCCURS	31.4.3.
2.5	BRUSHITE	MNCL.	2.33	PHO	NEEDLELK,ETC. VITR,PEARLY. 2 CLVGS. SOL-HCL. PHOSPHATE DEPOSITS	39.2.1.1
2.5	SYNGENITE	MNCL.	2.58-2.60	SUF	CRUSTS, LAMELLAR AGGS, PRISM. VITR. CLVGS. SOL-H2O(IN PART WITH FM 29.3.1. OF GYPSUM). RARE-MARINE SALT DEPOSITS, PROD OF VOLCANIC ACTION	
2.5	FERRINATRITE	HEXA.	2.55-2.61	SUF	FIBR AGGS, PRISM, MSV. VITR. CLVGS. SPLINTERY. SCNDRY WITH OTHER IRON SULFATES IN ARID REGIONS	29.4.1.
2.5	HIERATITE	ISOM. A	2.67	HAL	CUBO-OCTAH, STALACTITIC. VITR. OCTAH CLVG. SOL-H2O. FUMAROLIC DEPS	11.4.1.1
2.5	PARAGONITE	MNCL.	2.85	PSI	NA-MICA. DISTINGUISHED FROM MUSCOVITE BY CHEM OR X-RAY ANAL. WIDE- W370 PREAD IN SCHISTS AND PHYLLITES, ALSO IN VEINS	3-31

NONMETALLIC COLORLESS

H	Mineral	Sys	SG	Cl	Habit / Properties	Ref
2.5	CRYOLITE	MNCL	2.96-2.98	HAL	MSV, GRNLR, GRSY, UNEVN, THEMOLUM. SOL-H2O(SLIGHTLY), RARE-PEGS	11.5.1.
2.5	ELPASOLITE	ISOM.	3.00	HAL	MSV, CUBO-OCTAH, VITR. UNEVN. RARE-CRYOLITE-BEARING PEGIEL PASO COUNTY, COLORADO)	11.5.2.
2.5	SZOMOLNOKITE	MNCL.	3.03-3.07	SUF	BIPYRAMIDAL,STALAC.VITR. CONCH.SOL-H2O(SLOW). EFFLOR-ACID SOLUTION	29.6.2.2
2.5	MINNESOTAITE	MNCL.	3.0-3.1	PSI	ESSENTIALLY FE-ANALOGUE OF TALC. FLKS, NEEDLES. FE-FMS(MINNESOTA)	M365 3-124
2.5	PYROCHROITE	HEXA.	3.23-3.27	OXD	TBLR, RHOMB, FOL MSV, PEARLY-DULL, BASAL CLVG(FLEX), SOL-HCL, WITH OTHER MN MINERALS IN LOW TEMPERATURE HYDROTHERMAL VEINS	6.1.1.2
2.5	NANTOKITE	ISOM.	4.14	HAL	MSV, GRNLR, ADMN. DODEC CLVG, DCMP-H2O,SOL-HCL,HNO3,NH4OH. CU-DEPS	9.1.3.1
2.5	CLAUDETITE	MNCL.	4.15	OXD	PLTY, VITR-PEARLY, 1 PERF CLVG(FLEX). SOL-ALK SOLS. SCNDRY AS-MIN	4.4.3.
2.5	CHLORARGYRITE	ISOM.	5.56	HAL	CRUSTS,CUBIC, RESIN. DUCTILE. EXPOS-VIOLET-BRWN. SCNDRY IN AG DEPS	9.1.1.4
2.5	MARSHITE	ISOM.	5.68	HAL	YEL STRK. TETRA-STRIATED, ADMN. FLUO-RED. RARE-CU DEPS	9.1.3.3
2.5	COTUNNITE	ORTH. A	5.80	HAL	GRNLR, LATHLK, ADMN-SILKY, 1 CLVG. SOL-H2O,ACIDS. VOL SUBLIMATE, ALTERATION PRODUCT OF GALENA	9.2.6.
2.5	SCHULTENITE	MNCL.	5.94	ARS	TBLR, STRIATED. VITR 1 CLVG. RARE-WITH ANGLESITE,ETC.(TSUMEB)	37.1.2.
2.5	BROMARGYRITE	ISOM.	6.50	HAL	CRUSTS,CUBIC, RESIN. DUCTILE. SECONDARY IN SILVER DEPOSITS	9.1.1.5
2.5	MENDIPITE	ORTH.	7.24	HAL	FIBR, RADIATED, SILKY,RESIN. PRISM + OTHER CLVGS. SOL-HNO3. SCNDRY WITH CERUSSITE, ETC.	10.1.4.
2.5	GAYLUSSITE	MNCL.	1.99	CBT	TBLR,STRIATED. VITR. PRISM CLVG. EFF-ACIDS. WITH TRONA,ETC.	15.2.3.
2.5-3	TRONA	MNCL.	2.14	CBT	FIBR,ELONGATE. VITR, 1 PERF,OTHER CLVGS. ALK TASTE.SOL-H2O, EFF-ACIDS. SALINE LAKE DEPS, EFFLORESCENT ON SOIL IN ARID REGIONS	13.1.4.
2.5-3	KAINITE	MNCL.	2.15	SUF	GRNLR, TBLR. VITR 1 CLVG. SALTY,BITTER. IN MARINE POTASH DEPS	31.4.1.
2.5-3	LEONITE	MNCL.	2.20	SUF	ANHEDRAL, TBLR. WAXY. CONCH. BITTER. SCNDRY IN MARINE SALINE DEPS	29.3.5.2
2.5-3	WHEWELLITE	MNCL.	2.23	CCC	HEART-SHAPED XLS, VITR,PEARLY. CLVGS. CONCH. SOL-ACIDS. ORGANIC ASSOCIATION, VEINS	50.1.1.
2.5-3	BLOEDITE	MNCL.	2.22-2.28	SUF	PRISM, GRNLR. VITR. CONCH. FAINTLY SALTY + BITTER. EVAPORITE DEPS	29.3.5.1
2.5-3	LOEWEITE	UNKN.	2.37-2.42	SUF	ANHEDRAL GRAINS. VITR. CONCH. BITTER, DISSEM IN MARINE SALINE DEPS	29.3.4.
2.5-3	FOSHALLASITE	UNKN.	2.5	SSI	PERF TBLR CLVG. SAME COMPOSITION AS AFWILLITE. RARE-CALCITE VNS	W478 N
2.5-3	THENARDITE	ORTH.	2.66	SUF	DIPYRAMIDAL, CRUSTS. VITR. CLVGS. FAINTLY SALTY. SALINE LAKE DEPS, EFFLORESCENCES, FUMAROLIC	28.2.4.
2.5-3	LANTHANITE	ORTH.	2.61-2.74	CBT	PLTY-TBLR, GRNLR, PEARLY, MICACEOUS CLVG. SOL-ACIDS. COATING ON CERITE, ALLANITE, ETC.	15.2.10.
2.5-3	CRYOLITHIONITE	ISOM.	2.77	HAL	DODEC, VITR. DODEC CLVG. SOL-H2SO4(+HF). WITH CRYOLITE	11.2.4.
2.5-3	GLAUBERITE	MNCL.	2.75-2.85	SUF	TBLR, PRISM, DIPYRAMIDAL-STRIATED, VITR,PEARLY. CLVGS. CONCH. BRTL. SALTY TASTE.SOL-HCL.SALINE DEPS.DISSEM IN CLASTIC SEDS, FUMAROLIC	28.4.2.2
2.5-3	SERICITE	MNCL.	2.77-2.88	PSI	FINE-GRAINED WHITE (NA OR K) MICA. DEUTERIC-IG RKS, META ROCKS	M369 3-14
2.5-3	VALENTINITE	ORTH.	5.76	OXD	PRISM, TBLR, MSV, ADMN. PRISM CLVG. SOL-HCL. SCNDRY AFTER SB-MINS	4.4.4.
2.5-3	ANGLESITE	ORTH.	6.37-6.39	SUF	TBLR,PRISM,MSV,ETC. ADMN, CLVGS, CONCH, BRTL, SOME FLUO-YEL. SOL-HNO3(DIFFICULTLY), SECONDARY — BY OXIDATION OF GALENA	28.3.1.3

NONMETALLIC COLORLESS

HARDNESS	NAME	XL. SYS.	SPECIFIC GRAVITY	CHEM. CLASS	REMARKS	REFERENCES
2.5-3	LEADHILLITE	MNCL.	6.55	CBT	PSEUDOHEX, GRNLR, RESIN, PEARLY, 1 CLVG. FLUO-YEL. EFF-HNO3, SCNDRY IN OXIDIZED ZONES OF LEAD DEPOSITS	17.1.3.
2.5-3	MATLOCKITE	TETR.	7.12	HAL	TBLR, PLTY AGGS, ETC. ADMN. 1 CLVG. SOL-HNO3, DCMP-H2SO4. PB-DEPS	10.1.6.1
2.5-3.5	SODA ALUM	ISOM.	1.67	SUF	CRUSTS, VITR. CONCH. SWEET,ASTRINGENT. REPORTED OCCURRENCES NEED VERIFICATION	29.5.5.2
2.5-3.5	MENDOZITE	MNCL.	1.73-1.77	SUF	EXPOS-WHT + TURBID. FIBR MASSES.3 CLVGS. SOL-H2O. REPORTED NATURAL OCCURRENCES NEED VERIFICATION	29.5.4.1
2.5-3.5	TAMARUGITE	MNCL.	2.07	SUF	TBLR, FIBR, GRNLR. VITR. 1 CLVG. SWEET, ASTRINGENT. OXIDATION PROD OF SULFIDES IN AL- + ALK-RICH ENVIRONMENTS UNDER ARID CONDITIONS	29.5.3.1
2.5-3.5	ENGLISHITE	MNCL.	2.64-2.66	PHO	SUBPARALLEL PLTS. VITR,PEARLY.1 CLVG(CURVED PLTS). IN ALTERED VAR-ISCITE NODULES	42.6.8.
2.5-4	MUSCOVITE	MNCL.	2.77-2.88	PSI	K-MICA. COMMON RK-FORMING MIN-E.G., ACID IG, PEGS, GREISENS, CNTCT META-LS METASOMATIZED(FLUORINE) ZONES, REGIONAL META-ARGILLACEOUS ROCKS	M367 3-11
2.5-4	LEPIDOLITE	MNCL.	2.80-2.90	PSI	LI-MICA, TABULAR XLS, AGGS OF SMALL FLAKES. DISTINGUISHED FROM ROSE-COLORED MUSCOVITE BY CHEM OR X-RAY ANAL. RARE, LI-BEARING PEGS	M370 3-85
3	INDERITE	TRCL.	1.86	BOR	ACICULAR-NODULAR AGGS. VITR,DULL. CLVGS. SOL-HCL. EVAPORITE DEPS	25.1.15.
3	HYDROCALUMITE	MNCL.	2.15	OXD	MSV. VITR, PEARLY. BASAL + POOR CLVGS. SOL-HCL. RARE-CNTCT META-LS	6.2.4.
3	FLUELLITE	ORTH.	2.14-2.17	HAL	DIPYRAMIDAL. VITR. 1 + POOR PYRAMIDAL CLVGS. VNS, PEGS	11.5.9.
3	RINNEITE	HEXA.	2.35	HAL	EXPOS-BRWN. GRNLR. SILKY. CONCH. ASTRINGENT. SCNDRY-SED SALT DEPS	11.4.3.
3	METAVAUXITE	MNCL.	2.35	PHO	PRISM,ACICULAR,//-RADIAL AGGS. VITR-SILKY. WX PROD(I.E.,SUPERGENE)	42.8.5.
3	PARAVAUXITE	TRCL.	2.36	PHO	PRISM,SUBPARALLEL-RADIAL AGGS. VITR,PEARLY. 1 CLVG. CONCH. WITH VAUXITE, ETC. IN TIN VEINS	42.8.6.
3	DAWSONITE	ORTH.	2.44	CBT	ACICULAR CRUSTS VITR-SILKY. CLVG. EFF-ACIDS. LO-T HYDROTHERMAL MIN	16.2.1.
3	MOOREITE	MNCL.	2.47	SUF	PLTY.VITR.1 CLVG. SOL-ACIDS. WITH ZINCITE,ETC.(STERLING HILL,N.J.)	31.1.3.
3	FERRUCCITE	ORTH.	2.50	HAL	TBLR XLS. 3 CLVGS. BITTER-ACID TASTE. FUMAROLIC SUBLIMATE.	11.2.3.
3	SHORTITE	ORTH.	2.60	CBT	WEDGE-SHAPED XLS. VITR. 1 CLVG. FLUO-AMBER. DCMP-H2O. WITH TRONA	14.3.2.
3	ARCANITE	ORTH.	2.66	SUF	TBLR-CYCLIC(6) TWINS, 2 CLVGS. RARE-IN PINE RR TIE(SANTA ANA MINE, ORANGE COUNTY, CALIFORNIA)	28.2.1.2
3	CALCITE	HEXA.	2.71	CBT	VARIOUS TRIGONAL XL HABITS, MSV. VITR-PEARLY, RHOMB CLVG. SOME IS FLUO,PHOSPHO,THERMOLUM. EFF-DILUTE HCL,WDSPRD-VNS,DIVERSE ROCKS	14.1.1.1
3	APHTHITALITE	HEXA.	2.65-2.71	SUF	TRIGONAL-TBLR, MAMMILLARY,MSV. VITR. CLVGS. CONCH. SOL-H2O, ACIDS. FUMAROLIC, OCEANIC AND LACUSTRINE SALINE DEPOSITS	28.2.2.
3	ZEOPHYLLITE	HEXA.	2.76	PSI	SPHERES, RADIAL STRUCTURE. BASAL CLVG. WITH ZEOLITES IN AMYGDULES	W359 N
3	PACHNOLITE	MNCL.	2.98	HAL	PRISM-STRIATED, MSV. VITR.1 CLVG. SOL-H2SO4.AT CRYOLITE LOCALITIES	11.5.3.
3	CAHNITE	TETR.	3.16	BOR	TWINS-E.G. CRUCIFORM. VITR. PRISM CLVG. SOL-HCL. VNS CUTTING MAIN ORE BODY AT FRANKLIN, NEW JERSEY	27.1.2.
3	ROEBLINGITE	ORTH.	3.43	NSI	DENSE FIBR MASSES. GELAT. CONTACT META DEPS (FRANKLIN, N. J.)	W512 N
3	GANOMALITE	HEXA.	5.74	SSI	PRISM XLS + CLVG. GELAT. CONSTITUTES SER WITH NASONITE. LANGBAN	W478 N
3	LARSENITE	ORTH.	5.9	NSI	STRIATED PRISMS. ADMN. GELAT. RARE(FRANKLIN, NEW JERSEY)	W504 N

NONMETALLIC COLORLESS

H	Mineral	Crystal	SG	Streak/Luster	Description	Ref
3	BARYSILITE	ORTH.	6.72	SSI	TARN-EXPOS. TBLR, CURVED LAMELLAR. BASAL CLVG. GELAT-HNO3. FRNKLN	W476 N
3 -3.5	TEEPLEITE	TETR.	2.08	BOR	TBLR. GLASSY, EXPOS-DULL. SOL-H2O(TO ALK SOLUTION).WITH TRONA	26.1.3.
3 -3.5	NEMBERYITE	ORTH.	2.10	PHO	EQUANT. VITR. 2 CLVGS. SOL-HCL,H2O(SCARCELY). GUANO DEPOSITS	39.2.3.
3 -3.5	LAUMONTITE (ZEOLITE)	MNCL.	2.2 -2.3	TSI	PERF CLVGS-3 DIRECTIONS IN 1 ZONE. GELAT. CAVITIES IN SEVERAL RKS	W342 4-401
3 -3.5	PIRSSONITE	ORTH.	2.35	CBT	PRISM, TBLR. VITR. PYROELECT. EFF-ACIDS. WITH TRONA,ETC.-SED DEPS.	15.2.2.
3 -3.5	HANKSITE	HEXA.	2.56	SUF	STRIATED PRISMS. VITR-DULL. 1 CLVG. SALTY. FLUO-YEL(LW). SOL-H2O. EFF-DILUTE ACIDS. SALINE LAKE DEPOSITS	32.1.1.
3 -3.5	HOPEITE	ORTH.	3.00-3.10	PHO	TBLR,PRISM,TUFTED AGGS,CRUSTS. VITR,PEARLY. CLVGS. SOL-DILUTE HCL. WITH HEMIMORPHITE, ETC.	40.2.11.
3 -3.5	PHOSPHOPHYLLITE	MNCL.	3.08-3.13	PHO	THICK TBLR. VITR. CLVGS. FLUO-VIOLET(SW). SOL-ACIDS. PEG-WITH FE, MN PHOSPHATES AND SPHALERITE	40.2.12.
3 -3.5	REDDINGITE	ORTH.	3.0 -3.2	PHO	MSV,GRNLR,FIBR,TBLR. VITR. 1 CLVG. UNEVN. SOL-ACIDS. PEGMATITES	40.2.5.1
3 -3.5	CELESTITE	ORTH.	3.96-3.98	SUF	TYPICALLY COLORLESS-BLUISH. TBLR,ETC. VITR-PEARLY. CLVGS. SOME FLUO, THERMOLUM. SOL-H2O(SLIGHTLY).SED RKS(ESP GYPSUM + LS),VNS	28.3.1.2
3 -3.5	WITHERITE	ORTH.	4.29	CBT	PSEUDOHEX, GRBLR,ETC. VITR. FLUO(ETC.)-LK ARAGONITE.SOL-HCL. VNS	14.1.3.2
3 -3.5	BARITE	ORTH.	4.50	SUF	TBLR, MSV,ETC. VITR-PEARLY. CLVGS. SOME FLUO, PHOSPHO, THERMOLUM. SOME FETID WHEN RUBBED. COMMON-VNS,SED RKS, CAVITIES IN IG RKS	28.3.1.1
3 -3.5	DESCLOIZITE	ORTH.	6.1 -6.3	VAN	VARIABLE-XLS TO COMPACT MSV. GRSY,COMMONLY COLOR ZONED. SOL-ACIDS. OXIDIZED ZONES OF VANADIUM-BEARING ORE DEPOSITS	41.5.2.1
3 -3.5	CERUSSITE	ORTH.	6.53-6.57	CBT	TWINNED CLUSTERS, MSV. ADMN-RESIN. CLVGS. FLUO-YELIX-RAYS +(LW)UL- TRA VIOLET). OXIDIZED ZONES OF LEAD DEPOSITS	14.1.3.4
3 -4	FAUJASITE (ZEOLITE)	ISOM.	1.91-1.93	TSI	OCTAHEDRAL XLS + CLVG. DCMP-ACID. RARE-LIMBURGITE(BADEN)	W333 4-400
3 -4	MORDENITE (ZEOLITE)	ORTH.	2.12-2.15	TSI	STRIATED PRISMS, ACICULAR, COTTON-LK AGGS, ZEOL-ASSOC. AUTHIGENIC	W339 4-401
3 -4	GISMONDINE (ZEOLITE)	ORTH.	2.1 -2.3	TSI	PSEUDOTETRAG BIPYRAMIDS. 1 CLVG. GELAT. LEUCITE RKS, ALTER OF PLAG	W339 4-401
3 -4	WAVELLITE	ORTH.	2.36	PHO	RADIAL FIBR,STELLATE,CRUSTS. VITR-PEARLY,CLVGS. SOL-ACIDS. SCNDRY- WDSPRD-OPENINGS IN DIVERSE ROCKS,LIMONITE/PHOSPHATE ROCKS,ETC.	42.7.4.
3 -4	WEINSCHENKITE	MNCL.	3.27	PHO	CRUSTS,RADIAL FIBR-ROSETTES. SILKY. CLVGS. THERMOLUM.SOL-HOT ACID. WITH LIMONITE OF NEAR-SURFACE FORMATION	40.3.3.
3 -4	MOLYBDOPHYLLITE	HEXA.	4.72	SSI	FOLIATED MASSES. 1 PERF CLVG(BASAL). WITH HAUSMANNITE AT LANGBAN	W479 N
3 -4	FIEDLERITE	MNCL.	5.88	HAL	LATHLK. ADMN. 1 CLVG. ATTKD-H2O,SOL-HNO3. RARE-ACTION OF SEA WATER ON ANCIENT LEAD SLAGS (LAURIUM, GREECE)	10.1.10.
3 -5.5	ZEOLITE		1.9 -2.45	TSI	GROUP OF MIN FAMILIES WITH SOMEWHAT RELATED APPEARANCES, OCCURRENCES, AND COMPOSITIONS, MANY ARE INDISTINGUISHABLE MEGASCOPICALLY. DIVERSE HABITS(E.G., BUNDLES, FIBROUS), PEARLY-GLASSY, SUBCONCH. FLUO-ORANGE OR YEL-GREEN. CONTINUOUS, IN PART REVERSIBLE, DEHYDRATION(I.E.,WATER CAN BE EXPELLED WITH-OUT DESTROYING XL STRUCTURE). SCNDRY IN OPEN SPACES; COMMONLY TWO OR MORE OCCUR TOGETHER IN CAVITIES IN BASALTIC RKS - THIS OCCURRENCE IS REFERRED TO HEREIN AS THE ZEOLITE ASSOCIATION (ABBREVIATED "ZEOL-ASSOC").	W330 4-351
3.5	LAURIONITE	ORTH. A	6.24	HAL	TBLR. ADMN,PEARLY. CLVG. SOL-HOT H2O, HNO3. SCNDRY AFTER PB-MINS	10.1.7.
3.5	THAUMASITE	HEXA.	1.91	NSI	FILIFORM-MSV. HARDENS ON EXPOS. ATTACKED-ACIDS. SCNDRY-CAVITIES	W179 N
3.5	INDERBORITE	MNCL.	2.00	BOR	COARSE AGGS. VITR. 1 CLVG. SOL-HCL, H2O. BORATE DEPOSITS	25.1.11.
3.5	PROBERTITE	MNCL.	2.14	BOR	RADIAL GROUPS OF NEEDLES OR LATHS. VITR. PRISM CLVG. SOL-ACIDS. BORATE DEPOSITS	25.1.6.

NONMETALLIC COLORLESS

HARDNESS	NAME	XL. SYS.	SPECIFIC GRAVITY	CHEM. CLASS	REMARKS	REFERENCES
3.5	HYDROMAGNESITE	MNCL.	2.24	CBT	TUFTS, CRUSTS, MSV. VITR-EARTHY. EFF-ACIDS VNS IN ULTRABASIC IGS	16.1.13.
3.5	MINYULITE	ORTH.	2.45	PHO	RADIAL AGGS OF NEEDLES. SILKY. SOL-HOT ACIDS. ALTER PROD OF GLAU-CONITE	42.8.3.
3.5	SULFOHALITE	ISOM.	2.50	SUF	DODEC, OCTAH. VITR. CONCH. SALTY. SOL-H2O(SLOWLY). SALINE DEPOSITS	30.1.6.
3.5	KIESERITE	MNCL.	2.57	SUF	MSV,GRNLR. VITR. CLVGS. FRIABLE. SOL-H2O(SLOW). MARINE SALINE DEPS	29.6.2.1
3.5	SCHAIRERITE	HEXA.	2.61	SUF	RHOMB. VITR. CONCH. SOL-H2O(SLOW). WITH TRONA,ETC.-EVAPORITE DEPS	30.1.5.
3.5	VANTHOFFITE	UNKN.	2.69	SUF	MSV, GRNLR. VITR-PEARLY. UNEVN. FRIABLE. MARINE SALT DEPOSITS	28.4.1.
3.5	POLYHALITE	TRCL.	2.78	SUF	MSV-FIBR, FOL. VITR. 1 CLVG. DCMP-H2O(↓GYP). MARINE SALINE DEPS	29.4.2.
3.5	FLUOBORITE	HEXA.	2.85-2.98	BOR	ACICULAR, FELTED AGGS. 1 POOR CLVG. CNTCT METASOMATIC DEPOSITS	26.1.1.
3.5	ANHYDRITE	ORTH.	2.98	SUF	MSV, GRNLR, FIBR. VITR-PEARLY. 2 CLVGS. SOL-ACIDS. SED ROCKS	28.3.2.
3.5	HUREAULITE	MNCL.	3.15-3.20	PHO	TBLR,MSV,SCALY,FIBR. VITR. 1 CLVG. SOL-ACIDS. WITH ALTERED LITHI-OPHILITE AND/OR TRIPHYLITE	39.1.3.
3.5	BRANDTITE	MNCL.	3.67	ARS	STOUT PRISM,RADIAL FIBR GRPS. VITR. 1 CLVG. SOL-DILUTE ACIDS. VNS	40.2.4.2
3.5	CHALCOCYANITE	ORTH.	3.60-3.70	SUF	TBLR. HYGROSCOPIC TO BLUE PENTAHYDRATE OR CHALCANTHITE. FUMAROLIC	28.3.3.
3.5	STRONTIANITE	ORTH.	3.64-3.78	CBT	SPEAR-LK, MSV. VITR. CLVGS. FLUO(ETC.)-LK ARAGONITE. SOL-HCL. VNS	14.1.3.3
3.5	ADAMITE	ORTH.	4.32-4.48	ARS	ELONGATE,CRUSTS,RADIAL AGGS. VITR. CLVGS. UNEVN. SOL-DILUTE ACIDS. SOME FLUO YEL.OXIDIZED ZONES OF DEPS WITH ZN- + AS-RICH MINERALS	41.6.5.3
3.5	HYDROCERUSSITE	HEXA.	6.80	CBT	SCALY,TBLR.ADMN,PEARLY.BASAL CLVG. EFF-ACIDS. SCNDRY AFTER PB MINS	16.1.12.
3.5-4	HEULANDITE (ZEOLITE)	MNCL.	2.1 -2.2	TSI	PLTY GRP. TBLR-BROAD CENTRAL PORTIONS(COFFIN-LIKE). ZEOL-ASSOC	W347 4-377
3.5-4	STILBITE (ZEOLITE)	MNCL.	2.1 -2.2	TSI	PLTY GRP. SHEAF-LIKE AGGS. 1 GOOD CLVG. DCMP-HCL. ZEOL-ASSOC	W345 4-377
3.5-4	NORTHUPITE	ISOM.	2.38	CBT	OCTAH. VITR. CONCH. EFF-ACIDS,DCMP-HOT H2O. SALINE CLAY DEPS	16.2.2.
3.5-4	OVERITE	ORTH.	2.53	PHO	PLTY,SUBPARALLEL PLTY AGGS. VITR.2 CLVGS. SOL-HOT HNO3. IN ALTERED VARISCITE NODULES	42.8.12.
3.5-4	CREEDITE	MNCL.	2.71-2.73	HAL	PRISM, RADIATING AGGS. VITR.1 CLVG. SOL-ACIDS(SLOWLY).FLUORITE VNS	12.1.1.
3.5-4	METASTRENGITE	MNCL.	2.76	PHO	TBLR,BOTRYOIDAL,RADIAL FIBR CRUSTS. VITR. 2 CLVGS. SOL-HCL. WITH STRENGITE	40.3.2.2
3.5-4	LANGBEINITE	ISOM.	2.83	SUF	NODULAR, DISSEM. VITR.CONCH. SOL-H2O(SLOWLY). MARINE SALTS	28.4.3.1
3.5-4	DOLOMITE	HEXA.	2.84-2.86	CBT	RHOMB-CURVED, GRNLR. VITR-PEARLY. RHOMB CLVG. SOME FLUO. EFF-WARM HCL IF POWDERED. SEDIMENTARY ROCKS, CRYSTALS IN CAVITIES, ETC.	14.2.1.1
3.5-4	STRENGITE	ORTH.	2.5 -2.87	PHO	TBLR,PSEUDO-OCTAH,BOTRYOIDAL AGGS,ETC. WAXY. 2 CLVGS. SOL-HCL. ALTERATION PRODUCT OF FE-BEARING PHOSPHATES	40.3.1.2
3.5-4	CHIOLITE	TETR.	3.00	HAL	GRNLR. 1 PERF. OTHER CLVGS. RESEMBLES CRYOLITE. IN CRYOLITE PEGS	11.5.8.
3.5-4	SCORODITE	ORTH.	3.28	ARS	PYRAMIDAL,ETC.-AGGS,CRUSTS,ETC. VITR. CLVGS. SOL-ACIDS. SCNDRY-AFTER ARSENOPYRITE AND OTHER AS-BEARING MINERALS	40.3.1.3

NONMETALLIC COLORLESS

H	Mineral	Crystal	SG	Class	Description	Ref
3.5	GORDONITE	TRCL.	2.23	PHO	SHEA-FLK AGGS. VITR,PEARLY. CLVGS. CONCH. SOL-ACIDS. IN VARISCITE NODULES	42.8.8.
3.5-4	PARAHOPEITE	TRCL.	3.31	PHO	ELONGATE,TUFTS. VITR,PEARLY. 1 CLVG. SCNDRY-WITH HEMIMORPHITE,ETC.	40.2.10.
3.5-4	TARBUTTITE	TRCL.	4.12	PHO	CRUSTS,SHEA-FLK AGGS OF ROUNDED,STRIATED,EQUANT XLS. VITR,PEARLY. 1 CLEAVAGE. OXIDIZED ZONES OF ZINC DEPOSITS	41.6.7.
3.5-4	LAUTARITE	MNCL.	4.59	IOD	RADIAL AGGS-STRIATED PRISMS.CLVGS.SOL-HCL. GYPSUM BANDS-NITER DEPS	21.1.1.
3.5-4	PYROMORPHITE	HEXA.	7.00-7.08	PHO	PRISM,GLBL,ETC. RESIN. RHOMB CLVG. UNEVN. SOL-HNO3. OXIDIZED ZONES OF LEAD DEPOSITS	41.7.2.1
3.5-4	MIMETITE	HEXA.	7.24	ARS	PRISM,ACICULAR,GLBL,ETC. RESIN. RHOMB CLVG. UNEVN. SOL-HNO3. OXI-DIZED ZONES OF LEAD DEPOSITS	41.7.2.2
3.5-4.5	MAGNESITE	HEXA.	2.98-3.02	CBT	GRNLR, RHOMB. VITR. RHOMB CLVG. FLUO-GREEN,BLUE. SOL-WARM HCL. AL-TERATION OF MG-RICH ROCKS (E.G., SERPENTINE)	14.1.1.2
3.5-6	CLINTONITE (BRITTLE MICA)	MNCL. 3	3.1	PSI	WITH TALC-CHLORITE SCHIST, WITH SPINEL IN METASOMATIZED CALC RKS. DISTINGUISHED FROM XANTHOPHYLLITE BY OPTICAL OR X-RAY ANALYSES	W391 3-99
3.5-6	XANTHOPHYLLITE (BRITTLE MICA)	MNCL. 3	3.1	PSI	WITH TALC-CHLORITE SCHIST, WITH HUMITES IN METASOMATIZED CALC RKS. DISTINGUISHED FROM CLINTONITE BY OPTICAL AND/OR X-RAY ANALYSES	W391 3-99
3.5-7	BRITTLE MICA	MNCL.	2.6-2.85	PSI	GROUP OF MORE OR LESS RELATED MINS THAT RESEMBLE MICAS BUT CLEAVE TO BRTL FLKS AND HAVE HARDNESSES OF 3.5-4 VS. 2-3(MICA) ON CLVG FLKS AND 6 VS. 4 ⊥ TO FLKS. WITH CORUNDUM + DIASPORE IN EMERY DEPS, CHLORITE + MICA SCHISTS, SKARN	W392 3-95
4	WEDDELLITE	TETR.	1.93-1.95	CCC	PYRAMIDAL,AGGS. SUBCONCH. ORGANIC ASSOC, BOTTOM MUDS(WEDDELL SEA)	50.1.2.
4	EPISTILBITE (ZEOLITE)	MNCL.	2.1-2.3	TSI	PLTY GRP. SHEAF-LK + SPHERICAL AGGS. DCMP-HCL.PIEZOELECTRIC. ZEOL. ASSOC	W346 4-377
4	MONTGOMERYITE	MNCL.	2.48-2.58	PHO	LATHS,SUBPARALLEL PLTY AGGS. VITR. 2 CLVGS. IN ALTERED VARISCITE	42.8.11.
4	AFWILLITE	MNCL.	2.63	NSI	ELONGATE XLS. BASAL CLVG. SOL-HCL. DIAMOND PIPES + AFTER SPURRITE	W478 N
4	FLUORITE	ISOM.	3.18	HAL	CUBIC, MSV. VITR. OCTAH CLVG. FLUO-VIOLET. DCMP-H2SO4. WDSPRD-VNS, CAVITIES, DISSEMINATED IN IGNEOUS ROCKS, ETC.	9.2.1.
4	BARYTOCALCITE	MNCL.	3.66-3.71	CBT	PRISM-STRIATED, MSV. VITR. CLVG. WEAKLY FLUO. SOL-HCL. LO-T DEPS	14.2.3.
4	KALIBORITE	MNCL.	2.13	BOR	XL AGGS, GRNLR. VITR. CLVGS. SOL-ACIDS. SALINE EVAPORITE DEPOSITS	25.1.22.
4-.5	PHILLIPSITE (ZEOLITE)	MNCL.	2.2	TSI	PENETRATION TWINS, SPHERULITES. 2 GOOD CLVGS., GELAT. DISTINGUISHED FROM HARMOTOME BY OPTICAL METHODS. ZEOL-ASSOC,DEEP-SEA RED CLAYS	W343 4-386
4-.5	SULFOBORITE	ORTH.	2.38-2.45	BOR	PRISM. PRISM + BASAL CLVGS. BRTL. SOL-ACIDS,DCMP-H2O. SALT DEPS	27.1.3.
4-.5	VARISCITE	ORTH.	2.2-2.57	PHO	NODULES,CRUSTS-SOME OPALINE, WAXY. 2 CLVGS. SPLINTERY-CONCH. SOL-ALKALIES. NEAR-SURFACE DEPOSITION IN CAVITIES(E.G.,BRECCIAS)	40.3.1.1
4-.5	JEZEKITE	MNCL.	2.94	PHO	PRISM-STRIATED,FIBR CRUSTS,AGGS. VITR. 2 CLVGS. DCMP-HOT H2SO4. DRUSES IN GRANITE (SAXONY, GERMANY)	41.2.3.
4-.5	ALSTONITE	ORTH.	3.67-3.71	CBT	PSEUDOHEX, STRIATED. VITR. FLUO-YEL(LW). SOL-HCL. LO-T DEPS	14.2.2.
4-.5	JARLITE	MNCL.	3.78-3.93	HAL	TBLR, RADIAL AGGS. VITR. SOL-ALCL3 SOLUTION. RARE-IVIGTUT,GREENLD.	11.5.5.
4-.5	AUSTINITE	ORTH.	4.13	ARS	BLADED,FIBR CRUSTS + NODULES. SUBADMN,SILKY. 1 CLVG. SCNDRY-RARE	41.5.1.3
4-.5	SMITHSONITE	HEXA.	4.40-4.45	CBT	BOTRYOIDAL, MSV. VITR-PEARLY. RHOMB CLVG. SOME FLUO-GREEN, BLUE. EFF-ACIDS. SCNDRY-OXIDIZED ZONES OF ZINC DEPOSITS	14.1.1.6
4-5	FRIEDELITE	HEXA.	3.06-3.19	PSI	TBLR, MSV. BASAL CLVG. MN-RICH ORE DEPS (E.G., FRNKLN, N. J.)	W359 N
4-5	SVABITE	HEXA.	3.5-3.8	ARS	SHORT PRISM,MSV. VITR.CLVG. SOL-DILUTE ACIDS.RARE-E.G.,FRNKLN,N.J.	41.7.3.1
4.5	CHABAZITE (ZEOLITE)	HEXA.	2.05-2.10	TSI	RHOMBS(NEARLY CUBES). RHOMB CLVG. DCMP-HCL.ZEOL-ASSOC, RK CAVITIES	W334 4-387
4.5	GMELINITE (ZEOLITE)	HEXA.	2.0-2.2	TSI	PRISM CLVG. DCMP-HCL. ZEOL-ASSOC	W335 4-387

NONMETALLIC COLORLESS

HARDNESS	NAME	XL. SYS.	SPECIFIC GRAVITY	CHEM. CLASS	REMARKS	REFERENCES
4.5	LEVYNITE (ZEOLITE)	ORTH.	2.0 -2.2	TSI	CHEM OR X-RAY ANAL DISTINGUISH CHABAZITE +LEVYNITE.RARE-ZEOL-ASSOC	W335 4-387
4.5	COLEMANITE	MNCL.	2.42-2.43	BOR	EQUANT, GRNLR. VITR. 2 CLVGS. SOL-HOT HCL. EVAPORITE BORATE DEPS	25.1.9.
4.5	HARMOTOME (ZEOLITE)	MNCL.	2.41-2.47	TSI	HABIT LK PHILLIPSITE(ABOVE), + CRUCIFM TWINS. DCMP-HCL. ZEOL-ASSOC, ALSO WITH AL-WXING PRODS,IN GNEISSES, + WITH MN-MINERALIZATION	W344 4-386
4.5	RALSTONITE	ISOM.	2.56-2.62	HAL	OCTAH. VITR. OCTAH CLVG. DCMP-HF. RARE AT SOME CRYOLITE LOCALITIES	11.5.10.
4.5	WOODHOUSEITE	HEXA.	3.01	PHO	PSEUDOCUBIC,TBLR -CURVED,STRIATED. VITR,PEARLY. 1 CLVG. VUGS,ETC.	43.1.1.5
4.5	FILLOWITE	MNCL.	3.43	PHO	GRNLR,PSEUDORHOMB. GRSY. 1 CLVG. SOL-ACIDS. PEG-WITH TRIPLOIDITE, ETC.	40.2.2.
4.5	HINSDALITE	HEXA.	3.65	PHO	TBLR,PSEUDOCUBIC,MSV,GRNLR. VITR.1 CLVG. RARE-VN(HINSDALE CO,COLO)	43.1.1.3
4.5	SYNADELPHITE	TRCL.	3.57-3.79	ARS	COMMONLY COLOR ZONED. PSEUDOPRISMS, MSV. VITR. 1 CLVG. UNEVN. SOL-ACIDS. RARE WITH OTHER MANGANESE MINERALS	41.1.5.
4.5	CORDYLITE	HEXA.	4.31	CBT	PRISM, DIPYRAMIDAL. GRSY,PEARLY. 1 CLVG. SOL-ACIDS. ALK PEGS	16.2.7.
4.5	CARACOLITE	UNKN.	5.0 -5.2	SUF	CRUSTS OF PSEUDOHEX XLS. VITR. RARE-WITH BINDHEIMITE,ETC.(CHILE)	30.1.3.
4.5	ALAMOSITE	MNCL.	6.49	ISI	FIBR. ADMN. CLVG ⊥ LENGTH. GELAT-HNO3. VNS, ORE-DEP (ALAMOS,MEXICO)	W455 N
4.5-5	APOPHYLLITE	TETR.	2.33-2.37	PSI	VITR SQ PRISMS, PEARLY BASES. BASAL + PRISM CLVG. DCMP-HCL(SILICA RESIDUE). SECONDARY IN CAVITIES, COMMONLY WITH ZEOLITES, ETC.	W394 3-258
4.5-5	AUGELITE	MNCL.	2.70	PHO	TBLR,ACICULAR,MSV. VITR,PEARLY. CLVGS. SOL-HCL(SLOW). VNS,ETC.	41.6.8.
4.5-5	SCAWTITE	MNCL.	2.77	ISI	PLATY. VITR. 1 PERF, 1 POOR CLVG. ATTKD-HCL. CNTCT META-LIMESTONE	W394 N
4.5-5	PECTOLITE	TRCL.	2.86-2.90	ISI	RADIATING AGGREGATES. 2 PERF CLVGS. MOST FLUO-ORANGE(LW). WITH ZEO-LITES IN CAVITIES	W460 2-176
4.5-5	WOLLASTONITE	TRCL.	2.87-3.09	ISI	TBLR, MSV. CLVGS-84°+ 96°.DCMP-HCL. FLUO-YEL-ORANGE. CNTCT META-LS	W456 2-167
4.5-5	GOYAZITE	HEXA.	3.26	PHO	RHOMB-STRIATED,ROUNDED GRAINS. GRSY,PEARLY.1 CLVG.SOL-ACIDS(SLOW). DIAMOND SANDS, CREVICES	41.5.8.3
4.5-5	SCHEELITE	TETR.	6.08-6.12	WOS	DIPYRAMIDAL,MSV,GRNLR,VITR.CLVGS,UNEVN.DCMP-HCL,HNO3,FLUO-BLUE(X-,SW ULTRAVI-OLET--,+ CATHODE-RAYS),THERMOLUM. HI-T VNS,PEGS,GREISSENS,CNTCT META RKS,ETC.	48.1.3.1
4.5-5.5	LEWISTONITE	HEXA.	3.08	PHO	FIBR CRUSTS,PRISMS-SUB // SHEAFLK GROUPS.CLVGS. WITH DEHRNITE(FAIR-FIELD, UTAH)	41.7.5.
4.5-5.5	DEHRNITE	HEXA.	3.04-3.09	PHO	BOTRYOIDAL CRUSTS,PRISMS-SUB //SHEAFLK GROUPS. VITR. 1 CLVG. SOL-ACIDS. WITH OTHER PHOSPHATE MINERALS	41.7.4.
4.5-5.5	LEGRANDITE	MNCL.	3.96-4.06	ARS	RADIATING AGGS OF PRISMS. 1 CLVG. RARE-1 SPECIMEN(WITH SPHALERITE)	42.7.1.
5	NATROLITE (ZEOLITE)	ORTH.	2.20-2.26	TSI	FIBR GRP. SQ NEEDLES. PRISM CLVG. GELAT. PYROELECTRIC. ZEOL-ASSOC	W340 4-358
5	MESOLITE (ZEOLITE)	MNCL.	2.25-2.27	TSI	FIBR GRP. TUFTS OF HAIR-LIKE XLS. GELAT. PYROELECTRIC. ZEOL-ASSOC	W341 4-358
5	SCOLECITE (ZEOLITE)	MNCL.	2.25-2.29	TSI	FIBR GRP. THIN STRIATED PRISMS. PRISM CLVG. GELAT. ZEOL-ASSOC, AL-SO HYDROTHERMAL + IN METAMORPHOSED CALCAREOUS ROCKS	W341 4-358
5	GONNARDITE (ZEOLITE)	ORTH.	2.2 -2.4	TSI	FIBR GRP. RADIATED SPHERICAL MASSES. GELAT. RARE. ZEOL-ASSOCIATION	W337 4-359
5	STERRETTITE	ORTH.	2.44	PHO	PRISM-SOME STRIATED. VITR. CLVGS. IN ALTERED VARISCITE NODULES	42.7.5.

NONMETALLIC COLORLESS

H	Mineral	Crystal	SG	Group	Notes	Ref
5	HILGARDITE	MNCL.	2.71	BOR	TBLR, HEMIMORPHIC. VITR. 2 CLVGS. SALT DEPS(E.G., LOUISIANA)	26.1.8.
5	PARAHILGARDITE	TRCL.	2.71	BOR	INTERGROWN WITH HILGARDITE(SEE). SALT DEPS(E.G., LOUISIANA)	26.1.9.
5	WARDITE	TETR.	2.81-2.87	PHO	PYRAMIDAL-STRIATED;CRUSTS-RADIAL FIBR,ETC. VITR.1 CLVG. SOL-ACIDS (DIFFICULTLY).WITH MILLISITE IN VARISCITE-AFTER AMBLYGONITE--PEGS	42.5.4.
5	PSEUDOWOLLASTONITE	TRCL.	2.91	ISI	PSEUDO-ORTHORHOMBIC. EQUANT. BASAL CLVG. HI-T THERMAL META-CALC RK	W456 2-167
5	SPURRITE	MNCL.	3.	NSI	2 CLVGS AT 79° EFF +GELAT-HCL. ALTERS TO AFWILLITE. CNTCT META LS	W516 N
5	HYDROXYLAPATITE	HEXA.	2.9 -3.1	PHO	APATITE GROUP(SEE FOR PROPS.) RARE-TALC SCHIST	41.7.1.3
5	CARBONATE-APATITE	HEXA.	2.9 -3.1	PHO	APATITE GROUP(SEE FOR PROPS.)RARE-NODULES(PHOSPHATE RK),GUANO,ETC.	41.7.1.4
5	WHITLOCKITE	HEXA.	3.12	PHO	RHOMB, GRNLR. VITR. UNEVN. SOL-DILUTE ACIDS. PEGS, PHOSPHATE ROCKS	38.3.1.
5	SELLAITE	TETR.	3.15	HAL	PRISM, ACICULAR, FIBR. CLVGS. DCMP-H2SO4. VNS WITH NATIVE S, ETC.	9.2.2.
5	CHLORAPATITE	HEXA. A	3.17	PHO	APATITE GROUP(SEE FOR PROPS.) RARE-VNS IN GABBRO, METEORITES	41.7.1.2
5	FLUORAPATITE	HEXA. A	3.18	PHO	APATITE GROUP(SEE FOR PROPS.) COMMON-IGS, METAS, MARINE SEDS, ETC.	41.7.1.1
5	APATITE	HEXA.	2.9 -3.2	PHO	GROUP NAME. MOST TYPES INDISTINGUISHABLE MEGASCOPICALLY. HEX PRISMS-COMMONLY ROUNDED,MSV,GRNLR. VITR. SOL-ACIDS. SOME FLUO--YEL-ORANGE. IG RKS,VNS,PEGS	41.7.1.
5	SVANBERGITE	HEXA.	3.22	PHO	RHOMB,PSEUDOCUBIC,GRNLR,VITR-ADMN,1 CLVG. DIVERSE META RKS,PLACERS	43.1.1.4
5	HEMIMORPHITE	ORTH.	3.45	SSI	HEMIMORPHIC XLS, SHEAFLK, MAMMILLARY, ETC. 1 PERF + 1 POOR CLVG. GELAT. PYROELECTRIC.SOME FLUO-PALE ORANGE(LONG WAVE). ZINC DEPOSITS	W481 N
5	CALAMINE	ORTH.	3.45	SSI	(=HEMIMORPHITE)	W481 N
5	TITANITE(=SPHENE)	MNCL.	3.45-3.55	NSI	DOUBLE WEDGE-SHAPED XLS, MSV-GRNLR. ADAMANTINE. ALTERS TO LEUCOX-	W525 1-69
5	SPHENE	MNCL.	3.45-3.55	NSI	ENE. DCMP-H2SO4. COMMON ACCESSORY CONSTITUENT OF IG + META ROCKS	W525 1-69
5	ADELITE	ORTH.	3.70-3.76	ARS	MSV,ELONGATE. RESIN. CONCH. SOL-DILUTE ACIDS. MN DEPOSITS	41.5.1.1
5	SANBORNITE	ORTH.	4.19	PSI	1 PERF, 1 POOR CLVG. POLYSYNTHETIC TWINNING. WITH QTZ,CELSIAN,ETC.	W358 N
5 -5.5	THOMSONITE (ZEOLITE)	ORTH.	2.10-2.39	TSI	FIBR GRP. RADIATED SPHERICAL MASSES. GELAT. PYROELECT. ZEOL-ASSOC	W336 4-359
5 -5.5	EDINGTONITE (ZEOLITE)	TETR.	2.7 -2.8	TSI	FIBR GRP. PSEUDOTETRAGONAL. PRISM CLVG. GELAT. PYROELECTRIC. ZEOL-ASSOC, ORE DEPOSITS	W338 4-359
5 -5.5	HERDERITE	MNCL.	2.95	PHO	PRISM,TBLR,BOTRYOIDAL-RADIAL FIBR. VITR. CLVG. SOL-ACIDS. PEGS-LATE STAGE	41.5.4.1
5 -5.5	DATOLITE	MNCL.	2.96-3.00	NSI	TINTED WHT-COLORLESS. GLASSY. GELAT. SCNDRY WITH ZEOLITES-CAVITIES	W355 1-171
5 -5.5	HYDROXYL-HERDERITE	MNCL.	3.01	PHO	PRISM,TBLR,BOTRYOIDAL-RADIAL FIBR. VITR. CLVG. SOL-ACIDS. PEGS-LATE STAGE	41.5.4.2
5 -5.5	HYALOTEKITE	ORTH.	3.8	ISI	MSV, COARSELY XLINE. 2 CLVGS-90°. ORE DEP(LANGBAN, SWEDEN)	W401 N
5 -6	VISHNEVITE	HEXA.	2.32-2.42	TSI	SULFATE-RICH END MEMBER OF CANCRINITE FAMILY. OCCURRENCE LIKE CANCRINITE. CARBONATE-RICH MEMBERS EFFERVESCE AS WELL AS GELATINIZE WITH HCL	WN 4-310
5 -6	DAVYNE	HEXA.	2.42-2.5	TSI	VARIETY OF CANCRINITE (ACCORDING TO SOME). RARE-VESUVIUS, ITALY	W354 N

NONMETALLIC COLORLESS

HARDNESS	NAME	XL. SYS.	SPECIFIC GRAVITY	CHEM. CLASS	REMARKS	REFERENCES
5 -6	CANCRINITE	HEXA.	2.42-2.51	TSI	MSV (RARE PRISM XLS). PRISM CLVG. EFF-HCL. NEPHELINE-BEARING RKS	W354 4-310
5 -6	MARIALITE (SCAPOLITE)	TETR.	2.50-2.62	TSI	ME-0-20. PURE MARIALITE UNKNOWN IN NATURE	W352 4-321
5 -6	DIPYRE (SCAPOLITE)	TETR.	2.57-2.69	TSI	ME-20-50.	W352 4-321
5 -6	MIZZONITE (SCAPOLITE)	TETR.	2.67-2.74	TSI	ME-50-80.	W352 4-322
5 -6	SCAPOLITE	TETR.	2.50-2.78	TSI	GROUP NAME. MOST TYPES INDISTINGUISHABLE MEGASCOPICALLY BUT S.G. MAY BE INDICATIVE. LONG,STRIATED PRISMS + AGGS OF COARSE XLS. SUBCONCH. POOR C-VG GIVING IRREGULAR, STRIATED-APPEARING SURFACES. MARIALITE IS INSOL; MEIONITE IS DCMP IN HCL. MOST FLUO--YEL-RED(ULTRAVIOLET,LW). CNTCT META CALC RKS,PEGS	W352 4-321
5 -6	(WERNERITE) (SCAPOLITE)	TETR.	2.50-2.78	TSI	GROUP NAME NO LONGER USED	W352 4-322
5 -6	MEIONITE (SCAPOLITE)	TETR.	2.73-2.78	TSI	ME-80-100. PURE MEIONITE UNKNOWN IN NATURE	W352 4-321
5 -6	AKERMANITE	TETR.	2.94	SSI	ISOMORPHOUS WITH GEHLENITE + MELILITE. RARE IN NATURE, IN SLAGS	W473 1-236
5 -6	CUSPIDINE	MNCL.	2.95	SSI	SPEAR-SHAPED XLS. BASAL CLVG. SOL-HNO3. CONTACT METAMORPHIC ZONES	W480 N
5 -6	GEHLENITE	TETR.	3.04	SSI	SHORT SQ PRISMS. VITR-RESIN.CONCH.GELAT. CNTCT META-IMPURE C03-RKS	W473 1-236
5 -6	ENSTATITE (ORTHOPYROXENE)	ORTH.	3.21	ISI	MG PYROXENE. WITH OLIVINE IN BASIC + ULTRABASIC IG RKS. GRANULITES	W405 2-8
5 -6	HARDYSTONITE	TETR. A	3.40	SSI	GEHLENITE GROUP. GRNLR. SOME FLUO-PURPLE(SHORT-WAVE). RARE(FRNKLN)	W473 N
5 -6	TREMOLITE (AMPHIBOLE)	MNCL.	3.02-3.44	ISI	META SI-BEARING LS + DOLO, LOW-GRADE META ULTRABASIC IGNEOUS RKS	W434 2-249
5 -7	PYROXENE		2.96-3.96	ISI	GROUP NAME. SEVERAL END AND OTHER MEMBERS WITH NAMES UNSETTLED FOR SOME. MANY MEGASCOPICALLY INDISTINGUISHABLE; OCCURRENCE MAY BE INDICATIVE. ORTH + MNCL SERIES.SHORT PRISMS.PRISM CLVG AT 87° + 92°. ALTERS TO AMPHIBOLE.IG + META RKS	W402 2-1
5 -7	OLIVINE	ORTH.	3.22-4.39	NSI	GRP NAME FOR MINS OF FAYALITE-FORSTERITE-TEPHROITE SERIES. MONTI-CELLITE + GLAUCOCHROITE. GRNLR MASSES. DISSEM. CONCH. GELAT. COMMONLY ALTERED TO SERPENTINE. BASIC + ULTRABASIC IG RKS. CNTCT META DOLOMITIC LS. METEORITES	W498 1-2
5.5	EPIDIDYMITE	ORTH.	2.55	PSI	EQUANT-ELONGATE PLTS. 2 PERF CLVGS. ALK PEGS(NORWAY + GREENLAND)	W444 N
5.5	HARSTIGITE	ORTH.	3.05	SSI	SHORT PRISMATIC CRYSTALS. RARE-WITH GARNET + RHODONITE AT MN-DEPS	W479 N
5.5	MONTICELLITE	ORTH.	3.08-3.27	NSI	COLORLESS-GRAY. CA-MG OLIVINE. CONTACT META OF SILICEOUS DOLO-LS	W502 1-41
5.5	CLINOHEDRITE	MNCL.	3.33	SSI	CLINOHEDRAL. 1 PERF CLVG. WITH WILLEMITE, ETC. AT FRANKLIN, N.J.	W482 N
5.5	PERICLASE	ISOM.	3.55-3.57	OXD	OCTAH, ROUNDED GRAINS. CUBIC,OCTAH CLVGS. CNTCT META MARBLES(HI-T)	4.2.1.1
5.5-6	WAIRAKITE	MNCL.	2.26	TSI	CA-ANALCITE. PSEUDOCUBIC. RARE-HYDROTHERMALLY ALTERED SSINEW ZEALAND)	WN 4-342
5.5-6	MILARITE	HEXA.	2.57	CSI	PRISMATIC. GLASSY. RARE-PEGMATITE NEAR GLETSCH, SWITZERLAND	W257 N
5.5-6	NEPHELINE	HEXA.	2.56-2.67	TSI	TBLR,PRISM,VITR-GRSY.GOOD PRISM.POOR BASAL CLVGS.GELAT.FLUO-ORANGE.DISTINGUISH-ED FROM KALSILITE + KALIOPHILITE BY X-RAY. NA-RICH ALK IG AND NEARBY META ROCKS	W254 4-231
5.5-6	MONTEBRASITE	TRCL.	2.98	PHO	AMBLYGONITE SERIES(SEE FOR PROPERTIES + OCCURRENCES.)	41.5.5.2

110

NONMETALLIC COLORLESS

5.5-6	NATROMONTEBRASITE	TRCL.	3.04-3.1	PHO	AMBLYGONITE SERIES(SEE FOR PROPERTIES + OCCURRENCES.)	41.5.5.3
● 5.5-6	AMBLYGONITE	TRCL.	3.11	PHO	EQUANT, LARGE MASSES. VITR,PEARLY. CLVGS. SOL-ACIDS(DIFFICULTY). LI-, PO4-RICH PEGS	41.5.5.1
● 5.5-6	ANATASE	TETR.	3.90	OXD	SUBMET, PALE YEL STRK. ACUTE PYRAMIDAL. ADMN. BASAL + PYRAMIDAL CLVG. CONVERTS TO RUTILE ON HEATING. VNS,ACCESS-IG RKS,DETRITAL	4.5.2.
5.5-6	NORDENSKIOLDINE	HEXA.	4.20	BOR	TBLR, LENSLK XLS. VITR,PEARLY. BASAL CLVG. DCMP-HCL. ALK PEG	24.1.8.
5.5-6.5	OPAL	TETR.	1.8-2.25	SIL	SUBMICROXLINE QTZ + H2O-RESIN-PEARLY. MSV,MAMILLARY CRUSTS,CONCH.SOL-HF +CAUS-TIC ALKS.SOME FLUO-YEL-GREEN. INCRUSTATIONS, IN CAVITIES(ESP IN VOLCANIC RKS)	D-111 287
● 5.5-6.5	DIOPSIDE (CLINOPYROXENE)	MNCL.	3.22-3.38	ISI	SOME LIGHT ONES FLUO-BLUE. WHT-PALE GREEN. META CALC-SIL RKS	W411 2-42
5.5-7.5	NARSARSUKITE	TETR.	2.75	PSI	TBLR. PRISM CLVG. SOL-HF. IN PEGS ASSOC WITH + CAVITIES IN ALK IGS	W358 N
6	EUDIDYMITE	MNCL.	2.55	PSI	TBLR. LAMELLAR TWINNING. 1 PERF, 1 POOR CLVG. CAVITIES IN SYENITE-RARE	W445 N
● 6	ADULARIA	TRCL.	2.56	TSI	MUCH IS TRANSPARENT. PRISM. CAVITIES IN META RKS. MOONSTONE IS VAR	W303 4-7
6	LEIFITE	HEXA.	2.57	TSI	STRIATED ACICULAR XLS. PRISM CLVG. RARE-PEGINARSARSUK, GREENLAND)	W355 N
6	BERTRANDITE	ORTH.	2.6	ISI	PLATY. 1 PERF, OTHER CLVGS. PEGMATITES, TYPICALLY WITH BERYL	W479 N
6	KALIOPHILITE	HEXA.	2.61	TSI	METASTABLE PHASE OF KALSIO4.RARE-IN BLOCKS EJECTED WITH K-RICH LAVA	W257 4-232
6	SANIDINE	MNCL.	2.56-2.62	TSI	TBLR. GLASSY. OPTICALLY DISTINCT. VOLCANIC ROCKS	W305 4-7
6	KALSILITE	HEXA.	2.59-2.63	TSI	K CHEM-ANALOGUE OF NEPH.DISTINGUISHED BY X-RAY,K-RICH,SI-POOR VOLS	W257 4-231
6	LAWSONITE	ORTH.	3.05-3.10	SSI	PRISM, TBLR.CLVG-2 PERF, 1 POOR. LO-GRADE(E.G.,GLAUCOPHANE)SCHISTS	W482 1-221
6	MERMINITE	MNCL.	3.15	NSI	MULTIPLE TWINNING COMMON. 1 PERF CLVG. CONTACT META ZONES	W505 N
6	STOKESITE	ORTH.	3.19	SSI	PYRAMIDAL.1 PERF,1 POOR CLVG. RARE, WITH AXINITE(ST.JUST,CORNWALL)	W454 N
6	CLINOENSTATITE (CLINOPYROXENE)	MNCL.	3.19	ISI	HI-T FORM. RARE IN IGNEOUS ROCKS, IN METEORITES.	W408 N
● 6	JADEITE (CLINOPYROXENE)	MNCL.	3.24-3.43	ISI	RELATIVELY UNCOMMON. TYPICALLY WITH ALBITE, HI-T + ALSO LO-GRADE METAMORPHIC ROCKS	W411 2-99
● 6	EPIDOTE	MNCL.	3.38-3.49	SSI	GROUP NAME, NO GENERALLY ACCEPTED NOMENCLATURE FOR GROUP MEMBERS. CLOSELY RELATED TO ZOISITE + ALLANITE. PISTACHIO-BLACKISH GREEN-BROWNISH. LONG THIN GROOVED XLS, GRNLR MASSES. VNS, OTHER CAVITIES, REGIONAL META + IG ROCKS	W448 1-193
6	LAVENITE	MNCL.	3.5	NSI	PRISMATIC,TBLR. 1 GOOD CLVG. SOL-HCL(SLOWLY). ALKALIC IGNEOUS RKS	W517 N
● 6	ANORTHOCLASE (FELDSPAR)	TRCL.	2.56-2.62	TSI	DISTINGUISHED FROM ORTHOCLASE + MICROCLINE-OPTICALLY. NA-RICH VOLS	W311 4-7
● 6	ALBITE (PLAGIOCLASE)	TRCL.	2.57-2.63	TSI	AN0-10.. SEE FELDSPAR. IG, META RKS, PEGS. MOONSTONE +CLEAVELANDITE ARE VARIETIES	W312 4-94
● 6	MICROCLINE (K-FELDSPAR)	TRCL.	2.56-2.63	TSI	DISTINGUISHED FROM ORTHOCLASE OPTICALLY. PLUTONIC IG RKS, PEGS	W308 4-7
● 6	ORTHOCLASE (K-FELDSPAR)	TRCL.	2.55-2.63	TSI	DISTINGUISHED FROM MICROCLINE OPTICALLY. PLUTONIC IG RKS, PEGS	W303 4-7
● 6	OLIGOCLASE (PLAGIOCLASE)	TRCL.	2.62-2.67	TSI	AN10-30.. SEE FELDSPAR. PLUTONIC IG RKS, FEW PEGS. SUNSTONE IS VAR.	W312 4-95
● 6	ANDESINE (PLAGIOCLASE)	TRCL.	2.64-2.69	TSI	AN30-50.. SEE FELDSPAR. INTERMEDIATE SI-CONTENT IGNEOUS ROCKS	W312 4-95
● 6	LABRADORITE (PLAGIOCLASE)	TRCL.	2.68-2.72	TSI	AN50-70..BLUISH REFLECTIONS. SEE FELDSPAR. BASIC IGNEOUS ROCKS	W312 4-95

111

NONMETALLIC COLORLESS

HARDNESS	NAME	XL. SYS.	SPECIFIC GRAVITY	CHEM. CLASS	REMARKS	REFERENCES
6 -6.5	BYTOWNITE (PLAGIOCLASE)	TRCL.	2.71-2.75	TSI	AN70-90. GELAT. SEE FELDSPAR. CALCIUM-RICH IGNEOUS ROCKS	W312 4-95
6 -6.5	ANORTHITE (PLAGIOCLASE)	TRCL.	2.70-2.76	TSI	AN90-100. GELAT. SEE FELDSPAR. CNTCT META CALCAREOUS RKS	W312 4-94
6 -6.5	CELSIAN (BA-FELDSPAR)	MNCL.	3.10-3.39	TSI	SEE FELDSPAR TYPICALLY CLEAVABLE MASSES.GELAT.MN-RICH VNS + RKS	W307 4-166
6 -6.5	HYALOPHANE (BA-FELDSPAR)	MNCL.	2.58-2.82	TSI	BA EQUIVALENT OF ADULARIA. PRISM, MSV. MN-RICH VEINS + ROCKS	W304 4-166
6 -6.5	PARACELSIAN (BA-FELDSPAR)	MNCL.	3.31-3.32	TSI	PSEUDO-ORTHORHOMBIC. TOPAZ-LK XLS. MN-RICH VEINS AND ROCKS	W307 4-167
6 -6.5	FELDSPAR		2.55-3.39	TSI	INCLUDES PLAGIOCLASE(NA-CA)SERIES, ALKALI(K-NA) AND BA FELDSPARS. MOST EASILY DISTINGUISHED BY NONMEGASCOPIC MEANS. MONOCLINIC ONES HAVE 2 CLVGS AT 90° AND SIMPLE(IF ANY) TWINNING; TRICLINIC ONES HAVE 2 CLVGS UP TO 4° OFF 90° AND TYPICALLY HAVE POLYSYNTHETIC TWINNING (GIVING STRIATED APPEARANCE ON SOME CLVG SURFACES OF PLAGIOCLASE). PEGMATITES; IG, META; AND SOME SED ROCKS	W261 4-1
6 -6.5	BENITOITE	HEXA.	3.65	CSI	UNEVENLY COLORED. TBLR. SOL-HF. FLUO-BLUE(SW).FROM 1 NATROLITE VN	W453 N
6 -7	BERLINITE	HEXA.	2.64	PHO	MSV,GRNLR,VITR. CONCH. RARE-WITH AUGELITE AT FE-MINE(KRISTIANSTAD, SWEDEN)	38.4.3.
6 -7	MULLITE	ORTH.	3.03-3.16	NSI	INDISTINGUISHABLE MEGASCOPICALLY FROM SILLIMANITE. HI-T META-RARE	W401 1-124
6 -7	CASSITERITE	TETR.	6.99	OXD	SUBMET. RADIAL CONCRETIONARY MASSES. ADMN-DULL. HI-T VNS, GREISENS	4.5.1.5
6.5	XONOTLITE	MNCL.	2.71	ISI	EXPOS-COLOR MAY FADE. ELONGATE XLS. CLVG // LENGTH. CNTCT META LS	W455 N
6.5	POLLUCITE	ISOM.	2.9	TSI	CUBES, MSV. SLOWLY DCMPD-HCL. CAVITIES IN GRANITIC RKS + PEGS	W253 4-340
6.5	KOTOITE	ORTH.	3.10	BOR	GRNLR;DISSEM. VITR. PRISM CLVG. SOL-WARM HCL,H2SO4.CNTCT META DOLO	24.1.5.
6.5	JEREMEJEVITE	HEXA.	3.28	BOR	PRISMS. VITR. CONCH. RARE IN RESIDUUM ABOVE GRANITE(E.SIBERIA)	24.1.7.
6.5	CLINOZOISITE	MNCL.	3.21-3.38	SSI	CALCIUM-ALUMINUM END-MEMBER OF EPIDOTE GROUP. SAME OCCUR	W448 1-193
6.5	BADDELEYITE	MNCL.	5.4 -6.02	OXD	FLAT PRISM. VITR. BASAL + 2 POOR CLVGS. DCMP-HOT H2SO4. PLACERS, ASSOCIATED WITH ULTRAMAFIC IGNEOUS ROCKS	5.1.1.
6.5-7	SPODUMENE	MNCL.	3.03-3.22	ISI	COMMONLY FLUO(+PHOSPHORESCES)-ORANGE. THERMOLUM. LI-BEARING PEGS	W418 2-92
6.5-7	DIASPORE	ORTH.	3.3 -3.5	OXD	PLTY, FOL-MSV. VITR, PEARLY. WITH CORUNDUM IN EMERY ROCK	7.1.2.1
6.5-7,7.5	SILLIMANITE	ORTH.	3.23-3.27	NSI	PRISMATIC XLS, FIBR MASSES. VITR. 1 CLVG. HI-GRADE META ROCKS	W520 1-121
7	TRIDYMITE	ORTH.	2.25-2.27	SIL	PLTY MICROXLS. VITR-PEARLY. CONCH. SOL-BOILING NaCO3. SILICIC VOLS	D-111 264
7	QUARTZ	HEXA.	2.65	SIL	DOUBLY TERMINATED HEX PRISMS STRIATED ⊥ LENGTH(LO-T). BIPYRAMIDS (HI-T). MSV,VITR.CONCH.SOL-HF.WDSPRD IN MOST RK TYPES + VEINS	D-111 9
7	ZUNYITE	ISOM.	2.88	SSI	TETRAHEDRA. OCTAH CLVG. ALTERED IGNEOUS RKS + ALUMINOUS SHALES	W482 N
7	DANBURITE	ORTH.	3.0	TSI	PRISM. 1 POOR CLVG. VITR. GELAT(AFTER IGN). PEGS, WITH FELDSPAR IN DOLOMITE	W258 N

NONMETALLIC COLORLESS

Hardness	Mineral	Crystal System, S.G.	Class	Description	Ref
7	ELBAITE	HEXA. 3.03-3.10	CSI	NA- LI-TOURMALINE. INFUSIBLE. TYPICALLY PINK AND/OR GREEN. PEGS	W465 1-300
7 ●	TOURMALINE	HEXA. 3.03-3.25	CSI	GROUP NAME. VARIETIES COMMONLY DISTINGUISHABLE MEGASCOPICALLY ON BASIS OF COLOR(S). LENGTHWISE STRIATED PRISMATIC XLS WITH CROSS-SECTIONS THAT RESEMBLE SPHERICAL TRIANGLES. SOME XLS EXHIBIT BOTH LENGTHWISE + CONCENTRIC COLOR ZONING. VITR-RESIN. BRTL. ELECTRICALLY CHARGED ON HEATING + COOLING. MG-RICH VARIETIES FLUO-YEL (ULTRAVIOLET - SW). PEGS, HI-T VNS, IG AND META ROCKS	W465 1-300
7	BARYLITE	ORTH. 4.	SSI	PLATY XLS. 2 GOOD + 1 POOR CLVGS(PINACOIDAL). RARE-FRANKLIN, N.J.	W476 N
7 -7.5	BORACITE	ORTH. 2.91-2.97	BOR	CUBIC,ETC.,GRNLR,FIBR. VITR. SOL-HCL(SLOWLY). SALT-GYPSUM BEDS	26.1.7.
7.5	HAMBERGITE	ORTH. 2.36	BOR	PRISM-STRIATED. VITR. 2 CLVGS. SOL-HF. ALKALIC PEGMATITES	26.1.2.
7.5 ●	COESITE	MNCL. 2.91-2.95	SIL	TRANSPARENT. GYPSUM-LIKE HABIT. SOL-HF(<QTZ).IMPACT(METEOR) CRATERS	D-111 310
7.5 ●	PHENAKITE	HEXA. 2.98	NSI	RHOMBOHEDRAL(FLAT)-PRISMATIC. VITREOUS. PEGMATITES	W496 N
7.5	EUCLASE	MNCL. 3.0 -3.1	PSI	PRISM XLS. 1 GOOD, 2 POOR CLVGS. PEGS, PLACERS, GRANITES, SCHISTS	M357 N
7.5 ●	ZIRCON	TETR. 4.6 -4.7	NSI	TERMINATED PRISMS. ADMN. SOME FLUO-YEL-ORANGE. ACCESS-IG, SANDS	W494 1-59
7.5-8	BERYL	HEXA. 2.66-2.83	CSI	GROOVED PRISM. POOR BASAL CLVG. SOME FLUO WEAKLY-YEL. PEGMATITES	W463 1-256
7.5-8	SPINEL	ISOM. 3.55	OXD	GROUP NAME. SOME HAVE DISTINCTIVE COLORS OR OTHER HAND-SPECIMEN PROPS. OCTAH. GRNLR, GLASSY. CONCH. SOME FLUO-RED TO YEL-GREEN. IG.META(ESP CALC RKS),PEGS	7.2.1.1
7.5-8.5	SWEDENBORGITE	HEXA. 4.29	ANT	PRISM. 1 CLVG. SUBCONCH. RARE-SKARN(LANGBAN, SWEDEN)	44.2.2.
8	RHODIZITE	ISOM. 3.31-3.38	BOR	DODEC. VITR. OCTAH CLVG. WITH RED TOURMALINE IN PEGMATITES-RARE	24.1.6.
8 ●	TOPAZ	ORTH. 3.49-3.57	NSI	STRIATED PRISMS. PERF BASAL CLVG. CNTCT ZONES, GREISENS, PEGS	W509 1-145
9 ●	CORUNDUM	HEXA. 4.0 -4.1	OXD	STEEP PYRAMIDAL, PRISM. BASAL PARTING.WDSPRD IN SI-DEFICIENT RKS	4.4.1.1
10 ●	DIAMOND	ISOM. 3.50-3.53	ELE	OCTAHEDRAL. ADMN-GRSY. BRTL. OCTAH CLVG. ULTRABASIC IGS + PLACERS	1.2.4.1

NONMETALLIC WHITE

HARDNESS	NAME	XL. SYS.	SPECIFIC GRAVITY	CHEM. CLASS	REMARKS	REFERENCES
1	SASSOLITE	TRCL.	1.46-1.50	OXD	SCALES. PEARLY. BASAL CLVG(FLEX). BITTER,SALTY TASTE. HOT SPRINGS	6.2.1.
1	HOERNESITE	MNCL.	2.57-2.73	ARS	PRISM,RADIAL FOL. PEARLY. 2 CLVGS(FLEX). SOL-ACIDS. DIVERSE OCCUR.	40.2.17.
1	GLAUCOCERINITE	UNKN.	2.75	SUF	MARTY MASSES), RADIAL + CONCENTRIC. WAXY. WITH SMITHSONITE,ETC.(LAU-31.1.2. RIUM SPECIMEN)	
1	TALC	MNCL.	2.58-2.83	PSI	FOL MSV. BASAL CLVG. SECTILE. SOAPY FEEL. PEARLY. SCHIST, META IG	M364 3-121
1	LAWRENCITE	HEXA. A	3.16	HAL	MSV. BASAL CLVG. SOL-H2O. DLQSNT. METEORIC. VOL SUBLIMATE	9.2.3.1
1	ANTHOINITE	UNKN. A	4.5 -4.7	WOS	CHALK-LK MASSES. ATTKD-HCL,HNO3(SLOW). PLACERS,QTZ-FERBERITE VEINS	49.1.7.
1	PARALAURIONITE	MNCL.	6.15	HAL	LATHLK, SUBADMN. 1 CLVG. SOL-HNO3. XLS BENDIGLIDING). SCNDRY AFTER LEAD MINERALS	10.1.8.
1	NATRON	MNCL.	1.48	CBT	GRNLR CRUSTS. VITR. CLVGS. ALK TASTE. IN SOLUTIONS, EFFLORESCENCES	15.1.6.
1-1.5	THERMONATRITE	ORTH. A	2.26	CBT	CRUST-EFFLOR. VITR. 1 CLVG. ALK TASTE. RARE-WITH TRONA-ARID SOILS	15.1.1.
1-1.5	EPISTOLITE	MNCL.	2.89	NSI	PLATY XLS. 1 PERF. 1 GOOD CLVG. RARE IN PEG-JULIANEHAAB,GREENLAND	W508 N
1-2	BISCHOFITE	MNCL.	1.60	HAL	GRNLR, PRISM. VITR-DULL. CONCH. DLQSNT. BITTER TASTE. SALINE DEPS	9.2.8.
1-2	ALUMINITE	UNKN.	1.66-1.82	SUF	SUBTRANSLUCENT. NODULAR MASSES, FIBR., DULL. FRIABLE. SOL-ACIDS. VNS,CONCRETIONS FMED BY ACTION OF SULFATE SOLUTIONS ON AL-SILICATES	31.4.5.
1-2	ILLITE (CLAY)	MNCL.	2.64-2.69	PSI	ILL-DEFINED. ALTER PROD FORMED UNDER ALK CONDITIONS. AUTHIGENIC	WN 3-213
1-2	MONTMORILLONITE (CLAY)	MNCL.	2.2 -2.7	PSI	IRREG+HEX-SHAPED PLTS. "SWELLING CLAY." ALTER-HI MG +CA+LO K RKS (ALKALINE CONDITIONS). SOILS, BENTONITES, ETC.	W398 3-226
1-2	CELADONITE	MNCL.	2.6 -2.9	PSI	PHYSICALLY + CHEMICALLY LIKE(=?) GLAUCONITE. CAVITIES IN BASALTS	M378 3-217
1-2	HYDROMUSCOVITE	MNCL.	2.6 -2.9	PSI	FINE GRAINED. ALTER(UNDER ALK CONDITIONS) OF MICA, FELDSPAR, ETC.	M369 3-213
1-2	GUEMBELITE	MNCL.	2.6 -2.9	PSI	MG-BEARING HYDROMUSCOVITE. (SAME GENERAL PROPERTIES + MODE OF FM. AS HYDROMUSCOVITE)	3-215
1-2	HYDROPHLOGOPITE	MNCL.	2.6 -2.9	PSI	FMED UNDER ALK CONDITIONS. CLASTIC + DIAGENIC-SEDS, HYDROTHERMAL	WN 3-213
1-2	PYROPHYLLITE	MNCL.	2.65-2.90	PSI	LAMELLAR. CLVG-INELAST PLTS. PARTLY DCMP-H2SO4. VNS, FEW SCHISTS	M361 3-115
1-2	BEIDELLITE (CLAY)	MNCL.	2.	PSI	"SWELLING CLAY." ALTER-HI MG +CA+LO K RKS(ALK CONDITIONS). RARE-GOUGE (COLORADO), BENTONITES	W398 3-226
1-2	STEVENSITE (CLAY)	MNCL.	2.	PSI	POSSIBLY MONTMORILLONITIC "SWELLING CLAY." ALTER PROD OF PECTOLITE	M461 3-231
1-2	HECTORITE (CLAY)	MNCL.	2.	PSI	ELONGATE,MONTMORILLONITIC CLAY(SEE).IN ULTRABASICS +AFTER ZEOLITES	WN 3-226
1-2	SAPONITE (CLAY)	MNCL.	2.	PSI	IRREG PLTS, FIBR, MSV. SWELLING CLAY. ALTER-HI MG +CA+LO K RKS (ALK CONDITIONS). AS NODULES AND OTHER FMS IN AMYGDULES, IN VEINS	W399 3-226
1-2	HYDROBIOTITE	MNCL.	2.6 -3.3	PSI	FMED UNDER ALK CONDITIONS. CLASTIC + DIAGENIC-SEDS, HYDROTHERMAL	M396 3-213
1-3	CLAY	MNCL.	1.85-3.0	PSI	NAME GIVEN SEVERAL HYDROUS AL-SILICATES WITH ALKALIES OR ALKALINE EARTHS ± MG OR FE. NAMES UNSETTLED. MOST DISTINGUISHED BY NONMEGASCOPIC MEANS. FINE-GRAINED AGGS(COMPACT,MEALY,ETC.). EARTHY-PEARLY,UNCTUOUS,PLASTIC WHEN WET.DEHYDRATED ON HEATING. SOME ARE ABSORBENT. MXING PRODS,HYDROTHERMAL ALTER, DETRITAL, DIAGENIC	W398 3-191

NONMETALLIC WHITE

Hardness	Mineral	Crystal	S.G.	Class	Description	Ref
1.5	WATER	HEXA.	.92	OXD	LIQUID ABOVE 0°C.(32°F)+ BELOW 100°C.(212°F). S.G. FOR ICE	4.1.3.
1.5	TESCHEMACHERITE	ORTH. A	1.57	CBT	COMPACT MASSES. CLVG. BRTL. SOL-H2O. EFF-ACIDS. GUANO DEPS	13.1.3.
1.5	AMMONIA ALUM	ISOM.	1.65	SUF	FIBR, CRUSTS. VITR-SILKY. CONCH. SWEET,ASTRINGENT. IN COALY BEDS	29.5.5.3
1.5	APJOHNITE	MNCL.	1.78	SUF	FIBR. SILKY. SOL-H2O. EFFLORESCENCES IN SHELTERED PLACES	29.7.3.3
1.5	PICKERINGITE	MNCL.	1.73-1.79	SUF	FIBR. VITR. ASTRINGENT. EFFLORESCENCES IN SHELTERED PLACES	29.7.3.1
1.5	HALOTRICHITE	MNCL.	1.89	SUF	FIBR. VITR. ASTRINGENT. EFFLORESCENCES IN SHELTERED PLACES	29.7.3.2
1.5	FELSOBANYAITE	ORTH.	2.33	SUF	RADIAL AGGS LAMELLAR XLS. PEARLY. 2 CLVGS. WITH MARCASITE,STIBNITE AND BARITE (FELSOBANYA, HUNGARY)	31.2.3.
1.5	ISOCLASITE	MNCL.	2.92	PHO	PRISM,FIBR. VITR-PEARLY. 1 CLVG. SOL-ACIDS. RARE-WITH CHALCEDONY + DOLOMITE (JOACHIMOV, BOHEMIA)	42.4.5.
1.5	SZMIKITE	MNCL.	3.15	SUF	STALAC. SPLINERY. RARE-EFFLOR(FELSOBANYA, ROUMANIA)	29.6.2.3
1.5	ARSENOLITE	ISOM.	3.86-3.88	OXD	WHITE-COLOR TINTED. OCTAH, CRUSTS. VITR-SILKY. OCTAH CLVG. ASTRIN-GENT-SWEETISH TASTE. SCNDRY AFTER ARSENOPYRITE + OTHER AS MINS.	4.4.2.1
1.5	CALOMEL	TETR.	7.15	HAL	EXPOS-DARKENS. TBLR, MSV. ADMN. PRISM CLVG. PLASTIC.FLUO-RED. SOL-AQ-REG. SCNDRY AFTER SEVERAL MERCURY MINERALS	9.1.5.
1.5-2	MIRABILITE	MNCL.	1.49	SUF	PRISM, ACICULAR, MSV, CRUSTS. VITR. CLVGS. COOL, THEN SALTY + BIT-TER. LOSES H2O IN DRY AIR TO POWDER. EVAPORITE DEPS, EFFLORESCENCES	29.2.2.
1.5-2	SALAMMONIAC	ISOM.	1.53	HAL	SKELETAL, DENDRITIC. VITR.OCTAH CLVG. SALTY TASTE. VOL SUBLIMATE	9.1.2.
1.5-2	ALUNOGEN	TRCL.	1.77	SUF	FIBR MASSES. VITR-SILKY. SHARP, ACID TASTE. EFFLOR IN FES2-BEARING ROCKS	29.8.6.
1.5-2	SODA-NITER	HEXA.	2.24-2.29	NIT	INCRUST, RHOMB. VITR. RHOMB + OTHER CLVGS. COOLING TASTE. IN ARID SOILS IN PROTECTED LOCATIONS	18.1.1.
1.5-2.5	BILINITE	MNCL.	1.88	SUF	FIBR-RADIAL. SOL-H2O. RARE-ALTER OF FES2 IN LIGNITE(SCHMAZ,BOHEMIA)	29.7.3.5
1.5-2.5	LUENEBERGITE	MNCL.	2.05	BOR	NODULES, PSEUDOHEX TABLETS. PRISM CLVG. EVAPORITE DEPOSITS	27.1.1.
1.5-2.5	ANNABERGITE	MNCL. C	3.23	ARS	XLINE COATINGS,EARTHY. ADMN-DULL. CLVGS. SOL-ACIDS. OXIDATION PROD OF COBALT AND NICKEL ARSENIDES	40.2.15.3
2	TAYLORITE	UNKN.		SUF	COMPACT LUMPS. PUNGENT + BITTER TASTE. SOL-H2O,ACIDS. GUANO DEPS	28.2.1.3
2	DIETRICHITE	MNCL.	1.8	SUF	FIBR. SILKY. SOL-H2O. RARE-EFFLORESCENCES IN SHELTERED PLACES	29.7.3.4
2	STERCORITE	TRCL.	1.62	PHO	XLINE MASSES, NODULES. VITR. SOL-H2O. GUANO DEPOSITS	39.1.1.
2	STRUVITE	ORTH.	1.71	PHO	DIVERSE XLS,MOST HEMIMORPHIC.VITR.2 CLVGS. SOL-ACIDS, H2O(SLIGHT-LY). GUANO DEPOSITS, URINARY SEDIMENT, ETC.	40.1.1.
2	KIROVITE	MNCL.	1.76-1.87	SUF	DISTINGUISHED FROM MELANTERITE BY NONMEGASCOPIC MEANS. SAME OCCUR	29.6.8.3
2	INYOITE	MNCL.	1.88	BOR	TBLR, GRNLR. VITR. 2 CLVGS. SOL-HOT H2O, ACIDS. EVAPORITE DEPOSITS	25.1.13.
2	MELANTERITE	MNCL.	1.84-1.90	SUF	STALAC, CRUSTS,ETC. VITR. CLVGS. SWEET,ASTRINGENT,METALLIC TASTE. EXPOS(DRY AIR)--YEL-WHT + OPQ. EFFLORESCENCE -BY OXIDATION OF SUFS	29.6.8.1
2	PISANITE	MNCL.	1.95	SUF	DISTINGUISHED FROM MELANTERITE BY NONMEGASCOPIC MEANS. SAME OCCUR	29.6.8.2
2	SYLVITE	ISOM.	1.98-2.00	HAL	CUBIC, GRNLR. VITR. CUBIC CLVG. BITTER TASTE. SED EVAPORITE DEPS	9.1.1.2
2	HYDROTALCITE	HEXA.	2.03-2.09	OXD	PLTY, FOL-MSV. WAXY,PEARLY. BASAL CLVG(FLEX). EFF-HCL. ALTER PROD OF SPINEL	6.1.5.1
2	MANASSEITE	HEXA.	2.00-2.10	OXD	FOL-MSV. GRSY. BASAL CLVG(FLEX). SOL-HCL. WITH HYDROTALCITE + SER-PENTINE	6.1.6.1

115

NONMETALLIC WHITE

HARDNESS	NAME	XL. SYS.	SPECIFIC GRAVITY	CHEM. CLASS	REMARKS	REFERENCES
2	NITER	ORTH.	2.10-2.11	NIT	CRUSTS,GRNLR. VITR. CLVGS. SALTY, COOLING TASTE. EFFLOR-CAVES,ETC.	18.1.2.
2	MEYERHOFFERITE	TRCL.	2.12	BOR	ELONGATE, FIBR. VITR,SILKY. CLVGS. SOL-ACIDS. COLEMANITE DEPOSITS	25.1.12.
2	BRUGNATELLITE	HEXA.	2.14	OXD	FLKS, FOL-MSV. PEARLY. BASAL CLVG. EFF-HCL. INCRUST-CRACKS IN SER-PENTINE	6.1.7.
2	HALITE	ISOM.	2.17	HAL	CUBIC, GRNLR. VITR. CUBIC CLVG. SALTY TASTE. SED EVAPORITE DEPS	9.1.1.1
2	RHOMBOCLASE	ORTH.	2.23	SUF	THIN TBLR, STALACTITIC. SUBVITR. BASAL(FLEX),PRISM CLVGS. CONCH-FIBR. SOL-ACIDS,H2O(SLOWLY). RARE-ALTER PROD OF SULFIDES	29.1.1.
2	GYPSUM	MNCL.	2.31-2.32	SUF	TBLR, MSV. SUBVITR-PEARLY. 1 PERF CLVG(FLEX,INELAST), SOL-HCL. SEDIMENTARY BEDDED DEPOSITS, FUMAROLIC, EFFLORESCENCES, ETC.	29.6.3.
2	VEATCHITE	MNCL.	2.58-2.69	BOR	PLTY, FIBR. VITR, PEARLY. 2 CLVGS. RARE-BORATE DEP(LANG, CALIF.)	25.1.8.
2	GEARKSUTITE	MNCL.	2.77	HAL	MSV, NODULAR, NEEDLES. DULL. SOL-DILUTE ACID. ALK PEGS + IG RKS	11.5.6.
2	THOMSENOLITE	MNCL.	2.98	HAL	PRISM, OPALINE CRUSTS. VITR. BASAL,PYRAMIDAL CLVGS. SOL-H2SO4. AT CRYOLITE LOCALITIES	11.5.4.
2	DUNDASITE	UNKN.	3.25	CBT	RADIATING XLS,MATTED CRUSTS. VITR-SILKY. 1 CLVG. EFF-ACIDS. SCNDRY WITH CERUSSITE, ETC.	16.2.3.
2	TELLURITE	ORTH.	5.88-5.92	OXD	ACICULAR, LATHS-STRIATED. 1 PERF CLVG. SOL-HCL,HNO3,ALKALIES. OXI-DATION PRODUCT OF NATIVE TELLURIUM AND TELLURIDES	4.5.4.
2 -2.5	DICKITE (CLAY)	MNCL.	2.62	PSI	ALTER(ACID CONDITIONS)OF FELDSPARS, ETC. HYDROTHERMAL WITH QUARTZ, SULFIDES, ETC.	W363 3-194
2 -2.5	MELLITE	TETR.	1.64	CCC	MSV,GRNLR,CRUSTS,NODULAR, RESIN,PRISM CLVG. CONCH. SOL-HNO3. FLUO-BLUE(SW). SECONDARY IN OPEN SPACES IN LIGNITE	50.2.1.
2 -2.5	EPSOMITE	ORTH. A	1.68	SUF	CRUSTS. VITR-EARTHY. CLVGS. BITTER, SALTY. METALLIC. COMMON EFFLOR	29.6.9.1
2 -2.5	BORAX	MNCL.	1.71-1.72	BOR	HABIT-LK PYROXENE. VITR-EARTHY. CLVGS. SWEETISH. SOL-H2O(ITO ALK SOLUTION). BORAX DEPOSITS (SALINE LAKE DEPOSITS, ETC.)	25.1.4.
2 -2.5	KALINITE	MNCL.	1.75	SUF	FIBR. MORE DATA NEEDED ON PHYSICAL PROPERTIES.	29.5.4.2
2 -2.5	POTASH ALUM	ISOM.	1.76	SUF	MSV, STALAC. VITR. OCTAH CLVG. CONCH. SWEET,ASTRINGENT. EFFLOR	29.5.5.1
2 -2.5	ETTRINGITE	HEXA.	1.77	SUF	PRISM. PRISMATIC CLVG. SOL-ACIDS,DCMP-H2O TO ALK SOLUTION. META LS	31.3.2.
2 -2.5	MORENOSITE	ORTH.	1.95	SUF	CRUSTS. VITR,1 CLVG. CONCH. METALLIC,ASTRINGENT. EFFLOR-AT NI-SUL-FIDE DEPOSITS	29.6.9.3
2 -2.5	GOSLARITE	ORTH.	1.98	SUF	CRUSTS. VITR-SILKY. 1 CLVG. ASTRINGENT,METALLIC,NAUCEOUS. EFFLOR-TYPICALLY BY ALTERATION OF SPHALERITE	29.6.9.2
2 -2.5	SEPIOLITE (CLAY)	ORTH.	2.08	PSI	POROUS(FLOATS ON H2O WHEN DRY)-COMPACT. SMOOTH FEEL. GELAT. WITH SERP OR OPAL DESERT SOILS, CLAYEY SEDS WITH SALINES, LAKE SEDS. MEERSCHAUM IS A VARIETY	W444 N
2 -2.5	HALLOYSITE (CLAY)	MNCL.	2.0 -2.2	PSI	ALTER(ACID CONDITIONS) OF FELDSPARS, ETC. SEDS + MXED PEGS, ETC.	W400 3-194
2 -2.5	BOBIERRITE	MNCL. A	2.20	PHO	ACICULAR OR FIBR XLS,GLBL,ETC. VITR. 1 CLVG. SOL-ACIDS. GUANO,ETC.	40.2.16.
2 -2.5	NACRITE (CLAY)	MNCL.	2.45-2.55	PSI	ALTER(ACID CONDITIONS)OF FELDSPARS, ETC. HYDROTHERMAL WITH QUARTZ, SULFIDES, ETC., ALSO WITH CRYOLITE, ETC.	W363 3-194
2 -2.5	KAOLINITE (CLAY)	TRCL.	2.60-2.68	PSI	ALTER(ACID CONDITIONS) OF FELDSPARS, ETC. HYDROTHERMAL + WXING	W362 3-194
2 -2.5	PHARMACOLITE	MNCL.	2.53-2.73	ARS	FIBERS. VITR,PEARLY. 1 CLVG(FLEX). SOL-ACIDS. OXIDIZED AS-DEPS	39.2.1.2
2 -2.5	VILLIAUMITE	ISOM.	2.79	HAL	MSV, GRNLR. VITR. CUBIC CLVG. SOL-H2O. CAVITIES IN ALK IG RKS	9.1.1.3

NONMETALLIC WHITE

2 -2.5	HAIDINGERITE	ORTH.	2.85	ARS	BOTRYOIDAL, FIBR COATINGS. VITR,PEARLY, 1 CLVG(FLEX). SOL-ACIDS. OXIDIZED PARTS OF ARSENIC DEPOSITS	39.2.2.
2 -2.5	HYDROZINCITE	MNCL.	3.5 -4.0	CBT	MSV,INCRUST, SILKY-DULL.1 CLVG. FLUO-BLUE,LILAC. SOL-ACIDS.SCNDRY WITH OTHER ZINC MINERALS	16.1.3.
2 -2.5	SENARMONTITE	ISOM.	5.50	OXD	OCTAH, GRNLR. RESIN-ADMN. OCTAH CLVG.SOL-HCL. SCNDRY AFTER SB-MINS	4.4.2.2
2 -2.5	TELLURIUM	HEXA.	6.1 -6.3	ELE	OPQ.SN-WHT, GRAY STRK. SOL-H2SO4 TO RED SULFITE. VEINS	1.2.2.3
2 -2.5	LANARKITE	MNCL.	6.92	SUF	MSV, ELONGATE. ADMN,PEARLY, CLVGS(THIN LAMINAE-FLEX), FLUO-YEL(X-RAYS + ULTRAVIOLET RADIATION). SOL-WARM HNO3. RARE-LEAD DEPOSITS	30.2.1.
2 -3	ALLOPHANE (CLAY)	AMOR.	1.85-1.89	PSI	INCRUST, "BOOKS." GELAT, ALTER(ACID CONDITIONS) OF FELDSPARS, ETC. IN COAL, ORE DEPOSITS, AND VEINS	W531 3-194
2 -3	ROESSLERITE	MNCL. A	1.94	ARS	HAIR(LK CRUSTS. VITR-DULL. CLVG. SOL-DILUTE HCL,H2O(SLIGHT). OXI-DIZED ZONES OF ARSENIC DEPOSITS.	39.2.5.1
2 -3	VASHEGYITE	UNKN.	1.96	PHO	SUBTRANSLUCENT. MSV,MICROXLINE-FIBR.DULL.SOL-ACIDS. WITH VARISCITE	42.9.3.
2 -3	BIANCHITE	MNCL.	2.07	SUF	CRUSTS. VITR. SOL-H2O. RARE-EFFLOR ON MINE WALLS(IN JULIAN ALPS)	29.6.6.2
2 -3	HYDROBORACITE	MNCL.	2.17	BOR	LAMELLAR,FIBR. VITR,SILKY. 2 CLVGS. SOL-ACIDS. EVAPORITE DEPOSITS	25.1.10.
2 -3	PITTICITE	UNKN.	2.2 -2.5	ARS	MSV,BOTRYOIDAL,OPALINE CRUSTS,ETC. DULL-GRSY. SOL-ACIDS. OXIDATION ZONES OF AS-BEARING DEPOSITS. DEPOSITED FROM MINE + SPRING WATERS	43.2.6.
2 -3	FIBROFERRITE	ORTH.	1.84-2.52	SUF	FIBR CRUSTS. SILKY. 1 CLVG. DCMP-H2O. SCNDRY-BY OXIDATION OF FES2	31.6.6.
2 -3	RADIOPHYLLITE	UNKN.	2.53	PSI	SPHERES, RADIAL STRUCTURE. WITH PHILLIPSITE-FISSURES IN PHONOLITE	WN N
2 -3	COOKEITE (CLAY)	MNCL.	2.69	PSI	LI-BEARING CHLOR. WITH TOURMALINE IN PEGS, PROBABLY AS ALTER PROD	W378 3-147
2 -3	SHERIDANITE (CHLORITE)	MNCL.	2.65-2.80	PSI	INTERMEDIATE BETWEEN CLINOCHLORE + CORUNDOPHILITE. RARE-META ROCKS	WN 3-137
2 -3	TINTICITE	ORTH.	2.7 -2.9	PHO	DENSE EARTHY-PORCELANEOUS MASSES.SCNDRY-WITH JAROSITE,LIMONITE,ETC	42.8.4.
2 -3	CLINOCHLORE (CHLORITE)	MNCL.	2.6 -3.0	PSI	HEX-OUTLINE (COMMONLY SLENDER)XLS.MSV. VERY COMMON,E.G.IN TALC RKS	W383 3-137
2 -3	PENNINITE (CHLORITE)	MNCL.	2.6 -3.1	PSI	PSEUDORHOMB XLS. WITH MAGNETITE + CHONDRODITE(SERP RKS), SCHISTS	W383 3-137
2 -3	CHLORITE	MNCL.	2.6 -3.3	PSI	GROUP NAME. MEMBER NAMES UNSETTLED. ONLY FEW RARE VARIETIES MEGASCOPICALLY DIS-TINGUISHABLE. DISSEM FLKS. PEARLY, 1 PERF CLVG TO FLEX,INELAST FOLIA. DCMP-IN BOILING H2SO4. SCNDRY-HYDROTHERMAL ALTER, LO-GRADE META, DETRITAL, AUTHIGENIC	W381 3-131
2 -3	PHOSGENITE	TETR.	6.13	CBT	PRISM, MSV. ADMN, CLVGS(PERCUSSION FIGURE ON BASAL PLANE) FLUO-YEL (LW ULTRAVIOLET, CATHODE+ X-RAYS). EFF-HNO3., SCNDRY AFTER PB MINS	16.1.7.
2 -3	BEYERITE	TETR.	6.50-6.56	CBT	PLTY, MSV. VITR(XLS), CONCH. EFF-ACIDS. SCNDRY-WITH BISMUTITE	16.2.5.
2 -3	MICA	MNCL.	2.4 -3.4	PSI	GROUP NAME. SOME SPECIES DISTINGUISHABLE ON BASIS OF COLOR. SERIES ARE NOT COM-PLETE. HEX-SHAPED XLS,DISSEM. PERF CLVG TO FLEX,ELAST FLKS. PERCUSSION + PRES-SURE FIGURES MAY BE IMPOSED ON FLKS. CHIEF OCCURRENCES ARE NOTED UNDER SPECIES	W365 3-1
2 -4	BAUXITE	2	3.5	OXD	FIELD TERM FOR MATERIALS RICH IN HYDROUS ALUMINUM OXIDES.	6.2.3.
2.5	BISMOCLITE	TETR.	7.12	HAL	IN SERIES WITH DAUBREEITE. MSV, SCALY, GRSY-DULL, 1 CLVG. PLASTIC. SOL-ACIDS. SCNDRY-ALTER PROD OF BISMUTHINITE OR NATIVE BISMUTH	10.1.6.2
2.5	DAUBREEITE	TETR. C	7.56	HAL	IN SERIES WITH BISMOCLITE,WHICH SEE.SCNDRY AFTER BISMUTHINITE + BI	10.1.6.3
2.5	OXAMMITE	ORTH.	1.4 -1.6	CCC	LAMELLAR,POWDERY. 1 CLVG. WITH MASCAGNITE IN GUANO DEPOSITS(PERU)	50.1.4.

117

NONMETALLIC WHITE

HARDNESS	NAME	XL. SYS.	SPECIFIC GRAVITY	CHEM. CLASS	REMARKS	REFERENCES
● 2.5	CARNALLITE	ORTH.	1.60	HAL	GRNLR, PSEUDOHEX PRISM. GRSY. CONCH. DLQSNT. BITTER. SED SALT DEPS	11.1.2.
2.5	LANSFORDITE	MNCL.	1.69	CBT	STALACTITIC, PRISM. VITR. EXPOS-DULL. 2 CLVGS. EFF-ACIDS. EFFLOR	15.1.5.
2.5	NESQUEHONITE	ORTH.	1.85	CBT	PRISM-STRIATED, RADIAL AGGS. VITR. PYRAMID. BASAL CLVGS. SOL-CO2 CHARGED H2O. LOW TEMPERATURE DEPOSITION IN OPEN SPACES	15.1.2.
● 2.5	KERNITE	MNCL.	1.91	BOR	EQUANT-STRIATED,MSV. VITR,SATINY. CLVGS. SOL-HOT H2O,ACIDS. BORATE DEPOSITS	25.1.2.
2.5	ULEXITE	TRCL.	1.96	BOR	NODULAR, CRUSTS. SILKY-VITR. CLVGS. DCMP-HOT H2O. PLAYA SALT LAKES	25.1.7.
2.5	CRYPTOHALITE	ISOM.	2.00	HAL	MAMMILLARY-DENDRITIC. VITR. OCTAH CLVG. SALTY TASTE. VOL SUBLIMATE	11.4.1.2
2.5	ARTINITE	MNCL.	2.01-2.03	CBT	ACICULAR CRUSTS, FIBR. VITR,SILKY. 2 CLVGS. EFF-ACIDS. VNS IN UL-TRABASIC IGNEOUS ROCKS	16.1.10.
2.5	PICROMERITE	MNCL. A	2.03	SUF	PRISM, CRUSTS. VITR. BITTER. FUMAROLIC, MARINE SALINE DEPS-AS BEDS	29.3.7.1
2.5	PYROAURITE	HEXA.	2.10-2.14	OXD	TBLR-STRIATED. WAXY,PEARLY. BASAL CLVG(FLEX). EFF-HCL. LO-T VNS	6.1.5.3
2.5	SJOGRENITE	HEXA.	2.08-2.14	OXD	PLTS-STRIATED. WAXY,PEARLY. BASAL CLVG(FLEX).EFF-HCL. RARE-LANGBAN	6.1.6.3
2.5	BARARITE	HEXA.	2.15	HAL	TBLR, MAMMILLARY, DENDRITIC. VITR. 1 CLVG. SALTY. VOL SUBLIMATE	11.4.2.2
● 2.5	CHRYSOTILE (SERPENTINE)	MNCL.	2.2	PSI	FIBROUS, SOME IS ASBESTIFORM. SECONDARY, COMMONLY IN VEINS	W379 3-170
2.5	NAHCOLITE	MNCL.	2.21	CBT	PRISM, FRIABLE AGGS. VITR. CLVGS. SOL-H2O. WITH TRONA	13.1.1.
2.5	ALUMOHYDROCALCITE	MNCL.	2.23	CBT	CHALKY MASSES. 2 CLVGS. SOL-ACIDS,DCMP-BOILING H2O. WITH ALLOPHANE	16.2.4.
● 2.5	BRUCITE	HEXA.	2.38-2.40	OXD	WHT STRK. TBLR, FOL MSV. WAXY-PEARLY. BASAL CLVG(FLEX). SOL-ACIDS. LOW TEMPERATURE VEINS	6.1.1.1
2.5	CALCIOFERRITE	UNKN.	2.53	PHO	FOL NODULAR MASSES. PEARLY. CLVGS. RARE-NODULES IN CLAY(BAVARIA)	42.8.9.
2.5	LIZARDITE (SERPENTINE)	MNCL.	2.5 -2.6	PSI	SCALES, MASSIVE. SECONDARY (MASSIVE MATRIX MATERIALS)	WN 3-170
2.5	SYNGENITE	MNCL.	2.58-2.60	SUF	CRUSTS, LAMELLAR AGGS, PRISM. VITR. CLVGS. SOL-H2O(IN PART WITH FM OF GYPSUM).RARE-MARINE SALT DEPS, PRODUCT OF VOLCANIC ACTION	29.3.1.
2.5	FERRINATRITE	HEXA.	2.55-2.61	SUF	FIBR AGGS, PRISM, MSV. VITR. CLVGS. SPLINTERY. SCNDRY WITH OTHER IRON SULFATES IN ARID REGIONS	29.4.1.
2.5	DONBASSITE (CHLORITE)	TRCL.	2.63	PSI	PEARLY FLKS (RESEMBLES PYROPHYLLITE). RARE (DONETZ BASIN)	W512 N
2.5	HIERATITE	ISOM. A	2.67	HAL	CUBO-OCTAH, STALACTITIC. VITR. OCTAH CLVG. SOL-H2O. FUMAROLIC DEPS	11.4.1.1
2.5	PARAGONITE	MNCL.	2.85	PSI	NA-MICA. DISTINGUISHED FROM MUSCOVITE BY CHEM OR X-RAY ANAL. WIDE-SPREAD IN SCHISTS AND PHYLLITES, ALSO IN VEINS	W370 3-31
● 2.5	CRYOLITE	MNCL.	2.96-2.98	HAL	MSV, GRNLR. GRSY. UNEVN. THEMOLUM. SOL-H2O(SLIGHTLY). RARE-PEGS	11.5.1.
2.5	MINNESOTAITE	MNCL.	3.0 -3.1	PSI	ESSENTIALLY FE-ANALOGUE OF TALC. FLKS, NEEDLES. FE-FMS(MINNESOTA)	M365 3-124
2.5	FORBESITE	UNKN.	3.13	ARS	NEEDS VERIFICATION.	39.2.4.
2.5	NANTOKITE	ISOM.	4.14	HAL	MSV, GRNLR. ADMN. DODEC CLVG. DCMP-H2O,SOL-HCL,HNO3,NH4OH. CU-DEPS	9.1.3.1
2.5	CLAUDETITE	MNCL.	4.15	OXD	PLTY. VITR-PEARLY. 1 PERF CLVG(FLEX). SOL-ALK SOLS. SCNDRY AS-MIN	4.4.3.

NONMETALLIC WHITE

Hardness	Mineral	Crystal System	SG	Class	Description	Ref
2.5	COTUNNITE	ORTH. A	5.80	HAL	GRNLR, LATHLK, ADMN-SILKY. 1 CLVG. SOL-H2O,ACIDS. VOL SUBLIMATE, ALTERATION PRODUCT OF GALENA	9.2.6.
2.5	MENDIPITE	ORTH.	7.24	HAL	FIBR, RADIATED, SILKY,RESIN. PRISM + OTHER CLVGS. SOL-HNO3. SCNDRY WITH CERUSSITE, ETC.	10.1.4.
2.5-3	CHLOROCALCITE	ORTH.		HAL	PSEUDOCUBIC,TBLR. CUBELK CLVG. DLQSNT. BITTER TASTE. VOL SUBLIMATE	11.1.1.
2.5-3	GAYLUSSITE	MNCL.	1.99	CBT	TBLR,STRIATED. VITR. PRISM CLVG. EFF-ACIDS. WITH TRONA,ETC.	15.2.3.
2.5-3	ZINCALUMINITE	HEXA.	2.26	SUF	CRUSTS, PLTY. SOL-ACIDS + ALKALIES. SCNDRY AT ZINC DEPSILAURIUM)	31.1.7.
2.5-3	THENARDITE	ORTH.	2.66	SUF	DIPYRAMIDAL, CRUSTS. VITR. CLVGS. FAINTLY SALTY. SALINE LAKE DEPS, EFFLORESCENCES, FUMAROLIC	28.2.4.
2.5-3	LANTHANITE	ORTH.	2.61-2.74	CBT	PLTY-TBLR, GRNLR. PEARLY. MICACEOUS CLVG. SOL-ACIDS. COATING ON CERITE, ALLANITE, ETC.	15.2.10.
2.5-3	CRYOLITHIONITE	ISOM.	2.77	HAL	DODEC. VITR. DODEC CLVG. SOL-H2SO4(↑HF). WITH CRYOLITE	11.2.4.
2.5-3	VALENTINITE	ORTH.	5.76	OXD	PRISM, TBLR, MSV. ADMN. PRISM CLVG. SOL-HCL. SCNDRY AFTER SB-MINS	4.4.4.
2.5-3	ANGLESITE	ORTH.	6.37-6.39	SUF	TBLR,PRISM,MSV,ETC. ADMN. CLVGS. CONCH. BRTL. SOME FLUO-YEL. SOL-HNO3(DIFFICULTY). SCNDRY - BY OXIDATION OF GALENA	28.3.1.3
2.5-3	LEADHILLITE	MNCL.	6.55	CBT	PSEUDOHEX, GRNLR. RESIN, PEARLY. 1 CLVG. FLUO-YEL. EFF-HNO3. SCNDRY IN OXIDIZED ZONES OF LEAD DEPOSITS	17.1.3.
2.5-3	MENDOZITE	MNCL.	1.73-1.77	SUF	EXPOS-WHT + TURBID. FIBR MASSES.3 CLVGS. SOL-H2O. REPORTED NATURAL OCCURRENCES NEED VERIFICATION	29.5.4.1
2.5-3.5	GIBBSITE	MNCL.	2.3 -2.42	OXD	TBLR, MAMMILLARY, PEARLY-VITR. 1 CLVG. CLAYLK ODOR WHEN DAMP. WX-ING PRODUCT. BAUXITE DEPOSITS, LOW TEMPERATURE VEINS	6.2.2.
2.5-3.5	SERPENTINE		2.5 -2.6	PSI	GROUP NAME. NOMENCLATURE FOR MEMBERS IS UNSETTLED. MSV-ASBESTIFM, WAXY-GRSY. GELAT, SOME FLUO-CREAM YEL. ALTER PROD AFTER MG-SILICATES(E.G.,OLIVINE). VNS	W381 3-170
2.5-3.5	ANTIGORITE (=SERPENTINE)	MNCL.	2.65	PSI	COLORS COMMONLY VARIEGATED OR MOTTLED. PLATY. SECONDARY	W381 3-170
2.5-3.5	BISMUTITE	TETR.	6.1 -7.7	CBT	GRAY STRK. MSV, RADIAL FIBR CRUSTS. VITR-DULL. EFF-ACIDS. SCNDRY AFTER BISMUTH MINERALS	16.1.8.
3	KURNAKOVITE	MNCL.	1.85	BOR	GRNLR. VITR. 1 POOR CLVG. SOL-WARM ACIDS. BORATE DEPOSITS	25.1.14.
3	INDERITE	TRCL.	1.86	BOR	ACICULAR-NODULAR AGGS. VITR,DULL. CLVGS. SOL-HCL. EVAPORITE DEPS	25.1.15.
3	FLUELLITE	ORTH.	2.14-2.17	HAL	DIPYRAMIDAL. VITR. 1 + POOR PYRAMIDAL CLVGS. VNS, PEGS	11.5.9.
3	METAVAUXITE	MNCL.	2.35	PHO	PRISM,ACICULAR,∥-RADIAL AGGS. VITR-SILKY. WX PRODII.E.,SUPERGENE)	42.8.5.
3	PARAVAUXITE	TRCL.	2.36	PHO	PRISM,SUBPARALLEL-RADIAL AGGS. VITR,PEARLY. 1 CLVG. CONCH. WITH VAUXITE, ETC. IN TIN VEINS	42.8.6.
3	DAWSONITE	ORTH.	2.44	CBT	ACICULAR CRUSTS VITR-SILKY. CLVG. EFF-ACIDS. LO-T HYDROTHERMAL MIN	16.2.1.
3	FERRUCCITE	ORTH.	2.50	HAL	TBLR XLS. 3 CLVGS. BITTER-ACID TASTE. FUMAROLIC SUBLIMATE.	11.2.3.
3	RIVERSIDEITE(=CRESTMORITE)	UNKN.	2.6	ISI	FIBR, MSV. SILKY. DCMP-HCL. VNS WITH CALCITE + VESUVIANTE –RARE	W421 N
3	ARCANITE	ORTH.	2.66	SUF	TBLR-CYCLIC(6) TWINS. 2 CLVGS. RARE-IN PINE RR TIE(SANTA ANA MINE, ORANGE COUNTY, CALIFORNIA)	28.2.1.2
3	TORREYITE	UNKN.	2.67	SUF	MSV,GRNLR,FOL. VITR-PEARLY. 1 CLVG. SOL-ACIDS. VNSISTERLING HILL, NEW JERSEY)	31.1.4.
3	CALCITE	HEXA.	2.71	CBT	VARIOUS TRIGONAL XL HABITS. MSV. VITR-PEARLY. RHOMB CLVG. SOME IS FLUO,PHOSPHO,THERMOLUM. EFF-DILUTE HCL. WDSPRD-VNS,DIVERSE ROCKS	14.1.1.1

119

NONMETALLIC WHITE

HARDNESS	NAME	XL. SYS.	SPECIFIC GRAVITY	CHEM. CLASS	REMARKS	REFERENCES
3	APHTHITALITE	HEXA.	2.65-2.71	SUF	TRIGONAL-TBLR, MAMMILLARY,MSV, VITR,CLVGS, CONCH. SOL-H2O, ACIDS. FUMAROLIC, OCEANIC + LACUSTRINE SALINE DEPOSITS	28.2.2.
3	PACHNOLITE	MNCL.	2.98	HAL	PRISM-STRIATED, MSV, VITR.1 CLVG. SOL-H2SO4.AT CRYOLITE LOCALITIES	11.5.3.
3	BOEHMITE	ORTH.	3.01-3.06	OXD	PISOLITES, TBLR. 1 CLVG. IN BAUXITE DEPOSITS	6.1.2.2
3	SPENCERITE	MNCL.	3.14	PHO	TBLR,MSV,STALAC,COLUMNAR-PLTY. PEARLY-VITR. 2 CLVGS. SOL-ACIDS. RARE-WITH OTHER SECONDARY ZINC MINERALS	42.4.4.
3	CAHNITE	TETR.	3.16	BOR	TWINS-E.G.; CRUCIFORM, VITR. PRISM CLVG. SOL-HCL. VNS CUTTING MAIN ORE BODY AT FRANKLIN, NEW JERSEY	27.1.2.
3	LOSEYITE	MNCL.	3.27	CBT	LATHS-SUBPARALLEL TO RADIAL AGGS. IN CALCITE VNLETS AT FRNKLN,N.J.	16.1.1.
3	OLIVENITE	ORTH.	3.9 -4.46	ARS	ELONGATE,GLBL FIBR,MSV,ETC. ADMN-SILKY. CLVGS. CONCH. SOL-ACIDS + NH4OH. SECONDARY-OXIDIZED ZONES OF ORE DEPOSITS	41.6.5.1
3	LARSENITE	ORTH.	5.9	NSI	STRIATED PRISMS. ADMN. GELAT. RARE(FRANKLIN, NEW JERSEY)	M504 N
3 -3.5	TEEPLEITE	TETR.	2.08	BOR	TBLR. GLASSY, EXPOS-DULL. SOL-H2OITO ALK SOLUTION).WITH TRONA	26.1.3.
3 -3.5	LAUMONTITE (ZEOLITE)	MNCL.	2.2 -2.3	TSI	PERF CLVGS-3 DIRECTIONS IN 1 ZONE. GELAT. CAVITIES IN SEVERAL RKS	W342 4-401
3 -3.5	PIRSSONITE	ORTH.	2.35	CBT	PRISM, TBLR. VITR. PYROELECT. EFF-ACIDS. WITH TRONA,ETC.-SED DEPS.	15.2.2.
3 -3.5	PRICEITE	TRCL.	2.42	BOR	NODULES. EARTHY. CONCH. SOL-ACIDS. HOT SPRINGS, SOLFATARIC LAGOONS	25.1.5.
3 -3.5	SZAIBELYITE	ORTH.	2.62	BOR	IN SERIES WITH SUSSEXITE(SEE). VNS, SALT DEPS, SKARN ZONES	26.1.5.2
3 -3.5	HOPEITE	ORTH.	3.00-3.10	PHO	TBLR,PRISM,TUFTED AGGS,CRUSTS. VITR,PEARLY. CLVGS. SOL-DILUTE HCL. WITH HEMIMORPHITE, ETC.	40.2.11.
3 -3.5	REDDINGITE	ORTH.	3.0 -3.2	PHO	MSV,GRNLR,FIBR,TBLR. VITR. 1 CLVG. UNEVN. SOL-ACIDS. PEGMATITES	40.2.5.1
3 -3.5	SUSSEXITE	ORTH.	3.30	BOR	IN SERIES WITH SZAIBELYITE. FIBR(INFLEX),DENSE-CHALKY AGGS. SILKY-DULL. SOL-ACIDS(SLOWLY). VNS (E.G., FRANKLIN, NEW JERSEY)	26.1.5.1
3 -3.5	CELESTITE	ORTH.	3.96-3.98	SUF	TYPICALLY COLORLESS-BLUISH. TBLR,ETC. VITR-PEARLY. CLVGS. SOME FLUO,THERMOLUM. SOL-H2O(SLIGHTLY). SED RKS(ESP GYPSUM + LS),VNS	28.3.1.2
3 -3.5	WITHERITE	ORTH.	4.29	CBT	PSEUDOHEX, GLBLR,ETC. VITR. FLUO(ETC.)-LK ARAGONITE.SOL-HCL. VNS	14.1.3.2
3 -3.5	BARITE	ORTH.	4.50	SUF	TBLR, MSV,ETC. VITR-PEARLY. CLVGS. SOME FLUO, PHOSPHO, THERMOLUM. SOME FETID WHEN RUBBED. COMMON-VNS,SED RKS,CAVITIES IN IG ROCKS	28.3.1.1
3 -3.5	CERUSSITE	ORTH.	6.53-6.57	CBT	TWINNED CLUSTERS. MSV. ADMN-RESIN. CLVGS. FLUO-YEL(X-RAYS +(LM)UL-TRA VIOLET). OXIDIZED ZONES OF LEAD DEPOSITS	14.1.3.4
3 -4	KERSTENITE	UNKN.		SLI	BOTRYOIDAL, FIBR. GRSY-VITR. 1 CLVG. ALTER PROD OF ZORGITE-RARE	34.1.3.
3 -4	MORDENITE (ZEOLITE)	ORTH.	2.12-2.15	TSI	STRIATED PRISMS, ACICULAR, COTTON-LK AGGS. ZEOL-ASSOC, AUTHIGENIC	W339 4-401
3 -4	EVANSITE	UNKN.	1.8 -2.2	PHO	MSV,COLLOFM COATINGS. VITR. CONCH. BRTL. SOL-ACIDS. WITH LIMONITE	42.2.6.
3 -4	GISMONDINE (ZEOLITE)	ORTH.	2.1 -2.3	TSI	PSEUDOTETRAG BIPYRAMIDS. 1 CLVG. GELAT. LEUCITE RKS, ALTER OF PLAG	W339 4-401
3 -4	WAVELLITE	ORTH.	2.36	PHO	RADIAL FIBR,STELLATE,CRUSTS. VITR-PEARLY.CLVGS. SOL-ACIDS. SCNDRY-WDSPRD-OPENINGS IN DIVERSE ROCKS,LIMONITE,PHOSPHATE ROCKS, ETC.	42.7.4.
3 -4	DIADOCHITE	TRCL.	2.0 -2.4	PHO	NODULAR,PLTY,GEL-LK,COLLOFM. DULL-WAXY. UNEVN-CONCH. SOL-ACIDS. IN GOSSANS AND AS EFFLORESCENCES	43.2.4.
3 -4	TRUSCOTTITE	HEXA.	2.34-2.45	PSI	MEGASCOPICALLY INDISTINGUISHABLE FROM GYROLITE. LAMELLAR. PERF AL CLVG. SOL-ACID. WITH APOPHYLLITE AND ZEOLITES	BAS-W395 N

NONMETALLIC WHITE

Hardness	Mineral	Crystal	SG	Class	Habit/Properties	Occurrence	Ref
3 -4	GYROLITE	HEXA.	2.34-2.45	PSI	BASAL CLVG. SOL-ACID. WITH APOPHYLLITE AND ZEOLITES	MEGASCOPICALLY INDISTINGUISHABLE FROM TRUSCOTTITE. LAMELLAR. PERF	W395 N 40.3.2.1
3 -4	METAVARISCITE	MNCL.	2.54	PHO	LATHLK, GRNLR. VITR. 1 CLVG. SOL-ALKALIES. WITH VARISCITE		40.3.3.
3 -4	WEINSCHENKITE	MNCL.	3.27	PHO	CRUSTS,RADIAL FIBR,ROSETTES, SILKY, CLVGS. THERMOLUM.SOL-HOT ACID. WITH LIMONITE OF NEAR-SURFACE FORMATION		40.3.3.
3 -4	MIXITE	HEXA.	3.79	ARS	HAIRLK-STRIATED,CONCENTRIC FIBR MASSES. DULL. WITH OTHER BI-MINS		42.5.7.
3 -4	FIEDLERITE	MNCL.	5.88	HAL	LATHLK. ADMN. 1 CLVG. ATTKD-H2O,SOL-HNO3. RARE-ACTION OF SEA WATER ON ANCIENT LEAD SLAGS (LAURIUM, GREECE)		10.1.10.
3 -5.5	ZEOLITE		1.9 -2.45	TSI	GROUP OF MIN FAMILIES WITH SOMEWHAT RELATED APPEARANCES, OCCURRENCES, AND COMPOSITIONS. MANY ARE INDISTINGUISHABLE MEGASCOPICALLY. DIVERSE HABITS(E.G., BUNDLES, FIBROUS), PEARLY-GLASSY, SUBCONCH. SOME FLUO-ORANGE OR YEL-GREEN. CONTINUOUS, IN PART REVERSIBLE, DEHYDRATION(I.E.,WATER CAN BE EXPELLED WITH- OUT DESTROYING XL STRUCTURE). SCNDRY IN OPEN SPACES; COMMONLY TWO OR MORE OCCUR TOGETHER IN CAVITIES IN BASALTIC RKS — THIS OCCURRENCE IS REFERRED TO HEREIN AS THE ZEOLITE ASSOCIATION (ABBREVIATED "ZEOL-ASSOC").		W330 4-351
3.5	LAURIONITE	ORTH. A	6.24	HAL	TBLR. ADMN,PEARLY. CLVG. SOL-HOT H2O, HNO3. SCNDRY AFTER PB-MINS		10.1.7.
3.5	INDERBORITE	MNCL.	2.00	BOR	COARSE AGGS. VITR. 1 CLVG. SOL-HCL, H2O. BORATE DEPOSITS		25.1.11.
3.5	GINORITE	MNCL.	2.09	BOR	PLTY. DENSE MASSES.1 CLVG. RARE-CALCITE VEINS IN SS(TUSCANY,ITALY)		25.1.19.
3.5	GORDONITE	TRCL.	2.23	PHO	SHEAFLK AGGS. VITR,PEARLY. CLVGS. CONCH. SOL-ACIDS. IN VARISCITE NODULES		42.8.8.
3.5	HYDROMAGNESITE	MNCL.	2.24	CBT	TUFTS, CRUSTS, MSV. VITR-EARTHY. EFF-ACIDS VNS IN ULTRABASIC IGS		16.1.13.
3.5	MINYULITE	ORTH.	2.45	PHO	RADIAL AGGS OF NEEDLES. SILKY. SOL-HOT ACIDS. ALTER PROD OF GLAU- CONITE		42.8.3.
3.5	SEARLESITE	MNCL.	2.45	SSI	PRISM XLS, SPHERULITES OF RADIATING FIBERS. 1 GOOD CLVG. DCMP-HCL		W421 N
3.5	KIESERITE	MNCL.	2.57	SUF	MSV,GRNLR. VITR. CLVGS. FRIABLE. SOL-H2O(SLOW). MARINE SALINE DEPS		29.6.2.1
3.5	BURKEITE	ORTH.	2.57	SUF	TBLR,PLTY AGGS,ETC. GRSY. CONCH.SOL-H2O. WITH TRONA,ETC. BORAX DEP		32.14.
3.5	HOWLITE	MNCL.	2.53-2.59	BOR	NODULAR, DENSE-SCALY. SUBVITR. GELAT. IN EVAPORITE DEPOSITS		25.1.16.
3.5	POLYHALITE	TRCL.	2.78	SUF	MSV-FIBR, FOL, VITR. 1 CLVG. DCMP-H2O(+GYP). MARINE SALINE DEPS		29.4.2.
3.5	ANAPAITE	TRCL.	2.81	PHO	TBLR,ROSETTELK AGGS, CRUSTS. VITR. 2 CLVGS. SOL-HCL,HNO3. DIVERSE		40.2.9.
3.5	MONETITE	TRCL. A	2.93	PHO	CRUSTS,AGGS. VITR. 3 CLVGS. EXPOS-HYDRATES TO BRUSHITE. GUANO DEPS		37.1.1.
3.5	FLUOBORITE	HEXA.	2.85-2.98	BOR	ACICULAR, FELTED AGGS. 1 POOR CLVG. CNTCT METASOMATIC DEPOSITS		26.1.1.
3.5	FAIRFIELDITE	TRCL.	3.08	PHO	FOLIATED AGGS, PRISM. PEARLY-GRSY. CLVGS. SOL-ACIDS. PEGMATITES		40.2.3.1
3.5	LIME	ISOM.	3.3	OXD	POSSIBLY OCCURS NATURALLY IN CALC ROCKS WITHIN LAVA AT VESUVIUS		4.2.1.5
3.5	BRANDTITE	MNCL.	3.67	ARS	STOUT PRISM,RADIAL FIBR GRPS. VITR. 1 CLVG. SOL-DILUTE ACIDS. VNS		40.2.4.2
3.5	STRONTIANITE	ORTH.	3.64-3.78	CBT	SPEAR-LK, MSV. VITR. CLVGS. FLUO(ETC.)-LK ARAGONITE. SOL-HCL. VNS		14.1.3.3
3.5	RHABDOPHANE	HEXA.	3.94-4.01	PHO	RADIATED FIBR MAMMILLARY CRUSTS. GRSY. UNEVN. SOL-HCL. WITH LIMO- NITE		40.3.5.
3.5	ADAMITE	ORTH.	4.32-4.48	ARS	ELONGATE,CRUSTS,RADIAL AGGS. VITR. CLVGS. UNEVN. SOL-DILUTE ACIDS. SOME FLUO YEL.OXIDIZED ZONES OF DEPS WITH ZN- + AS-RICH MINERALS		41.6.5.3
3.5	HYDROCERUSSITE	HEXA.	6.80	CBT	SCALY,TBLR.ADMN,PEARLY,BASAL CLVG. EFF-ACIDS. SCNDRY AFTER PB MINS		16.1.12.

NONMETALLIC WHITE

HARDNESS	NAME	XL. SYS.	SPECIFIC GRAVITY	CHEM. CLASS	REMARKS	REFERENCES
3.5	GEORGIADESITE	MNCL.	7.1	ARS	PSEUDOHEX PLTS,STRIATED. RESIN. SOL-HNO3. RARE-1 SPECIMEN(LAURIUM)	41.3.3.
3.5-4	FERRIERITE (ZEOLITE)	ORTH.	2.15	TSI	PLTY GRP. LATH-LK. RARE-IN VNS WITH CHALCEDONY IN BASALT (KAMLOOPS LAKE, BRITISH COLUMBIA)	W337 4-382
3.5-4	DACHIARDITE (ZEOLITE)	MNCL.	2.16	TSI	PLTY GRP. SECTOR(8) TWINNED XLS. 2 CLVGS. DCMP-HCL. RARE-PEG(ELBA)	W348 4-382
3.5-4	HEULANDITE (ZEOLITE)	MNCL.	2.1-2.2	TSI	PLTY GRP. TBLR-BROAD CENTRAL PORTIONS(COFFIN-LIKE). ZEOL-ASSOC	W347 4-377
● 3.5-4	STILBITE (ZEOLITE)	MNCL.	2.1-2.2	TSI	PLTY GRP. SHEAF-LIKE AGGS. 1 GOOD CLVG. DCMP-HCL. ZEOL-ASSOC	W345 4-377
3.5-4	BREWSTERITE (ZEOLITE)	MNCL.	2.45	TSI	PLTY GRP. ELONGATE,STRIATED PRISMS. 1 CLVG. DCMP-HCL. ZEOL-ASSOC	W347 4-383
3.5-4	TYCHITE	ISOM.	2.46-2.59	CBT	OCTAH.VITR.CONCH.SOL-DILUTE ACIDS. RARE-CLAY BED,BORAX LAKE,CALIF	17.1.1.
3.5-4	CREEDITE	MNCL.	2.71-2.73	HAL	PRISM, RADIATING AGGS. VITR.1 CLVG. SOL-ACIDS(SLOWLY).FLUORITE VNS	12.1.1.
● 3.5-4	DOLOMITE	HEXA.	2.84-2.86	CBT	RHOMB-CURVED, GRNLR, VITR-PEARLY. RHOMB CLVG. SOME FLUO. EFF-WARM HCL IF POWDERED. SEDIMENTARY RKS, CRYSTALS IN CAVITIES, ETC.	14.2.1.1
● 3.5-4	ALUNITE	HEXA.	2.6-2.9	SUF	GRNLR, MSV. VITR,PEARLY. CLVGS. SOL-H2SO4. "ALUNITIZED" ROCKS	30.2.4.1
3.5-4	NATROALUNITE	HEXA.	2.6-2.9	SUF	GRNLR, MSV. VITR,PEARLY. CLVGS. SOL-H2SO4. RARE,BUT VARIED OCCUR.	30.2.4.2
● 3.5-4	ARAGONITE	ORTH.	2.94-2.95	CBT	ACICULAR, RADIAL FIBR, STALACTITIC.VITR. CLVGS. FLUO(ULTRA-VIOLET, X-RAYS,ELECTRON BEAMS), THERMOLUM. EFF-ACIDS. LO-T DEPOSITION	14.1.3.1
3.5-4	KUTNAHORITE	HEXA.	3.0	CBT	DISTINGUISHED FROM DOLOMITE(SEE) BY NONMEGASCOPIC METHODS. RARE	14.2.1.3
3.5-4	CHIOLITE	TETR.	3.00	HAL	GRNLR. 1 PERF, OTHER CLVGS. RESEMBLES CRYOLITE. IN CRYOLITE PEGS	11.5.8.
● 3.5-4	ANKERITE	HEXA.	3.01-3.03	CBT	DISTINGUISHED FROM DOLOMITE(SEE) BY NONMEGASCOPIC METHODS. VNS	14.2.1.2
3.5-4	MANSFIELDITE	ORTH.	3.03	ARS	CELLULAR MASSES,CRUSTS.VITR. CLVGS. SOL-ACIDS. RARE-WITH SCORODITE	40.3.1.4
3.5-4	POWELLITE	TETR.	4.21-4.25	MBS	PYRAMIDAL,TBLR,MSV-FOL,POWDER,CRUSTS.SUBADMN-PEARLY. CLVGS. UNEVN. FLUO-YEL. DCMP-HCL,HNO3. ALTER PROD OF MOLYBDENITE	48.1.3.2
3.5-4	MIMETITE	HEXA.	7.24	ARS	PRISM,ACICULAR,GLBL,ETC. RESIN. RHOMB CLVG. UNEVN. SOL-HNO3. OXI-DIZED ZONES OF LEAD DEPOSITS	41.7.2.2
3.5-4.5	MAGNESITE	HEXA.	2.98-3.02	CBT	GRNLR, RHOMB. VITR. RHOMB CLVG. FLUO-GREEN,BLUE. SOL-WARM HCL. AL-TERATION OF MG-RICH ROCKS (E.G., SERPENTINE)	14.1.1.2
● 3.5-4.5	SIDERITE	HEXA.	3.95-3.97	CBT	TARN-IRID. MSV. RHOMB,ETC. VITR.RHOMB CLVG. SOL-HOT ACID. SED. VNS	14.1.1.3
4	WEDDELLITE	TETR.	1.93-1.95	CCC	PYRAMIDAL.AGGS. SUBCONCH. ORGANIC ASSOC. BOTTOM MUDS(WEDDELL SEA)	50.1.2.
4	EPISTILBITE (ZEOLITE)	MNCL.	2.1-2.3	TSI	PLTY GRP. SHEAF-LK + SPHERICAL AGGS. DCMP-HCL.PIEZOELECTRIC. ZEOL-ASSOC	W346 4-377
4	LEUCOPHANE	ORTH.	2.96	SSI	TBLR. 2 PERF CLVGS. TWINNING COMMON(SOME PENETRATION). RARE-PEGS	W476 N
● 4	FLUORITE	ISOM.	3.18	HAL	CUBIC, MSV. VITR. OCTAH CLVG. FLUO-VIOLET. DCMP-H2SO4. WDSPRD-VNS, CAVITIES, DISSEMINATED IN IGNEOUS ROCKS, ETC.	9.2.1.
4	BARYTOCALCITE	MNCL.	3.66-3.71	CBT	PRISM-STRIATED, MSV. VITR. CLVG. WEAKLY FLUO. SOL-HCL. LO-T DEPS	14.2.3.
4	NASONITE	HEXA.	5.43	SSI	GRNLR. PRISM CLVG. GELAT. CONSTITUTES SER WITH GANOMALITE. FRNKLN	W478 N
4 -4.5	KALIBORITE	MNCL.	2.13	BOR	XL AGGS, GRNLR. VITR. CLVGS. SOL-ACIDS. SALINE EVAPORITE DEPOSITS	25.1.22.

NONMETALLIC WHITE

H	Mineral	Cryst	SG	Sp.Gr.	Class	Description	Ref
4 -4.5	PHILLIPSITE (ZEOLITE)	MNCL.	2.2	TSI	PENETRATION TWINS, SPHERULITES, 2 GOOD CLVGS. GELAT. DISTINGUISHED FROM HARMOTOME BY OPTICAL METHODS. ZEOL-ASSOC,DEEP-SEA RED CLAYS	W343 4-386
4 -4.5	JEZEKITE	MNCL.	2.94	PHO	PRISM;STRIATED FIBR CRUSTS,AGGS. VITR. 2 CLVGS. DCMP-HOT H2SO4. DRUSES IN GRANITE (SAXONY, GERMANY)	41.2.3.
4 -4.5	ALSTONITE	ORTH.	3.67-3.71	CBT	PSEUDOHEX, STRIATED. VITR. FLUO-YEL(LW). SOL-HCL. LO-T DEPS	14.2.2.
4 -4.5	JARLITE	MNCL.	3.78-3.93	HAL	TBLR, RADIAL AGGS. VITR. SOL-ALCL3 SOLUTION. RARE-IVIGTUT,GREENLD.	11.5.5.
4 -4.5	AUSTINITE	ORTH.	4.13	ARS	BLADED,FIBR CRUSTS + NODULES. SUBADMN,SILKY. 1 CLVG. SCNDRY-RARE	41.5.1.3
4 -4.5	SMITHSONITE	HEXA.	4.40-4.45	CBT	BOTRYOIDAL, MSV. VITR-PEARLY. RHOMB CLVG. SOME FLUO-GREEN, BLUE. EFF-ACIDS. SECONDARY - OXIDIZED ZONES OF ZINC DEPOSITS	14.1.1.6
4 -4.5	BINDHEIMITE	ISOM.	4.6 -5.6	ANT	DENSE-EARTHY CRUSTS,NODULAR-CONCENTRIC,SOME IS OPALINE. RESIN-DULL DCMP-HNO3, HCL. OXIDATION ZONES OF ANTIMONY-BEARING LEAD DEPOSITS	44.1.1.1
4 -4.5	SVABITE	HEXA.	3.5 -3.8	ARS	SHORT PRISM,MSV. VITR.CLVG. SOL-DILUTE ACIDS.RARE-E.G.,FRNKLN,N.J.	41.7.3.1
4 -5	XENOTIME	TETR.	4.4 -5.1	PHO	PRISM + PYRAMIDAL. VITR. 1 CLVG. ACID + ALK IG RKS + PEGS,DETRITAL	38.4.1.
4 -5	FLUOCERITE	HEXA.	5.93-6.14	HAL	FRESH-YEL. MSV, TBLR. RESIN,PEARLY. BASAL CLVG. SOL-H2SO4. PEGS	9.3.2.
4 -5	CERVANTITE	ORTH. A	6.64	OXD	MSV, FIBR, ACICULAR. GRSY-PEARLY. OXIDATION PROD OF SB-MINS	4.5.6.
4 -5.5	STIBICONITE	ISOM.	5.58	OXD	MSV, POWDER, CRUSTS. EARTHY-GLASSY. OXIDATION PROD OF SB-MINS	4.5.7.
4.5	CHABAZITE (ZEOLITE)	HEXA.	2.05-2.10	TSI	RHOMBS(NEARLY CUBES). RHOMB CLVG. DCMP-HCL.ZEOL-ASSOC. RK CAVITIES	W334 4-387
4.5	GMELINITE (ZEOLITE)	HEXA.	2.0 -2.2	TSI	PRISM CLVG. DCMP-HCL. ZEOL-ASSOC	W335 4-387
4.5	LEVYNITE (ZEOLITE)	ORTH.	2.0 -2.2	TSI	CHEM OR X-RAY ANAL DISTINGUISH CHABAZITE +LEVYNITE.RARE-ZEOL-ASSOC	W335 4-387
4.5	COLEMANITE	MNCL.	2.42-2.43	BOR	EQUANT, GRNLR. VITR. 2 CLVGS. SOL-HOT HCL. EVAPORITE BORATE DEPS	25.1.9.
4.5	HARMOTOME (ZEOLITE)	MNCL.	2.41-2.47	TSI	HABIT LK PHILLIPSITE(ABOVE), + CRUCIFM TWINS. DCMP-HCL. ZEOL-ASSOC, ALSO WITH AL-WXING PRODS,IN GNEISSES, + WITH MN-MINERALIZATION	W344 4-386
4.5	RALSTONITE	ISOM.	2.56-2.62	HAL	OCTAH. VITR. OCTAH CLVG. DCMP-HF. RARE AT SOME CRYOLITE LOCALITIES	11.5.10.
4.5	DAVISONITE	UNKN.	2.85	PHO	BOTRYOIDAL CRUSTS-FIBR. 1 CLVG. IN CAVITIES IN VARISCITE-RARE	42.5.3.
4.5	BAKERITE	UNKN.	2.88	BOR	NODULES. DENSE-LK UNGLAZED PORCELAIN. BORATE DEPS, ZEOLITE-ASSOC.	25.1.17.
4.5	PROSOPITE	MNCL.	2.88-2.90	HAL	TBLR, MSV, GRNLR. VITR-DULL. PYRAMIDAL CLVG. DCMP-H2SO4. VNS, GREI SENS, PEGMATITES - RARE	11.5.7.
4.5	WOODHOUSEITE	HEXA.	3.01	PHO	PSEUDOCUBIC, TBLR -CURVED,STRIATED. VITR,PEARLY. 1 CLVG. VUGS,ETC.	43.1.1.5
4.5	LACROIXITE	MNCL.	3.13	PHO	XLS. VITR-RESIN. CLVGS. SOL-HCL,H2SO4. RARE-DRUSES IN GRANITE(SAX- ONY, GERMANY)	41.2.1.
4.5	HEDYPHANE	HEXA.	5.82	ARS	PRISM,TBLR,MSV. GRSY. RHOMB CLVG. SOL-HNO3. RARE-E.G.,FRNKLN,N.J.	41.7.3.2
4.5	ALAMOSITE	MNCL.	6.49	ISI	FIBR. ADMN. CLVG ⊥ LENGTH. GELAT-HNO3. VNS, ORE-DEP (ALAMOS,MEXICO)	W455 N
4.5-5	APOPHYLLITE	TETR.	2.33-2.37	PSI	VITR SQ PRISMS, PEARLY BASES. BASAL + PRISM CLVG. DCMP-HCL(SILICA RESIDUE). SECONDARY IN CAVITIES, COMMONLY WITH ZEOLITES, ETC.	W394 3-258
4.5-5	AUGELITE	MNCL.	2.70	PHO	TBLR,ACICULAR,MSV. VITR,PEARLY. CLVGS. SOL-HCL(SLOW). VNS,ETC.	41.6.8.

123

NONMETALLIC WHITE

HARDNESS	NAME	XL. SYS.	SPECIFIC GRAVITY	CHEM. CLASS	REMARKS	REFERENCES
4.5-5	PECTOLITE	TRCL.	2.86-2.90	ISI	RADIATING AGGREGATES. 2 PERF CLVGS. MOST FLUO-ORANGE(LW). WITH ZEO-LITES IN CAVITIES.	W460 2-176
4.5-5	PARAWOLLASTONITE	MNCL.	2.92	ISI	DISTINGUISHED FROM WOLLASTONITE BY X-RAY ANAL. CONTACT META LS-RARE	W455 2-167
4.5-5	WOLLASTONITE	TRCL.	2.87-3.09	ISI	TBLR, MSV. CLVGS-84°+ 96°.DCMP-HCL. FLUO-YEL-ORANGE. CNTCT META-LS	W456 2-167
4.5-5	PLUMBOGUMMITE	HEXA.	4.01	PHO	CONCENTRIC BOTRYOIDAL. GUM-LIKE LUSTER. UNEVN. SOL-HOT ACIDS. SECONDARY - IN LEAD DEPOSITS WITH PYROMORPHITE	41.5.8.1
4.5-5	SCHEELITE	TETR.	6.08-6.12	WOS	DIPYRAMIDAL,MSV,GRNLR,VITR.CLVGS.UNEVN,DCMP-HCL,HNO3,FLUO-BLUE(X-SW ULTRAVI-OLET-, + CATHODE-RAYS),THERMOLM. HI-T VNS,PEGS,GREISSENS,CNTCT META RKS,ETC.	48.1.3.1
4.5-5.5	LEWISTONITE	HEXA.	3.08	PHO	FIBR CRUSTS,PRISMS-SUB//SHEAFLK GROUPS.CLVGS. WITH DEHRNITE(FAIR-FIELD, UTAH)	41.7.5.
4.5-5.5	DEHRNITE	HEXA.	3.04-3.09	PHO	BOTRYOIDAL CRUSTS,PRISMS-SUB//-SHEAFLK GROUPS. VITR. 1 CLVG. SOL-ACIDS. WITH OTHER PHOSPHATE MINERALS	41.7.4.
5	NATROLITE (ZEOLITE)	ORTH.	2.20-2.26	TSI	FIBR GRP. SQ NEEDLES. PRISM CLVG. GELAT. PYROELECTRIC. ZEOL-ASSOC	W340 4-358
5	MESOLITE (ZEOLITE)	MNCL.	2.25-2.27	TSI	FIBR GRP. TUFTS OF HAIR-LIKE XLS. GELAT. PYROELECTRIC. ZEOL-ASSOC	W341 4-358
5	SCOLECITE (ZEOLITE)	MNCL.	2.25-2.29	TSI	FIBR GRP. THIN STRIATED PRISMS. PRISM CLVG. GELAT. ZEOL-ASSOC, AL-SO HYDROTHERMAL + IN METAMORPHOSED CALCAREOUS ROCKS	W341 4-358
5	OKENITE	TRCL.	2.28-2.33	ISI	FIBR, TBLR, MSV. 1 PERF CLVG-ELAST. GELAT. IN AMYGDULES IN BASALT	W358 N
5	GONNARDITE (ZEOLITE)	ORTH.	2.2 -2.4	TSI	FIBR GRP. RADIATED SPHERICAL MASSES. GELAT. RARE. ZEOL-ASSOCIATION	W337 4-359
5	COERULEOLACTITE	UNKN.	2.57-2.69	PHO	MICROXLINE FIBR CRUSTS + BOTRYOIDAL AGGS. CONCH. SOL-ACIDS + ALKS. RARE-WITH LIMONITE AND/CR WAVELLITE	42.7.3.
5	CRANDALLITE	HEXA.	2.78-2.92	PHO	CONCENTRIC NODULAR AGGS,ETC. VITR-DULL. 1 CLVG. SOL-ACIDS(DIFFI-CULTLY). SECONDARY - DIVERSE OCCURRENCES	41.5.8.4
5	HYDROXYLAPATITE	HEXA.	2.9 -3.1	PHO	APATITE GROUP(SEE FOR PROPS.) RARE-TALC SCHIST	41.7.1.3
5	CARBONATE-APATITE	HEXA.	2.9 -3.1	PHO	APATITE GROUP(SEE FOR PROPS.)RARE-NODULES(PHOSPHATE RK),GUANO,ETC.	41.7.1.4
5	WHITLOCKITE	HEXA.	3.12	PHO	RHOMB, GRNLR. VITR. UNEVN. SOL-DILUTE ACIDS. PEGS, PHOSPHATE ROCKS	38.3.1.
5	SELLAITE	TETR.	3.15	HAL	PRISM, ACICULAR, FIBR. CLVGS. DCMP-H2SO4. VNS WITH NATIVE S, ETC.	9.2.2.
5	CHLORAPATITE	HEXA.	A 3.17	PHO	APATITE GROUP(SEE FOR PROPS.) RARE-VNS IN GABBRO, METEORITES	41.7.1.2
5	FLUORAPATITE	HEXA.	A 3.18	PHO	APATITE GROUP(SEE FOR PROPS.) COMMON-IGS, METAS, MARINE SEDS, ETC.	41.7.1.1
5	APATITE	HEXA.	2.9 -3.2	PHO	GROUP NAME. MOST TYPES INDISTINGUISHABLE MEGASCOPICALLY. HEX PRISMS-COMMONLY ROUNDED,MSV,GRNLR. VITR. SOL-ACIDS. SOME FLUO-YEL-ORANGE. IG RKS,VNS,PEGS	41.7.1.
5	HEMIMORPHITE	ORTH.	3.45	SSI	HEMIMORPHIC XLS, SHEAFLK, MAMMILLARY, ETC. 1 PERF + 1 POOR CLVG. GELAT. PYROELECTRIC. SOME FLUO-PALE ORANGE(LONG WAVE). ZINC DEPS	W481 N
5	CALAMINE	ORTH.	3.45	SSI	(=HEMIMORPHITE)	W481 N
5	FERMORITE	HEXA.	3.52	ARS	MSV,GRNLR. GRSY. EVN FRACTURE. SOL-ACIDS. VNS IN MN ORE(INDIA)	41.7.6.
5	THOMSONITE (ZEOLITE)	ORTH.	2.10-2.39	TSI	FIBR GRP. RADIATED SPHERICAL MASSES. GELAT. PYROELECT. ZEOL-ASSOC	W336 4-359
5	EDINGTONITE (ZEOLITE)	TETR.	2.7 -2.8	TSI	FIBR GRP. PSEUDOTETRAGONAL. PRISM CLVG. GELAT. PYROELECTRIC. ZEOL-ASSOC, ORE DEPOSITS	W338 4-359
5 -5.5	HERDERITE	MNCL.	2.95	PHO	PRISM,TBLR,BOTRYOIDAL-RADIAL FIBR. VITR. CLVG. SOL-ACIDS. PEGS-LATE STAGE	41.5.4.1

NONMETALLIC WHITE

H	Mineral	Xl Sys	S.G.	Clvg	Color, Luster, Other Properties	Ref
5 -5.5	DATOLITE	MNCL.	2.96-3.00	NSI	TINTED WHT-COLORLESS. GLASSY. SCNDRY WITH ZEOLITES-CAVITIES	W355 1-171
5 -5.5	HYDROXYL-HERDERITE	MNCL.	3.01	PHO	PRISM,TBLR,BOTRYOIDAL-RADIAL FIBR. VITR. CLVG. SOL-ACIDS. PEGS-LATE STAGE	41.5.4.2
5 -5.5	MONAZITE	MNCL.	4.6 -5.4	PHO	TBLR-STRIATED. WAXY-ADMN. CLVGS. UNEVN. DCMP-ACIDS(SLOW). ACCESS IN GRANITIC IG + GNEISSIC RKS,PEGS,DETRITAL SEDIMENTARY ROCKS	38.4.2.
5 -6	VISHNEVITE	HEXA.	2.32-2.42	TSI	SULFATE-RICH END MEMBER OF CANCRINITE FAMILY. OCCURRENCE LIKE CANCRINITE.	WN 4-310
5 -6	CANCRINITE	HEXA.	2.42-2.51	TSI	MSV (RARE PRISM XLS). PRISM CLVG. EFF-HCL. NEPHELINE-BEARING RKS	W354 4-310
5 -6	MARIALITE (SCAPOLITE)	TETR.	2.50-2.62	TSI	ME-0-20. PURE MARIALITE UNKNOWN IN NATURE	W352 4-321
5 -6	DIPYRE (SCAPOLITE)	TETR.	2.57-2.69	TSI	ME-20-50.	W352 4-321
5 -6	MIZZONITE (SCAPOLITE)	TETR.	2.67-2.74	TSI	ME-50-80.	W352 4-322
5 -6	SCAPOLITE	TETR.	2.50-2.78	TSI	GROUP NAME. MOST TYPES INDISTINGUISHABLE MEGASCOPICALLY BUT S.G. MAY BE INDICATIVE. LONG,STRIATED PRISMS + AGGS OF COARSE XLS. SUBCONCH. POOR CLVG GIVING IRREGULAR, STRIATED-APPEARING SURFACES. MARIALITE IS INSOL; MEIONITE IS DCMP IN HCL. MOST FLUO—YEL-RED(ULTRAVIOLET,LW). CNTCT META CALC RKS,PEGS	W352 4-321
5 -6	(WERNERITE) (SCAPOLITE)	TETR.	2.50-2.78	TSI	GROUP NAME NO LONGER USED	W352 4-322
5 -6	MEIONITE (SCAPOLITE)	TETR.	2.73-2.78	TSI	ME-80-100. PURE MEIONITE UNKNOWN IN NATURE	W352 4-321
5 -6	AMPHIBOLE		3.0 -3.57	ISI	GROUP NAME. SEVERAL END AND OTHER MEMBERS WITH NAMES UNSETTLED FOR SOME, MANY MEGASCOPICALLY INDISTINGUISHABLE BUT SOME COLORS AND OCCURRENCES ARE INDICATIVE. ORTH AND MNCL SERIES. LONG SLENDER XLS. PRISM CLVG AT 56° + 124°. SOL-HF(SLOWLY). ALTERS TO BIOTITE +/OR CHLORITE. IGNEOUS AND METAMORPHIC RKS, VNS	W422 2-203
5 -7	PYROXENE		2.96-3.96	ISI	GROUP NAME. SEVERAL END AND OTHER MEMBERS WITH NAMES UNSETTLED FOR SOME. MANY MEGASCOPICALLY INDISTINGUISHABLE; OCCURRENCE MAY BE INDICATIVE. ORTH + MNCL SERIES.SHORT PRISMS.PRISM CLVG AT 87° + 92°. ALTERS TO AMPHIBOLE.IG + META RKS	W402 2-1
5.5	ANALCIME (ZEOLITE)	ISOM.	2.24-2.29	TSI	TRAPEZOHEDRA, MSV, GRNLR. POOR CUBIC CLVG. GELAT. AMYGDULES-BASALT	W333 4-338
5.5	HILLEBRANDITE	MNCL.	2.69	NSI	RADIATING FIBR. PRISM CLVG. SOL-HCL. CNTCT META CALCAREOUS ROCKS	W506 N
5.5	BAVENITE	ORTH.	2.74	SSI	FIBR-RADIATING PRISM XLS. 1 GOOD, 1 POOR CLVG. RARE-GRANITIC PEGS	W259 N
5.5	MILLISITE	TETR.	2.83	PHO	FIBR,CHALCEDONIC(LK CRUSTS. CLVG(?). WITH WARDITE IN VARISCITE	42.5.5.
5.5	LEHIITE	UNKN.	2.89	PHO	AGGS-SUBPARALLEL FIBERS. WITH WARDITE,ETC. IN ALTERED VARISCITE	42.5.6.
5.5	BITYITE	MNCL.	3.	PSI	SIMILAR TO COMMON MICAS, DISTINGUISHED OPTICALLY. LI-BEARING PEGS	W391 N
5.5	CLINOHEDRITE	MNCL.	3.33	SSI	CLINOHEDRAL. 1 PERF CLVG. WITH WILLEMITE, ETC. AT FRANKLIN, N.J.	W482 N
5.5	PERICLASE	ISOM.	3.55-3.57	OXD	OCTAH, ROUNDED GRAINS. CUBIC,OCTAH CLVGS. CNTCT META MARBLES(HI-T)	4.2.1.1
5.5	WILLEMITE	HEXA	3.9 -4.1	NSI	PRISMS, MSV. MAY FLUO-GRN. SOME PHOS + TRIBOLUM. GELAT. ZINC DEPS	W497 N
5.5-6	LEUCITE	TETR.	2.47-2.50	TSI	TRAPEZOHEDRA(STRIATED FACES). DODEC CLVG. DCMP-HCL. K-RICH ALK VOL	W251 4-276
5.5-6	HAUYNE	ISOM.	2.44-2.50	TSI	SODALITE(E)SE) GROUP PROPS. NEPH-BEARING IG RKS. DISSOLVE-HND3, EVAPORATE SLOWLY, GYPSUM CRYSTALS REMAIN.	W350 4-289
5.5-6	NEPHELINE	HEXA.	2.56-2.67	TSI	TBLR,PRISM.VITR-GRSY.GOOD PRISM,POOR BASAL CLVGS.GELAT.FLUO-ORANGE.DISTINGUISHED FROM KALSILITE + KALIOPHILITE BY X-RAY. NA-RICH ALK IG AND NEARBY META ROCKS	W254 4-231

NONMETALLIC WHITE

HARDNESS	NAME	XL. SYS.	SPECIFIC GRAVITY	CHEM. CLASS	REMARKS	REFERENCES
5.5-6	BERYLLONITE	MNCL.	2.81	PHO	TBLR, COMPLEX. VITR.,PEARLY. CLVGS. CONCH. SOL-ACIDS(SLOW). PEGS	38.1.6.
5.5-6	MONTEBRASITE	TRCL.	2.98	PHO	AMBLYGONITE SERIES(SEE FOR PROPERTIES + OCCURRENCES.)	41.5.5.2
5.5-6	LAZULITE	MNCL.	3.08	PHO	SUBTRANSLUCENT. ACUTE PYRAMIDAL,MSV,GRNLR. VITR. CLVGS. SOL-HOT ACIDS(SLOWLY).PEGS,QTZ VNS,SI-RICH ALUMINOUS META(HI-GRADE) RKS	41.8.1.1
5.5-6	NATROMONTEBRASITE	TRCL.	3.04-3.1	PHO	AMBLYGONITE SERIES(SEE FOR PROPERTIES + OCCURRENCES.)	41.5.5.3
5.5-6	AMBLYGONITE	TRCL.	3.11	PHO	EQUANT, LARGE MASSES. VITR.,PEARLY. CLVGS. SOL-ACIDS(DIFFICULTY). LI-, PO4-RICH PEGS	41.5.5.1
5.5-6	SCORZALITE	MNCL.	3.38	PHO	IN SERIES WITH LAZULITE(SEE FOR PROPS). PEGMATITES	41.8.1.2
5.5-6	ANTHOPHYLLITE (AMPHIBOLE)	ORTH.	2.85-3.57	ISI	WITH CORDIERITE IN GNEISSES, WITH TALC IN META ULTRABASIC IG RKS	W425 2-211
5.5-6	GEDRITE (AMPHIBOLE)	ORTH.	2.85-3.57	ISI	RELATED IN COMPOSITION + OCCURRENCE TO ANTHOPHYLLITE	W425 2-211
5.5-6.5	OPAL	TETR.	1.8 -2.25	SIL	SUBMICROXLINE QTZ + H2O.RESIN-PEARLY, MSV.MAMILLARY CRUSTS,CONCH.SOL-HF +CAUS- TIC ALKS.SOME FLUO—YEL-GREEN. INCRUSTATIONS, IN CAVITIES(ESP IN VOLCANIC RKS)	D-111 287
5.5-6.5	DIOPSIDE (CLINOPYROXENE)	MNCL.	3.22-3.38	ISI	SOME LIGHT ONES FLUO—YEL-BLUE. WHT-PALE GREEN. META CALC-SIL RKS	W411 2-42
5.5-6.5	SALITE (CLINOPYROXENE)	MNCL.	3.2 -3.6	ISI	INTERMEDIATE BETWEEN DIOPSIDE + HEDENBERGITE. META CA-RICH SED + IGNEOUS ROCKS	W411 2-42
5.5-6.5	FERROSALITE (CLINOPYROXENE)	MNCL.	3.2 -3.6	ISI	INTERMEDIATE BETWEEN DIOPSIDE + HEDENBERGITE. OCCURS LIKE SALITE	W411 2-42
5.5-7	KYANITE	TRCL.	3.53-3.65	NSI	HARDNESS-5 ⊥ TO + 7 // TO LENGTH OF BLADE-LIKE XLS. GNEISS +SCHIST	W527 1-137
6	ADULARIA (K-FELDSPAR)	TRCL.	2.56	TSI	MUCH IS TRANSPARENT. PRISM. CAVITIES IN META RKS. MOONSTONE IS VAR	W303 4-7
6	SANIDINE (K-FELDSPAR)	MNCL.	2.56-2.62	TSI	TBLR. GLASSY. OPTICALLY DISTINCT. VOLCANIC ROCKS	W305 4-7
6	KALSILITE	HEXA.	2.59-2.63	TSI	K CHEM-ANALOGUE OF NEPH.DISTINGUISHED BY X-RAY.K-RICH,SI-POOR VOLS	W257 4-231
6	BANALSITE	ORTH.	3.06	TSI	RESEMBLES FELDSPARS. DCMP-HCL. RARE-MN-RICH VEIN(WALES)	W260 4-168
6	LAWSONITE	ORTH.	3.05-3.10	SSI	PRISM. TBLR.CLVG-2 PERF, 1 POOR. LO-GRADE(E.G.,GLAUCOPHANE)SCHISTS	W482 1-221
6	JADEITE (CLINOPYROXENE)	MNCL.	3.24-3.43	ISI	RELATIVELY UNCOMMON, TYPICALLY WITH ALBITE, HI-T + ALSO LO-GRADE METAMORPHIC ROCKS	W411 2-99
6 -6.5	ANORTHOCLASE (FELDSPAR)	TRCL.	2.56-2.62	TSI	DISTINGUISHED FROM ORTHOCLASE + MICROCLINE-OPTICALLY. NA-RICH VOLS	W311 4-7
6 -6.5	ALBITE (PLAGIOCLASE)	TRCL.	2.57-2.63	TSI	ANO-10. SEE FELDSPAR. IG, META RKS, PEGS. MOONSTONE +CLEAVELANDITE ARE VARIETIES	W312 4-94
6 -6.5	MICROCLINE (K-FELDSPAR)	TRCL.	2.56-2.63	TSI	DISTINGUISHED FROM ORTHOCLASE OPTICALLY. PLUTONIC IG RKS, PEGS	W308 4-7
6 -6.5	ORTHOCLASE (K-FELDSPAR)	TRCL.	2.55-2.63	TSI	DISTINGUISHED FROM MICROCLINE OPTICALLY. PLUTONIC IG RKS, PEGS	W303 4-7
6 -6.5	PERTHITE (FELDSPARS)	TRCL.	2.56-2.65	TSI	MEGASCOPICALLY, MICROSCOPICALLY,OR SUBMICROSCOPICALLY INTERDIGITA- TED MICROCLINE OR ORTHOCLASE + PLAGIOCLASE(TYPICALLY ALBITE)	M299 4-7
6 -6.5	OLIGOCLASE (PLAGIOCLASE)	TRCL.	2.62-2.67	TSI	AN10-30. SEE FELDSPAR. PLUTONIC IG RKS, FEW PEGS. SUNSTONE IS VAR.	W312 4-95
6 -6.5	ANDESINE (PLAGIOCLASE)	TRCL.	2.64-2.69	TSI	AN30-50. SEE FELDSPAR. INTERMEDIATE SI-CONTENT IGNEOUS ROCKS	W312 4-95
6 -6.5	LABRADORITE (PLAGIOCLASE)	TRCL.	2.68-2.72	TSI	AN50-70. BLUISH REFLECTIONS. SEE FELDSPAR. BASIC IGNEOUS ROCKS	W312 4-95

NONMETALLIC WHITE

Hardness	Mineral	System	S.G.	Habit/Properties	Group	Notes	Ref
6 –6.5	BYTOWNITE (PLAGIOCLASE)	TRCL.	2.71–2.75	AN70–90. GELAT. SEE FELDSPAR. CALCIUM-RICH IGNEOUS ROCKS	TSI		W312 4-95
6 –6.5	ANORTHITE (PLAGIOCLASE)	TRCL.	2.70–2.76	AN90–100. GELAT. SEE FELDSPAR. CNTCT META CALCAREOUS RKS	TSI		W312 4-94
6 –6.5	HYALOPHANE (BA-FELDSPAR)	MNCL.	2.58–2.82	BA EQUIVALENT OF ADULARIA. PRISM, MSV. MN-RICH VEINS + ROCKS	TSI		W304 4-166
6 –6.5	PREHNITE	ORTH.	2.90–2.95	ROSETTES-TBLR XLS. 1 CLVG. SOL-HCL(SLOW). CAVITIES(WITH ZEOLITES)	SSI		W359 3-263
6 –6.5	FELDSPAR		2.55–3.39	INCLUDES PLAGIOCLASE(NA-CAISERIES, ALKALI(K-NA) AND BA FELDSPARS. MOST EASILY DISTINGUISHED BY NONMEGASCOPIC MEANS. MONOCLINIC ONES HAVE 2 CLVGS AT 90° AND SIMPLE(IF ANY) TWINNING; TRICLINIC ONES HAVE 2 CLVGS UP TO 4° OFF 90° AND TYPICALLY HAVE POLYSYNTHETIC TWINNING (GIVING STRIATED APPEARANCE ON SOME CLVG SURFACES OF PLAGIOCLASE). PEGMATITES; IG, META, AND SOME SED ROCKS	TSI		W261 4-1
6 –6.5	CELSIAN (BA-FELDSPAR)	MNCL.	3.10–3.39	SEE FELDSPAR. TYPICALLY CLEAVABLE MASSES.GELAT.MN-RICH VNS + RKS	TSI		W307 4-166
6 –7	MULLITE	ORTH.	3.03–3.16	INDISTINGUISHABLE MEGASCOPICALLY FROM SILLIMANITE. HI-T META-RARE	NSI		W401 1-124
6 –7	CASSITERITE	TETR.	6.99	SUBMET, RADIAL CONCRETIONARY MASSES. ADMN-DULL. HI-T VNS, GREISENS	OXD		4.5.1.5
6 –7.5	GROSSULAR	ISOM. A	3.59	CA-AL GARNET. EQUANT XLS. CONTACT + REGIONAL META CALCAREOUS RKS	NSI		W489 1-93
6 –7.5	HYDROGROSSULAR	ORTH.	3.13–3.59	GROSSULAR +H2O. SLOWLY SOL-HCL + HNO3. META CALCAREOUS ROCKS	NSI		W493 1-104
6 –7.5	GARNET	ISOM.	3.58–4.32	GROUP NAME. DOMINANT MOLECULE DETERMINES NAME. DISTINGUISHED BY S.G. + NONMEGA-SCOPIC MEANS.EQUANT XLS,MSV.VITR-DULL,DODEC PARTING. SLIGHTLY- TO IN-SOL IN HF	SIL		W483 1-77
6.5	CRISTOBALITE	TETR.	2.32–2.34	MILKY-WHT.EQUANT XLS—COMMONLY SKELETAL. WITH TRIDYMITE IN SIL VOLS	SIL		D-111 273
6.5	PETALITE	MNCL.	2.41–2.42	FOL, MSV. 1 PERF + 1 GOOD-CLVGS AT 38.5°. LI-BEARING PEGMATITES	TSI		W260 4-271
6.5	AGATE	HEXA.	2.57–2.64	BANDED OR VARIEGATED.CHALCEDONY(MICROXLINE QTZ). NODULES IN BASALT	SIL		D-111 210
6.5	XONOTLITE	MNCL.	2.71	EXPOS—COLOR MAY FADE. ELONGATE XLS. CLVG // LENGTH. CNTCT META LS	ISI		W455 N
6.5	LEUCOSPHENITE	MNCL.	3.05	PLATY, ELONGATE XLS. 1 GOOD CLVG. DCMP-HF. RARE(NARSARSUK, GREENLAND)	SSI		W455 N
6.5	KORNERUPINE	ORTH.	3.27–3.34	PRISM MASSES. PRISM CLVG. WITH CORDIERITE + SAPPHIRINE IN SCHISTS	NSI		W523 N
6.5–7	SPODUMENE	MNCL.	3.03–3.22	COMMONLY FLUO(+PHOSPHORESCES)—ORANGE. THERMOLUM. LI-BEARING PEGS	ISI		W418 2-92
6.5–7	DIASPORE	ORTH.	3.3 –3.5	PLTY, FOL-MSV. VITR, PEARLY. WITH CORUNDUM IN EMERY ROCK	OXD		7.1.2.1
6.5–7.5	ANDALUSITE	ORTH.	3.13–3.16	BLUNT, NEARLY SQ PRISMS. COMMONLY ALTERED. CONTACT +REGIONAL META	NSI		W521 1-129
6.5–7.5	SILLIMANITE	ORTH.	3.23–3.27	PRISMATIC XLS, FIBR MASSES. VITR. 1 CLVG. HI-GRADE META ROCKS	NSI		W520 1-121
7	TRIDYMITE	ORTH.	2.25–2.27	PLTY MICROXLS. VITR-PEARLY. CONCH. SOL-BOILING NaCO3. SILICIC VOLS	SIL		D-111 264
7	ELPIDITE	ORTH.	2.58	FIBR-MSV. PRISM CLVG. IN ALKALIC IGNEOUS ROCKSIE.G., OSLO EKERITE)	SSI		W454 N
7	QUARTZ	HEXA.	2.65	DOUBLY TERMINATED HEX PRISMS STRIATED ⊥ LENGTH(LO-T), BIPYRAMIDS (HI-T),MSV.VITR.CONCH.SOL-HF.WDSPRD IN MOST ROCK TYPES + VEINS	SIL		D-111 9
7	UVITE	HEXA.	3.05	CA- MG-TOURMALINE. TYPICALLY WHITE. RARE	CSI		W465 N

NONMETALLIC WHITE

HARDNESS	NAME	XL. SYS.	SPECIFIC GRAVITY	CHEM. CLASS	REMARKS	REFERENCES
7	TOURMALINE	HEXA.	3.03-3.25	CSI	GROUP NAME. VARIETIES COMMONLY DISTINGUISHABLE MEGASCOPICALLY ON BASIS OF COLOR(S). LENGTHWISE STRIATED PRISMATIC XLS WITH CROSS-SECTIONS THAT RESEMBLE SPHERICAL TRIANGLES. SOME XLS EXHIBIT BOTH LENGTHWISE + CONCENTRIC COLOR ZONING. VITR-RESIN. BRTL. ELECTRICALLY CHARGED ON HEATING + COOLING. MG-RICH VARIETIES FLUO-YEL (ULTRAVIOLET - SW). PEGS, HI-T VNS, IG AND META ROCKS	W465 1-300
7	BARYLITE	ORTH.	4.	SSI	PLATY XLS. 2 GOOD + 1 POOR CLVGS(PINACOIDAL). RARE-FRANKLIN, N.J.	W476 N
7 -7.5	BORACITE	ORTH.	2.91-2.97	BOR	CUBIC,ETC.,GRNLR,FIBR. VITR. SOL-HCL(SLOWLY). SALT-GYPSUM BEDS	26.1.7.
7.5	HAMBERGITE	ORTH.	2.36	BOR	PRISM-STRIATED. VITR. 2 CLVGS. SOL-HF. ALKALIC PEGMATITES	26.1.2.
7.5	COESITE	MNCL.	2.91-2.95	SIL	TRANSPARENT. GYPSUM-LIKE HABIT.SOL-HF(<QTZ).IMPACT(METEOR) CRATERS	D-111 310
7.5-8	BERYL	HEXA.	2.66-2.83	CSI	GROOVED PRISM. POOR BASAL CLVG.SOME FLUO WEAKLY-YEL. PEGMATITES	W463 1-256
8	RHODIZITE	ISOM.	3.31-3.38	BOR	DODEC. VITR. OCTAH CLVG. WITH RED TOURMALINE IN PEGMATITES-RARE	24.1.6.
8	TOPAZ	ORTH.	3.49-3.57	NSI	STRIATED PRISMS. PERF BASAL CLVG. CNTCT ZONES, GREISENS, PEGS	W509 1-145
9	BROMELLITE	HEXA.	3.02	OXD	PRISM.PRISM CLVG.SOL-CONC ACIDS. RARE-CALCITE VN IN SKARN(LANGBAN)	4.2.2.2
10	DIAMOND	ISOM.	3.50-3.53	ELE	OCTAHEDRAL. ADMN-GRSY. BRTL. OCTAH CLVG. ULTRABASIC IGS + PLACERS	1.2.4.1

NONMETALLIC GRAY

HARDNESS	NAME	XL. SYS.	SPECIFIC GRAVITY	CHEM. CLASS	REMARKS	REFERENCES
1	SASSOLITE	TRCL.	1.46-1.50	OXD	SCALES. PEARLY. BASAL CLVG(FLEX). BITTER,SALTY TASTE. HOT SPRINGS	6.2.1.
1	GLAUCOCERINITE	UNKN.	2.75	SUF	WARTY MASSES, RADIAL + CONCENTRIC. WAXY. WITH SMITHSONITE,ETC.(LAU-RIUM SPECIMEN)	(LAU-31.1.2.
1	SILLENITE	ISOM.	8.80	OXD	GRNLR. WAXY-DULL. RARE-SCNDRY AFTER BISMUTITE(DURANGO,MEXICO)	4.4.8.
1 -1.5	NATRON	MNCL.	1.48	CBT	GRNLR CRUSTS. VITR. CLVGS. ALK TASTE. IN SOLUTIONS, EFFLORESCENCES	15.1.6.
1 -1.5	THERMONATRITE	ORTH.	2.26	CBT	CRUST-EFFLOR. VITR. 1 CLVG. ALK TASTE. RARE-WITH TRONA-ARID SOILS	15.1.1.
1 -1.5	EPISTOLITE	MNCL.	2.89	NSI	PLATY XLS. 1 PERF, 1 GOOD CLVG. RARE IN PEG-JULIANEHAAB,GREENLAND	W508 N
1 -2	STEVENSITE (CLAY)	MNCL.	3.	PSI	POSSIBLY MONTMORILLONITIC "SWELLING CLAY." ALTER PROD OF PECTOLITE	W461 3-231
1 -3	CLAY	MNCL.	1.85-3.0	PSI	NAME GIVEN SEVERAL HYDROUS AL-SILICATES WITH ALKALIES OR ALKALINE EARTHS ± MG OR FE. NAMES UNSETTLED. MOST DISTINGUISHED BY NONMEGASCOPIC MEANS. FINE-GRAINED AGGS(COMPACT,MEALY,ETC.). EARTHY-PEARLY,UNCTUOUS.PLASTIC WHEN WET,DEHYDRATED ON HEATING. SOME ARE ABSORBENT. MXING PRODS,HYDROTHERMAL ALTER, DETRITAL, DIAGENIC	W398 3-191
1.5	IODARGYRITE	HEXA.	5.69	HAL	EXPOS-YEL. MSV. PRISM-HEMIMORPHIC. RESIN. BASAL CLVG(FLEX). DCMP-H2SO4, HNO3, KI SOLUTION. OXIDIZED ZONES OF SILVER DEPOSITS	9.1.4.
1.5	CALOMEL	TETR.	7.15	HAL	EXPOS-DARKENS. TBLR. MSV. ADMN. PRISM CLVG. PLASTIC-FLUO-RED. SOL-AQ-REG. SECONDARY AFTER SEVERAL MERCURY MINERALS	9.1.5.
1.5-2	SALAMMONIAC	ISOM.	1.53	HAL	SKELETAL, DENDRITIC. VITR.OCTAH CLVG. SALTY TASTE. VOL SUBLIMATE	9.1.2.
1.5-2	SODA-NITER	HEXA.	2.24-2.29	NIT	INCRUST, RHOMB. VITR. RHOMB + OTHER CLVGS. COOLING TASTE. IN ARID SOILS IN PROTECTED LOCATIONS	18.1.1.
1.5-2.5	ANNABERGITE	MNCL. C	3.23	ARS	XLINE COATINGS,EARTHY. ADMN-DULL. CLVGS. SOL-ACIDS. OXIDATION PROD OF COBALT AND NICKEL ARSENIDES	40.2.15.3
2	SYLVITE	ISOM.	1.98-2.00	HAL	CUBIC. GRNLR. VITR. CUBIC CLVG. BITTER TASTE. SED EVAPORITE DEPS	9.1.1.2
2	MANASSEITE	HEXA.	2.00-2.10	OXD	FOL-MSV. GRSY. BASAL CLVG(FLEX). SOL-HCL. WITH HYDROTALCITE + SER-PENTINE	6.1.6.1
2	NITER	ORTH.	2.10-2.11	NIT	CRUSTS,GRNLR. VITR. CLVGS. SALTY. COOLING TASTE. EFFLOR-CAVES,ETC.	18.1.2.
2	HALITE	ISOM.	2.17	HAL	CUBIC, GRNLR. VITR. CUBIC CLVG. SALTY TASTE. SED EVAPORITE DEPS	9.1.1.1
2	RHOMBOCLASE	ORTH.	2.23	SUF	THIN TBLR, STALACTITIC. SUBVITR. BASAL(FLEX),PRISM CLVGS. CONCH-FIBR. SOL-ACIDS.H2O(SLOWLY). RARE-ALTER PROD OF SULFIDES	29.1.1.
2	GYPSUM	MNCL.	2.31-2.32	SUF	TBLR, MSV. SUBVITR-PEARLY. 1 PERF CLVG(FLEX),INELAST). SOL-HCL. SEDIMENTARY BEDDED DEPS, FUMAROLIC, EFFLORESCENCES, ETC.	29.6.3.
2	BORAX	MNCL.	1.71-1.72	BOR	HABIT-LK PYROXENE. VITR-EARTHY. CLVGS. SWEETISH. SOL-H2O(TO ALK SOLUTION). BORAX DEPOSITS (SALINE LAKE DEPOSITS, ETC.)	25.1.4.
2 -2.5	MASCAGNITE	ORTH.	1.77	SUF	CRUSTS, STALACTITIC. VITR-DULL. 1 CLVG. SHARP,BITTER. FUMAROLIC	28.2.1.1
2 -2.5	PHARMACOLITE	MNCL.	2.53-2.73	ARS	FIBERS. VITR,PEARLY. 1 CLVG(FLEX). SOL-ACIDS. OXIDIZED AS-DEPS	39.2.1.2
2 -2.5	HYDROZINCITE	MNCL.	3.5 -4.0	CBT	MSV,INCRUST. SILKY-DULL,1 CLVG. FLUO-BLUE,LILAC. SOL-ACIDS.SCNDRY WITH OTHER ZINC MINERALS	16.1.3.
2 -2.5	SENARMONTITE	ISOM.	5.50	OXD	OCTAH, GRNLR. RESIN-ADMN. OCTAH CLVG.SOL-HCL. SCNDRY AFTER SB-MINS	4.4.2.2
2 -2.5	TELLURIUM	HEXA.	6.1 -6.3	ELE	OPQ.SN-WHT, GRAY STRK. SOL-H2SO4 TO RED SULFITE. VEINS	1.2.2.3
2 -2.5	LANARKITE	MNCL.	6.92	SUF	MSV, ELONGATE. ADMN,PEARLY. CLVGS(THIN LAMINAE-FLEX). FLUO-YEL(X-RAYS + ULTRAVIOLET RADIATION). SOL-WARM HNO3. RARE-LEAD DEPOSITS	30.2.1.

NONMETALLIC GRAY

HARDNESS	NAME	XL. SYS.	SPECIFIC GRAVITY	CHEM. CLASS	REMARKS	REFERENCES
2 -2.5	CINNABAR	HEXA.	8.09	SLD	SCARLET STRK. ADMN-DULL. NEAR VOLS + HOT SPRINGS-VNS + INCRUSTS	2.6.9.
2 -3	PITTICITE	UNKN.	2.2 -2.5	ARS	MSV,BOTRYOIDAL,OPALINE CRUSTS,ETC. DULL-GRSY. SOL-ACIDS. OXIDATION ZONES OF AS-BEARING DEPOSITS. DEPOSITED FROM MINE + SPRING WATERS	43.2.6.
2 -3	FIBROFERRITE	ORTH.	1.84-2.52	SUF	FIBR CRUSTS. SILKY. 1 CLVG. DCMP-H2O. SCNDRY-BY OXIDATION OF FES2	31.6.6.
2 -3	RHIPIDOLITE (CHLORITE)	MNCL.	2.88-3.08	PSI	INTERMEDIATE BETWEEN SHERIDANITE + DAPHNITE. FELDSPAR SCHISTS, VNS	W383 3-137
2 -3	CHLORITE	MNCL.	2.6 -3.3	PSI	GROUP NAME. MEMBER NAMES UNSETTLED. ONLY FEW RARE VARIETIES MEGASCOPICALLY DISTINGUISHABLE. DISSEM FLKS. PEARLY. 1 PERF CLVG TO FLEX,INELAST FOLIA. DCMP-IN BOILING H2SO4. SCNDRY-HYDROTHERMAL ALTER, LO-GRADE META, DETRITAL, AUTHIGENIC	W381 3-131
2 -3	LAMPROPHYLLITE	ORTH.	3.45-3.54	NSI	ELONGATE PLTS. 1 GOOD, 1 POOR CLVG. NEPHELINE SYENITES	W527 N
2 -3	PHOSGENITE	TETR.	6.13	CBT	PRISM, MSV. ADMN. CLVGS(PERCUSSION FIGURE ON BASAL PLANE) FLUO-YEL (LW ULTRAVIOLET, CATHODE - X-RAYS). EFF-HNO3. SCNDRY AFTER PB MINS	16.1.7.
2 -3	BEYERITE	TETR.	6.50-6.56	CBT	PLTY, MSV. VITR(XLS). CONCH. EFF-ACIDS. SCNDRY-WITH BISMUTITE	16.2.5.
2 -4	MICA	MNCL.	2.4 -3.4	PSI	GROUP NAME. SOME SPECIES DISTINGUISHABLE ON BASIS OF COLOR. SERIES ARE NOT COMPLETE. HEX-SHAPED XLS,DISSEM, PERF CLVG TO FLEX,ELAST FLKS. PERCUSSION + PRESSURE FIGURES MAY BE IMPOSED ON FLKS. CHIEF OCCURRENCES ARE NOTED UNDER SPECIES	W365 3-1
2 -4	BAUXITE	2	3.5	OXD	FIELD TERM FOR MATERIALS RICH IN HYDROUS ALUMINUM OXIDES.	6.2.3.
2.5	BISMOCLITE	TETR.	7.12	HAL	IN SERIES WITH DAUBREEITE. MSV, SCALY, GRSY-DULL. 1 CLVG. PLASTIC. SOL-ACIDS..SCNDRY - ALTER PROD OF BISMUTHINITE OR NATIVE BISMUTH	10.1.6.2
2.5	DAUBREEITE	TETR. C	7.56	HAL	IN SERIES WITH BISMOCLITE,WHICH SEE.SCNDRY AFTER BISMUTHINITE + BI	10.1.6.3
2.5	CRYPTOHALITE	ISOM.	2.00	HAL	MAMMILLARY,DENDRITIC. VITR. OCTAH CLVG. SALTY TASTE. VOL SUBLIMATE	11.4.1.2
2.5	PICROMERITE	MNCL. A	2.03	SUF	PRISM, CRUSTS. VITR. BITTER. FUMAROLIC. MARINE SALINE DEPS-AS BEDS	29.3.7.1
2.5	CHRYSOTILE (SERPENTINE)	MNCL.	2.2	PSI	FIBROUS, SOME IS ASBESTIFORM. SECONDARY, COMMONLY IN VEINS	W379 3-170
2.5	NAHCOLITE	MNCL.	2.21	CBT	PRISM, FRIABLE AGGS. VITR. CLVGS. SOL-H2O. WITH TRONA	13.1.1.
2.5	ALUMOHYDROCALCITE	MNCL.	2.23	CBT	CHALKY MASSES. 2 CLVGS. SOL-ACIDS,DCMP-BOILING H2O. WITH ALLOPHANE	16.2.4.
2.5	CHALCOALUMITE	TRCL.	2.29	SUF	BOTRYOIDAL,LATHLK.DULL-VITR. CLVGS. SOL-DILUTE ACIDS(SLOW). CRUSTS ON LIMONITE STALACTITES (BISBEE, ARIZONA)	31.1.9.
2.5	BRUCITE	HEXA.	2.38-2.40	OXD	WHT STRK. TBLR, FOL MSV. VITR. CLVGS. MAXY-PEARLY. BASAL CLVG(FLEX). SOL-ACIDS. LOW TEMPERATURE VEINS	6.1.1.1
2.5	FERRINATRITE	HEXA.	2.55-2.61	SUF	FIBR AGGS, PRISM, MSV. VITR. CLVGS. SPLINTERY. SCNDRY WITH OTHER IRON SULFATES IN ARID REGIONS	29.4.1.
2.5	MIERATITE	ISOM. A	2.67	HAL	CUBO-OCTAH, STALACTITIC. VITR. OCTAH CLVG. SOL-H2O. FUMAROLIC DEPS	11.4.4.1
2.5	FORBESITE	UNKN.	3.13	ARS	NEEDS VERIFICATION.	39.2.4.
2.5	TEINEITE	ORTH.	3.80	TEL	CRUSTS, AGGS, PRISM. 3 CLVGS. SOL-HCL(TO BLUE),HNO3(TO YEL). OXIDATION PRODUCT OF TETRAHEDRITE AND NATIVE TELLURIUM	33.1.2.
2.5	NANTOKITE	ISOM.	4.14	HAL	MSV, GRNLR. ADMN. DODEC CLVG. DCMP-H2O.SOL-HCL,HNO3,NH4OH. CU-DEPS	9.1.3.1
2.5	CHLORARGYRITE	ISOM.	5.56	HAL	CRUSTS,CUBIC. RESIN. DUCTILE. EXPOS-VIOLET-BRWN. SCNDRY IN AG DEPS	9.1.1.4
2.5	BROMARGYRITE	ISOM.	6.50	HAL	CRUSTS,CUBIC. RESIN. DUCTILE. SECONDARY IN SILVER DEPOSITS	9.1.1.5
2.5	MENDIPITE	ORTH.	7.24	HAL	FIBR, RADIATED, SILKY,RESIN. PRISM + OTHER CLVGS. SOL-HNO3. SCNDRY WITH CERUSSITE, ETC.	10.1.4.

NONMETALLIC GRAY

Hardness	Mineral	System	S.G.		Description	Ref.
2.5	FINNEMANITE	HEXA.	7.27	ASI	CRUSTS-PRISM XLS. SUBADMN. RHOMB CLVG. IN CREVICES IN HEMATITE AT LANGBAN, SWEDEN	46.1.3.
2.5-3	GAYLUSSITE	MNCL.	1.99	CBT	TBLR,STRIATED. VITR. PRISM CLVG. EFF-ACIDS. WITH TRONA,ETC.	15.2.3.
2.5-3	TRONA	MNCL.	2.14	CBT	FIBR,ELONGATE. VITR. 1 PERF,OTHER CLVGS. ALK TASTE,SOL-H2O, EFF-ACIDS. SALINE LAKE DEPS, EFFLORESCENT ON SOILS IN ARID REGIONS	13.1.4.
2.5-3	KAINITE	MNCL.	2.15	SUF	GRNLR, TBLR. VITR. 1 CLVG. SALTY,BITTER. IN MARINE POTASH DEPS	31.4.1.
2.5-3	THENARDITE	ORTH.	2.66	SUF	DIPYRAMIDAL, CRUSTS, VITR. CLVGS. FAINTLY SALTY. SALINE LAKE DEPS, EFFLORESCENCES, FUMAROLE DEPOSITS	28.2.4.
2.5-3	GLAUBERITE	MNCL.	2.75-2.85	SUF	TBLR, PRISM, DIPYRAMIDAL-STRIATED. VITR,PEARLY. CLVGS. CONCH. BRTL. SALTY TASTE. SOL-HCL. SALINE LAKE DEPS,DISSEM IN CLASTIC SEDS, FUMAROLIC	28.4.2.
2.5-3	VALENTINITE	ORTH.	5.76	OXD	PRISM, TBLR, MSV. ADMN. PRISM CLVG. SOL-HCL. SCNDRY AFTER SB-MINS	4.4.4.
2.5-3	ANGLESITE	ORTH.	6.37-6.39	SUF	TBLR,PRISM,MSV,ETC. ADMN. CLVGS. CONCH. BRTL. SOME FLUO-YEL. SOL-HNO3(DIFFICULTLY). SECONDARY - BY OXIDATION OF GALENA	28.3.1.3
2.5-3	LEADHILLITE	MNCL.	6.55	CBT	PSEUDOHEX, GRNLR. RESIN, PEARLY. 1 CLVG. FLUO-YEL. EFF-HNO3. SCNDRY17.1.3. IN OXIDIZED ZONES OF LEAD DEPOSITS	17.1.3.
2.5-3	WULFENITE	TETR.	6.5 -7.0	MBS	TBLR,DIPYRAMIDAL,MSV,GRNLR, RESIN-ADMN. CLVGS,UNEVN. DCMP-HCL, HNO3,SOL-H2SO4-ALKS.SECONDARY-OXIDIZED ZONES OF PB-, MO-DEPOSITS	48.1.4.1
2.5-3	STOLZITE	TETR.	7.9 -8.34	WOS	DIPYRAMIDAL-STRIATED, RESIN, CLVGS. CONCH. SOL-HCL+YEL). WITH LIMONITE, ETC.-OXIDIZED DEPOSITS WITH TUNGSTEN MINERALS	48.1.4.2
2.5-3	RASPITE	MNCL.	8.46	WOS	TBLR-STRIATED. ADMN.1 CLVG. DCMP-HCL(YEL). WITH STOLZITE(MXED VNS)	48.1.5.
2.5-3.5	GIBBSITE	MNCL.	2.3 -2.42	OXD	TBLR, MAMMILLARY, PEARLY-VITR. 1 CLVG. CLAYLK ODOR WHEN DAMP. WX-ING PRODUCT, BAUXITE DEPOSITS, LOW TEMPERATURE VEINS	6.2.2.
2.5-3.5	SERPENTINE		2.5 -2.6	PSI	GROUP NAME. NOMENCLATURE FOR MEMBERS IS UNSETTLED. MSV-ASBESTIFM. WAXY-GRSY. GELAT. SOME FLUO-CREAM YEL. ALTER PROD AFTER MG-SILICATES(E.G.,OLIVINE). VNS	W381 3-170
2.5-3.5	ANTIGORITE (SERPENTINE)	MNCL.	2.65	PSI	COLORS COMMONLY VARIEGATED OR MOTTLED. PLATY. SECONDARY	W381 3-170
2.5-3.5	BISMUTITE	TETR.	6.1 -7.7	CBT	GRAY STRK. MSV, RADIAL FIBR CRUSTS. VITR-DULL. EFF-ACIDS. SCNDRY AFTER BISMUTH MINERALS	16.1.8.
2.5-4	ZINNWALDITE	MNCL.	2.90-3.02	PSI	LI-,FE-MICA. WITH LI-BEARING MINS-PEGS, CASSITERITE-BEARING VEINS	W371 3-92
3	CALCITE	HEXA.	2.71	CBT	VARIOUS TRIGONAL XL HABITS. MSV. VITR-PEARLY. RHOMB CLVG. SOME IS FLUO,PHOSPHO,THERMOLUM. EFF-DILUTE HCL. WDSPRD-VNS,DIVERSE ROCKS	14.1.1.1
3	APHTHITALITE	HEXA.	2.65-2.71	SUF	TRIGONAL-TBLR, MAMMILLARY,MSV. VITR. CLVGS. CONCH. SOL-H2O. ACIDS. FUMAROLIC, OCEANIC + LACUSTRINE SALINE DEPOSITS	28.2.2.
3	CHURCHITE	MNCL.	3.14	PHO	FAN-SHAPED GROUPS OF COLUMNAR XLS. VITR,PEARLY. 1 CLVG. CONCH. RARE - COATINGS AT COPPER DEPOSIT	40.3.4.
3	OLIVENITE	ORTH.	3.9 -4.46	ARS	ELONGATE,GLBL FIBR,MSV,ETC. ADMN-SILKY. CLVGS. CONCH. SOL-ACIDS + NH4OH. SECONDARY - OXIDIZED ZONES OF ORE DEPOSITS	41.6.5.1
3	GANOMALITE	HEXA.	5.74	SSI	PRISM XLS + CLVG. GELAT. CONSTITUTES SER WITH NASONITE. LANGBAN	W478 N
3	DUFTITE	ORTH. C	6.98	ARS	SUBTRANSLUCENT. AGGS CURVED XLS. VITR-DULL. WITH AZURITE(TSUMEB)	41.5.1.4
3 -3.5	PIRSSONITE	ORTH.	2.35	CBT	PRISM, TBLR. VITR. PYROELECT. EFF-ACIDS. WITH TRONA,ETC.-SED DEPS.	15.2.2.
3 -3.5	HANKSITE	HEXA.	2.56	SUF	STRIATED PRISMS. VITR-DULL. 1 CLVG. SALTY. FLUO-YEL(LW). SOL-H2O. EFF-DILUTE ACIDS. SALINE LAKE DEPOSITS	32.1.1.
3 -3.5	HOPEITE	ORTH.	3.00-3.10	PHO	TBLR,PRISM,TUFTED AGGS,CRUSTS. VITR,PEARLY. CLVGS. SOL-DILUTE HCL. WITH HEMIMORPHITE, ETC.	40.2.11.
3 -3.5	MITHERITE	ORTH.	4.29	CBT	PSEUDOHEX, GLBLR,ETC. VITR. FLUO(ETC.)-LK ARAGONITE.SOL-HCL. VNS	14.1.3.2

131

NONMETALLIC GRAY

HARDNESS	NAME	XL. SYS.	SPECIFIC GRAVITY	CHEM. CLASS	REMARKS	REFERENCES
3 -3.5	BARITE	ORTH.	4.50	SUF	TBLR, MSV, ETC. VITR-PEARLY. CLVGS. SOME FLUO, PHOSPHO, THERMOLUM. SOME FETID WHEN RUBBED. COMMON-VNS, SED RKS, CAVITIES IN IGNEOUS RKS	28.3.1.1
3 -3.5	LAUTITE	ORTH.	4.8 -5.0	SLD	SUBMET. OPQ. BLK-GRAY(RED TINGE). SOL-HNO3. VNS-WITH NATIVE AS, ETC.	2.9.5.5
3 -3.5	CERUSSITE	ORTH.	6.53-6.57	CBT	TWINNED CLUSTERS, MSV. ADMN-RESIN. CLVGS. FLUO-YEL(X-RAYS +(LW)UL- TRA VIOLET), OXIDIZED ZONES OF LEAD DEPOSITS	14.1.3.4
3 -5.5	ZEOLITE		1.9 -2.45	TSI	GROUP OF MIN FAMILIES WITH SOMEWHAT RELATED APPEARANCES, OCCURRENCES, AND COMPOSITIONS. MANY ARE INDISTINGUISHABLE MEGASCOPICALLY. DIVERSE HABITS(E.G., BUNDLES, FIBROUS). PEARLY-GLASSY. SUBCONCH. SOME FLUO-ORANGE OR YEL-GREEN. CONTINUOUS, IN PART REVERSIBLE, DEHYDRATION(I.E., WATER CAN BE EXPELLED WITH- OUT DESTROYING XL STRUCTURE). SCNDRY IN OPEN SPACES; COMMONLY TWO OR MORE OCCUR TOGETHER IN CAVITIES IN BASALTIC RKS - THIS OCCURRENCE IS REFERRED TO HEREIN AS THE ZEOLITE ASSOCIATION (ABBREVIATED "ZEOL-ASSOC").	W330 4-351
3.5-7	BRITTLE MICA	MNCL.	2.6 -3.2	PSI	GROUP OF MORE OR LESS RELATED MINS THAT RESEMBLE MICAS BUT CLEAVE TO BRTL FLKS AND HAVE HARDNESSES OF 3.5-4 VS. 2-3(MICA) ON CLVG FLKS AND 6 VS. 4 ⊥ TO FLKS. WITH CORUNDUM + DIASPORE IN EMERY DEPS, CHLORITE + MICA SCHISTS, SKARN	W392 3-95
3.5	GORDONITE	TRCL.	2.23	PHO	SHEAFLK AGGS. VITR, PEARLY. CLVGS. CONCH. SOL-ACIDS. IN VARISCITE NODULES	42.8.8.
3.5	SULFOHALITE	ISOM.	2.50	SUF	DODEC, OCTAH. VITR. CONCH. SALTY. SOL-H2O(SLOWLY). SALINE DEPOSITS	30.1.6.
3.5	KIESERITE	MNCL.	2.57	SUF	MSV, GRNLR. VITR. CLVGS. FRIABLE. SOL-H2O(SLOW). MARINE SALINE DEPS	29.6.2.1
3.5	BURKEITE	ORTH.	2.57	SUF	TBLR, PLTY AGGS, ETC. GRSY. CONCH. SOL-H2O. WITH TRONA, ETC. BORAX DEP	32.1.4.
3.5	POLYHALITE	TRCL.	2.78	SUF	MSV-FIBR, FOL. VITR. 1 CLVG. DCMP-H2O(ⓁGYP). MARINE SALINE DEPS	29.4.2.
3.5	WEBERITE	ORTH.	2.96-2.97	HAL	GRAINS(SOME INCLUDED IN CRYOLITE). VITR. POOR CLVGS. SOL-AGCL3 SO- LUTION, H2O(SLIGHTLY). RARE-IVIGTUT, GREENLAND	11.5.11.
3.5	ANHYDRITE	ORTH.	2.98	SUF	MSV, GRNLR, FIBR. VITR-PEARLY. 2 CLVGS. SOL-ACIDS. SED ROCKS	28.3.2.
3.5	HUREAULITE	MNCL.	3.15-3.20	PHO	TBLR, MSV, SCALY, FIBR. VITR. 1 CLVG. SOL-ACIDS. WITH ALTERED LITHI- OPHILITE AND/OR TRIPHYLITE	39.1.3.
3.5	LIME	ISOM.	3.3	OXD	POSSIBLY OCCURS NATURALLY IN CALC ROCKS WITHIN LAVA AT VESUVIUS	4.2.1.5
3.5	CHLOROPHOENICITE	MNCL.	3.46	ARS	LONG PRISM, STRIATED. VITR, PEARLY. 1 CLVG. SOL-ACIDS. FRNKLN, N.J.	41.1.4.1
3.5	STRONTIANITE	ORTH.	3.64-3.78	CBT	SPEAR-LK, MSV. VITR. CLVGS. FLUO(ETC.)-LK ARAGONITE. SOL-HCL. VNS	14.1.3.3
3.5	VRBAITE	ORTH.	5.27-5.33	SST	SUBMET. DK BLUISH GRAY, DK RED WITH TRANSMITTED LIGHT. STRK-RED WITH YEL TINGE. TBLR, PYRAMIDAL. 1 CLVG. SOL-HNO3, H2SO4. WITH REALGAR	3.8.13.
3.5	HYDROCERUSSITE	HEXA.	6.80	CBT	SCALY, TBLR, ADMN, PEARLY, BASAL CLVG. EFF-ACIDS. SCNDRY AFTER PB MINS	16.1.12.
3.5-4	HEULANDITE (ZEOLITE)	MNCL.	2.1 -2.2	TSI	PLTY GRP. TBLR-BROAD CENTRAL PORTIONS(COFFIN-LIKE). ZEOL-ASSOC	W347 4-377
3.5-4	STILBITE (ZEOLITE)	MNCL.	2.1 -2.2	TSI	PLTY GRP. SHEAF-LIKE AGGS. 1 GOOD CLVG. DCMP-HCL. ZEOL-ASSOC	W345 4-377
3.5-4	NORTHUPITE	ISOM.	2.38	CBT	OCTAH. VITR. CONCH. EFF-ACIDS, DCMP-HOT H2O. SALINE CLAY DEPS	16.2.2.
3.5-4	KOLBECKITE	MNCL.	2.39	PHO	PRISM. VITR-PEARLY. 1 CLVG. BRTL. DCMP-ACIDS. IN QTZ-WOLFRAMITE VN	43.2.7.
3.5-4	LANGBEINITE	ISOM.	2.83	SUF	NODULAR, DISSEM. VITR. CONCH. SOL-H2O(SLOWLY). MARINE SALTS	28.4.3.1
3.5-4	DOLOMITE	HEXA.	2.84-2.86	CBT	RHOMB-CURVED, GRNLR. VITR-PEARLY. RHOMB CLVG. SOME FLUO. EFF-WARM HCL IF POWDERED. SED RKS, XLS IN CAVITIES, ETC.	14.2.1.1
3.5-4	ALUNITE	HEXA.	2.6 -2.9	SUF	GRNLR, MSV. VITR, PEARLY. CLVGS. SOL-H2SO4. "ALUNITIZED" ROCKS	30.2.4.1

132

NONMETALLIC GRAY

Hardness	Mineral	Crystal	SG	Luster	Properties	Ref
3.5-4	NATROALUNITE	HEXA.	2.6-2.9	SUF	GRNLR, MSV. VITR,PEARLY. CLVGS. SOL-H2SO4. RARE,BUT VARIED OCCUR	30.2.4.2
3.5-4	ARAGONITE	ORTH.	2.94-2.95	CBT	ACICULAR, RADIAL FIBR, STALACTITIC,VITR. CLVGS. FLUO(ULTRA-VIOLET, X-RAYS,ELECTRON BEAMS),THERMOLUM. EFF-ACIDS. LOW TEMPERATURE DEPS	14.1.3.1
3.5-4	MANSFIELDITE	ORTH.	3.03	ARS	CELLULAR MASSES,CRUSTS,VITR. CLVGS. SOL-ACIDS. RARE-WITH SCORODITE	40.3-1.4
3.5-4	SCORODITE	ORTH.	3.28	ARS	PYRAMIDAL,ETC.-AGGS,CRUSTS,ETC. VITR. CLVGS. SOL-ACIDS. SCNDRY-AFTER ARSENOPYRITE AND OTHER AS-BEARING MINERALS	40.3.1.3
3.5-4	LAUBMANNITE	ORTH.	3.33	PHO	SUBTRANSLUCENT. CRUSTS. VITR-SILKY. CLVG. RARE WITH LIMONITE	41.4.6.
3.5-4	RHODOCHROSITE	HEXA.	3.70	CBT	GRNLR, BOTRYOIDAL. VITR-PEARLY, RHOMB CLVG. EFF-WARM ACID. VNS	14.1.1.4
3.5-4	POWELLITE	TETR.	4.21-4.25	MBS	PYRAMIDAL,TBLR,MSV-FOL,POWDER,CRUSTS,SUBADMN-PEARLY. CLVGS. UNEVN. FLUO-YEL. DCMP-HCL. ALTER PROD OF MOLYBDENITE	48.1.3.2
3.5-4	NADORITE	ORTH.	7.02	ATI	TBLR,PRISM,SUBPARALLEL-DIVERGENT GROUPS. RESIN-ADMN. CLVG. DIVERSE	46.1.4.
3.5-4	PYROMORPHITE	HEXA.	7.00-7.08	PHO	PRISM,GLBL,ETC. RESIN. RHOMB CLVG. UNEVN. SOL-HNO3. OXIDIZED ZONES OF LEAD DEPOSITS	41.7.2.1
3.5-4.5	MAGNESITE	HEXA.	2.98-3.02	CBT	GRNLR, RHOMB. VITR. RHOMB CLVG. FLUO-GREEN,BLUE. SOL-WARM HCL. AL-TERATION OF MG-RICH ROCKS (E.G., SERPENTINE)	14.1.1.2
3.5-4.5	SIDERITE	HEXA.	3.95-3.97	CBT	TARN-IRID. MSV. RHOMB,ETC. VITR,RHOMB CLVG. SOL-HOT ACID. SED, VNS	14.1.1.3
3.5-6	MARGARITE (BRITTLE MICA)	MNCL.	3.1	PSI	AGGS OF LAMELLAE. ATTKD-H2SO4. WITH DIASPORE + CORUNDUM-EMERY DEPS	W392 3-95
4	EPISTILBITE (ZEOLITE)	MNCL.	2.1-2.3	TSI	PLTY GRP. SHEAF-LK + SPHERICAL AGGS. DCMP-HCL,PIEZOELECTRIC. ZEOL-ASSOC	W346 4-377
4	FLUORITE	ISOM.	3.18	HAL	CUBIC, MSV. VITR. OCTAH CLVG. FLUO-VIOLET, DCMP-H2SO4. WDSPRD-VNS, CAVITIES, DISSEMINATED IN IGNEOUS ROCKS, ETC.	9.2.1.
4	BARYTOCALCITE	MNCL.	3.66-3.71	CBT	PRISM-STRIATED, MSV. VITR. CLVG. WEAKLY FLUO. SOL-HCL. LO-T DEPS	14.2.3.
4	SPHEROCOBALTITE	HEXA.	4.13	CBT	PINK STRK. MSV-RADIAL STRUCTURE. VITR. EFF-HOT ACID. VNS	14.1.1.5
4	MANGANITE	MNCL.	4.32-4.34	OXD	SUBMET. RED-BRWN STRK. STRIATED PRISMS. 3 CLVGS.SOL-HCL. LO-T VNS, SECONDARY, ETC.	6.1.3.
4-.5	PHILLIPSITE (ZEOLITE)	MNCL.	2.2	TSI	PENETRATION TWINS, SPHERULITES. 2 GOOD CLVGS. GELAT. DISTINGUISHED FROM HARMOTOME BY OPTICAL METHODS. ZEOL-ASSOC,DEEP-SEA RED CLAYS	W343 4-386
4-.5	ALSTONITE	ORTH.	3.67-3.71	CBT	PSEUDOHEX. STRIATED. VITR. FLUO-YEL(LW). SOL-HCL. LO-T DEPS	14.2.2.
4-.5	JARLITE	MNCL.	3.78-3.93	HAL	TBLR, RADIAL AGGS. VITR. SOL-AlCl3 SOLUTION. RARE-IVIGTUT,GREENLD.	11.5.5.
4-.5	ANCYLITE	ORTH.	3.95	CBT	PSEUDO-OCTAH,CRUSTS,VITR. SPLINTERY. SOL-ACIDS. DRUSES IN ALK PEGS	16.2.10.
4-.5	SMITHSONITE	HEXA.	4.40-4.45	CBT	BOTRYOIDAL, MSV. VITR-PEARLY. RHOMB CLVG. SOME FLUO-GREEN, BLUE. EFF-ACIDS. SECONDARY - OXIDIZED ZONES OF ZINC DEPOSITS	14.1.1.6
4-.5	BINDHEIMITE	ISOM.	4.6 -5.6	ANT	DENSE-EARTHY CRUSTS,NODULAR-CONCENTRIC,SOME IS OPALINE. RESIN-DULL. DCMP-HNO3, HCL OXIDATION ZONES OF ANTIMONY-BEARING LEAD DEPOSITS	44.1.1.1
4-.5	ROOSEVELTITE	MNCL.	6.86	ARS	BOTRYOIDAL CRUSTS. ADMN. SOL-ACIDS. WOOD-TIN VNS ACID LAVA FLOWS	38.4.4.
4-.5	WOLFRAMITE	MNCL.	7.31	WOS	SUBMET. TARN-IRID. PRISM-EQUANT--STRIATED,MSV-GRNLR,ETC. 1 CLVG. DCMP-AQ-REG,H2SO4 OR HCL(SLOWLY).DIVERSE-VNS,CNTCT META, PLACERS	48.1.1.2
4-5	TRIPHYLITE	ORTH.	3.34-3.50	PHO	MSV, ANHEDRAL. VITR. CLVGS. UNEVN. SOL-ACIDS. PEGMATITES	38.1.1.1
4-5	SVABITE	HEXA.	3.5 -3.8	ARS	SHORT PRISM,MSV. VITR.CLVG. SOL-DILUTE ACIDS.RARE-E.G.,FRNKLN,N.J.	41.7.3.1
4-5	STAINIERITE	ORTH.	4.13-4.47	OXD	SUBMET. OPQ. MAMMILLARY. UNEVN. SOL-HCL. OXIDATION PROD OF CO-MINS	6.1.4.

133

NONMETALLIC GRAY

HARDNESS	NAME	XL. SYS.	SPECIFIC GRAVITY	CHEM. CLASS	REMARKS	REFERENCES
4, -5	XENOTIME	TETR.	4.4-5.1	PHO	PRISM + PYRAMIDAL. VITR. 1 CLVG. ACID + ALK IG RKS + PEGS.DETRITAL	38.4.1.
4.5	COLEMANITE	MNCL.	2.42-2.43	BOR	EQUANT. GRNLR. VITR. 2 CLVGS. SOL-HOT HCL. EVAPORITE BORATE DEPS	25.1.9.
4.5	HARMOTOME (ZEOLITE)	MNCL.	2.41-2.47	TSI	HABIT LK PHILLIPSITE(ABOVE). + CRUCIFM TWINS. DCMP-HCL. ZEOL-ASSOC ALSO WITH AL-WXING PRODS.IN GNEISSES, + WITH MN-MINERALIZATION	W344 4-386
4.5	DIDYMOLITE	MNCL.	2.71	SSI	SUBTRANSLUCENT. TWINNED XLS. 2 FAIR CLVGS. SOL-HF. CNTCT META LS	W455 N
4.5	PROSOPITE	MNCL.	2.88-2.90	HAL	TBLR. MSV. GRNLR. VITR-DULL. PYRAMIDAL CLVG. DCMP-H2SO4. VNS. GREI SENS. PEGMATITES - RARE	-11.5.7.
4.5	SYNCHISITE	HEXA.	3.90	CBT	ACUTE RHOMB, TBLR-HEMIMORPHIC. GRSY. 1 CLVG. SOL-ACIDS. ALK PEGS	16.2.8.
4.5	PARISITE	HEXA.	4.33-4.39	CBT	STRIATED BIPYRAMIDS. VITR.PEARLY. SOL-HOT ACIDS. VNS, ALK PEGS	16.2.6.
4.5	CARACOLITE	UNKN.	5.0 -5.2	SUF	CRUSTS OF PSEUDOHEX XLS. VITR. RARE-WITH BINDHEIMITE.ETC.(CHILE)	30.1.3.
4.5	CESAROLITE	UNKN.	5.29	OXD	SUBMET. STL-GRAY. FRIABLE MASSES(RESEMBLES COKE).SOL-HCL. RARE- CAVITIES IN GALENA(SIDI-AMOR-BEN-SALEM,TUNIS)	7.7.1.3
4.5	EULYTINE	ISOM.	6.6	NSI	MINUTE TETRAHEDRA. POOR DODEC CLVG. GELAT. WITH BISMUTH + QTZ	W494 N
4.5	BISMITE	MNCL.	8.64-9.22	OXD	GRAY-YEL STRK. MSV. GRNLR. VITR-DULL. OXIDATION PROD OF BI-MINS	4.4.7.
4.5	PARAWOLLASTONITE	MNCL.	2.92	ISI	DISTINGUISHED FROM WOLLASTONITE BY X-RAY ANAL. CONTACT META LS- RARE	W455 2-167
4.5-5	WOLLASTONITE	TRCL.	2.87-3.09	ISI	TBLR. MSV. CLVGS-84° + 96°.DCMP-HCL. FLUO-YEL-ORANGE. CNTCT META-LS	W456 2-167
4.5-5	PLUMBOGUMMITE	HEXA.	4.01	PHO	CONCENTRIC BOTRYOIDAL. GUM-LIKE LUSTER. UNEVN. SOL-HOT ACIDS. SECONDARY - IN LEAD DEPOSITS WITH PYROMORPHITE	41.5.8.1
4.5-5	CORONADITE	TETR.	5.44	OXD	SUBMET. DPQ. BRWN-GRAY STRK. MSV-BOTRYOIDAL. OXIDIZED ZONE-MN DEPS	7.7.1.1
4.5-5	SCHEELITE	TETR.	6.08-6.12	MOS	DIPYRAMIDAL.MSV.GRNLR.VITR.CLVGS.UNEVN.DCMP-HCL.HNO3.FLUO-BLUE(X-SW ULTRAVI- OLET-, + CATHODE-RAYS).THERMOLM. HI-T VNS,PEGS,GREISSENS,CNTCT META RKS,ETC.	48.1.3.1
4.5-5.5	DEHRNITE	HEXA.	3.04-3.09	PHO	BOTRYOIDAL CRUSTS,PRISMS-SUB/-SHEAFLK GROUPS. VITR. 1 CLVG. SOL- ACIDS. WITH OTHER PHOSPHATE MINERALS	41.7.4.
4.5-6	MONIMOLITE	ISOM.	5.9 -7.3	ANT	OCTAH,CUBIC. GRSY. OCTAH CLVG. CONCH-SPLINTERY. RARE-E.G.,LANGBAN	44.1.2.
5	NATROLITE (ZEOLITE)	ORTH.	2.20-2.26	TSI	FIBR GRP. SQ NEEDLES. PRISM CLVG. GELAT. PYROELECTRIC. ZEOL-ASSOC	W340 4-358
5	MESOLITE (ZEOLITE)	MNCL.	2.25-2.27	TSI	FIBR GRP. TUFTS OF HAIR-LIKE XLS. GELAT. PYROELECTRIC. ZEOL-ASSOC	M341 4-358
5	SCOLECITE (ZEOLITE)	MNCL.	2.25-2.29	TSI	FIBR GRP. THIN STRIATED PRISMS. PRISM CLVG. GELAT. ZEOL-ASSOC, AL- SO HYDROTHERMAL + IN METAMORPHOSED CALCAREOUS ROCKS	M341 4-358
5	CRANDALLITE	HEXA.	2.78-2.92	PHO	CONCENTRIC NODULAR AGGS,ETC. VITR-DULL. 1 CLVG. SOL-ACIDS(DIFFI- CULTY). SECONDARY - DIVERSE OCCURRENCES	41.5.8.4
5	DELTAITE (MIXTURE ?)	HEXA.	2.90-3.00	PHO	FIBR CRUSTS,PRISM. 1 CLVG. RARE-WITH CRANDALLITE(FAIRFIELD, UTAH)	41.5.8.5
5	HYDROXYLAPATITE	HEXA.	2.9 -3.1	PHO	APATITE GROUP(SEE FOR PROPS.) RARE-TALC SCHIST	41.7.1.3
5	CARBONATE-APATITE	HEXA.	2.9 -3.1	PHO	APATITE GROUP(SEE FOR PROPS.)RARE-NODULES(PHOSPHATE RK),GUANO,ETC.	41.7.1.4
5	WHITLOCKITE	HEXA.	3.12	PHO	RHOMB. GRNLR. VITR. UNEVN. SOL-DILUTE ACIDS. PEGS, PHOSPHATE ROCKS	38.3.1.
5	CHLORAPATITE	HEXA. A	3.17	PHO	APATITE GROUP(SEE FOR PROPS.) RARE-VNS IN GABBRO, METEORITES	41.7.1.2
5	FLUORAPATITE	HEXA. A	3.18	PHO	APATITE GROUP(SEE FOR PROPS.) COMMON-IGS, METAS, MARINE SEDS, ETC.	41.7.1.1

NONMETALLIC GRAY

H	Mineral	Crystal	S.G.	Luster	Notes	Ref
5	APATITE	HEXA.	2.9 -3.2	PHO	GROUP NAME. MOST TYPES INDISTINGUISHABLE MEGASCOPICALLY. HEX PRISMS-COMMONLY ROUNDED,MSV,GRNLR, VITR. SOL-ACIDS. SOME FLUO-ORANGE,IG RKS,VNS,PEGS	41.7.1.
5	HEMIMORPHITE	ORTH.	3.45	SSI	HEMIMORPHIC XLS, SHEAFLK, MAMMILLARY, ETC. 1 PERF + 1 POOR CLVG. GELAT,PYROELECTRIC. SOME FLUO-PALE ORANGE(LONG WAVE). ZINC DEPS	W481 N
5	CALAMINE	ORTH.	3.45	SSI	(=HEMIMORPHITE)	W481 N
5	ADELITE	ORTH.	3.70-3.76	ARS	MSV,ELONGATE. RESIN. CONCH. SOL-DILUTE ACIDS. MN DEPOSITS	41.5.1.1
5	TILASITE	MNCL.	3.75-3.79	ARS	ELONG,MSV,RESIN. CLVGS + PARTINGS. SOL-HCL,HNO3,WITH OTHER MN-MINS	41.5.6.
5	PLUMBOFERRITE	HEXA.	6.07	OXD	RED STRK. TBLR. 1 CLVG. SOL-HCL. RARE-MN MINS(VERMLAND, SWEDEN)	7.3.2.
5 -5.5	MAGNERITE	MNCL.	3.15	PHO	NEARLY OPQ-ALTERED. PRISM STRIATED,MSV. VITR. CLVGS. UNEVN-SPLIN-TERY. SOL-ACIDS. DIVERSE(E.G., VEINS AND VOLCANIC ROCKS)	41.6.1.
5 -5.5	HYALOTEKITE	ORTH.	3.8	ISI	MSV, COARSELY XLINE. 2 CLVGS-90°. ORE DEP(LANGBAN, SWEDEN)	W401 N
5 -5.5	YTTROTANTALITE	ORTH.	5.5 -5.9	OXD	SUBMET, GRAY STRK. PRISM. VITR. METAMICT. RARE-PEGS	8.1.4.
5 -6	VISHNEVITE	HEXA.	2.32-2.42	TSI	SULFATE-RICH END MEMBER OF CANCRINITE FAMILY. OCCURRENCE LIKE CANCRINITE. CARBONATE-RICH MEMBERS EFFERVESCE AS WELL AS GELATINIZE WITH HCL	MN 4-310
5 -6	CANCRINITE	HEXA.	2.42-2.51	TSI	MSV (RARE PRISM XLS). PRISM CLVG. EFF-HCL. NEPHELINE-BEARING RKS	W354 4-310
5 -6	MARIALITE (SCAPOLITE)	TETR.	2.50-2.62	TSI	ME-0-20. PURE MARIALITE UNKNOWN IN NATURE	W352 4-321
5 -6	DIPYRE (SCAPOLITE)	TETR.	2.57-2.69	TSI	ME-20-50.	W352 4-321
5 -6	MIZZONITE (SCAPOLITE)	TETR.	2.67-2.74	TSI	ME-50-80.	W352 4-322
5 -6	SCAPOLITE	TETR.	2.50-2.78	TSI	GROUP NAME. MOST TYPES INDISTINGUISHABLE MEGASCOPICALLY BUT S.G. MAY BE IN-DICATIVE. LONG,STRIATED PRISMS + AGGS OF COARSE XLS. SUBCONCH. POOR CLVG GIVING IRREGULAR, STRIATED-APPEARING SURFACES. MARIALITE IS INSOL. MEIONITE IS DCMP IN HCL. MOST FLUO-YEL-RED(ULTRAVIOLET,LW). CNTCT META CALC RKS,PEGS	W352 4-321
5 -6	(WERNERITE) (SCAPOLITE)	TETR.	2.50-2.78	TSI	GROUP NAME NO LONGER USED	W352 4-322
5 -6	MEIONITE (SCAPOLITE)	TETR.	2.73-2.78	TSI	ME-80-100. PURE MEIONITE UNKNOWN IN NATURE	W352 4-321
5 -6	TURQUOIS	TRCL.	2.6 -2.84	PHO	XLS-RARE,CRUSTS,STALAC. WAXY. 2 CLVGS. SOL-HCL(DIFFICULTLY). WXED ALUMINOUS ROCKS (ESPECIALLY IN ARID AREAS)	42.6.2.1
5 -6	AKERMANITE	TETR.	2.94	SSI	ISOMORPHOUS WITH GEHLENITE + MELILITE. RARE IN NATURE, IN SLAGS	W473 1-236
5 -6	GEHLENITE	TETR. A	3.04	SSI	SHORT SQ PRISMS. VITR-RESIN.CONCH.GELAT. CNTCT META-IMPURE CO3-RKS	W473 1-236
5 -6	ENSTATITE (ORTHOPYROXENE)	ORTH.	3.21	ISI	MG PYROXENE. WITH OLIVINE IN BASIC + ULTRABASIC IG RKS, GRANULITES	W405 2-8
5 -6	ROSENBUSCHITE	TRCL.	3.3	NSI	PRISMATIC, ACICULAR. 1 PERF CLVG. SOL-HCL. NEPHELINE SYENITE-RARE	W518 N
5 -6	CROSSITE (AMPHIBOLE)	MNCL.	3.1 -3.4	ISI	INTERMEDIATE BETWEEN RIEBECKITE + GLAUCOPHANE. SCHISTS, LK GLAUCO-PHANE	W440 2-334
5 -6	CROCIDOLITE (AMPHIBOLE)	MNCL.	3.0 -3.4	ISI	FIBR RIEBECKITE. IN VEINS IN METAMORPHOSED BEDDED IRON-STONES	W440 2-334
5 -6	HARDYSTONITE	TETR. A	3.40	SSI	GEHLENITE GROUP, GRNLR. SOME FLUO-PURPLE(SHORT-WAVE). RARE(FRNKLN)	W473 N
5 -6	TREMOLITE (AMPHIBOLE)	MNCL.	3.02-3.44	ISI	META SI-BEARING LS + DOLO, LOW-GRADE META ULTRABASIC IGNEOUS RKS	W434 2-249
5 -6	URALITE (AMPHIBOLE)	MNCL.	3.02-3.45	ISI	SECONDARY AFTER PYROXENE, ESPECIALLY IN IGNEOUS ROCKS	W436 N
5 -6	ARFVEDSONITE (AMPHIBOLE)	MNCL.	3.50	ISI	PLUTONIC ALKALIC IGNEOUS MASSES AND ASSOCIATED PEGMATITE MASSES	W439 2-364

NONMETALLIC GRAY

HARDNESS	NAME	XL. SYS.	SPECIFIC GRAVITY	CHEM. CLASS	REMARKS	REFERENCES
5 -6	AMPHIBOLE		3.0 -3.57	ISI	GROUP NAME. SEVERAL END AND OTHER MEMBERS WITH NAMES UNSETTLED FOR SOME. MANY MEGASCOPICALLY INDISTINGUISHABLE BUT SOME COLORS AND OCCURRENCES ARE INDICATIVE. ORTH AND MNCL SERIES. LONG SLENDER XLS. PRISM CLVG AT 56° + 124°. SOL-HF(SLOWLY). ALTERS TO BIOTITE +/OR CHLORITE. IGNEOUS AND METAMORPHIC RKS, VNS	W422 2-203
5 -6	PSILOMELANE	ORTH.	4.70-4.72	OXD	SUBMET. DPQ. MSV. MAMMILLARY. SOL-HCL. SCNDRY(E.G.,RESIDUAL DEPS)	6.1.8.
5 -6	ILMENITE	HEXA.	4.68-4.76	OXD	SUBMET. OPQ. TBLR. VNS + DISSEM DEPS ASSOC WITH GABBROS + DIORITES	4.4.1.3
5 -6	HEMATITE	HEXA.	5.26	OXD	MET-EARTHY. STRK-RED. PLTY,GRNLR. SOL-HCL. SED FE-FMS, RARE IN VNS	4.4.1.2
5 -7	EKMANNITE	HEXA.	2.79	PSI	MICACEOUS. RARE IN ORE DEPOSITS-E.G., AT GRYTHYTTAN, SWEDEN	W361 N
5 -7	PYROXENE		2.96-3.96	ISI	GROUP NAME. SEVERAL END AND OTHER MEMBERS WITH NAMES UNSETTLED FOR SOME. MANY MEGASCOPICALLY INDISTINGUISHABLE; OCCURRENCE MAY BE INDICATIVE. ORTH + MNCL SERIES.SHORT PRISMS.PRISM CLVG AT 87° + 92°. ALTERS TO AMPHIBOLE.IG + META RKS	W402 2-1
5 -7	OLIVINE	ORTH.	3.22-4.39	NSI	GRP NAME FOR MINS OF FAYALITE-FORSTERITE + FAYALITE-TEPHROITE SERIES, MONTI-CELLITE + GLAUCOCHROITE, GRNLR MASSES, DISSEM. CONCH. GELAT. COMMONLY ALTERED TO SERPENTINE. BASIC + ULTRABASIC IG RKS, CNTCT META DOLOMITIC LS, METEORITES	W498 1-2
5.5	ANALCIME (ZEOLITE)	ISOM.	2.24-2.29	TSI	TRAPEZOHEDRA. MSV. GRNLR. POOR CUBIC CLVG. GELAT. AMYGDULES-BASALT	W333 4-338
5.5	MILLISITE	TETR.	2.83	PHO	FIBR,CHALCEDONIC(LK CRUSTS. CLVG(?). WITH WARDITE IN VARISCITE	42.5.5.
5.5	LEHIITE	UNKN.	2.89	PHO	AGGS-SUBPARALLEL FIBERS. WITH WARDITE,ETC. IN ALTERED VARISCITE	42.5.6.
5.5	MONTICELLITE	ORTH.	3.08-3.27	NSI	COLORLESS-GRAY. CA-MG OLIVINE. CONTACT META OF SILICEOUS DOLO-LS	W502 1-41
5.5	AUGITE (CLINOPYROXENE)	MNCL.	3.23-3.52	ISI	THE COMMON PYROXENE OF SUB-ALKALIC IGNEOUS RKS(E.G., GABBROS)	W411 2-109
5.5	PERICLASE	ISOM.	3.55-3.57	OXD	OCTAH. ROUNDED GRAINS. CUBIC,OCTAH CLVGS. CNTCT META MARBLES(HI-T)	4.2.1.1
5.5	ALLEGHANYITE	MNCL.	4.02	NSI	SIMILAR TO OLIVINE + HUMITE MINS. WITH MN-MINS IN VNS IN META RKS	W516 1-35
5.5	PEROVSKITE	MNCL.	3.97-4.05	OXD	SUBMET. CUBIC, OCTAH. GRNLR. ADMN. 1 CLVG. DCMP-HF, H2SO4.BASIC IG	7.4.2.1
5.5	ARIZONITE (MIXTURE ?)	MNCL.	4.25	OXD	SUBMET. STL-GRAY. BRWN STRK. MSV. DCMP-HCL. RARE-PEG(25 MI. SE OF HACKBERRY, ARIZONA)	8.2.1.
5.5	CERITE	MNCL.	4.65-4.91	SSI	SHORT PRISM, GRNLR. GELAT. WITH ALLANITE ETC. IN GNEISS + GRANITE	W507 N
5.5-6	SODALITE	ISOM.	2.27-2.33	TSI	DODEC. MSV. POOR DODEC CLVG. GELAT. SOME FLUO-YEL-ORANGE. IN NEPH-ELINE-BEARING RKS. DISSOLVE-HNO3.EVAPORATE SLOWLY, HALITE REMAINS	W348 4-289
5.5-6	LEUCITE	TETR.	2.47-2.50	TSI	TRAPEZOHEDRA(STRIATED FACES). DCMP-HCL. K-RICH ALK VOL	W251 4-276
5.5-6	HAUYNE	ISOM.	2.44-2.50	TSI	SODALITE(SEE) GROUP PROPS. NEPH-BEARING IG RKS. DISSOLVE-HNO3. EVAPORATE SLOWLY, GYPSUM CRYSTALS REMAIN	W350 4-289
5.5-6	NEPHELINE	HEXA.	2.56-2.67	TSI	TBLR,PRISM,VITR-GRSY,GOOD PRISM.POOR BASAL CLVGS.GELAT.FLUO-ORANGE.DISTINGUISH-ED FROM KALSILITE +LEUCOPHYLLITE BY X-RAY. NA-RICH ALK IG AND NEARBY META ROCKS	W254 4-231
5.5-6	MONTEBRASITE	TRCL.	2.98	PHO	AMBLYGONITE SERIES(SEE FOR PROPERTIES + OCCURRENCES.)	41.5.5.2
5.5-6	NATROMONTEBRASITE	TRCL.	3.04-3.1	PHO	AMBLYGONITE SERIES(SEE FOR PROPERTIES + OCCURRENCES.)	41.5.5.3
5.5-6	AMBLYGONITE	TRCL.	3.11	PHO	EQUANT, LARGE MASSES. VITR,PEARLY. CLVGS. SOL-ACIDS(DIFFICULTY).	41.5.5.1
5.5-6	ANTHOPHYLLITE (AMPHIBOLE)	ORTH.	2.85-3.57	ISI	WITH CORDIERITE IN GNEISSES, WITH TALC IN META ULTRABASIC IG RKS	W425 2-211

NONMETALLIC GRAY

Hardness	Mineral		Crystal	SG	Code	Description	Ref
5.5-6	GEDRITE	(AMPHIBOLE)	ORTH.	2.85-3.57	ISI	RELATED IN COMPOSITION + OCCURRENCE TO ANTHOPHYLLITE	W425 2-211
5.5-6	ANATASE		TETR.	3.90	OXD	SUBMET. PALE YEL STRK. ACUTE PYRAMIDAL, ADMN. BASAL + PYRAMIDAL CLVG. CONVERTS TO RUTILE ON HEATING. VNS,ACCESS-IG RKS, DETRITAL	4.5.2.
5.5-6	ILVAITE		ORTH.	3.8 -4.1	SSI	SUBTRANSLUCENT. STRIATED PRISMS. 2 CLVGS. GELAT. CNTCT ZNS, SCNDRY	W511 N
5.5-6.5	OPAL			1.8 -2.25	SIL	SUBMICROXLINE QTZ + H2O-RESIN-PEARLY. MSV,MAMILLARY CRUSTS.CONCH.SOL-HF +CAUSTIC ALKS.SOME FLUO--YEL-GREEN. INCRUSTATIONS, IN CAVITIES(ESP IN VOLCANIC RKS)	D-111 287
5.5-6.5	FERGUSONITE		TETR. C	5.38	OXD	SUBMET. IN SERIES WITH FORMANITE, PRISM, COMMONLY HEMIHEDRAL, IR-REGULAR MASSES. VITR. METAMICT. DCMP-HF. GRANITIC PEGMATITES	8.1.3.1
5.5-6.5	FORMANITE		TETR. C	7.03	OXD	SUBMET IN SERIES WITH FERGUSONITE,WHICH SEE. PLACERS(W. AUSTRALIA)	8.1.3.2
5.5-7	KYANITE		TRCL.	3.53-3.65	NSI	HARDNESS-5 $\perp$ TO + 7 $\parallel$ TO LENGTH OF BLADE-LIKE XLS. GNEISS +SCHIST	W527 1-137
6	SANIDINE	(K-FELDSPAR)	MNCL.	2.56-2.62	TSI	TBLR. GLASSY. OPTICALLY DISTINCT. VOLCANIC ROCKS	W305 4-7
6	KALSILITE		HEXA.	2.59-2.63	TSI	K CHEM-ANALOGUE OF NEPH.DISTINGUISHED BY X-RAY.K-RICH,SI-POOR VOLS	W257 4-231
6	GLAUCOPHANE	(AMPHIBOLE)	MNCL.	3.08-3.30	ISI	WITH LAWSONITE, ETC. IN SCHISTS + GNEISSES (COMMONLY NA-RICH ONES)	W439 2-333
6	ZOISITE		ORTH.	3.15-3.37	SSI	DISTINGUISHED FROM CLINOZOISITE BY OPTICAL OR X-RAY ANAL. IG,META(ESP CALC RKS),PEGS	W446 1-185
6	DANALITE		ISOM.	3.28-3.44	TSI	HELVITE(SEE) GRP PROPS. FLESH-RED TO GRAY. OCTAH. GRANITES, SKARNS	M351 4-303
6	EPIDOTE		MNCL.	3.38-3.49	SSI	GROUP NAME, NO GENERALLY ACCEPTED NOMENCLATURE FOR GROUP MEMBERS. CLOSELY RELATED TO ZOISITE + ALLANITE. PISTACHIO-BLACKISH GREEN-BROWNISH. LONG THIN GROOVED XLS, GRNLR MASSES. VNS, OTHER CAVITIES, REGIONAL META + IG ROCKS	W448 1-193
6	PISTACITE		MNCL.	3.3 -3.49	SSI	NAME USED BY SOME FOR INTERMEDIATE COPOSITION MEMBERS-EPIDOTE GRP	W448 1-184
6	JOHANNSENITE	(CLINOPYROXENE)	MNCL.	3.44-3.55	ISI	MN-ANALOG OF DIOPSIDE-HEDENBERGITE. SKARNS, CNTCT ZONES. VEINS	W411 2-75
6	TEPHROITE		ORTH.	3.78-4.1	NSI	FAYALITE-TEPHROITE (FE-MN)OLIVINE SERIES HAS NAME KNEBELITE APPLIED TO INTERMEDIATE MEMBERS. OCCUR IN META RKS + FE-MN ORE DEPS	W498 1-34
6	MAGNETOPLUMBITE		HEXA.	5.52	OXD	SUBMET. OPQ. BRWN STRK. PYRAMIDAL, BASAL CLVG. MAG. SOL-HCL(SLOW). RARE - WITH MANGANOPHYLLITE (LANGBAN, SWEDEN)	7.3.3.
6	ANORTHOCLASE	(FELDSPAR)	TRCL.	2.56-2.62	TSI	DISTINGUISHED FROM ORTHOCLASE + MICROCLINE-OPTICALLY. NA-RICH VOLS	W311 4-7
6	ALBITE	(PLAGIOCLASE)	TRCL.	2.57-2.63	TSI	AN0-10. SEE FELDSPAR. IG, META RKS, PEGS. MOONSTONE +CLEAVELANDITE ARE VARIETIES	M312 4-94
6	MICROCLINE	(K-FELDSPAR)	TRCL.	2.56-2.63	TSI	DISTINGUISHED FROM ORTHOCLASE OPTICALLY. PLUTONIC IG RKS, PEGS	W308 4-7
6	ORTHOCLASE	(K-FELDSPAR)	TRCL.	2.55-2.63	TSI	DISTINGUISHED FROM MICROCLINE OPTICALLY. PLUTONIC IG RKS, PEGS	W303 4-7
6	PERTHITE	(FELDSPARS)	TRCL.	2.56-2.65	TSI	MEGASCOPICALLY, MICROSCOPICALLY,OR SUBMICROSCOPICALLY INTERDIGITATED MICROCLINE OR ORTHOCLASE + PLAGIOCLASE(TYPICALLY ALBITE)	W299 4-7
6	OLIGOCLASE	(PLAGIOCLASE)	TRCL.	2.62-2.67	TSI	AN10-30. SEE FELDSPAR. PLUTONIC IG RKS, FEW PEGS. SUNSTONE IS VAR.	M312 4-95
6	ANDESINE	(PLAGIOCLASE)	TRCL.	2.64-2.69	TSI	AN30-50. SEE FELDSPAR. INTERMEDIATE SI-CONTENT IGNEOUS ROCKS	M312 4-95
6	LABRADORITE	(PLAGIOCLASE)	TRCL.	2.68-2.72	TSI	AN50-70. BLUISH REFLECTIONS. SEE FELDSPAR. BASIC IGNEOUS ROCKS	M312 4-95
6	BYTOWNITE	(PLAGIOCLASE)	TRCL.	2.71-2.75	TSI	AN70-90. GELAT. SEE FELDSPAR. CALCIUM-RICH IGNEOUS ROCKS	M312 4-95
6	ANORTHITE	(PLAGIOCLASE)	TRCL.	2.70-2.76	TSI	AN90-100. GELAT. SEE FELDSPAR. CNTCT META CALCAREOUS RKS	M312 4-94
6	PREHNITE		ORTH.	2.90-2.95	SSI	ROSETTES-TBLR XLS. 1 CLVG. GELAT. SOL-HCL(SLOW). CAVITIES(WITH ZEOLITES)	W359 3-263

NONMETALLIC GRAY

HARDNESS	NAME	XL. SYS.	SPECIFIC GRAVITY	CHEM. CLASS	REMARKS	REFERENCES
6 -6.5	FELDSPAR		2.55-3.39	TSI	INCLUDES PLAGIOCLASE(NA-CA)SERIES, ALKALI(K-NA) AND BA FELDSPARS. MOST EASILY DISTINGUISHED BY NONMEGASCOPIC MEANS. MONOCLINIC ONES HAVE 2 CLVGS AT 90° AND SIMPLE(IF ANY) TWINNING; TRICLINIC ONES HAVE 2 CLVGS UP TO 4° OFF 90° AND TYPICALLY HAVE POLYSYNTHETIC TWINNING (GIVING STRIATED APPEARANCE ON SOME CLVG SURFACES OF PLAGIOCLASE). PEGMATITES; IG, META, AND SOME SED ROCKS	M261 4-1
6 -6.5	BRAUNITE	TETR.	4.72-4.83	NSI	SUBMET, OPQ. PYRAMIDAL, GRNLR. PYRAMIDAL CLVG. GELAT. MAG. WITH OTHER MN MINERALS - VEINS, WEATHERING PRODUCT DEPOSITS, ETC.	W60 N
6 -7	BERLINITE	HEXA.	2.64	PHO	MSV,GRNLR,VITR. CONCH. RARE-WITH AUGELITE AT FE-MINE(KRISTIANSTAD, SWEDEN)	38.4.3.
6 -7	MULLITE	ORTH.	3.03-3.16	NSI	INDISTINGUISHABLE MEGASCOPICALLY FROM SILLIMANITE. HI-T META-RARE	M401 1-124
6 -7	THORTVEITITE	MNCL.	3.58	SSI	GRAYISH GREEN. TAPERED, PRISM. PRISM CLVG. RARE-GRANITIC PEGS	W477 N
6 -7	CASSITERITE	TETR.	6.99	OXD	SUBMET. RADIAL CONCRETIONARY MASSES. ADMN-DULL. HI-T VNS, GREISENS	4.5.1.5
6 -7.5	HYDROGROSSULAR	ORTH.	3.13-3.59	NSI	GROSSULAR +H2O. SLOWLY SOL-HCL + HNO3. META CALCAREOUS ROCKS	M493 1-104
6 -7.5	GARNET	ISOM.	3.58-4.32	NSI	GROUP NAME. DOMINANT MOLECULE DETERMINES NAME. DISTINGUISHED BY S.G. + NONMEGASCOPIC MEANS.EQUANT XLS,MSV,VITR-DULL.DODEC PARTING. SLIGHTLY- TO IN-SOL IN HF	W483 1-77
6.5	PETALITE	MNCL.	2.41-2.42	TSI	FOL, MSV. 1 PERF + 1 GOOD-CLVGS AT 38.5°. LI-BEARING PEGMATITES	M260 4-271
6.5	CHRYSOPRASE	HEXA.	2.57-2.64	SIL	APPLE-GREEN CHALCEDONY(MICROFIBR QTZ). SECONDARY-VNS IN SERPENTINE	D-111 218
6.5	AGATE	HEXA.	2.57-2.64	SIL	BANDED OR VARIEGATED-CHALCEDONY(MICROXLINE QTZ). NODULES IN BASALT	D-111 210
6.5	CHALCEDONY	HEXA.	2.57-2.64	SIL	MICROXLINE + MICROFIBR QTZ. MAMMILLARY. LO-T VEINS +OTHER CAVITIES	D-111 195
6.5	CLINOZOISITE	MNCL.	3.21-3.38	SSI	CALCIUM-ALUMINUM END-MEMBER OF EPIDOTE GROUP. SAME OCCUR	W448 1-193
6.5	KNEBELITE	ORTH.	3.96-4.25	NSI	FAYALITE-TEPHROITE (FE-MN)OLIVINE SERIES HAS NAME KNEBELITE APPLIED TO INTERMEDIATE MEMBERS. OCCUR IN META RKS + FE-MN ORE DEPS	M498 1-34
6.5	THORIANITE	ISOM.	9.7	OXD	SUBMET. CUBIC. HORNLK. POOR BASAL CLVG. ROACTV. PLACERS, IN SERPENTINE OF CONTACT METAMORPHOSED ZONES	5.1.2.2
6.5-7	BLOODSTONE	HEXA.	2.57-2.65	SIL	SUBTRANSLUCENT. GREEN-RED SPOTS. CHALCEDONY(MICROCRYSTALLINE QTZ)	D-111 219
6.5-7	HELIOTROPE	HEXA.	2.57-2.65	SIL	SUBTRANSLUCENT. GREEN-RED SPOTS. CHALCEDONY(MICROCRYSTALLINE QTZ)	D-111 219
6.5-7	SPODUMENE	MNCL.	3.03-3.22	ISI	COMMONLY FLUO(+PHOSPHORESCES)-ORANGE. THERMOLUM. LI-BEARING PEGS	W418 2-92
6.5-7	AXINITE	TRCL.	3.26-3.36	CSI	STRIATED WEDGE-SHAPED XLS. 1 GOOD CLVG. SOL-HCL(SLOW). CNTCT META	W505 1-320
6.5-7	DIASPORE	ORTH.	3.3 -3.5	OXD	PLTY, FOL-MSV. VITR, PEARLY. WITH CORUNDUM IN EMERY ROCK	7.1.2.1
6.5-7.5	ANDALUSITE	ORTH.	3.13-3.16	NSI	BLUNT, NEARLY SQ PRISMS. COMMONLY ALTERED. CONTACT +REGIONAL META	M521 1-129
7	QUARTZ	HEXA.	2.65	SIL	DOUBLY TERMINATED HEX PRISMS STRIATED ⊥ LENGTH(LO-T), BIPYRAMIDS (HI-T),MSV.VITR.CONCH.SOL-HF.WDSPRD IN MOST ROCK TYPES + VEINS	D-111 9
7	BLUE QUARTZ	HEXA.	2.65	SIL	GRSY. MSV. META + IG RKS. PEGMATITES-COMMONLY WITH TI-BEARING MINS	D-111 189
7	CORDIERITE	ORTH.	2.53-2.78	CSI	COLOR DIFFERS WITH DIRECTION. DISSEM GRNS. 1 GOOD CLVG. META RKS	W470 1-268

NONMETALLIC GRAY

7 -7.5	BORACITE	ORTH. 2.91-2.97	BOR	CUBIC,ETC.,GRNLR,FIBR. VITR. SOL-HCL(SLOWLY). SALT-GYPSUM BEDS		26.1.7.
7.5	HAMBERGITE	ORTH. 2.36	BOR	PRISM-STRIATED. VITR. 2 CLVGS. SOL-HF. ALKALIC PEGMATITES		26.1.2.
7.5	SAPPHIRINE	MNCL. 3.40-3.58	OXD	TBLR, DISSEM. VITR. HI-T META RKS WITH CORUNDUM, SPINEL, ETC.		7.3.1.2
7.5	ZIRCON	TETR. 4.6 -4.7	NSI	TERMINATED PRISMS. ADMN. SOME FLUO-YEL-ORANGE. ACCESS-IG, SANDS		W494 1-59
8	RHODIZITE	ISOM. 3.31-3.38	BOR	DODEC. VITR. OCTAH CLVG. WITH RED TOURMALINE IN PEGMATITES-RARE		24.1.6.
8	TOPAZ	ORTH. 3.49-3.57	NSI	STRIATED PRISMS. PERF BASAL CLVG. CNTCT ZONES, GREISENS, PEGS		W509 1-145
10	DIAMOND	ISOM. 3.50-3.53	ELE	OCTAHEDRAL. ADMN-GRSY. BRTL. OCTAH CLVG. ULTRABASIC IGS + PLACERS		1.2.4.1

139

NONMETALLIC BLACK

HARDNESS	NAME	XL. SYS.	SPECIFIC GRAVITY	CHEM. CLASS	REMARKS	REFERENCES
1.5-2	VIVIANITE	MNCL.	2.67-2.69	PHO	PRISM, ROUNDED GLBL, VITR-DULL CLVGS, FIBR-FLEX, DARKENS ON EXPOSURE. SOL-ACIDS, SECONDARY OF DIVERSE OCCURRENCES	40.2.15.1
1.5-2.5	CHRYSOCOLLA	UNKN.	2.3 -2.5	CSI	MSV, VITR-EARTHY, CONCH, DCMP-HCL. OXIDE ZONES CU-DEPS IN SIL RKS	W420 N
1.5-4.5	ARSENIOSIDERITE	TETR.	3.60	ARS	STRK-YEL. SUBTRANSLUCENT, RADIAL FIBR AGGS, SUBMET-SILKY, 1 CLVG. SOL-HOT ACIDS. SECONDARY - DIVERSE OCCURRENCES	42.6.4.
2 -3	IANTHINITE	ORTH.		OXD	SUBMET. BRWN-VIOLET STRK. PLTY, TBLR. BASAL CLVG. ALTER AFTER URANINITE	5.3.5.
2 -3	PITTICITE	UNKN.	2.2 -2.5	ARS	MSV,BOTRYOIDAL,OPALINE CRUSTS,ETC. DULL-GRSY. SOL-ACIDS. OXIDATION ZONES OF AS-BEARING DEPOSITS, DEPOSITED FROM MINE + SPRING WATERS	43.2.6.
2 -3	DELESSITE (CHLORITE)	MNCL.	2.7 -2.9	PSI	COMMONLY AS REPLACEMENT OF BIOTITE + OTHER FERROMAGNESIAN MINERALS	W383 3-137
2 -3	SYMPLESITE	TRCL.	3.01	ARS	SPHERICAL AGGS-FIBR. VITR,PEARLY. CLVGS. SOL-ACIDS. SCNDRY AS-DEPS	40.2.15.5
2 -3	CHLORITE	MNCL.	2.6 -3.3	PSI	GROUP NAME. MEMBER NAMES UNSETTLED. ONLY FEW RARE VARIETIES MEGASCOPICALLY DISTINGUISHABLE. DISSEM FLKS. PEARLY, 1 PERF CLVG TO FLEX,INELAST FOLIA. DCMP-IN BOILING H2SO4. SCNDRY-HYDROTHERMAL ALTER, LO-GRADE META, DETRITAL, AUTHIGENIC	W381 3-131
2 -4	MICA	MNCL.	2.4 -3.4	PSI	GROUP NAME. SOME SPECIES DISTINGUISHABLE ON BASIS OF COLOR. SERIES ARE NOT COMPLETE, HEX-SHAPED XLS,DISSEM. PERF CLVG TO FLEX,ELAST FLKS. PERCUSSION + PRESSURE FIGURES MAY BE IMPOSED ON FLKS. CHIEF OCCURRENCES ARE NOTED UNDER SPECIES	M365 3-1
2 -6.5	WAD		2.8 -4.4	OXD	FIELD TERM FOR HYDROUS MN-OXIDES OF UNKNOWN IDENTITY. (MN-DEPS)	4.5.1.3
2.5	MITRIDATITE	UNKN.		PHO	MSV,NODULES,CRUSTS. DULL-RESIN. FRIABLE-DENSE. SOL-HOT ACIDS. OOLITIC SEDIMENTARY IRON ORES - RARE (SOUTHERN RUSSIA)	42.6.6.
2.5	CRYOLITE	MNCL.	2.96-2.98	HAL	MSV, GRNLR. GRSY. UNEVN. THEMOLUM. SOL-H2O(SLIGHTLY). RARE-PEGS	11.5.1.
2.5	PYROCHROITE	HEXA.	3.23-3.27	OXD	TBLR, RHOMB, FOL MSV. PEARLY-DULL. BASAL CLVG(FLEX). SOL-HCL. WITH OTHER MN MINERALS IN LOW TEMPERATURE HYDROTHERMAL VEINS	6.1.1.2
2.5	MELANOVANADITE	MNCL.	3.48	VAS	RED-BRWN STRK. SUBTRANSLUCENT. DIVERGENT AGGS OF ROUNDED,STRIATED, ELONGATE PRISMS.NEARLY SUBMET.1 CLVG.SOL-ACIDS.WITH PASCOITE, ETC.	47.1.17.
2.5	QUENSELITE	MNCL.	6.84	OXD	SUBMET. OPQ, BRWNISH STRK. TBLR-STRIATED. 1 CLVG(FLEX). SOL-DILUTE ACIDS. RARE-WITH CALCITE + BARITE IN CREVICES(LANGBAN, SWEDEN)	7.4.1.
2.5	FINNEMANITE	HEXA.	7.27	ASI	CRUSTS-PRISM XLS. SUBADMN. RHOMB CLVG. IN CREVICES IN HEMATITE AT LANGBAN, SWEDEN	46.1.3.
2.5	EGLESTONITE	ISOM.	8.33-8.45	HAL	EXPOS-DK BRWN, BLK. YEL STRK. MSV, CRUSTS, DODEC. RESIN. CONCH. DCMP-ACIDS(CALOMEL RESIDUE). MERCURY DEPOSITS	10.1.1.
2.5	CORVUSITE	UNKN.	2.82	OXD	OPQ. MSV. SOL-ACIDS. WDSPRD IN UTAH-COLO U-ORE AREA-IMPREGNATING SS	4.6.2.
2.5-3	BIOTITE	MNCL.	2.7 -3.3	PSI	MG-,FE-MICA. BLEACHED + ATTKD-H2SO4.COMMON-E.G.,ACID IG,PEGS, META	W373 3-55
2.5-3	SIDEROPHYLLITE	MNCL.	2.8 -3.4	PSI	FE-MICA. RARE IN PEGMATITES	W373 3-65
2.5-3	CLINOCLASE	MNCL.	4.38	ARS	ELONGATE,ROSETTES,CRUSTS,FIBR, VITR,PEARLY, 1 CLVG. SOL-ACIDS + NH4OH. WITH OLIVENITE + OTHER SECONDARY COPPER MINERALS	41.3.1.
2.5-3	VAUQUELINITE	MNCL.	6.02	CHR	WEDGE-SHAPED,MAMMILLARY FIBR. ADMN-RESIN. UNEVN. SOL-HNO3(PARTLY). SECONDARY - WITH CROCOITE, ETC.	36.1.1.
2.5-3.5	SERPENTINE		2.5 -2.6	PSI	GROUP NAME. NOMENCLATURE FOR MEMBERS IS UNSETTLED. MSV-ASBESTIFM, WAXY-GRSY. GELAT. SOME FLUO-CREAM YEL. ALTER PROD AFTER MG-SILICATES(E.G.,OLIVINE), VNS	W381 3-170
2.5-3.5	HULSITE	ORTH.	4.28	BOR	SUBMET-VITR. TBLR. PRISM CLVG. SOL-ACIDS. RARE-CNTCT META LS	24.1.3.
2.5-3.5	BISMUTITE	TETR.	6.1 -7.7	CBT	GRAY STRK. MSV, RADIAL FIBR CRUSTS. VITR-DULL. EFF-ACIDS. SCNDRY AFTER BISMUTH MINERALS	16.1.8.
2.5-5	GUMMITE		3.9 -6.4	OXD	FIELD TERM FOR HYDROUS OXIDES OF URANIUM OF UNKNOWN IDENTITY	5.2.1.

NONMETALLIC BLACK

Hardness	Mineral	Crystal	SG	Streak	Description	Formula Ref
3	CALCITE	HEXA.	2.71	CBT	VARIOUS TRIGONAL XL HABITS, MSV. VITR-PEARLY, RHOMB CLVG. SOME IS FLUO, PHOSPHO, THERMOLUM. EFF-DILUTE HCL. WDSPRD-VNS, DIVERSE RKS	14.1.1.1
3	VOLTAITE	ISOM.	2.6-2.8	SUF	SUBTRANSLUCENT, CUBIC, OCTAH, MSV. GRNLR. RESIN, CONCH. SOL-ACIDS. DCMP-H2O TO ACID SOLUTION. SOLFATARAS, FUMAROLES, EFFLORESCENCES	29.5.2.
3	HEMAFIBRITE	ORTH.	3.65	ARS	PRISM, RADIAL FIBR AGGS, VITR.CLVGS. SOL-ACIDS. WITH SCNDRY MN-MINS	42.2.3.
3	PARATACAMITE	HEXA.	3.74	HAL	GRNLR, CRUSTS, VITR, RHOMB CLVG. SOL-ACIDS. SCNDRY AFTER TENORITE, ETC.	10.1.12.
3	DOLEROPHANITE	MNCL.	4.17	SUF	SUBTRANSLUCENT. YELLOWISH STRK. ELONGATE. SOL-HNO3. VOL SUBLIMATE	30.2.2.
3	CADMIUM OXIDE	ISOM.	8.1-8.2	OXD	RED-ORANGE BRWN WITH TRANSMITTED LIGHT, OCTAH. SOL-DILUTE ACIDS. COATING ON HEMIMORPHITE (GENARUTTA, SARDINIA)	4.2.1.4
3	HISINGERITE	AMOR.	3.	PSI	RESINOUS. CONCH. 1 POOR CLVG. DCMP-HCL. SCNDRY(E.G., AFTER OLIVINE)	W400 N
3 -3.5	ATACAMITE	ORTH.	3.75-3.78	HAL	APPLE GREEN STRK. PRISM-STRIATED, GRNLR, FIBR. VITR-ADMN. CLVGS. SOL-ACIDS. SCNDRY AFTER CU MINS, ESP UNDER SALINE-ARID CONDITIONS	10.1.11.1
3 -3.5	LAUTITE	ORTH.	4.8 -5.0	SLD	SUBMET. OPQ. BLK-GRAYIRED TINGE). SOL-HNO3. VNS-WITH NATIVE AS, ETC. TRA VIOLET). OXIDIZED ZONES OF LEAD DEPOSITS	2.9.5.5
3 -3.5	BOLEITE	TETR.	5.05	HAL	GREEN-BLUE STRK. PSEUDOCUBIC. VITR-PEARLY. CLVGS. SOL-HNO3. SCNDRY	10.21.
3 -3.5	DESCLOIZITE	ORTH.	6.1 -6.3	VAN	VARIABLE-XLS TO COMPACT MSV. GRSY.COMMONLY COLOR ZONED. SOL-ACIDS. OXIDIZED ZONES OF VANADIUM-BEARING ORE DEPOSITS	41.5.2.1
3 -3.5	CERUSSITE	ORTH.	6.53-6.57	CBT	TWINNED CLUSTERS, MSV. ADMN-RESIN. CLVGS. FLUO-YEL(X-RAYS +(LW)UV-	14.1.3.4
3 -4	WAVELLITE	ORTH.	2.36	PHO	RADIAL FIBR, STELLATE, CRUSTS. VITR-PEARLY.CLVGS. SOL-ACIDS. SCNDRY- WDSPRD-OPENINGS IN DIVERSE ROCKS, LIMONITE, PHOSPHATE RKS, ETC.	42.7.4.
3 -4	STILPNOMELANE	MNCL.	2.59-2.96	PSI	MICACEOUS(BUT 2ND CLVG⊥ 1ST). SOL-HF+H2SO4(1/1). LO-GRD META RKS	W390 3-103
3 -4	DIXENITE	ORTH.	4.2	NSI	RED-TRANSMITTED LIGHT. SCALY. BASAL CLVG. DCMP-HCL. RARE-ORE DEPS	W236 N
3.5	CRONSTEDTITE (SERPENTINE ?)	HEXA.	3.45	PSI	FINE GRAINED. MORE COMMON THAN GREENALITE, IN IRON FORMATIONS	W389 3-164
3.5	HEMATOLITE	HEXA.	3.49	ARS	SUBMET. TBLR, RHOMB, STRIATED, VITR. 1 CLVG. SOL-ACIDS. WITH BARITE, JACOBSITE, ETC. IN VNS IN LIMESTONE (NORDMARK, SWEDEN)	41.1.3.
3.5	KALKOWSKITE	UNKN.	3.98-4.04	OXD	SUBMET. REDDISH STRK. FIBR PLTS, WAXY. DCMP-HCL.RARE-SCHIST(BRAZIL)	8.2.2.
3.5-4	WARWICKITE	ORTH.	3.34-3.36	BOR	SUBMET-PEARLY, BLUE-BLK STRK. SLENDER PRISMS, ROUNDED ENDS. DCMP- H2SO4. RARE – CONTACT METAMORPHOSED LIMESTONE	24.1.4.
3.5-4	BROCHANTITE	MNCL.	3.97	SUF	ACICULAR, CRUSTS. VITR. 1 CLVG. SOL-ACIDS. OXIDE ZONES OF CU DEPS	30.1.1.
3.5-4	WURTZITE	HEXA.	3.98	SLD	BRWN STRK. HEMIMORPH PYRAMIDAL. RESIN. SOL-HCL WITH SPHAL. S-VNS	2.6.4.1
3.5-4	ALABANDITE	ISOM.	3.9 -4.04	SLD	SUBMET. GRN STRK, TARN-BRWN. PERF CUBIC CLVG. SOL-HCL. SULFIDE VNS	2.6.1.4
3.5-4	MALACHITE	MNCL.	3.6 -4.05	CBT	INCRUST, MAMMILLARY. VITR-DULL. CLVGS. SOL-ACIDS. OXIDIZED ZONES-CU DEPOSITS	16.1.6.
3.5-4	POWELLITE	TETR.	4.21-4.25	MBS	PYRAMIDAL, TBLR, MSV-FOL, POWDER, CRUSTS-SUBADMN-PEARLY. CLVGS. UNEVN. FLUO-YEL. DCMP-HCL, HNO3. ALTER PROD OF MOLYBDENITE	48.1.3.2
3.5-4	CUPRITE	ISOM.	6.14	OXD	SUBMET(SOME). OCTAH, HAIRLK, MSV. COMMON-OXIDIZED ZONES OF CU-DEPS	4.1.1.
3.5-4,5	DUFRENITE	MNCL.	3.1 -3.34	PHO	BOTRYOIDAL-RADIAL, CRUSTS. VITR-SILKY. CLVGS. SOL-DILUTE ACIDS. WITH LIMONITE IN GOSSANS	41.6.9.
3.5-4,5	BEUDANTITE	HEXA.	4 -4.3	ARS	RHOMB, PSEUDOCUBIC. VITR. 1 CLVG. SOL-HCL. SCNDRY-DIVERSE	43.1.1.1
3.5-4,5	ARMANGITE	HEXA.	4.43	ASI	PRISM. 1 CLVG. SOL-HCL. RARE-CALCITE-BARITE VNS(LANGBAN, SWEDEN)	45.1.1.

NONMETALLIC BLACK

HARDNESS	NAME	XL. SYS.	SPECIFIC GRAVITY	CHEM. CLASS	REMARKS	REFERENCES
4	MELANOSTIBIAN	UNKN.		ATI	RED STRK. FOL-MSV,STRIATED XLS. OPQ. PRISM, BASAL CLVGS. SOL-HCL. NEEDS VERIFICATION	46.1.5.
4	FLUORITE	ISOM.	3.18	HAL	CUBIC, MSV. VITR. OCTAH CLVG. FLUO-VIOLET. DCMP-H2SO4. WDSPRD-VNS, CAVITIES, DISSEMINATED IN IGNEOUS ROCKS, ETC.	9.2.1.
4	STEENSTRUPINE	HEXA.	3.4 -3.47	NSI	RHOMBOHEDRAL. RARE-E.G., JULIANEHAAB AREA, GREENLAND	W509 N
4	LIBETHENITE	ORTH.	3.97	PHO	STRIATED PRISM. VITR-GRSY. 2 CLVGS. CONCH. SOL-ACIDS + NH4OH. OXI- DIZED ZONES OF ORE DEPS., COMMONLY WITH MALACHITE, LIMONITE, ETC.	41.6.5.2
4	SPHEROCOBALTITE	HEXA.	4.13	CBT	PINK STRK. MSV-RADIAL STRUCTURE. VITR. EFF-HOT ACID. VNS	14.1.1.5
4	MANGANITE	MNCL.	4.32-4.34	OXD	SUBMET. RED-BRWN STRK. STRIATED PRISMS. 3 CLVGS.SOL-HCL. LO-T VNS, SECONDARY, ETC.	6.1.3.
4	AMPANGABEITE	ORTH.	3.36-4.64	OXD	PRISM, RADIAL AGGS. HORNLK. CONCH. DCMP-ACIDS. K-RICH PEGMATITES	8.4.3.
4	VANDENBRANDEITE	TRCL.	5.03	OXD	LATHLK, MSV. BASAL CLVG. SOL-ACIDS TO GREEN-YEL SOLUTION. SCNDRY	5.3.4.
4 -4.5	WOLFRAMITE	MNCL.	7.31	MOS	SUBMET. TARN-IRID. PRISM-EQUANT--STRIATED,MSV-GRNLR,ETC. 1 CLVG. DCMP-AQ-REG, H2SO4 OR HCL(SLOWLY).DIVERSE-VNS,CNTCT META, PLACERS	48.1.1.2
4 -5	STAINIERITE	ORTH.	4.13-4.47	OXD	SUBMET. OPQ. MAMMILLARY. UNEVN. SOL-HCL. OXIDATION PROD OF CO-MINS	6.1.4.
4 -5.5	LIMONITE		2.7 -4.3	OXD	FIELD TERM FOR HYDROUS IRON OXIDES OF UNKNOWN IDENTITIES.	7.1.3.
4 -5.5	BETAFITE	ISOM.	3.7 -5.	OXD	SUBMET. OCTAH. WAXY. CONCH. METAMICT. DCMP-ACIDS. GRANITIC PEGS	8.4.1.
4.5	FRONDELITE	ORTH.	3.3 -3.49	PHO	SUBTRANSLUCENT. RADIAL FIBR BOTRYOIDAL;SOME HAVE CONCENTRIC COLOR BANDING). VITR-DULL.CLVGS,UNEVEN,SOL-HCL.SCNDRY-PO4 MINS IN PEGS	41.6.6.1
4.5	ROCKBRIDGEITE	ORTH.	3.3 -3.49	PHO	IN SERIES WITH FRONDELITE(SEE FOR PROPS + OCCUR.) WITH LIMONITE	41.6.6.2
4.5	SYNADELPHITE	TRCL.	3.57-3.79	ARS	COMMONLY COLOR ZONED. PSEUDOPRISMS, MSV. VITR. 1 CLVG. UNEVN. SOL- ACIDS. RARE WITH OTHER MANGANESE MINERALS	41.1.5.
4.5	CORNWALLITE	UNKN.	4.17	ARS	RADIAL FIBR BOTRYOIDAL CRUSTS.DATA LACKING.WITH OLIVENITE + TENOR- ITE	42.3.1.
4.5	BRANNERITE	UNKN.	4.5 -5.43	OXD	BLK, EXPOS-BRWNISH YEL. PRISM. CONCH. RDACTV.DCMP-H2SO4.RARE-PLAC- ER (KELLY GULCH, IDAHO)	8.2.4.
4.5	FERBERITE	MNCL.	7.51	MOS	SUBMET. TARN-IRID. ELONGATE-STRIATED, WEDGE-SHAPED;BLDS,MSV.1 CLVG. DCMP-AQ-REG,H2SO4 OR HCL(SLOWLY).DIVERSE-VNS,CNTCT META, PLACERS	48.1.1.3
4.5-5	PSEUDOMALACHITE	MNCL.	4.00-4.40	PHO	CONCENTRIC,RADIAL FIBR BOTRYOIDAL. VITR. 1 CLVG. SOL-ACIDS. SCNDRY IN OXIDIZED ZONES OF COPPER DEPOSITS	41.4.3.
4.5-5	THORITE	TETR.	5.2 -5.4	NSI	TERMINATED PRISMS.PRISM CLVG. MOST IS ALTERED. RARE ACCESS-SYENITE	W496 N
4.5-5	CORONADITE	TETR.	5.44	OXD	SUBMET. OPQ. BRWN-GRAY STRK. MSV-BOTRYOIDAL. OXIDIZED ZONE-MN DEPS	7.7.1.1
5	HOMILITE	MNCL.	3.36	NSI	TABULAR XLS. GELAT. COMMONLY METAMICT. ALTERED RIMS. RARE IN PEGS	M356 N
5	RIEBECKITE (AMPHIBOLE)	MNCL.	3.02-3.42	ISI	QTZ-BEARING SODIC IG RKS. MAGNESIO-RIEB.-AUTHIGENIC IN SEDS,IN VNS	W439 2-333
5	KATOPHORITE (AMPHIBOLE)	MNCL.	3.20-3.50	ISI	RARE IN INTERMEDIATE TO BASIC ALKALIC IGNEOUS ROCKS	W440 2-359
5	DERBYLITE	ORTH.	4.53	ANT	SUBTRANSLUCENT. PRISM. RESIN. CONCH. RARE-CINNABAR-BEARING GRAVELS	44.2.1.
5	LORANSKITE	UNKN.	4.6	OXD	SUBMET. RESEMBLES SAMARSKITE. WITH WIIKITE-PEGIPITKARANTA,FINLAND, ILL DEFINED	8.1.7.
5	PAIGEITE	ORTH.	4.7	BOR	SUBTRANSLUCENT. IN SERIES WITH LUDWIGITE(SEE). HI-T CNTCT META	24.1.1.2

NONMETALLIC BLACK

H	Mineral	Crystal System	G	Streak/Luster	Remarks	Ref
5	TREVORITE	ISOM.	5.16	OXD	SUBMET. MAGNETITE SERIES OF SPINEL GROUP. BRWN STRK. GRNLR, MSV. DIFFICULTLY SOL-HCL. STRONGLY MAGNETIC. RARE-NI-DEPS, TRANSVAAL.	7.2.1.9
5	PLUMBOFERRITE	HEXA.	6.07	OXD	RED STRK. TBLR. 1 CLVG. SOL-HCL. RARE-MN MINS(VERMLAND, SWEDEN)	7.3.2.
5	BISMUTOTANTALITE	ORTH.	8.26	OXD	SUBMET. EXPOS-PINKISH YEL. PRISM. RARE-PEGMATITE(SW UGANDA)	8.1.9.
5 -5.5	ALLUAUDITE	UNKN.	3.4 -3.5	PHO	SUBTRANSLUCENT. MSV, GRNLR, FIBR-GLBL AGGS. 2 CLVGS. SOL-ACIDS. ALTERATION PRODUCT OF VARULITE-HUHNERKOBELITE (IN PEGMATITES)	38.1.4.1
5 -5.5	MANGAN-ALLUAUDITE	UNKN.	3.4 -3.5	PHO	SUBTRANSLUCENT. DISTINGUISHED FROM ALLUAUDITE BY CHEM ANALYSIS	38.1.4.2
5 -5.5	TRIPLITE	MNCL.	3.5 -3.9	PHO	SUBTRANSLUCENT. MSV. VITR-RESIN. 3 CLVGS. UNEVN. SOL-ACIDS. PEGS	41.6.2.
5 -5.5	GOETHITE	ORTH.	3.3 -4.29	OXD	SUBMET. YELLOWISH STRK. MSV, FIBR, ETC. SILKY-ADMN. SOL-HCL. WXING PRODUCT	7.1.2.2
5 -5.5	PYROCHLORE	ISOM.	4.45	OXD	OCTAH. VITR. SUBCONCH. DCMP-H2SO4(SLOW). ALK PEGS + IG RKS	8.1.1.1
5 -6	HOLMQUISTITE (AMPHIBOLE)	ORTH.	3.06-3.13	ISI	CONTAINS NOTABLE LI. NEAR CNTCTS BETWEEN LI-RICH PEGS +COUNTRY RKS	W440 2-230
5 -6	NEPTUNITE	MNCL.	3.19	CSI	BLK,RED IN SPLINTERS. PRISM. PERF PRISM CLVG-AT 80°. ALK SYENITE	W463 N
5 -6	TORENDRIKITE (AMPHIBOLE)	MNCL.	3.21	ISI	ALKALIC IGNEOUS ROCKS	W442 N
5 -6	KAERSUTITE (AMPHIBOLE)	MNCL.	3.2 -3.28	ISI	VOLCANICS (ESP ALKALIC ONES). SOME APPEARS TO BE SECONDARY	W437 2-321
5 -6	BASALTIC HORNBLENDE (AMPHIBOLE)	MNCL.	3.19-3.30	ISI	ALSO CALLED OXYHORNBLENDE. CHIEFLY IN VOLCANIC IGNEOUS ROCKS	W437 2-315
5 -6	CROSSITE (AMPHIBOLE)	MNCL.	3.1 -3.4	ISI	INTERMEDIATE BETWEEN RIEBECKITE + GLAUCOPHANE. SCHISTS, LK GLAUCO-PHANE	W440 2-334
5 -6	CROCIDOLITE (AMPHIBOLE)	MNCL.	3.0 -3.4	ISI	FIBR RIEBECKITE. IN VEINS IN METAMORPHOSED BEDDED IRON-STONES	W440 2-334
5 -6	FERROACTINOLITE (AMPHIBOLE)	MNCL.	3.02-3.44	ISI	REGIONAL + CONTACT METAMORPHISM, ESP OF SI-BEARING LS + DOLOMITE	W434 2-249
5 -6	TSCHERMAKITE (AMPHIBOLE)	MNCL.	3.02-3.45	ISI	KYANITE AMPHIBOLITES DERIVED FROM CALC OR DOLO SHALES OR ECLOGITE	W434 2-263
5 -6	FERROTSCHERMAKITE (AMPHIBOLE)	MNCL.	3.02-3.45	ISI	MORE OR LESS END-MEMBER COMPOSITION IN HORNBLENDE(SEE) SERIES	W435 2-263
5 -6	EDENITE (AMPHIBOLE)	MNCL.	3.02-3.45	ISI	MORE OR LESS END-MEMBER COMPOSITION IN HORNBLENDE(SEE) SERIES	W434 2-263
5 -6	FERROEDENITE (AMPHIBOLE)	MNCL.	3.02-3.45	ISI	MORE OR LESS END-MEMBER COMPOSITION IN HORNBLENDE(SEE) SERIES	W435 2-263
5 -6	COMMON HORNBLENDE (AMPHIBOLE)	MNCL.	3.02-3.45	ISI	IN MANY IGNEOUS RKS + METAMORPHIC RKS OF LOW TO HIGH GRADE	W430 2-263
5 -6	URALITE (AMPHIBOLE)	MNCL.	3.0 -3.5	ISI	SECONDARY AFTER PYROXENE, ESPECIALLY IN IGNEOUS ROCKS	W436 N
5 -6	HASTINGSITE (AMPHIBOLE)	MNCL.	3.50	ISI	INTERMEDIATE IGNEOUS ROCKS AND LOCALLY IN CONTACT ZONES	W434 2-266
5 -6	ARFVEDSONITE (AMPHIBOLE)	MNCL.	3.50	ISI	PLUTONIC ALKALIC IGNEOUS MASSES AND ASSOCIATED PEGMATITE MASSES	W439 2-364
5 -6	FERROHASTINGSITE (AMPHIBOLE)	MNCL.	3.50	ISI	ALKALIC GRANITES AND SYENITES AND ASSOCIATED PEGMATITES	W435 2-264
5 -6	AMPHIBOLE		3.0 -3.57	ISI	GROUP NAME. SEVERAL END AND OTHER MEMBERS WITH NAMES UNSETTLED FOR SOME. MANY MEGASCOPICALLY INDISTINGUISHABLE BUT SOME COLORS AND OCCURRENCES ARE INDICA-TIVE. ORTH AND MNCL SERIES. LONG SLENDER XLS. PRISM CLVG AT 56° + 124°. SOL-HF(SLOWLY). ALTERS TO BIOTITE +/OR CHLORITE. IGNEOUS AND METAMORPHIC RKS, VNS	W422 2-203

NONMETALLIC BLACK

HARDNESS	NAME	XL. SYS.	SPECIFIC GRAVITY	CHEM. CLASS	REMARKS	REFERENCES
5 -6	GEIKIELITE	HEXA.	4.05	OXD	SUBMET.BRWN-RED STRK. TBLR. RHOMB CLVG. RARE-GEM GRAVELS(CEYLON)	4.4.1.4
5 -6	HYDROHETAEROLITE	TETR.	4.6	OXD	SUBMET. BOTRYOIDAL, FIBR. CLVG // FIBERS. SOL-HCL. WITH CHALCOPHAN-ITE, ETC.	7.2.2.3
5 -6	PSILOMELANE	ORTH.	4.70-4.72	OXD	SUBMET. OPQ. MSV, MAMMILLARY. SOL-HCL. SCNDRY IE.G.,RESIDUAL DEPS)	6.1.8.
5 -6	ILMENITE	HEXA.	4.68-4.76	OXD	SUBMET. OPQ. TBLR. VNS + DISSEM DEPS ASSOC WITH GABBROS + DIORITES	4.4.1.3
5 -6	PRIORITE	ORTH.	4.85-5.05	OXD	SUBMET. IN SERIES WITH ESCHYNITE(SEE). RED-YEL STRK. GRANITIC PEGS	8.3.5.2
5 -6	ESCHYNITE	ORTH.	5.14-5.24	OXD	SUBMET. IN SERIES WITH PRIORITE. BLK-BRWN STRK. PRISM, MSV, WAXY, EXPOS-DULL. POOR CLVG. METAMICT.ALKIE.G.,NEPH SYENITE) PEGMATITES	8.3.5.1
5 -6	"SAMARSKITE"	UNKN.	5.67	OXD	SUBMET. DISTINGUISHED FROM SAMARSKITE ONLY BY X-RAY ANALYSIS	8.3.6.2
5 -6	SAMARSKITE	ORTH.	5.69	OXD	SUBMET. RED-BRWN TO BLK STRK. PRISM, MSV. VITR. RDACTV. PEGS	8.3.6.1
5 -6	ISHIKAWAITE	ORTH.	6.2 -6.4	OXD	OPQ. BRWN STRK. TBLR. WAXY. RARE-WITH SAMARSKITE-PEG(IWAKI, JAPAN). ILL DEFINED	8.1.6.
5 -6	URANINITE	ISOM.	10.63	OXD	SUBMET. OPQ. OCTAH,ETC., MSV.GRSY-DULL. RDACTV. PEGS, HI+MOD-T VNS	5.1.2.1
5 -6.5	MAGNESIUM ORTHITE	MNCL.	3.90	SSI	EXTREMELY SIMILAR TO ORTHITE MEGASCOPICALLY	W452 N
5 -6.5	ORTHITE (ALLANITE)	MNCL.	2.8 -4.2	SSI	SUBTRANSLUCENT, TABULAR, ACICULAR, MSV. PITCHY-RESIN. SUBCONCH. RADIOACTIVE. COMMONLY METAMICT. GELAT. ACCESS IN IG RKS, PEGS.	W451 1-211
5 -7	PYROXENE		2.96-3.96	ISI	GROUP NAME. SEVERAL END AND OTHER MEMBERS WITH NAMES UNSETTLED FOR SOME. MANY MEGASCOPICALLY INDISTINGUISHABLE; OCCURRENCE MAY BE INDICATIVE. ORTH + MNCL SERIES.SHORT PRISMS.PRISM CLVG AT 87° + 92°. ALTERS TO AMPHIBOLE.IG + META RKS	W402 2-1
5 -7	OLIVINE	ORTH.	3.22-4.39	NSI	GRP NAME FOR MINS OF FAYALITE-FORSTERITE + FAYALITE-TEPHROITE SERIES, MONTI-CELLITE + GLAUCOCHROITE. GRNLR MASSES, DISSEM. CONCH. GELAT. COMMONLY ALTERED TO SERPENTINE. BASIC + ULTRABASIC IG RKS, CNTCT META DOLOMITIC LS, METEORITES	W498 1-2
5.5	GRIPHITE	ISOM.	3.40	PHO	MSV. RESIN-VITR. UNEVN-CONCH. SOL-ACIDS. PEGMATITES	41.5.11.
5.5	AUGITE (CLINOPYROXENE)	MNCL.	3.23-3.52	ISI	THE COMMON PYROXENE OF SUB-ALKALIC IGNEOUS RKS(E.G., GABBROS)	W411 2-109
5.5	ENIGMATITE	TRCL.	3.74-3.85	ISI	ELONGATE XLS, 2 PERF CLVGS. ALK SYENITES, RARE IN BASALTIC VOLS	W477 N
5.5	NAGATELITE	MNCL.	3.91	SSI	CONSIDERED TO BE OF ORTHITE SERIES OF EPIDOTE GROUP. RARE IN PEGS	W452 N
5.5	PEROVSKITE	MNCL.	3.97-4.05	OXD	SUBMET. CUBIC, OCTAH. GRNLR. ADMN. 1 CLVG. DCMP-HF, H2SO4.BASIC IG	7.4.2.1
5.5	TSCHEFFKINITE	MNCL.	4.3 -4.65	SSI	SUBTRANSLUCENT. VELVET BLK. MSV. GELAT. ALTER PROD. (=CHEVKINITE)	W443 N
5.5	PERRIERITE	MNCL.	4.3 -4.7	SSI	DISTINGUISHED FROM TSCHEFFKINITE ONLY BY X-RAY ANAL-AFTER HEATING	WN N
5.5	ZIRKELITE	ISOM.	4.74	OXD	FLATTENED OCTAH. RESIN. CONCH. DCMP-HF. RARE-IG RK(JACUPIRANGITE)	7.6.2.
5.5	SCHETELIGITE	ORTH.	4.74	OXD	YEL-GRAY STRK. ROUGH XLS. CONCH. SOL-HF. RARE-PEG(IVELAND, NORWAY)	8.1.2.
5.5	HAUSMANNITE	TETR.	4.83-4.85	OXD	SUBMET. PSEUDO-OCTAH, MSV. BASAL CLVG. SOL-HOT HCL. HI-T MN DEPS	7.2.2.1
5.5	MANGANOSITE	ISOM.	5.36	OXD	GREEN, EXPOS-BLK. STRK-BRWN. OCTAH, GRNLR. CUBIC CLVG. FIBR BREAK. LANGBAN, SWEDEN AND FRANKLIN, NEW JERSEY	4.2.1.3
5.5	DJALMAITE	ISOM.	5.75-5.88	OXD	MODIFIED OCTAH. SUBCONCH. (=TA-BETAFITE?). RARE-GRANITIC PEGMATITE	8.4.2.

144

NONMETALLIC BLACK

Hardness	Mineral	Crystal System	SG	Luster	Description	Ref
5.5-6	BABINGTONITE	TRCL.	3.36	ISI	SMALL NEARLY EQUANT STRIATED XLS. BRILLIANT,GLASSY. 1 GOOD, 1 POOR CLVGS. FUSES TO BLK MAG GLBL. IN CAVITIES, COMMONLY WITH ZEOLITES	W462 N
5.5-6	TITANAUGITE (CLINOPYROXENE)	MNCL.	3.3 -3.6	ISI	AUGITE WITH NOTABLE TI-CONTENT. ALKALIC BASIC IGNEOUS ROCKS	W411 2-113
5.5-6	PYROXMANGITE	TRCL.	3.61-3.80	ISI	PINK, EXPOS-BRWN OR BLK. F-BLK MAG GLBL. META + METASOMATIC MN-RKS	W460 2-196
5.5-6	ANATASE	TETR.	3.90	OXD	SUBMET. PALE YEL STRK. ACUTE PYRAMIDAL. ADMN. BASAL + PYRAMIDAL CLVG. CONVERTS TO RUTILE ON HEATING. VNS. ACCESS-IG RKS, DETRITAL	4.5.2.
5.5-6	ILVAITE	ORTH.	3.8 -4.1	SSI	SUBTRANSLUCENT. STRIATED PRISMS. 2 CLVGS. GELAT. CNTCT ZNS, SCNDRY	W511 N
5.5-6	BROOKITE	ORTH.	4.08-4.20	OXD	GRAY TO YEL STRK.TBLR XLS. ADMN. POOR CLVGS. VNS, DISSEM, DETRITAL	4.5.3.
5.5-6	DELORENZITE	ORTH.	4.7	OXD	TBLR, LATHLK. RESIN. SUBCONCH. RDACTV. RARE-PEG(CRAVEGGIA, ITALY)	8.4.4.
5.5-6.5	OPAL		1.8 -2.25	SIL	SUBMICROXLINE QTZ + H2O.RESIN-PEARLY. MSV,MAMILLARY CRUSTS,CONCH.SOL-HF +CAUSTIC ALKS.SOME FLUO—YEL-GREEN. INCRUSTATIONS, IN CAVITIES(ESP IN VOLCANIC RKS)	D-111 287
5.5-6.5	SALITE (CLINOPYROXENE)	MNCL.	3.2 -3.6	ISI	INTERMEDIATE BETWEEN DIOPSIDE + HEDENBERGITE. META CA-RICH SED + IGNEOUS ROCKS	W411 2-42
5.5-6.5	FERROSALITE (CLINOPYROXENE)	MNCL.	3.2 -3.6	ISI	INTERMEDIATE BETWEEN DIOPSIDE + HEDENBERGITE. OCCURS LIKE SALITE	W411 2-42
5.5-6.5	DIALLAGE (CLINOPYROXENE)	MNCL.	3.2 -3.6	ISI	DIOPSIDE + AUGITE WITH GOOD PARTING OFTEN OCCUPIED BY MAGNETITE OR ILMENITE	W411 2-43
5.5-6.5	YTTRIALITE	MNCL.	4.3 -4.6	SSI	PRISMATIC, MASSIVE. RARE — WITH GADOLINITE, ETC. IN PEGMATITES	W477 N
5.5-6.5	MAGNESIOFERRITE	ISOM.	4.56-4.65	OXD	SUBMET. MAGNETITE SERIES OF SPINEL GROUP. BLK STRK. GRNLR, MSV. DIFFICULTLY SOL-HCL. STRONGLY MAGNETIC. FUMAROLE DEPOSITS	7.2.1.5
5.5-6.5	JACOBSITE	ISOM.	4.76	OXD	SUBMET. MAGNETITE SERIES OF SPINEL GROUP. BRWN STR. GRNLR, MSV. SOL-HCL. WEAKLY MAGNETIC. MN DEPS, E.G., LANGBAN, SWEDEN	7.2.1.8
5.5-6.5	MAGNETITE	ISOM.	5.18	OXD	SUBMET. MAGNETITE SERIES OF SPINEL GROUP. BLK STRK. OCTAH, GRNLR. OCTAH PARTING. DIFFICULTLY SOL-HCL. STRONGLY MAGNETIC. WDSPRD	7.2.1.6
5.5-6.5	FRANKLINITE	ISOM.	5.07-5.22	OXD	SUBMET. MAGNETITE SERIES OF SPINEL GROUP. RED-BRWN STRK. OCTAH. OCTAH PARTING.SOL-HCL.WEAKLY MAGNETIC.RARE—FRANKLIN, NEW JERSEY	7.2.1.7
5.5-6.5	FERGUSONITE	TETR. C	5.38	OXD	SUBMET. IN SERIES WITH FORMANITE. PRISM. COMMONLY HEMIHEDRAL. IR-REGULAR MASSES. VITR. METAMICT. DCMP-HF. GRANITIC PEGMATITES.	8.1.3.1
5.5-6.5	EUXENITE	ORTH.	4.9 -5.9	OXD	SUBMET. IN SERIES WITH POLYCRASE. STRK YEL.GRAY.RED-BRWN PRISM, MSV. GRSY. CONCH.DCMP-HOT HCL,HF,H2SO4. GRANITIC PEGMATITES	8.3.3.1
5.5-6.5	POLYCRASE	ORTH.	4.9 -5.9	OXD	SUBMET. IN SERIES WITH EUXENITE, WHICH SEE. PEGMATITES	8.3.3.2
5.5-6.5	FORMANITE	TETR. C	7.03	OXD	SUBMET.IN SERIES WITH FERGUSONITE,WHICH SEE. PLACERS(W. AUSTRALIA)	8.1.3.2
6	CLINOENSTATITE (CLINOPYROXENE)	MNCL.	3.19	ISI	HI-T FORM. RARE IN IGNEOUS ROCKS, IN METEORITES.	W408 N
6	RAMSAYITE	ORTH.	3.43	ISI	ACICULAR. 1 GOOD CLVG. VEINS IN IGNEOUS ROCKS(E.G.,KOLA PENINSULA)	W401 N
6	PIGEONITE (CLINOPYROXENE)	MNCL.	3.30-3.46	ISI	COMMON PYROXENE OF MANY DOLERITES(INCLUDING DIABASES) + BASALTS	W408 2-143
6	EPIDOTE	MNCL.	3.38-3.49	SSI	GROUP NAME. NO GENERALLY ACCEPTED NOMENCLATURE FOR GROUP MEMBERS. CLOSELY RELATED TO ZOISITE + ALLANITE. PISTACHIO-BLACKISH GREEN-BROWNISH. LONG THIN GROOVED XLS, GRNLR MASSES. VNS, OTHER CAVITIES, REGIONAL META + IG ROCKS	W448 1-193
6	PIEMONTITE	MNCL.	3.45-3.52	SSI	MN-BEARING EPIDOTE. COLOR GENERALLY DISTINCTIVE(DKR THAN THULITE)	W448 1-194
6	JOHANNSENITE (CLINOPYROXENE)	MNCL.	3.44-3.55	ISI	MN-ANALOG OF DIOPSIDE-HEDENBERGITE. SKARNS, CNTCT ZONES. VEINS	W411 2-75
6	AEGIRINE-AUGITE (CLINOPYROXENE)	MNCL.	3.40-3.55	ISI	TYPICAL PYROXENE OF ALKALIC ROCKS — ESPECIALLY SYENITIC ONES	W411 2-79
6	HEDENBERGITE (CLINOPYROXENE)	MNCL.	3.50-3.56	ISI	BRWNISH OR DK GREEN TO BLK. SKARNS, META FE-RICH SEDS, ORE DEPS	W411 2-42

NONMETALLIC BLACK

HARDNESS	NAME	XL. SYS.	SPECIFIC GRAVITY	CHEM. CLASS	REMARKS	REFERENCES
6	AEGIRINE (CLINOPYROXENE)	MNCL.	3.55-3.60	ISI	TYPICAL PYROXENE OF ALKALIC ROCKS - ESPECIALLY GRANITIC ONES	W411 2-79
6	PSEUDOBROOKITE	ORTH.	4.33-4.39	OXD	SUBMET. TBLR-STRIATED. GRSY. 1 CLVG. VOLCANIC AREAS FUMAROLIC,ETC.	7.5.1.
6	HETAEROLITE	TETR.	5.18	OXD	SUBMET. BRWN STRK. PSEUDO-OCTAH, MSV. BASAL CLVG. RARE-FRNKLN,N.J.	7.2.2.2
6	COLUMBITE	ORTH.	5.15-5.25	OXD	SUBMET. IN SERIES WITH TANTALITE. RED-BLK STRK. TARN-IRID. TBLR, PRISM. 2 CLEAVAGES. BRITTLE. GRANITIC PEGMATITES	8.3.2.1
6	MAGNETOPLUMBITE	HEXA.	5.52	OXD	SUBMET. OPQ. BRWN STRK. PYRAMIDAL BASAL CLVG. MAG. SOL-HCL(SLOW). RARE - WITH MANGANOPHYLITE (LANGBAN, SWEDEN)	7.3.3.
6 -6.5	RUTILE	TETR.	4.21-4.25	OXD	SUBMET. PRISM-STRIATED.ADMN.POOR CLVGS. VNS, ACCESS-META + IG RKS	4.5.1.1
6 -6.5	BRAUNITE	TETR.	4.72-4.83	NSI	SUBMET. OPQ. PYRAMIDAL. GRNLR. PYRAMIDAL CLVG. GELAT. MAG. WITH OTHER MN MINERALS - VEINS, WEATHERING PRODUCT DEPOSITS, ETC.	W60 N
6 -6.5	BIXBYITE	ISOM.	4.95	OXD	SUBMET. OPQ. CUBIC. OCTAH CLVG. SOL-HCL. IN CAVITIES IN RHYOLITE	4.4.5.
6 -6.5	SENAITE	HEXA.	5.30	OXD	SUBMET. BRWN-BLK STRK. ROUGH XLS. CONCH. DCMP-HF, H2SO4. PLACERS	4.4.1.6
6 -6.5	TAPIOLITE	TETR.	7.85-7.95	OXD	SUBMET. STRK-BRWNISH. PRISM. UNEVN. PEGMATITES, PLACERS	8.3.1.1
6 -6.5	MOSSITE	TETR.	7.85-7.95	OXD	SUBMET. POSSIBLE CB END-MEMBER OF TAPIOLITE SERIES(SEE). RARE-PEGS	8.3.1.2
6 -6.5	TANTALITE	ORTH.	7.90-8.00	OXD	SUBMET. IN SERIES WITH COLUMBITE, WHICH SEE. PEGMATITES	8.3.2.2
6 -7	CASSITERITE	TETR.	6.99	OXD	SUBMET. RADIAL CONCRETIONARY MASSES. ADMN-DULL. HI-T VNS, GREISENS	4.5.1.5
6 -7.5	PYROPE	ISOM. A	3.58	NSI	MG-AL GARNET. XLS ARE RARE. ALTERS TO KELYPHITE. ULTRABASIC IG RKS	W486 1-97
6 -7.5	ANDRADITE	ISOM. A	3.86	NSI	CA-FE,TI GARNET. F-MAG GLBL. CONTACT META CALCAREOUS RKS, ALK IGS	W489 1-89
6 -7.5	SPESSARTINE	ISOM. A	4.19	NSI	MN-AL GARNET. EQUANT CRYSTALS. MN-RICH META RKS + VEINS, PEGS	W486 1-99
6 -7.5	GARNET	ISOM. A	3.58-4.32	NSI	GROUP NAME. DOMINANT MOLECULE DETERMINES NAME. DISTINGUISHED BY S.G. + NONMEGA-SCOPIC MEANS.EQUANT XLS,MSV,VITR-DULL,DODEC PARTING, SLIGHTLY- TO IN-SOL IN HF	W483 1-77
6 -7.5	ALMANDINE	ISOM. A	4.32	NSI	FE-AL GARNET. F-MAG GLBL. META-ARGILLACEOUS RKS, RARE-IG RKS,PEGS	W486 1-85
6.5	AGATE	HEXA.	2.57-2.64	SIL	BANDED OR VARIEGATED.CHALCEDONY(MICROXLINE QTZ). NODULES IN BASALT	D-111 210
6.5	KNEBELITE	ORTH.	3.96-4.25	NSI	FAYALITE-TEPHROITE (FE-MN)OLIVINE SERIES HAS NAME KNEBELITE AP-PLIED TO INTERMEDIATE MEMBERS. OCCUR IN META RKS + FE-MN ORE DEPS	W498 1-34
6.5	LANGBANITE	HEXA.	4.6 -4.8	NSI	COMPLEX XLS. DIFFICULTY SOL-HCL. RARE-ORE DEPS, LANGBAN,SWEDEN	W529 N
6.5	POLYMIGNYTE	ORTH.	4.77-4.85	OXD	SUBMET. BRWN STRK. PRISM-STRIATED. 2 CLVGS. ALKALIC PEGMATITES	8.1.5.
6.5	MELANOTEKITE	ORTH.	5.7	NSI	PRISM XLS. 2 CLVGS. DCMP-HNO3. RARE IN LEAD ORE DEPOSITS	W525 N
6.5	BADDELEYITE	MNCL.	5.4 -6.02	OXD	FLAT PRISM. VITR. BASAL + 2 POOR CLVGS. DCMP-HOT H2SO4. PLACERS, ASSOCIATED WITH ULTRAMAFIC IGNEOUS ROCKS	5.1.1.
6.5	THORIANITE	ISOM.	9.7	OXD	SUBMET. CUBIC. HORNLK. POOR BASAL CLVG. RDACTV. PLACERS, IN SERPEN-TINE OF CONTACT METAMORPHIC ZONES	5.1.2.2
6.5-7	JASPER	HEXA.	2.57-2.65	SIL	SUBTRANSLUCENT. SOME IS VARIEGATED OR BANDED. MICROGRNLR. CAVITIES	D-111 224
6.5-7	GADOLINITE	MNCL.	4.0 -4.6	NSI	PRISMATIC. GELAT. COMMONLY METAMICT. WITH FLUORITE IN SOME PEGS	W356 N

NONMETALLIC BLACK

●	7	QUARTZ	HEXA.	2.65	SIL	DOUBLY TERMINATED HEX PRISMS STRIATED LENGTH(LO-T), BIPYRAMIDS (HI-T), MSV. VITR. CONCH. SOL-HF. WDSPRD IN MOST ROCK TYPES + VNS	D-111 181
●●	7	SMOKY QUARTZ	HEXA.	2.65	SIL	YEL-BRWN TO BRWNISH BLK. LOSSES COLOR-CA.225℃. HI-T VNS, PEGS	D-111 182
●	7	DRAVITE	HEXA.	3.03-3.15	CSI	NA— MG-TOURMALINE. FUSES AT 4. TYPICALLY BRWN. META CALCAREOUS RKS	W465 1-300
●	7	TOURMALINE	HEXA.	3.03-3.25	CSI	GROUP NAME. VARIETIES COMMONLY DISTINGUISHABLE MEGASCOPICALLY ON BASIS OF COLOR(S). LENGTHWISE STRIATED PRISMATIC XLS WITH CROSS-SECTIONS THAT RESEMBLE SPHERICAL TRIANGLES. SOME XLS EXHIBIT BOTH LENGTHWISE + CONCENTRIC COLOR ZONING. VITR-RESIN. BRTL. ELECTRICALLY CHARGED ON HEATING + COOLING. MG-RICH VARIETIES FLUO-YEL (ULTRAVIOLET — SW). PEGS, HI-T VNS, IG AND META ROCKS	W465 1-300
●	7	SCHORL	HEXA. 3.10-3.25		CSI	NA— FE-TOURMALINE. FUSES AT 5.5. TYPICALLY BLACK. IG RKS. SCHISTS	W465 1-300
●	7.5-8	SPINEL	ISOM.	3.55	OXD	GROUP NAME. SOME HAVE DISTINCTIVE COLORS OR OTHER HAND-SPECIMEN PROPS. OCTAH. GRNLR. GLASSY. CONCH. SOME FLUO—RED TO YEL-GREEN. IG,META(ESP CALC RKS),PEGS	7.2.1.1
	7.5-8	GALAXITE	ISOM.	4.03	OXD	SPINEL(SEE). GRNLR. RARE-WITH ALLEGHANYITE,ETC.-VNS(BALD KNOB,N.C)	7.2.1.4
	7.5-8	HERCYNITE	ISOM.	4.39	OXD	SPINEL(SEE). GRNLR. HI-T META(E.G., GRANULITES) + IGNEOUS ROCKS	7.2.1.2
	9.5	MOISSANITE	HEXA. 3.1 -3.21		ELE	SUBMET. TABULAR. CONCHOIDAL FRACT. METEOR. (ARTIF=CARBORUNDUM)	1.1.7.4
●	10	DIAMOND	ISOM. 3.50-3.53		ELE	OCTAHEDRAL. ADMN-GRSY. BRTL. OCTAH CLVG. ULTRABASIC IGS + PLACERS	1.2.4.1

147

NONMETALLIC BROWN

HARDNESS	NAME	XL. SYS.	SPECIFIC GRAVITY	CHEM. CLASS	REMARKS	REFERENCES
1	SASSOLITE	TRCL.	1.46-1.50	OXD	SCALES. PEARLY. BASAL CLVG(FLEX). BITTER,SALTY TASTE. HOT SPRINGS	6.2.1.
1	GLAUCOCERINITE	UNKN.	2.75	SUF	WARTY MASSES, RADIAL + CONCENTRIC. WAXY. WITH SMITHSONITE,ETC.ILAU-31.1.2. RIUM SPECIMEN	ILAU-31.1.2.
1	TALC	MNCL.	2.58-2.83	PSI	FOL MSV. BASAL CLVG. SECTILE. SOAPY FEEL. PEARLY. SCHIST, META IG	W364 3-121
1	MOLYSITE	HEXA. A	2.90	HAL	MSV. BASAL CLVG. DLQSNT. VOL SUBLIMATE	9.3.1.
1	SCACCHITE	HEXA.	2.98	HAL	MSV. BASAL CLVG. SOL-H2O. DLQSNT. WITH OTHER SALTS-VOL SUBLIME	9.2.3.2
1	LAWRENCITE	HEXA. A	3.16	HAL	MSV. BASAL CLVG. SOL-H2O. DLQSNT. METEORIC, VOL SUBLIMATE	9.2.3.1
1	PLUMBOJAROSITE	HEXA.	3.67	SUF	PLTY MASSES. SILKY-DULL. CLVG. SOL-ACIDS(SLOW). OXIDE ZONES OF PB ORES - ESPECIALLY IN ARID REGIONS	30.2.4.9
1 -2	VERMICULITE	MNCL.	2.2 -2.4	PSI	BRONZY. MICACEOUS, EXFOLIATES WITH SWELLING ON HEATING. ATTKD-ACID (SILICA RESID). ALTER PROD OF BIOTITE(ETC.)-HYDROTHERMAL OR WXING	W396 3-246
1 -2	HYDROPHLOGOPITE	MNCL.	2.6 -2.9	PSI	FMED UNDER ALK CONDITIONS. CLASTIC + DIAGENIC-SEDS, HYDROTHERMAL	WN 3-213
1 -2	PYROPHYLLITE	MNCL.	2.65-2.90	PSI	LAMELLAR. CLVG-INELAST PLTS. PARTLY DCMP-H2SO4. VNS, FEW SCHISTS	W361 3-115
1 -2	BEIDELLITE (CLAY)	MNCL.	3.	PSI	"SWELLING CLAY." ALTER-HI MG +CA,LO K RKS(ALK CONDITIONS). RARE- GOUGE (COLORADO), BENTONITES	W398 3-226
1 -2	STEVENSITE (CLAY)	MNCL.	3.	PSI	POSSIBLY MONTMORILLONITIC "SWELLING CLAY." ALTER PROD OF PECTOLITE	W461 3-231
1 -2	SAUCONITE (CLAY)	MNCL.	3.	PSI	FINE-GRAINED AGGS, VERMIFM, LAMELLAR,ETC. "SWELLING CLAY." ALTER- HI MG +CA,LO K RKS(ALK CONDITIONS). HYDROTHERMAL, SOILS, BENTONITE	W400 3-226
1 -2	HYDROBIOTITE	MNCL.	2.6 -3.3	PSI	FMED UNDER ALK CONDITIONS. CLASTIC + DIAGENIC-SEDS. HYDROTHERMAL	W396 3-213
1 -2.5	NONTRONITE (CLAY)	MNCL.	2.5 -3.0	PSI	SUBTRANSLUCENT. LATHS, MSV. GELAT. "SWELLING CLAY." ALTER-HI MG + CA,LO K RKS(ALK CONDITIONS). WITH OPAL AND/OR QTZ IN VNS, BENTONITE	W398 3-226
1 -3	CLAY	MNCL.	1.85-3.0	PSI	NAME GIVEN SEVERAL HYDROUS AL-SILICATES UNDER ALKALIES OR ALKALINE EARTHS ± MG OR FE. NAMES UNSETTLED. MOST DISTINGUISHED BY NONMEGASCOPIC MEANS. FINE-GRAINED AGGS(COMPACT,MEALY,ETC.). EARTHY-PEARLY,UNCTUOUS,PLASTIC WHEN WET.DEHYDRATED ON HEATING. SOME ARE ABSORBENT. WXING PRODS,HYDROTHERMAL ALTER, DETRITAL,DIAGENIC	W398 3-191
1.5	IODARGYRITE	HEXA.	5.69	HAL	EXPOS-YEL. MSV. PRISM-HEMIMORPHIC. RESIN. BASAL CLVG(FLEX). DCMP- H2SO4, HNO3, KI SOLUTION. OXIDIZED ZONES OF SILVER DEPOSITS	9.1.4.
1.5	CALOMEL	TETR.	7.15	HAL	EXPOS-DARKENS, TBLR, MSV. ADMN. PRISM CLVG. PLASTIC,FLUO-RED. SOL- AQ-REG. SECONDARY AFTER SEVERAL MERCURY MINERALS	9.1.5.
1.5 -2	SALAMMONIAC	ISOM.	1.53	HAL	SKELETAL, DENDRITIC. VITR.OCTAH CLVG. SALTY TASTE. VOL SUBLIMATE	9.1.2.
1.5 -2	HUMBOLDTINE	ORTH.	2.28	CCC	CAPILLARY,BOTRYOIDAL FIBR CRUSTS,GRNLR+DENSE. DULL-RESIN. CLVGS. SOL-ACIDS, ASSOCIATED WITH LIGNITE	50.1.3.
1.5 -2	SODA-NITER	HEXA.	2.24-2.29	NIT	INCRUST, RHOMB, VITR. RHOMB + OTHER CLVGS. COOLING TASTE. IN ARID SOILS IN PROTECTED LOCATIONS	18.1.1.
1.5 -2.5	LUENEBERGITE	MNCL.	2.05	BOR	NODULES, PSEUDOHEX TABLETS. PRISM CLVG. EVAPORITE DEPOSITS	27.1.1.
1.5 -2.5	SULFUR	ORTH.	2.07	ELE	CONCH-UNEVN. BRTL. BURNS-BLU FL + ACRID ODDR. SOL-CS2. VOLS + SED	1.2.3.1
1.5 -2.5	SIDERONATRITE	ORTH.	2.15-2.35	SUF	FIBR CRUSTS,NODULAR. 1 CLVG. DCMP-BOILING H2O. WITH OTHER SULFATES IN ARID REGIONS	31.5.3.
1.5 -2.5	CHRYSOCOLLA	UNKN.	2.3 -2.5	CSI	MSV. VITR-EARTHY. CONCH. DCMP-HCL. OXIDE ZONES CU-DEPS IN SIL RKS	W420 N
1.5 -4.5	ARSENIOSIDERITE	TETR.	3.60	ARS	STRK-YEL. SUBTRANSLUCENT. RADIAL FIBR AGGS. SUBMET-SILKY. 1 CLVG. SOLUBLE IN HOT ACIDS. SECONDARY - DIVERSE OCCURRENCES	42.6.4.
2	DIETRICHITE	MNCL.	1.8	SUF	FIBR. SILKY. SOL-H2O. RARE-EFFLORESCENCES IN SHELTERED PLACES	29.7.3.4

NONMETALLIC BROWN

H	Mineral	System	S.G.	Class	Description	Ref
2	STERCORITE	TRCL.	1.62	PHO	XLINE MASSES, NODULES. VITR. SOL-H2O. GUANO DEPOSITS	39.1.1.
2	STRUVITE	ORTH.	1.71	PHO	DIVERSE XLS.MOST HEMIMORPHIC.VITR.2 CLVGS. SOL-ACIDS, H2O(SLIGHT-LY). GUANO DEPOSITS, URINARY SEDIMENT, ETC.	40.1.1.
2	HYDROTALCITE	HEXA.	2.03-2.09	OXD	PLTY, FOL-MSV. WAXY,PEARLY. BASAL CLVG(FLEX). EFF-HCL. ALTER PROD OF SPINEL	6.1.5.1
2	MANASSEITE	HEXA.	2.00-2.10	OXD	FOL-MSV. GRSY. BASAL CLVG(FLEX). SOL-HCL. WITH HYDROTALCITE + SER-PENTINE	6.1.6.1
2	BRUGNATELLITE	HEXA.	2.14	OXD	FLKS, FOL-MSV. PEARLY. BASAL CLVG. EFF-HCL. INCRUST-CRACKS IN SER-PENTINE	6.1.7.
2	GYPSUM	MNCL.	2.31-2.32	SUF	TBLR, MSV. SUBVITR-PEARLY. 1 PERF CLVG(FLEX,INELAST). SOL-HCL. SEDIMENTARY BEDDED DEPOSITS, FUMAROLIC, EFFLORESCENCES, ETC.	29.6.3.
2	THOMSENOLITE	MNCL.	2.98	HAL	PRISM, OPALINE CRUSTS. VITR. BASAL,PYRAMIDAL CLVGS. SOL-H2SO4. AT CRYOLITE LOCALITIES	11.5.4.
2	MELLITE	TETR.	1.64	CCC	MSV,GRNLR,CRUSTS,NODULAR. RESIN,PRISM CLVG. CONCH. SOL-HNO3, FLUO-BLUE(SW). SECONDARY IN OPEN SPACES IN LIGNITE	50.2.1.
2	GOSLARITE	ORTH.	1.98	SUF	CRUSTS, VITR-SILKY. 1 CLVG. ASTRINGENT,METALLIC,NAUCEOUS. EFFLOR-TYPICALLY BY ALTERATION OF SPHALERITE	29.6.9.2
-2.5	KAOLINITE (CLAY)	TRCL.	2.60-2.68	PSI	ALTER(ACID CONDITIONS) OF FELDSPARS, ETC. HYDROTHERMAL + WXING	W362 3-194
-2.5	PHLOGOPITE	MNCL.	2.76-2.90	PSI	MG-MICA. ATTKD-HCL. CNTCT META-LS. ULTRABASIC IGNEOUS ROCKS	W373 3-42
-2.5	HYDROZINCITE	MNCL.	3.5 -4.0	CBT	MSV,INCRUST, SILKY-DULL.1 CLVG. FLUO-BLUE,LILAC. SOL-ACIDS.SCNDRY WITH OTHER ZINC MINERALS	16.1.3.
-2.5	CINNABAR	HEXA.	8.09	SLD	SCARLET STRK. ADMN-DULL. NEAR VOLS + HOT SPRINGS-VNS + INCRUSTS	2.6.9.
-3	ALLOPHANE (CLAY)	AMOR.	1.85-1.89	PSI	INCRUST, "BOOKS." GELAT. ALTER(ACID CONDITIONS) OF FELDSPARS, ETC. IN COAL, ORE DEPOSITS, AND VEINS	W531 3-194
-3	VASHEGYITE	UNKN.	1.96	PHO	SUBTRANSLUCENT. MSV,MICROXLINE-FIBR.DULL.SOL-ACIDS. WITH VARISCITE	42.9.3.
-3	BETA SULFUR	MNCL.	1.96-1.98	ELE	LIGHT YEL-COLORLESS PURE, BROWN IMPURE. THICK TABULAR. FUMAROLES	1.2.3.2
-3	RICHELLITE	UNKN.	2.	PHO	MSV COMPACT-FOL,RADIAL FIBR GLBLS. GRSY-HORNLK. SOL-ACIDS. RARE-WITH HALLOYSITE, ETC. (RICHELLE, BELGIUM)	42.6.7.
-3	PITTICITE	UNKN.	2.2 -2.5	ARS	MSV,BOTRYOIDAL,OPALINE DEPOSITS,ETC. DULL-GRSY. SOL-ACIDS. OXIDATION ZONES OF AS-BEARING DEPOSITS. DEPOSITED FROM MINE + SPRING WATERS	43.2.6.
-3	XANTHOXENITE	UNKN.	2.8 -2.97	PHO	MASSES + CRUSTS-PLTY XLS. DULL-WAXY. 1 CLVG. PHOSPHATE-RICH PEGS	42.8.10.
-3	GONYERITE	MNCL.	3.01	PSI	RADIAL AGGS WITH BARITE, ETC. IN HYDROTHERMAL VNS-E.G.,LANGBAN, SWEDEN	WN 3-146
-3	PENNINITE (CHLORITE)	MNCL.	2.6 -3.1	PSI	PSEUDORHOMB XLS. WITH MAGNETITE + CHONDRODITE(SERP RKS), SCHISTS	W383 3-137
-3	CHLORITE	MNCL.	2.6 -3.3	PSI	GROUP NAME. MEMBER NAMES UNSETTLED. ONLY FEW RARE VARIETIES MEGASCOPICALLY DIS-TINGUISHABLE, DISSEM FLKS. PEARLY. 1 PERF CLVG TO FLEX,INELAST FOLIA. DCMP- IN BOILING H2SO4. SCNDRY+HYDROTHERMAL ALTER, LO-GRADE META, DETRITAL, AUTHIGENIC	W381 3-131
-3	THURINGITE (CHLORITE)	MNCL.	2.8 -3.3	PSI	SCALES, MSV. IRON FORMATIONS, SCNDRY AFTER FERROMAG MINS, GNEISSES	W383 3-137
-3	GREENALITE (SERPENTINE)	MNCL.	2.5 -3.5	PSI	FINE GRAINED. RARE-IRON FORMATIONS (MESABI RANGE OF MINNESOTA)	W380 3-164
-3	LAMPROPHYLLITE	ORTH.	3.45-3.54	NSI	ELONGATE PLTS. 1 GOOD, 1 POOR CLVG. NEPHELINE SYENITES	W527 N
-3	BECQUERELITE	ORTH.	5.2	OXD	YEL STRK. TBLR-STRIATED, MSV. ADMN. BASAL CLVG. RDACTV. OXIDATION PRODUCT OF URANINITE	5.2.3.
-3	XANTHOCONITE	MNCL.	5.40-5.68	SST	STRK-ORANGE-YEL. TBLR, LATHLK. ADMN. 1 CLVG. WITH RUBY SILVERS	3.2.2.3

149

NONMETALLIC BROWN

HARDNESS	NAME	XL. SYS.	SPECIFIC GRAVITY	CHEM. CLASS	REMARKS	REFERENCES
2 -3	PHOSGENITE	TETR.	6.13	CBT	PRISM. MSV. ADMN. CLVGS(PERCUSSION FIGURE ON BASAL PLANE) FLUO-YEL (LW ULTRAVIOLET, CATHODE + X-RAYS). EFF-HNO3. SCNDRY AFTER PB MINS	16.1.7.
2 -3	TRIGONITE	MNCL.	6.1 -7.1	ASI	DOMATIC, VITR-ADMN, CLVGS, UNEVN. SOL-DILUTE ACIDS. RARE-WITH NA-TIVE LEAD + DIXENITE IN CREVICES IN DOLO-HAUSMANNITE ORE(LANGBAN)	45.1.2.
2 -3	HEMATOPHANITE	TETR.	7.70	OXD	SUBMET. YEL-RED STRK. TBLR, LAMELLAR AGGS. 1 CLVG. SOL-HCL, HNO3 RARE-WITH PLUMBOFERRITE, ETC. (VERMLAND, SWEDEN)	7.3.4.
2 -4	MICA	MNCL.	2.4 -3.4	PSI	GROUP NAME. SOME SPECIES DISTINGUISHABLE ON BASIS OF COLOR. SERIES ARE NOT COMPLETE. HEX-SHAPED XLS,DISSEM. PERF CLVG TO FLEX,ELAST FLKS. PERCUSSION + PRESSURE FIGURES MAY BE IMPOSED ON FLKS. CHIEF OCCURRENCES ARE NOTED UNDER SPECIES	W365 3-1
2 -4	BAUXITE		2	OXD	FIELD TERM FOR MATERIALS RICH IN HYDROUS ALUMINUM OXIDES.	6.2.3.
2 -6.5	WAD		2.8 -4.4	OXD	FIELD TERM FOR HYDROUS MN-OXIDES OF UNKNOWN IDENTITY. (MN-DEPS)	4.5.1.3
2.5	MITRIDATITE	UNKN.		PHO	MSV,NODULES,CRUSTS, DULL-RESIN. FRIABLE-DENSE. SOL-HOT ACIDS. OOLITIC SEDIMENTARY IRON ORES - RARE (SOUTHERN RUSSIA)	42.6.6.
2.5	PYROAURITE	HEXA.	2.10-2.14	OXD	TBLR-STRIATED. WAXY,PEARLY. BASAL CLVG(FLEX). EFF-MCL. LO-T VNS	6.1.5.3
2.5	SJOGRENITE	HEXA.	2.08-2.14	OXD	PLTS-STRIATED. WAXY,PEARLY. BASAL CLVG(FLEX).EFF-HCL. RARE-LANGBAN	6.1.6.3
2.5	NAHCOLITE	MNCL.	2.21	CBT	PRISM, FRIABLE AGGS. VITR. CLVGS. SOL-H2O. WITH TRONA	13.1.1.
2.5	AMARANTITE	TRCL.	2.19-2.29	SUF	STRK-YEL. ELONGATE,RADIATED AGGS. VITR. SOL-HCL,DCMP-H2O. SCNDRY-WITH COPIAPITE, ETC.	31.6.4.
2.5	BRUSHITE	MNCL.	2.33	PHO	NEEDELK,ETC. VITR,PEARLY. 2 CLVGS. SOL-HCL. PHOSPHATE DEPOSITS	39.2.1.1
2.5	METAVOLTINE	HEXA.	2.5	SUF	TBLR, GRNLR-SCALY AGGS. RESIN. 1 CLVG. PARTLY SOL-H2O,DCMP-DILUTE ACIDS(SLOWLY). WITH OTHER HYDROUS SULFATES IN DIVERSE ENVIRONMENTS	31.6.9.
2.5	PARABUTLERITE	ORTH.	2.55	SUF	PRISM-STRIATED. VITR. POOR PRISM CLVG. CONCH. SOL-DILUTE ACIDS. SCNDRY-ALTER PROD AFTER COPIAPITE, OXIDIZED ZONES OF PYRITIC VNS	31.6.3.
2.5	GUILDITE	MNCL.	2.72	SUF	YEL STRK. PSEUDOCUBIC XLS. VITR. 2 CLVGS. RARE-WITH OTHER SULFATES	31.6.8.
2.5	PHARMACOSIDERITE	TETR.	2.80	ARS	STRIATED CUBES,TETRA,GRNLR. ADMN.1 CLVG. UNEVN. SECTILE. OXIDATION PRODUCT OF OTHER ARSENIC MINERALS	42.9.1.
2.5	ROSCOELITE	MNCL.	2.97	PSI	V-MICA. RARE-E.G., ASSOCIATED WITH GOLD ORES	W369 3-14
2.5	CRYOLITE	MNCL.	2.96-2.98	HAL	MSV, GRNLR. GRSY. UNEVN. THEMOLUM. SOL-H2O(SLIGHTLY). RARE-PEGS	11.5.1.
2.5	SZOMOLNOKITE	MNCL.	3.03-3.07	SUF	BIPYRAMIDAL,STALAC.VITR. CONCH.SOL-H2O(SLOW). EFFLOR-ACID SOLUTION	29.6.2.2
2.5	BLAKEITE	UNKN.	3.1	TLI	STRK-YELLOWISH, MICROXLINE CRUSTS. DULL, FRIABLE. OXIDE ZONE OF VN WITH NATIVE TELLURIUM AND PYRITE	34.2.3.
2.5	PYROCHROITE	HEXA.	3.23-3.27	OXD	TBLR, RHOMB, FOL MSV. PEARLY-DULL. BASAL CLVG(FLEX). SOL-HCL. WITH OTHER MN MINERALS IN LOW TEMPERATURE HYDROTHERMAL VEINS	6.1.1.2
2.5	CHLORARGYRITE	ISOM.	5.56	HAL	CRUSTS,CUBIC. RESIN. DUCTILE. EXPOS-VIOLET-BRWN. SCNDRY IN AG DEPS	9.1.1.4
2.5	EGLESTONITE	ISOM.	8.33-8.45	HAL	EXPOS-DK BRWN, BLK. YEL STRK. MSV, CRUSTS, DODEC. RESIN. CONCH. DCMP-ACIDS(CALOMEL RESIDUE). MERCURY DEPOSITS	10.1.1.
2.5	TERLINGUAITE	MNCL.	8.73	HAL	EXPOS-OLIVE GREEN. PRISM, POWDERY. ADMN. 1 CLVG. SOL-ACIDS. RARE-MERCURY DEPOSIT (TERLINGUA, TEXAS)	10.1.2.
2.5	MINIUM	UNKN.	8.9 -9.2	OXD	MSV, EARTHY.GRSY-DULL. SOL-HCL(CL↑).DCMP-HNO3.ALTER PROD OF GALENA DEPOSITS	4.3.1.
2.5	MONTROYDITE	ORTH.	11.23	OXD	YEL-BRWN STRK. BENT, ELONGATE-EQUANT XLS, WORMLK-SPHERICAL AGGS. VITR,ICLVG.SECTILE.SOL-HCL,HNO3,KI,ETC. WITH OTHER MERCURY MINS	4.2.5.
2.5-3	WHEWELLITE	MNCL.	2.23	CCC	HEART-SHAPED XLS. VITR,PEARLY. CLVGS. CONCH. SOL-ACIDS. ORGANIC ASSOCIATION, VEINS	50.1.1.

150

NONMETALLIC BROWN

H	Mineral	Xl Sys	G	Streak	Habit, Cleavage, Fracture, etc.	Occ, Assoc, Alter	Ref
2.5-3	THENARDITE	ORTH.	2.66	SUF	DIPYRAMIDAL, CRUSTS, VITR. CLVGS. FAINTLY SALTY. SALINE LAKE DEPS. EFFLORESCENCES, FUMAROLIC		28.2.4.
2.5-3	CORVUSITE	UNKN.		OXD	OPQ. MSV. SOL-ACIDS. WDSPRD IN UTAH-COLO U-ORE AREA-IMPREGNATING SS		4.6.2.
2.5-3	SERICITE	MNCL.	2.77-2.88	PSI	FINE-GRAINED WHITE (NA OR K) MICA. DEUTERIC-IG RKS, META ROCKS		W369 3-14
2.5-3	BIOTITE	MNCL.	2.7 -3.3	PSI	MG-,FE-MICA. BLEACHED + ATTKD-H2SO4.COMMON-E.G.,ACID IG,PEGS, META RKS		W373 3-55
2.5-3	SIDEROPHYLLITE	MNCL.	2.8 -3.4	PSI	FE-MICA. RARE IN PEGMATITES		W373 3-65
2.5-3	VALENTINITE	ORTH.	5.76	OXD	PRISM, TBLR, MSV. ADMN. PRISM CLVG. SOL-HCL. SCNDRY AFTER SB-MINS		4.4.4.
2.5-3	VAUQUELINITE	MNCL.	6.02	CHR	WEDGE-SHAPED,MAMMILLARY FIBR, ADMN-RESIN. SOL-HNO3(PARTLY). SECONDARY - WITH CROCOITE, ETC.		36.1.1.
2.5-3	VANADINITE	HEXA.	6.88	VAN	PRISM,HAIRLK, RESIN. CONCH. SOME SHOW CONCENTRIC ZONING. SOL-HNO3 (YEL), HCL(GREEN). OXIDIZED ZONES OF LEAD DEPOSITS		41.7.2.3
2.5-3	WULFENITE	TETR.	6.5 -7.0	MBS	TBLR,DIPYRAMIDAL,MSV,GRNLR. RESIN-ADMN. CLVGS-UNEVN. DCMP-HCL, HNO3,SOL-H2SO4,ALKS. SCNDRY-OXIDIZED ZONES OF PB- + MO- DEPS		48.1.4.1
2.5-3	MATLOCKITE	TETR.	7.12	HAL	TBLR, PLTY AGGS, ETC. ADMN. 1 CLVG. SOL-HNO3, DCMP-H2SO4. PB-DEPS		10.1.6.1
2.5-3	STOLZITE	TETR.	7.9 -8.34	WOS	DIPYRAMIDAL-STRIATED. RESIN. CLVGS. CONCH. SOL-HCL(YEL). WITH LIMONITE,ETC.-OXIDIZED DEPOSITS WITH TUNGSTEN MINERALS		48.1.4.2
2.5-3	RASPITE	MNCL.	8.46	WOS	TBLR-STRIATED. ADMN.1 CLVG. DCMP-HCL(YEL). WITH STOLZITE(MXED VNS)		48.1.5.
2.5-3.5	SERPENTINE		2.5 -2.6	PSI	GROUP NAME. NOMENCLATURE FOR MEMBERS IS UNSETTLED. MSV-ASBESTIFM, WAXY-GRSY. GELAT. SOME FLUO-CREAM YEL. ALTER PROD AFTER MG-SILICATES(E.G.,OLIVINE), VNS		W381 3-170
2.5-3.5	JAROSITE	HEXA.	2.91-3.26	SUF	CRUSTS,GRNLR,ETC. RESIN.1 CLVG. SOL-HCL. ON FE-BEARING ORES + RKS		30.2.4.3
2.5-3.5	BISMUTITE	TETR.	6.1 -7.7	CBT	GRAY STRK. MSV, RADIAL FIBR CRUSTS. VITR-DULL. EFF-ACIDS. SCNDRY AFTER BISMUTH MINERALS		16.1.8.
2.5-4	DELVAUXITE	UNKN.	1.99-2.83	PHO	NODULES,BOTRYOIDAL CRUSTS. WAXY-VITR. DATA LACKING.		42.4.7.
2.5-4	MUSCOVITE	MNCL.	2.77-2.88	PSI	K-MICA. COMMON RK-FORMING MIN-E.G., ACID IG, PEGS, GREISENS, CNTCT METASOMATIZED(FLUORINE) ZONES, REGIONAL META-ARGILLACEOUS ROCKS		W367 3-11
2.5-4	ZINNWALDITE	MNCL.	2.90-3.02	PSI	LI-,FE-MICA. WITH LI-BEARING MINS-PEGS, CASSITERITE-BEARING VEINS		W371 3-92
2.5-5	GUMMITE		3.9 -6.4	OXD	FIELD TERM FOR HYDROUS OXIDES OF URANIUM OF UNKNOWN IDENTITY		5.2.1.
3	HOHMANNITE	TRCL.	2.2	SUF	GRNLR, PRISM, VITR. CLVGS. EXPOS-DEHYDRATES TO METAHOHMANNITE. SOL-HCL,DCMP-HOT H2O. SECONDARY-COMMONLY WITH COPIAPITE		31.6.5.
3	RINNEITE	HEXA.	2.35	HAL	EXPOS-BRWN. GRNLR. SILKY. CONCH. ASTRINGENT. SCNDRY-SED SALT DEPS		11.4.3.
3	CALCITE	HEXA.	2.71	CBT	VARIOUS TRIGONAL XL HABITS, MSV. VITR-PEARLY. RHOMB CLVG. SOME IS FLUO,PHOSPHO,THERMOLUM. EFF-DILUTE HCL.WDSPRD-VNS, DIVERSE ROCKS		14.1.1.1
3	BOEHMITE	ORTH.	3.01-3.06	OXD	PISOLITES, TBLR. 1 CLVG. IN BAUXITE DEPOSITS		6.1.2.2
3	NATROJAROSITE	HEXA.	3.18	SUF	CRUSTS, SCALY. VITR. 1 CLVG. CONCH. SOL-HCL(SLOW). WITH OTHER SUL-FATES FORMED BY OXIDATION OF PYRITE, ETC.		30.2.4.5
3	LOSEYITE	MNCL.	3.27	CBT	LATHS-SUBPARALLEL TO RADIAL AGGS. IN CALCITE VNLETS AT FRNKLN,N.J.		16.1.1.
3	ASTROPHYLLITE	TRCL.	3.3 -3.4	SSI	SUBMET-PEARLY. BRONZY-GLDN YEL. CLVG-BRTL FLKS. DCMP-HCL. ALK RKS		W480 N
3	HEMAFIBRITE	ORTH.	3.65	ARS	PRISM,RADIAL FIBR AGGS,VITR.CLVGS. SOL-ACIDS. WITH SCNDRY MN-MINS		42.2.3.
3	ARGENTOJAROSITE	HEXA.	3.66	SUF	SCALY. BRILLANT. 1 CLVG. RARE-SCNDRY WITH ANGLESITE,ETC.(ITINTIC)		30.2.4.6

NONMETALLIC BROWN

HARDNESS	NAME	XL. SYS.	SPECIFIC GRAVITY	CHEM. CLASS	REMARKS	REFERENCES
3	DOLEROPHANITE	MNCL.	4.17	SUF	SUBTRANSLUCENT. YELLOWISH STRK. ELONGATE. SOL-HNO3. VOL SUBLIMATE	30.2.2.
3	OLIVENITE	ORTH.	3.9-4.46	ARS	ELONGATE,GLBL FIBR,MSY,ETC. ADMN-SILKY. CLVGS. CONCH. SOL-ACIDS + NH4OH. SECONDARY-OXIDIZED ZONES OF ORE DEPOSITS	41.6.5.1
3 -3.5	TEEPLEITE	TETR.	2.08	BOR	TBLR. GLASSY, EXPOS-DULL. SOL-H2O(TO ALK SOLUTION).WITH TRONA	26.1.3.
3 -3.5	ROEMERITE	TRCL.	2.17	SUF	VIOLET OR YEL TINGE. CRUSTS, CUBICLK. OILY. 2 CLVGS. SALTY,ASTRIN- GENT. WITH COPIAPITE - BY OXIDATION OF PYRITE	29.7.2.
3 -3.5	LAUMONTITE (ZEOLITE)	MNCL.	2.2 -2.3	TSI	PERF CLVGS-3 DIRECTIONS IN 1 ZONE. GELAT. CAVITIES IN SEVERAL RKS	W342 4-401
3 -3.5	SZAIBELYITE	ORTH.	2.62	BOR	IN SERIES WITH SUSSEXITE(SEE). VNS, SALT DEPS, SKARN ZONES	26.1.5.2
3 -3.5	LANDESITE	UNKN.	3.03	PHO	OCTAHLK XLS (PSEUDOMORPHOUS). 2 CLVGS(AT 90°). ALTER PROD OF RED- DINGITE - IN PEGMATITE	40.2.6.
3 -3.5	PHOSPHOFERRITE	ORTH.	3.0 -3.2	PHO	MSV,GRNLR,FIBR,TBLR. VITR. 1 CLVG. UNEVN. SOL-ACIDS. PEGMATITES	40.2.5.2
3 -3.5	SUSSEXITE	ORTH.	3.30	BOR	IN SERIES WITH SZAIBELYITE. FIBR(INFLEX),DENSE-CHALKY AGGS. SILKY- DULL. SOL-ACIDS(SLOWLY). VEINS (E.G., FRANKLIN, NEW JERSEY)	26.1.5.1
3 -3.5	CELESTITE	ORTH.	3.96-3.98	SUF	TYPICALLY COLORLESS-BLUISH. TBLR,ETC. VITR-PEARLY. CLVGS. SOME FLUO,THERMOLUM. SOL-H2O(SLIGHTLY).SED RKS(ESP GYPSUM + LS). VNS	28.3.1.2
3 -3.5	WITHERITE	ORTH.	4.29	CBT	PSEUDOHEX, GLBLR,ETC. VITR. FLUO(ETC.)-LK ARAGONITE.SOL-HCL. VNS	14.1.3.2
3 -3.5	DESCLOIZITE	ORTH.	6.1 -6.3	VAN	VARIABLE-XLS TO COMPACT MSV. GRSY.COMMONLY COLOR ZONED. SOL-ACIDS. OXIDIZED ZONE OF VANADIUM-BEARING ORE DEPOSITS	41.5.2.1
3 -4	ARSENIOPLEITE	HEXA.		PHO	SUBTRANSLUCENT. YEL-BRWN STRK. MSV, GRNLR. RHOMB CLVG. CONCH. SOL- HCL, HNO3. VNS + NODULES WITH RHODONITE (ØREBRO, SWEDEN)	41.5.12.
3 -4	MORDENITE (ZEOLITE)	ORTH.	2.12-2.15	TSI	STRIATED PRISMS, ACICULAR, COTTON-LK AGGS. ZEOL-ASSOC, AUTHIGENIC	W339 4-401
3 -4	EVANSITE	UNKN.	1.8 -2.2	PHO	MSV,COLLOFM COATINGS. VITR. CONCH. BRTL. SOL-ACIDS. WITH LIMONITE	42.2.6.
3 -4	GISMONDINE (ZEOLITE)	ORTH.	2.1 -2.3	TSI	PSEUDOTETRAG BIPYRAMIDS. 1 CLVG. GELAT. LEUCITE RKS, ALTER OF PLAG	W339 4-401
3 -4	WAVELLITE	ORTH.	2.36	PHO	RADIAL FIBR,STELLATE,CRUSTS. VITR-PEARLY,CLVGS. SOL-ACIDS. SCNDRY- WDSPRD-OPENINGS IN DIVERSE ROCKS, LIMONITE, PHOSPHATE RKS, ETC.	42.7.4.
3 -4	CACOXENITE	HEXA.	2.2 -2.4	PHO	ACICULAR,TUFTS,RADIAL AGGS(FIBR CRUSTS. SILKY. SOL-ACIDS. SCNDRY- DIVERSE OCCURRENCES(WITH DUFRENITE, WAVELLITE, ETC.)	42.9.2.
3 -4	DIADOCHITE	TRCL.	2.0 -2.4	PHO	NODULAR,PLTY.GEL-LK,COLLOFM. DULL-WAXY. UNEVN-CONCH. SOL-ACIDS. IN GOSSANS AND AS EFFLORESCENCES	43.2.4.
3 -4	STILPNOMELANE	MNCL.	2.59-2.96	PSI	MICACEOUS(BUT 2ND CLVG⊥1ST). SOL-HF+H2SO4(1/1). LO-GRD META RKS	W390 3-103
3 -4	STILPNOCHLORANE	UNKN.	2.5 -3.0	PSI	YELLOWISH TO BRONZY RED. SCALES.	WN N
3 -4	FOURMARIERITE	ORTH.	6.05	OXD	TBLR-STRIATED. ADMN. BASAL CLVG. SOL-ACIDS. SCNDRY URANIUM MINERAL	5.3.1.
3 -5.5	ZEOLITE		1.9 -2.45	TSI	GROUP OF MIN FAMILIES WITH SOMEWHAT RELATED APPEARANCES, OCCURRENCES, AND COMPOSITIONS. MANY ARE INDISTINGUISHABLE MEGASCOPICALLY. DIVERSE HABITS(E.G., BUNDLES, FIBROUS), PEARLY-GLASSY, SUBCONCH. SOME FLUO-ORANGE OR YEL-GREEN. CONTINUOUS, IN PART REVERSIBLE, DEHYDRATION(I.E.,WATER CAN BE EXPELLED WITH- OUT DESTROYING XL STRUCTURE). SCNDRY IN OPEN SPACES; COMMONLY TWO OR MORE OCCUR TOGETHER IN CAVITIES IN BASALTIC RKS — THIS OCCURRENCE IS REFERRED TO HEREIN AS THE ZEOLITE ASSOCIATION (ABBREVIATED "ZEOL-ASSOC").	W330 4-351
3.5	BURKEITE	ORTH.	2.57	SUF	TBLR,PLTY AGGS,ETC. GRSY. CONCH.SOL-H2O. WITH TRONA,ETC. BORAX DEP	32.1.4.
3.5	BORICKITE	UNKN.	2.69-2.71	PHO	SUBTRANSLUCENT. COMPACT-SOME OPALINE. WAXY. SOL-HCL. RARE-GOSSAN	42.1.1.

NONMETALLIC BROWN

H	Mineral	Cryst	SG	Lus	Notes	Ref
3.5	BERMANITE	ORTH.	2.84	PHD	TBLR,FANLK + ROSETTELK AGGS, LAMELLAR MASSES,VITR.CLVGS,SOL-HNO3. SECONDARY — IN TRIPLITE (PEGMATITES)	42.8.1.
3.5	ANHYDRITE	ORTH.	2.98	SUF	MSV, GRNLR, FIBR. VITR-PEARLY. 2 CLVGS. SOL-ACIDS. SED ROCKS	28.3.2.
3.5	COLLINSITE	TRCL	2.99	PHD	RADIAL BLDS. SILKY. 2 CLVGS. SOL-ACIDS. VN IN ANDESITE WITH CARBO- NACEOUS MATERIALS	40.2.3.2
3.5	AKROCHORDITE	MNCL.	3.19	ARS	WARTLK AGGS. 2 CLVGS. SOL-H2SO4(PURPLE). RARE-LANGBAN, SWEDEN	42.3.3.
3.5	HUREAULITE	MNCL.	3.15-3.20	PHD	TBLR,MSV,SCALY,FIBR. VITR. 1 CLVG. SOL-ACIDS. WITH ALTERED LITHI- OPHILITE AND/OR TRIPHYLITE	39.1.3.
3.5	CRONSTEDTITE (SERPENTINE ?)	HEXA.	3.45	PSI	FINE GRAINED. MORE COMMON THAN GREENALITE. IN IRON FORMATIONS	W389 3-164
3.5	HEMATOLITE	HEXA.	3.49	ARS	SUBMET. TBLR,RHOMB,STRIATED. VITR. 1 CLVG. SOL-ACIDS. WITH BARITE, JACOBSITE,ETC. IN VNS IN LIMESTONE (NORDMARK, SWEDEN)	41.1.3.
3.5	CHALCOCYANITE	ORTH.	3.60-3.70	SUF	TBLR. HYGROSCOPIC TO BLUE PENTAHYDRATE OR CHALCANTHITE. FUMAROLIC	28.3.3.
3.5	STRONTIANITE	ORTH.	3.64-3.78	CBT	SPEAR-LK, MSV. VITR. CLVGS. FLUOIETC.)-LK ARAGONITE. SOL-HCL. VNS	14.1.3.3
3.5	RHABDOPHANE	HEXA.	3.94-4.01	PHD	RADIATED FIBR MAMMILLARY CRUSTS. GRSY. UNEVN. SOL-HCL. WITH LIMO- NITE	40.3.5.
3.5	KALKOWSKITE	UNKN.	3.98-4.04	OXD	SUBMET. REDDISH STRK. FIBR PLTS.WAXY. DCMP-HCL.RARE-SCHIST(BRAZIL)	8.2.2.
3.5	SCHAFARZIKITE	TETR.	4.3	ATI	BRWN STRK. SUBTRANSLUCENT. PRISM-STRIATED. CLVGS. RARE-SB-DEPOSITS	45.1.4.
3.5	ADAMITE	ORTH.	4.32-4.48	ARS	ELONGATE,CRUSTS,RADIAL AGGS. VITR. CLVGS. UNEVN. SOL-DILUTE ACIDS. SOME FLUO YEL.OXIDIZED ZONES OF DEPS WITH ZN — + AS-RICH MINERALS	41.6.5.3
3.5	CARMINITE	ORTH.	4.10-5.22	ARS	LATHLK,FIBR AGGS,ETC.VITR,PEARLY.SOL-HNO3,DCMP-HCL. SCNDRY-VN DEPS	41.8.3.
3.5	GEORGIADESITE	MNCL.	7.1	ARS	PSEUDOHEX PLTS,STRIATED. RESIN. SOL-HNO3. RARE-1 SPECIMEN(LAURIUM)	41.3.3.
3.5-4	HEULANDITE (ZEOLITE)	MNCL.	2.1-2.2	TSI	PLTY GRP. TBLR-BROAD CENTRAL PORTIONS(COFFIN-LIKE). ZEOL-ASSOC	W347 4-377
3.5-4	STILBITE (ZEOLITE)	MNCL.	2.1-2.2	TSI	PLTY GRP. SHEAF-LIKE AGGS. 1 GOOD CLVG. DCMP-HCL. ZEOL-ASSOC	W345 4-377
3.5-4	NORTHUPITE	ISOM.	2.38	CBT	OCTAH. VITR. CONCH. EFF-ACIDS,DCMP-HOT H2O. SALINE CLAY DEPS	16.2.2.
3.5-4	BREWSTERITE (ZEOLITE)	MNCL.	2.45	TSI	PLTY GRP. ELONGATE,STRIATED PRISMS. 1 CLVG. DCMP-HCL. ZEOL-ASSOC	W347 4-383
3.5-4	DOLOMITE	HEXA.	2.84-2.86	CBT	RHOMB-CURVED, GRNLR. VITR-PEARLY. RHOMB CLVG. SOME FLUO. EFF-WARM HCL IF POWDERED. SEDIMENTARY RKS, CRYSTALS IN CAVITIES, ETC.	14.2.1.1
3.5-4	ANKERITE	HEXA.	3.01-3.03	CBT	DISTINGUISHED FROM DOLOMITE(SEE) BY NONMEGASCOPIC METHODS. VNS	14.2.1.2
3.5-4	BERAUNITE	MNCL.	2.8-3.08	PHD	STRK-YEL,OLIVE. RADIATED FOL GLBLS,CRUSTS,ETC. VITR. 1 CLVG. SOL- ACIDS. SCNDRY-WITH OTHER PHOSPHATES + LIMONITE--FE DEPS + PEGS	42.7.2.
3.5-4	SCORODITE	ORTH.	3.28	ARS	PYRAMIDAL,ETC. AGGS,CRUSTS,ETC. VITR. CLVGS. SOL-ACIDS. SCNDRY- AFTER ARSENOPYRITE AND OTHER AS-BEARING MINERALS	40.3.1.3
3.5-4	LAUBMANNITE	ORTH.	3.33	PHD	SUBTRANSLUCENT. CRUSTS. VITR-SILKY. CLVG. RARE WITH LIMONITE	41.4.6.
3.5-4	WARWICKITE	ORTH.	3.34-3.36	BOR	SUBMET-PEARLY. BLUE-BLK STRK. SLENDER PRISMS, ROUNDED ENDS. DCMP- H2SO4. SCALY — CONTACT METAMORPHOSED LIMESTONE	24.1.4.
3.5-4	DICKINSONITE	MNCL.	3.38-3.41	PHD	TBLR,SCALY. VITR,PEARLY.1 CLVG. SOL-ACIDS. PEGS-WITH LITHIOPHILITE, ETC.	40.2.1.
3.5-4	RHODOCHROSITE	HEXA.	3.70	CBT	GRNLR, BOTRYOIDAL. VITR-PEARLY. RHOMB CLVG. EFF-WARM ACID. VNS	14.1.1.4

NONMETALLIC BROWN

HARDNESS	NAME	XL. SYS.	SPECIFIC GRAVITY	CHEM. CLASS	REMARKS	REFERENCES
3.5-4	WURTZITE	HEXA.	3.98	SLD	BRWN STRK. HEMIMORPH PYRAMIDAL. RESIN. SOL-HCL. WITH SPHAL, S-VNS	2.6.4.1
3.5-4	ALABANDITE	ISOM.	3.9-4.04	SLD	SUBMET.GRN STRK, TARN-BRWN. PERF CUBIC CLVG. SOL-HCL. SULFIDE VNS	2.6.1.4
3.5-4	TARBUTTITE	TRCL.	4.12	PHO	CRUSTS,SHEAFLK AGGS OF ROUNDED,STRIATED,EQUANT XLS. VITR-PEARLY. 1 CLEAVAGE. OXIDIZED ZONES OF ZINC DEPOSITS	41.6.7.
3.5-4	POWELLITE	TETR.	4.21-4.25	MBS	PYRAMIDAL,TBLR,MSV-FOL,PWDER,CRUSTS,SUBADMN-PEARLY. CLVGS. UNEVN. VITR-PEARLY. FLUO-YEL. DCMP-HCL,HNO3.ALTER PROD OF MOLYBDENITE	48.1.3.2
3.5-4	NADORITE	ORTH.	7.02	ATI	TBLR,PRISM,SUBPARALLEL-DIVERGENT GROUPS. RESIN-ADMN. CLVG. DIVERSE	46.1.4.
3.5-4	PYROMORPHITE	HEXA.	7.00-7.08	PHO	PRISM,GLBL,ETC. RESIN. RHOMB CLVG. UNEVN. SOL-HNO3. OXIDIZED ZONES OF LEAD DEPOSITS	41.7.2.1
3.5-4	MIMETITE	HEXA.	7.24	ARS	PRISM,ACICULAR,GLBL,ETC. RESIN. RHOMB CLVG. UNEVN. SOL-HNO3. OXI-DIZED ZONES OF LEAD DEPOSITS	41.7.2.2
3.5-4.5	MAGNESITE	HEXA.	2.98-3.02	CBT	GRNLR, RHOMB. VITR. RHOMB CLVG. FLUO-GREEN,BLUE. SOL-WARM HCL. AL-TERATION OF MG-RICH ROCKS (E.G., SERPENTINE)	14.1.1.2
3.5-4.5	DUFRENITE	MNCL.	3.1-3.34	PHO	BOTRYOIDAL-RADIAL,CRUSTS. VITR-SILKY. CLVGS. SOL-DILUTE ACIDS. WITH LIMONITE IN GOSSANS	41.6.9.
3.5-4.5	FERRI-SICKLERITE	ORTH.	3.2-3.4	PHO	SUBTRANSLUCENT. MSV. 1 CLVG. SOL-ACIDS. WXING PROD OF TRIPHLITE, LITHIOPHILITE (PEGMATITES)	38.1.3.1
3.5-4.5	SICKLERITE	ORTH.	3.2-3.45	PHO	SUBTRANSLUCENT. DISTINGUISHED BY CHEM ANAL FROM FERRI-SICKERLITE	38.1.3.2
3.5-4.5	SIDERITE	HEXA.	3.95-3.97	CBT	TARN-IRID. MSV. RHOMB,ETC. VITR.RHOMB CLVG. SOL-HOT ACID. SED, VNS	14.1.1.3
3.5-4.5	BEUDANTITE	HEXA.	4.3	ARS	RHOMB,PSEUDOCUBIC. VITR. 1 CLVG. SOL-HCL. SCNDRY-DIVERSE	43.1.1.1
3.5-6	CLINTONITE (BRITTLE MICA)	MNCL.	3.1	PSI	WITH TALC-CHLORITE SCHIST, WITH SPINEL IN METASOMATIZED CALC RKS. DISTINGUISHED FROM XANTHOPHYLLITE BY OPTICAL OR X-RAY ANALYSES	W391 3-99
3.5-6	XANTHOPHYLLITE (BRITTLE MICA)	MNCL.	3.1	PSI	WITH TALC-CHLORITE SCHIST, WITH HUMITES IN METASOMATIZED CALC RKS. DISTINGUISHED FROM CLINTONITE BY OPTICAL AND/OR X-RAY ANALYSES	W391 3-99
3.5-7	BRITTLE MICA	MNCL.	2.6-3.2	PSI	GROUP OF MORE OR LESS RELATED MINS THAT RESEMBLE MICAS BUT CLEAVE TO BRTL FLKS AND HAVE HARDNESSES OF 3.5-4 VS. 2-3(MICA) ON CLVG FLKS AND 6 VS. 4 ⊥ TO FLKS. WITH CORUNDUM + DIASPORE IN EMERY DEPS, CHLORITE + MICA SCHISTS, SKARN'	W392 3-95
4	WEDDELLITE	TETR.	1.93-1.95	CCC	PYRAMIDAL,AGGS. SUBCONCH. ORGANIC ASSOC. BOTTOM MUDS(WEDDELL SEA)	50.1.2.
4	EPISTILBITE (ZEOLITE)	MNCL.	2.1-2.3	TSI	PLTY GRP. SHEAF-LK + SPHERICAL AGGS. DCMP-HCL.PIEZOELECTRIC. ZEOL-ASSOC	W346 4-377
4	GANDPHYLLITE	MNCL.	2.84	PSI	SIMILAR TO STILPNOMELANE. STUBBY PRISMS. GELAT. RARE-WITH MN-ORES	W391 N
4	SALMONSITE	ORTH.	2.88	PHO	FIBR MASSES. 2 CLVGS≈AT 90°.ALTER PROD OF HUREAULITE IN PEGMATITE	40.2.8.
4	MOSANDRITE	MNCL.	2.93-3.03	NSI	REDDISH BRWN. LONG PRISM XLS. 1 CLVG. ALKALIC PEGMATITES-RARE	W517 N
4	FLUORITE	ISOM.	3.18	HAL	CUBIC, MSV, VITR. OCTAH CLVG. FLUO-VIOLET. DCMP-H2SO4. WDSPRD-VNS, CAVITIES, DISSEMINATED - IGNEOUS ROCKS, ETC.	9.2.1.
4	JOHNSTRUPITE	MNCL.	3.3	NSI	BRWNISH GREEN. STRIATED PRISM PLATES. 1 CLVG. SOL-HCL. ALK IG RKS	M516 N
4	STEENSTRUPINE	HEXA.	3.4-3.47	NSI	RHOMBOHEDRAL. RARE-E.G., JULIANEHAAB AREA, GREENLAND	M509 N
4	SARCOPSIDE	MNCL.	3.64-3.73	PHO	RED TO BROWNISH,EXPOS-DK BROWN,BLUE,VIOLET,GREEN,FIBR MASSES,SILKY. 2 CLVGS (∥ + ⊥ FIBERS). SOL-ACIDS. RARE-BPEGMATITES	41.6.4.
4	SPHEROCOBALTITE	HEXA.	4.13	CBT	PINK STRK. MSV-RADIAL STRUCTURE. VITR. EFF-HOT ACID. VNS	14.1.1.5
4	RETZIAN	ORTH.	4.15	ARS	PRISM,TBLR. VITR-GRSY. CONCH. SOL-ACIDS. RARE-WITH JACOBSITE IN BRAUNITE-BEARING DOLOMITE (NORDMARK, SWEDEN)	41.3.6.
4	CARYINITE	ORTH.	4.29	ARS	SUBTRANSLUCENT. MSV, GRNLR. GRSY, CLVGS. SOL-HNO3. VNS IN SKARN-LANGBAN, SWEDEN	38.2.2.

NONMETALLIC BROWN

Hardness	Mineral	System	S.G.	Class	Description	Ref
4	AMPANGABEITE	ORTH.	3.36-4.64	OXD	PRISM, RADIAL AGGS. HORNLK. CONCH. DCMP-ACIDS. K-RICH PEGMATITES	8.4.3.
4	WALTMERITE	UNKN.	5.32	CBT	YEL STRK. PRISM. VITR. CLVGS. EFF-ACIDS. RARE-SCNDRY(JOACHIMOV, BOHEMIA)	80-16.1.9.
4	PUCHERITE	ORTH. C	6.57	VAS	TBLR,ACICULAR,MSV,CRUSTS. VITR-ADMN.1 CLVG. SUBCONCH,SOL-HCL(RED). SECONDARY, WITH BISMUTITE AND BEYERITE	47.1.8.
4	HUEBNERITE	MNCL.	7.12	WOS	SUBMET. TARN-IRID. PRISM-STRIATED,-RADIATING XL GRPS.,1 CLVG. DCMP-AQ-REG,H2SO4 OR HCL(SLOWLY).,DIVERSE-VNS,CNTCT META,PLACERS	48.1.1.1
4-4.5	KALIBORITE	MNCL.	2.13	BOR	XL AGGS, GRNLR. VITR. CLVGS. SOL-ACIDS. SALINE EVAPORITE DEPOSITS	25.1.22.
4-4.5	HODGKINSONITE	MNCL.	3.91	NSI	PYRAMIDAL XLS. BASAL CLVG. GELAT. WITH WILLEMITE(ETC.)-FRNKLN,N.J. W515 N	
4-4.5	ANCYLITE	ORTH.	3.95	CBT	PSEUDO-OCTAH,CRUSTS.VITR. SPLINTERY. SOL-ACIDS. DRUSES IN ALK PEGS	16.2.10.
4-4.5	SMITHSONITE	HEXA.	4.40-4.45	CBT	BOTRYOIDAL, MSV. VITR-PEARLY. RHOMB CLVG. SOME FLUO-GREEN, BLUE. EFF-ACIDS. SECONDARY – OXIDIZED ZONES OF ZINC DEPOSITS	14.1.1.6
4-4.5	BASTNASITE	HEXA.	4.9-5.2	CBT	TBLR-GROOVED, GRNLR. VITR,PEARLY. CLVGS. SOL-HOT ACIDS. CNTCT META	16.2.9.
4-4.5	BINDHEIMITE	ISOM.	4.6-5.6	ANT	DENSE-EARTHY CRUSTS,NODULAR-CONCENTRIC.SOME IS OPALINE. RESIN-DULL. DCMP-HNO3, HCL. OXIDATION ZONES OF ANTIMONY-BEARING LEAD DEPOSITS	44.1.1.1
4-4.5	CLARKEITE	UNKN.	6.39	OXD	YEL-BRWN STRK. MSV. WAXY. CONCH. RDACTV. ALTER PROD OF URANINITE	5.2.2.
4-4.5	WOLFRAMITE	MNCL.	7.31	WOS	SUBMET. TARN-IRID. PRISM-EQUANT–STRIATED,MSV-GRNLR,ETC. 1 CLVG. DCMP-AQ-REG.H2SO4 OR HCL(SLOWLY).,DIVERSE-VNS,CNTCT META,PLACERS	48.1.1.2
4-5	PYROSMALITE	HEXA.	3.06-3.19	PSI	HEX PLTS. PERF BASAL CLVG. GELAT. RARE MN-,FE-DEPS (FRNKLN, N.J.)	W359 N
4-5	LITHIOPHILITE	ORTH.	3.50-3.58	PHO	MSV, ANHEDRAL. VITR. CLVGS. UNEVN. SOL-ACIDS. PEGMATITES	38.1.1.2
4-5	XENOTIME	TETR.	4.4-5.1	PHO	PRISM + PYRAMIDAL. VITR. 1 CLVG. ACID + ALK IG RKS + PEGS,DETRITAL	38.4.1.
4-5	FLUOCERITE	HEXA.	5.93-6.14	HAL	FRESH-YEL. MSV. TBLR. RESIN,PEARLY. BASAL CLVG. SOL-H2SO4. PEGS	9.3.2.
4-5	KASOLITE	MNCL.	5.96-6.46	NSI	ELONGATE XLS. 1 PERF,2 POOR CLVGS.GELAT. RDACTV. RARE-WITH CURITE	W530 N
4-5.5	LIMONITE	ISOM.	2.7-4.3	OXD	FIELD TERM FOR HYDROUS IRON OXIDES OF UNKNOWN IDENTIES.	7.1.3.
4-5.5	BETAFITE	ISOM.	3.7-5.	OXD	SUBMET. OCTAH. WAXY. CONCH. METAMICT. DCMP-ACIDS. GRANITIC PEGS	8.4.1.
4.5	COLEMANITE	MNCL.	2.42-2.43	BOR	EQUANT, GRNLR. VITR. 2 CLVGS. SOL-HOT HCL. EVAPORITE BORATE DEPS	25.1.9.
4.5	ROSCHERITE	MNCL.	2.92	PHO	6 OR 8 SIDED PRISMS,PLTY VERMICULAR AGGS. 2 CLVGS. SOL-ACIDS. PEG. DRUSES IN GRANITE	42.8.2.
4.5	FILLOWITE	MNCL.	3.43	PHO	GRNLR,PSEUDORHOMB. GRSY. 1 CLVG. SOL-ACIDS. PEG-WITH TRIPLOIDITE. ETC.	40.2.2.
4.5	FRONDELITE	ORTH.	3.3-3.49	PHO	SUBTRANSLUCENT. RADIAL FIBR BOTRYOIDAL.SOME HAVE CONCENTRIC COLOR BANDING). VITR-DULL. CLVGS,UNEVN,SOL-HCL,SCNDRY-PO4 MINS (PEGS)	41.6.6.1
4.5	ROCKBRIDGEITE	ORTH.	3.3-3.49	PHO	IN SERIES WITH FRONDELITE!SEE FOR PROPS + OCCUR.) WITH LIMONITE	41.6.6.2
4.5	SYNADELPHITE	TRCL.	3.57-3.79	ARS	COMMONLY COLOR ZONED. PSEUDOPRISMS, MSV. VITR. 1 CLVG. UNEVN. SOL-ACIDS. RARE WITH OTHER MANGANESE MINERALS	41.1.5.
4.5	ALLACTITE	MNCL.	3.83	ARS	ELONGATE,ROSETTELK AGGS. VITR. 1 CLVG. SOL-HCL. VNS(E.G.,LANGBAN)	41.2.4.
4.5	FLINKITE	ORTH.	3.87	ARS	FEATHERLK AGGS. VITR. SOL-ACL. RARE-VNS IN OREIPERSBERG, SWEDEN)	41.3.5.

NONMETALLIC BROWN

HARDNESS	NAME	XL. SYS.	SPECIFIC GRAVITY	CHEM. CLASS	REMARKS	REFERENCES
4.5	SYNCHISITE	HEXA.	3.90	CBT	ACUTE RHOMB, TBLR-HEMIMORPHIC. GRSY. 1 CLVG. SOL-ACIDS. ALK PEGS	16.2.8.
4.5	PARISITE	HEXA.	4.33-4.39	CBT	STRIATED BIPYRAMIDS. VITR,PEARLY. SOL-HOT ACIDS. VNS, ALK PEGS	16.2.6.
4.5	BRANNERITE	UNKN.	4.5 -5.43	OXD	BLK, EXPOS-BRWNISH YEL. PRISM. CONCH. RDACTV.DCMP-H2SO4.RARE-PLAC- ER (KELLY GULCH, IDAHO)	8.2.4.
4.5	HEDYPHANE	HEXA.	5.82	ARS	PRISM,TBLR,MSV. GRSY. RHOMB CLVG. SOL-HNO3. RARE-E.G.,FRNKLN, N.J.	41.7.3.2
4.5	EULYTINE	ISOM.	6.6	NSI	MINUTE TETRAHEDRA. POOR DDDEC CLVG. GELAT. WITH BISMUTH + QTZ	W494 N
4.5-5	ORIENTITE	ORTH.	3.05	NSI	TABULAR,RADIATING PRISMATIC. SOL-HCL(HOT). ORE DEPOSITS(E.G.,CUBA)	W507 N
4.5-5	SCHALLERITE	HEXA.	3.37	PSI	MSV. BASAL CLVG. DCMP-HCL. VEINLETS IN ORE BODY AT FRNKLN, N.J.	W359 N
4.5-5	TRIPLOIDITE	MNCL.	3.66	PHO	//-DIVERGENT FIBR-COLUMNAR AGGS,STRIATED PRISM. VITR-ADMN. CLVGS. UNEVN-SUBCONCH. SOL-ACIDS. PEGMATITES-PROBABLY SECONDARY	41.6.3.1
4.5-5	WOLFEITE	MNCL.	3.83	PHO	IN SERIES WITH TRIPLOIDITE(SEE FOR PROPERTIES + OCCURRENCE.)	41.6.3.2
4.5-5	PLUMBOGUMMITE	HEXA.	4.01	PHO	CONCENTRIC BOTRYOIDAL. GUM-LIKE LUSTER. UNEVN. SOL-HOT ACIDS. SECONDARY – IN LEAD DEPOSITS WITH PYROMORPHITE	41.5.8.1
4.5-5	THORITE	TETR.	5.2 -5.4	NSI	TERMINATED PRISMS.PRISM CLVG. MOST IS ALTERED. RARE ACCESS-SYENITE	W496 N
4.5-5	TRIPUHYITE	UNKN.	5.82	ANT	MICROXLINE AGGS. DATA LACKING. VNS IN DACITE, PLACERS.	44.1.3.
4.5-5	SCHEELITE	TETR.	6.08-6.12	WOS	DIPYRAMIDAL,MSV,GRNLR,VITR,CLVGS.UNEVN,DCMP-HCL,FLUO-BLUE(X-,SW ULTRAVI- OLET-, + CATHODE-RAYS),THERMOLM. HI-T VNS,PEGS,GREISSENS,CNTCT META RKS,ETC.	48.1.3.1
4.5-5	ROWEITE	ORTH.	2.90-2.94	BOR	LATHLK. POOR CLVG. BRTL. SOL-HCL. RARE-1 SPECIMEN(FRANKLIN, N.J.)	26.1.6.
4.5-6	MONIMOLITE	ISOM.	5.9 -7.3	ANT	OCTAH,CUBIC. GRSY. CONCH CLVG. CONCH-SPLINTERY. RARE-E.G.,LANGBAN	44.1.2.
5	OKENITE	TRCL.	2.28-2.33	ISI	FIBR, TBLR, MSV. 1 PERF CLVG-ELAST. GELAT. IN AMYGDULES IN BASALT	W358 N
5	GONNARDITE (ZEOLITE)	ORTH.	2.2 -2.4	TSI	FIBR GRP. RADIATED SPHERICAL MASSES. GELAT. RARE. ZEOL-ASSOCIATION	W337 4-359
5	HYDROXYLAPATITE	HEXA.	2.9 -3.1	PHO	APATITE GROUP(SEE FOR PROPS.) RARE-TALC SCHIST	41.7.1.3
5	CARBONATE-APATITE	HEXA.	2.9 -3.1	PHO	APATITE GROUP(SEE FOR PROPS.)RARE-NODULES(PHOSPHATE RK),GUANO,ETC.	41.7.1.4
5	CHLORAPATITE	HEXA.	3.17	PHO	APATITE GROUP(SEE FOR PROPS.) RARE-VNS IN GABBRO, METEORITES	41.7.1.2
5	FLUORAPATITE	HEXA.	3.18	PHO	APATITE GROUP(SEE FOR PROPS.) COMMON-IGS, METAS, MARINE SEDS, ETC.	41.7.1.1
5	APATITE	HEXA.	2.9 -3.2	PHO	GROUP NAME. MOST TYPES INDISTINGUISHABLE MEGASCOPICALLY. HEX PRISMS-COMMONLY ROUNDED,MSV,GRNLR. VITR. SOL-ACIDS. SOME FLUO—YEL-ORANGE. IG RKS,VNS,PEGS	41.7.1.
5	SVANBERGITE	HEXA.	3.22	PHO	RHOMB,PSEUDOCUBIC,GRNLR,VITR-ADMN. 1 CLVG. DIVERSE META RKS,PLACERS	43.1.1.4
5	CHILDRENITE	ORTH.	3.22-3.28	PHO	EQUANT-PLTY. VITR. 1 CLVG. UNEVN. SOL-ACIDS. DIVERSE-E.G.,VNS,PEGS	42.5.2.1
5	HOMILITE	MNCL.	3.36	NSI	TABULAR XLS. GELAT. COMMONLY METAMICT. ALTERED RIMS. RARE IN PEGS	W356 N
5	HEMIMORPHITE	ORTH.	3.45	SSI	HEMIMORPHIC XLS, SHEAFLK, MAMMILLARY, ETC. 1 PERF + 1 POOR CLVG. GELAT. PYROELECTRIC. SOME FLUO-PALE ORANGE(LONG WAVE). ZINC DEPS	W481 N
5	CALAMINE	ORTH.	3.45	SSI	(=HEMIMORPHITE)	W481 N

156

NONMETALLIC BROWN

H	Mineral	Crystal	G	Class	Description	Ref
5	RINKITE	MNCL.	3.46	NSI	PRISM, TBLR. 1 CLVG. ATTACKED BY H2SO4. ALKALIC SYENITES-RARE	W517 N
5	KATOPHORITE (AMPHIBOLE)	MNCL.	3.20-3.50	ISI	RARE IN INTERMEDIATE TO BASIC ALKALIC IGNEOUS ROCKS	W440 2-359
5	MAGNESIOKATOPHORITE (AMPHIBOLE)	MNCL.	3.20-3.50	ISI	RARE IN INTERMEDIATE AND BASIC ALKALIC IGNEOUS ROCKS	WN 2-359
5	TITANITE(=SPHENE)	MNCL.	3.45-3.55	NSI	DOUBLE WEDGE-SHAPED XLS, MSV-GRNLR. ADAMANTINE. ALTERS TO LEUCOX-	W525 1-69
5	SPHENE	MNCL.	3.45-3.55	NSI	ENE. DCMP-H2SO4. COMMON ACCESSARY CONSTITUENT OF IG + META ROCKS	W525 1-69
5	HUHNERKOBELITE	ORTH.	3.5 -3.6	PHO	MSV. GRNLR. VITR. 2 CLVGS. DISTINGUISHED FROM VARULITE BY CHEM ANAL. PEGS & HYDROTHERMAL ALTER OF TRIPHYLITE	38.1.1.3
5	VARULITE	ORTH.	3.5 -3.6	PHO	MSV. GRNLR. VITR. 2 CLVGS DISTINGUISHED FROM HUHNERKOBELITE BY CHEM ANAL. PEGS & HYDROTHERMAL ALTER OF TRIPHYLITE	38.1.1.4
5	GRAFTONITE	MNCL.	3.67-3.79	PHO	PRISM, MSV. VITR. UNEVN. SOL-ACIDS. WITH TRIPHYLITE IN PEGMATITES	38.3.2.
5	LEPIDOCROCITE	ORTH.	4.05-4.13	OXD	SUBMET. ORANGE STRK. SCALY, FIBR MSV. 3 CLVGS. WITH GOETHITE, ETC.	6.1.2.1
5	BECKELITE	HEXA.	4.15	NSI	CUBIC OR DODECAHEDRAL. CUBIC CLVG. SOL-ACIDS. SYENITE (RARE)	W520 N
5	TREVORITE	ISOM.	5.16	OXD	SUBMET. MAGNETITE SERIES OF SPINEL GROUP. BRWN STR. GRNLR, MSV. DIFFICULTLY SOL-HCL. STRONGLY MAGNETIC. RARE-NI DEPS, TRANSVAAL	7.2.1.9
5	MAGHEMITE	ISOM.	5.0 -5.2	OXD	MSV. MAG(STRONGLY). ALTER PROD OF MAGNETITE OR LEPIDOCROCITE	7.2.1.10
5	KENTROLITE	ORTH.	6.19	NSI	DK REDDISH BRWN. PRISM XLS + CLVG. SOL-HCL. MN-DEPS(SO.CHILE)-RARE	W523 N
5 -5.5	THOMSONITE (ZEOLITE)	ORTH.	2.10-2.39	TSI	FIBR GRP. RADIATED SPHERICAL MASSES. GELAT. PYROELECT. ZEOL-ASSOC	W336 4-359
5 -5.5	EDINGTONITE (ZEOLITE)	TETR.	2.7 -2.8	TSI	FIBR GRP. PSEUDOTETRAGONAL. PRISM CLVG. GELAT. PYROELECTRIC. ZEOL- ASSOC, ORE DEPOSITS	W338 4-359
5 -5.5	SCHIZOLITE	TRCL.	2.97-3.13	ISI	(MANGANOAN PECTOLITE). ALTER-BRWN. RADIATING AGGS. ALK SYENITES	W462 2-177
5 -5.5	ALLUAUDITE	UNKN	3.4 -3.5	PHO	SUBTRANSLUCENT. MSV, GRNLR, FIBR-GLBL AGGS. 2 CLVGS. SOL-ACIDS. ALTERATION PRODUCT OF VARULITE-HUHNERKOBELITE (PEGMATITES)	38.1.4.1
5 -5.5	MANGAN-ALLUAUDITE	UNKN.	3.4 -3.5	PHO	SUBTRANSLUCENT. DISTINGUISHED FROM ALLUAUDITE BY CHEM ANALYSIS	38.1.4.2
5 -5.5	KAINOSITE	ORTH.	3.34-3.61	CSI	SHORT PRISMS. 1 CLVG. EFF-HCL. ORE VEINS-RARE(E.G.,N. BURGESS,ONT.)	M463 N
5 -5.5	TRIPLITE	MNCL.	3.5 -3.9	PHO	SUBTRANSLUCENT. MSV. VITR-RESIN. 3 CLVGS. UNEVN. SOL-ACIDS. PEGS	41.6.2.
5 -5.5	GOETHITE	ORTH.	3.3 -4.29	OXD	SUBMET. YELLOWISH STRK. MSV, FIBR, ETC. SILKY-ADMN. SOL-HCL. WXING DEPOSITS	7.1.2.2
5 -5.5	PYROCHLORE	ISOM.	4.45	OXD	OCTAH. VITR. SUBCONCH. DCMP-H2SO4(SLOW). ALK PEGS + IG RKS	8.1.1.1
5 -5.5	MONAZITE	MNCL.	4.6 -5.4	PHO	TBLR-STRIATED. WAXY-ADMN. CLVGS. UNEVN. DCMP-ACIDS(SLOW). ACCESS IN GRANITIC IG + GNEISSIC RKS, PEGS, DETRITAL SEDIMENTS	38.4.2.
5 -5.5	MICROLITE	ISOM.	6.38-6.46	OXD	OCTAH. VITR. SUBCONCH. DCMP-H2SO4(SLOW).GRANITIC PEGS WITH TA MINS	8.1.1.2
5 -6	MARIALITE (SCAPOLITE)	TETR.	2.50-2.62	TSI	ME-0-20. PURE MARIALITE UNKNOWN IN NATURE	W352 4-321
5 -6	DIPYRE (SCAPOLITE)	TETR.	2.57-2.69	TSI	ME-20-50.	W352 4-321

157

NONMETALLIC BROWN

HARDNESS	NAME	XL. SYS.	SPECIFIC GRAVITY	CHEM. CLASS	REMARKS	REFERENCES
5 -6	MIZZONITE (SCAPOLITE)	TETR.	2.67-2.74	TSI	ME-50-80.	W352 4-322
5 -6	SCAPOLITE	TETR.	2.50-2.78	TSI	GROUP NAME. MOST TYPES INDISTINGUISHABLE MEGASCOPICALLY BUT S.G. MAY BE INDICATIVE. LONG,STRIATED PRISMS + AGGS OF COARSE XLS. SUBCONCH. POOR CLVG GIVING IRREGULAR, STRIATED-APPEARING SURFACES. MARIALITE IS INSOL; MEIONITE IS DCMP IN HCL. MOST FLUO--YEL-RED(ULTRAVIOLET,LW). CNTCT META CALC RKS,PEGS	W352 4-321
5 -6	(WERNERITE) (SCAPOLITE)	TETR.	2.50-2.78	TSI	GROUP NAME NO LONGER USED	W352 4-322
5 -6	MEIONITE (SCAPOLITE)	TETR.	2.73-2.78	TSI	ME-80-100. PURE MEIONITE UNKNOWN IN NATURE	W352 4-321
5 -6	AKERMANITE	TETR.	2.94	SSI	ISOMORPHOUS WITH GEHLENITE + MELILITE. RARE IN NATURE, IN SLAGS	W473 1-236
5 -6	GEHLENITE	TETR.	3.04	SSI	SHORT SQ PRISMS. VITR-RESIN.CONCH-GELAT. CNTCT META-IMPURE CO3-RKS	W473 1-236
5 -6	MELILITE	TETR.	2.95-3.05	SSI	SHORT PRISMS. VITR-RESIN. BASAL CLVG. GELAT. BASIC,CA-RICH ALK IG	W473 1-236
5 -6	PARGASITE (AMPHIBOLE)	MNCL.	3.05	ISI	META IMPURE CALC RKS + SKARNS. DISTINGUISHED FROM HBLDE OPTICALLY	W435 2-264
5 -6	ENSTATITE (ORTHOPYROXENE)	ORTH.	3.21	ISI	MG PYROXENE. WITH OLIVINE IN BASIC + ULTRABASIC IG RKS, GRANULITES	W405 2-8
5 -6	KAERSUTITE (AMPHIBOLE)	MNCL.	3.2 -3.28	ISI	VOLCANICS (ESP ALKALIC ONES), SOME APPEARS TO BE SECONDARY	W437 2-321
5 -6	BASALTIC HORNBLENDE (AMPHIBOLE)	MNCL.	3.19-3.30	ISI	ALSO CALLED OXYHORNBLENDE. CHIEFLY IN VOLCANIC IGNEOUS ROCKS	W437 2-315
5 -6	HARDYSTONITE	TETR. A	3.40	SSI	GEHLENITE GROUP. GRNLR. SOME FLUO-PURPLE(SHORT-WAVE). RARE(FRNKLN)	W473 N
5 -6	BRONZITE (ORTHOPYROXENE)	ORTH.	3.3 -3.45	ISI	SUBMET.(BRONZY). FS12-30. IN BASIC + ULTRABSIC IG RKS, SOME METAS	W405 2-9
5 -6	RICHTERITE (AMPHIBOLE)	MNCL.	2.97-3.45	ISI	THERMALLY META LS + SKARNS, HYDROTHERMAL VNS IN ALK IGNEOUS ROCKS	W435 2-352
5 -6	AMPHIBOLE		3.0 -3.57	ISI	GROUP NAME. SEVERAL END AND OTHER MEMBERS WITH NAMES UNSETTLED FOR SOME. MANY MEGASCOPICALLY INDISTINGUISHABLE BUT SOME COLORS AND OCCURRENCES ARE INDICATIVE. ORTH AND MNCL SERIES. LONG SLENDER XLS. PRISM CLVG AT 56° +124°. SOL-HF(SLOWLY). ALTERS TO BIOTITE +/OR CHLORITE. IGNEOUS AND METAMORPHIC RKS, VNS	W422 2-203
5 -6	HYPERSTHENE (ORTHOPYROXENE)	ORTH.	3.43-3.60	ISI	FS30-50. BASIC(E.G.,NORITE) + ULTRABASIC IG RKS, SOME META ROCKS	W405 2-9
5 -6	CUMMINGTONITE (AMPHIBOLE)	MNCL.	3.10-3.60	ISI	CNTCT META ZONES, ORE DEPS, + LESS COMMONLY REGIONAL META ROCKS	W427 2-234
5 -6	GRUNERITE (AMPHIBOLE)	MNCL.	3.10-3.60	ISI	TYPICALLY IN METAMORPHOSED IRON-RICH SEDIMENTARY ROCKS	W427 2-234
5 -6	FERROHYPERSTHENE (ORTHOPYROXENE)	ORTH.	3.59-3.75	ISI	FS50-70. BASIC + ULTRABASIC IGNEOUS ROCKS, SOME METAMORPHIC ROCKS	WN 2-9
5 -6	EULITE (ORTHOPYROXENE)	ORTH.	3.73-3.86	ISI	FS70-88. RARE IN GRANITE, IN FE-RICH DOLERITES + META-SED ROCKS	WN 2-9
5 -6	ORTHOFERROSILITE (ORTHOPYROXENE)	ORTH.	3.96	ISI	FE PYROXENE. RARE, IN SOME FE-RICH METAMORPHIC ROCKS	W405 2-8
5 -6	GEIKIELITE	HEXA.	4.05	OXD	SUBMET.BRWN-RED STRK. TBLR. RHOMB CLVG. RARE-GEM GRAVELS(CEYLON)	4.4.1.4
5 -6	HYDROMETAEROLITE	TETR.	4.6	OXD	SUBMET. BOTRYOIDAL, FIBR. CLVG // FIBERS. SOL-HCL. WITH CHALCOPHAN-ITE, ETC.	7.2.2.3
5 -6	PRIORITE	ORTH.	4.85-5.05	OXD	SUBMET. IN SERIES WITH ESCHYNITE(SEE). RED-YEL STRK. GRANITIC PEGS	8.3.5.2
5 -6	ESCHYNITE	ORTH.	5.14-5.24	OXD	SUBMET., IN SERIES WITH PRIORITE. BLK-BRWN STRK. PRISM, MSV, WAXY, EXPOS--DULL. POOR CLVG. METAMICT.ALK(E.G., NEPH SYENITE) PEGS	8.3.5.1
5 -6	"SAMARSKITE"	UNKN.	5.67	OXD	SUBMET. DISTINGUISHED FROM SAMARSKITE ONLY BY X-RAY ANALYSIS	8.3.6.2

158

NONMETALLIC BROWN

H	Mineral	SG	Class	Crystal	Properties	Ref
5 -6	SAMARSKITE	5.69	OXD	ORTH.	SUBMET. RED-BRWN TO BLK STRK. PRISM, MSV. VITR. RDACTV. PEGS	8.3.6.1
5 -6	URANINITE	10.63	OXD	ISOM.	SUBMET. OPQ. OCTAH,ETC., MSV.GRSY-DULL. RDACTV. PEGS, HI+MOD-T VNS	5.1.2.1
5 -6.5	MAGNESIUM ORTHITE	3.90	SSI	MNCL.	EXTREMELY SIMILAR TO ORTHITE MEGASCOPICALLY	W452 N
5 -6.5	ORTHITE (ALLANITE)	2.8 -4.2	SSI	MNCL.	SUBTRANSLUCENT. TABULAR, ACICULAR, MSV. PITCHY-RESIN. SUBCONCH. RADIOACTIVE. COMMONLY METAMICT. GELAT. ACCESS IN IG ROCKS, PEGS	W451 1-211
5 -7	PYROXENE	2.96-3.96	ISI		GROUP NAME. SEVERAL END AND OTHER MEMBERS WITH NAMES UNSETTLED FOR SOME. MANY MEGASCOPICALLY INDISTINGUISHABLE; OCCURRENCE MAY BE INDICATIVE. ORTH + MNCL SERIES,SHORT PRISMS,PRISM CLVG AT 87° + 92°. ALTERS TO AMPHIBOLE,IG + META RKS	W402 2-1
5 -7	OLIVINE	3.22-4.39	NSI	ORTH.	GRP NAME FOR MINS OF FAYALITE-FORSTERITE + FAYALITE-TEPHROITE SERIES, MONTI-CELLITE + GLAUCOCHROITE. GRNLR MASSES, DISSEM. CONCH. GELAT. COMMONLY ALTERED TO SERPENTINE. BASIC + ULTRABASIC IG RKS, CNTCT META DOLOMITIC LS, METEORITES	W498 1-2
5.5	NOSEAN	2.30-2.40	TSI	ISOM.	SODALITE(SEE) GROUP PROPS. IN PHONOLITE. DISSOLVE-HNO3, EVAPORATE SLOWLY,NO XLS REMAIN, ADD CACL2, HALITE + GYPSUM FORM	M350 4-289
5.5	GRIPHITE	3.40	PHO	ISOM.	MSV. RESIN-VITR. UNEVN-CONCH. SOL-ACIDS. PEGMATITES	41.5.11.
5.5	FERSMANITE	3.44	NSI	MNCL.	PSEUDOTETRAGONAL XLS. RARE - WITH NEPHELINE + ACMITE IN ALK RKS	W527 N
5.5	AUGITE (CLINOPYROXENE)	3.23-3.52	ISI	MNCL.	THE COMMON PYROXENE OF SUB-ALKALIC IGNEOUS RKS(E.G., GABBROS)	W411 2-109
5.5	HELLANDITE	3.35-3.70	NSI	MNCL.	TBLR. SOL-HCL. COMMONLY METAMICT. PEGS WITH ALLANITE, ETC.-RARE	W518 N
5.5	JOAQUINITE	3.89	SSI	ORTH.	TYPICALLY HONEY-YEL. TBLR-EQUANT XLS. WITH BENITOITE IN VEIN	W463 N
5.5	NAGATELITE	3.91	SSI	MNCL.	CONSIDERED TO BE OF ORTHITE SERIES OF EPIDOTE GROUP. RARE IN PEGS	W452 N
5.5	TARAMELLITE	3.92	ISI	ORTH.	REDDISH BRWN. FIBR. CLVG // FIBERS. CNTCT METAMORPHOSED LIMESTONE	W401 N
5.5	PEROVSKITE	3.97-4.05	OXD	MNCL.	SUBMET. CUBIC, OCTAH, GRNLR. ADMN. 1 CLVG. GELAT. RARE IN PEGMATITES	7.4.2.1
5.5	TRITOMITE	4.2	CSI	HEXA.	ACUTE TRIGONAL PYRAMIDS. 1 POOR CLVG. GELAT. RARE-LANGESUND, NORWAY	W509 N
5.5	TSCHEFFKINITE	4.3 -4.65	SSI	MNCL.	SUBTRANSLUCENT. VELVET BLK. MSV. GELAT. ALTER PROD. (=CHEVKINITE)	W443 N
5.5	PERRIERITE	4.3 -4.7	SSI	MNCL.	DISTINGUISHED FROM TSCHEFFKINITE ONLY BY X-RAY ANAL-AFTER HEATING	WN N
5.5	HAUSMANNITE	4.83-4.85	OXD	TETR.	SUBMET. PSEUDO-OCTAH, MSV. BASAL CLVG. SOL-HOT HCL. HI-T MN DEPS	7.2.2.1
5.5	CERITE	4.65-4.91	SSI	MNCL.	SHORT PRISM, GRNLR. GELAT. WITH ALLANITE ETC. IN GNEISS + GRANITE	M507 N
5.5	STIBIOCOLUMBITE	5.68	OXD	ORTH.	IN SERIES WITH STIBIOTANTALITE, WHICH SEE. RARE IN PEGMATITES	8.1.8.2
5.5	DJALMAITE	5.75-5.88	OXD	ISOM.	MODIFIED OCTAH. SUBCONCH. (=TA-BETAFITE?). RARE-GRANITIC PEGMATITE	8.4.2.
5.5	STIBIOTANTALITE	7.34	OXD	ORTH.	IN SERIES WITH STIBIOCOLUMBITE. YEL, YEL-BRWN STRK.PRISM-STRIATED. RESIN-ADMN. 1 GOOD 1 POOR CLVGS. SOL-HF. PYROELECT. PEGS, PLACERS	8.1.8.1
5.5-6	MONTEBRASITE	2.98	PHO	TRCL.	AMBLYGONITE SERIES(SEE FOR PROPERTIES + OCCURRENCES.)	41.5.5.2
5.5-6	NATROMONTEBRASITE	3.04-3.1	PHO	TRCL.	AMBLYGONITE SERIES(SEE FOR PROPERTIES + OCCURRENCES.)	41.5.5.3
5.5-6	AMBLYGONITE	3.11	PHO	TRCL.	EQUANT, LARGE MASSES. VITR,PEARLY. CLVGS. SOL-ACID(DIFFICULTY). LI-, PO4-RICH PEGS	41.5.5.1
5.5-6	BABINGTONITE	3.36	ISI	TRCL.	SMALL NEARLY EQUANT STRIATED XLS. BRILLIANT,GLASSY. 1 GOOD,1 POOR CLVGS. FUSES TO BLK MAG GLBL. IN CAVITIES, COMMONLY WITH ZEOLITES	W462 N
5.5-6	WOEHLERITE	3.42	NSI	MNCL.	PRISMATIC, TBLR. 1 GOOD CLVG. SOL-HCL. PEGS + SOME SILICIC IG RKS	W516 N

NONMETALLIC BROWN

HARDNESS	NAME	XL. SYS.	SPECIFIC GRAVITY	CHEM. CLASS	REMARKS	REFERENCES
5.5-6	ANTHOPHYLLITE (AMPHIBOLE)	ORTH.	2.85-3.57	ISI	WITH CORDIERITE IN GNEISSES, WITH TALC IN META ULTRABASIC IG RKS	W425 2-211
5.5-6	GEDRITE (AMPHIBOLE)	ORTH.	2.85-3.57	ISI	RELATED IN COMPOSITION + OCCURRENCE TO ANTHOPHYLLITE	W425 2-211
5.5-6	PYROXMANGITE	TRCL.	3.61-3.80	ISI	PINK, EXPOS-BRWN OR BLK. F-BLK MAG GLBL. META + METASOMATIC MN-RKS	W460 2-196
5.5-6	ANATASE	TETR.	3.90	OXD	SUBMET. PALE YEL STRK. ACUTE PYRAMIDAL. ADMN. BASAL + PYRAMIDAL CLVG. CONVERTS TO RUTILE ON HEATING. VNS, ACCESS-IG RKS, DETRITAL	4.5.2.
5.5-6	BROOKITE	ORTH.	4.08-4.20	OXD	GRAY TO YEL STRK.TBLR XLS. ADMN. POOR CLVGS. VNS, DISSEM, DETRITAL	4.5.3.
5.5-6.5	OPAL	TETR.	1.8 -2.25	SIL	SUBMICROXLINE QTZ + H2O.RESIN.PEARLY. MSV,MAMILLARY CRUSTS,CONCH.SOL-HF +CAUS- TIC ALKS.SOME FLUO-YEL-GREEN. INCRUSTATIONS, IN CAVITIES(ESP IN VOLCANIC RKS)	D-111 287
5.5-6.5	BUSTAMITE	TRCL.	3.32-3.43	ISI	PINK, EXPOS-FADES. 1 PERF +OTHER CLVGS. MN-ORE DEPS(ESP META ONES)	W457 2-191
5.5-6.5	SALITE (CLINOPYROXENE)	MNCL.	3.2 -3.6	ISI	INTERMEDIATE BETWEEN DIOPSIDE + HEDENBERGITE. META CA-RICH SED + IGNEOUS ROCKS	W411 2-42
5.5-6.5	FERROSALITE (CLINOPYROXENE)	MNCL.	3.2 -3.6	ISI	INTERMEDIATE BETWEEN DIOPSIDE + HEDENBERGITE. OCCURS LIKE SALITE	W411 2-42
5.5-6.5	RHODONITE	TRCL.	3.57-3.76	ISI	TBLR(COMMONLY ROUGH). MSV. ALTERS READILY TO BLK MN-OXIDES	W459 2-182
5.5-6.5	YTTRIALITE	MNCL.	4.3 -4.6	SSI	PRISMATIC, MASSIVE. RARE - WITH GADOLINITE, ETC. IN PEGMATITES	W477 N
5.5-6.5	MAGNESIOFERRITE	ISOM.	4.56-4.65	OXD	SUBMET. MAGNETITE SERIES OF SPINEL GROUP. BLK STRK. GRNLR, MSV. DIFFICULTY SOL-HCL. STRONGLY MAGNETIC. FUMAROLE DEPOSITS	7.2.1.5
5.5-6.5	JACOBSITE	ISOM.	4.76	OXD	SUBMET. MAGNETITE SERIES OF SPINEL GROUP. BRWN STR. GRNLR, MSV. SOL-HCL. WEAKLY MAGNETIC. MN DEPS, E.G., LANGBAN, SWEDEN	7.2.1.8
5.5-6.5	MAGNETITE	ISOM.	5.18	OXD	SUBMET. MAGNETITE SERIES OF SPINEL GROUP. BLK STRK. OCTAH, GRNLR. OCTAH PARTING. DIFFICULTY SOL-HCL. STRONGLY MAGNETIC. WIDESPREAD	7.2.1.6
5.5-6.5	FRANKLINITE	ISOM.	5.07-5.22	OXD	SUBMET. MAGNETITE SERIES OF SPINEL GROUP. RED-BRWN STRK. OCTAH. OCTAH PARTING. SOL-HCL. WEAKLY MAGNETIC. RARE-FRANKLIN, N.J.	7.2.1.7
5.5-6.5	FERGUSONITE	TETR. C	5.38	OXD	SUBMET. IN SERIES WITH FORMANITE. PRISM. COMMONLY HEMIHEDRAL, IR- REGULAR MASSES. VITR. METAMICT. DCMP-HF. GRANITIC PEGMATITES	8.1.3.1
5.5-6.5	ROMEITE	ISOM.	4.7 -5.4	ANT	OCTAH,MSV. GRSY. OCTAH CLVG. DIVERSE(E.G.,MN DEPS)	44.1.1.2
5.5-6.5	EUXENITE	ORTH.	4.9 -5.9	OXD	SUBMET. IN SERIES WITH POLYCRASE. STRK YEL,GRAY,RED-BRWN. PRISM, MSV., GRSY. CONCH. DCMP-HOT HCL, HF, H2SO4. GRANITIC PEGMATITES	8.3.3.1
5.5-6.5	POLYCRASE	ORTH.	4.9 -5.9	OXD	SUBMET. IN SERIES WITH EUXENITE, WHICH SEE. PEGMATITES	8.3.3.2
5.5-6.5	FORMANITE	TETR. C	7.03	OXD	SUBMET. IN SERIES WITH FERGUSONITE,WHICH SEE. PLACERS(W. AUSTRALIA)	8.1.3.2
6	SANIDINE (K-FELDSPAR)	MNCL.	2.56-2.62	TSI	TBLR, GLASSY. OPTICALLY DISTINCT. VOLCANIC ROCKS	W305 4-7
6	CATAPLEIITE	HEXA.	2.75	CSI	THIN PLATES. 2 PERF CLVGS. RARE-ALKALIC RKS, INCLUDING PEGMATITES	W454 N
6	GORCEIXITE	HEXA.	3.04-3.19	PHO	MICROXLINE GRAINS. VITR-DULL. SUBCONCH. IN DIAMANTIFEROUS SANDS	41.5.8.2
6	CLINOENSTATITE (CLINOPYROXENE)	MNCL.	3.19	ISI	HI-T FORM. RARE IN IGNEOUS ROCKS, IN METEORITES.	M408 N
6	PUMPELLYITE	MNCL.	3.18-3.23	SSI	FIBR-PLTY. 1 GOOD + 1 POOR CLVG. CAVITIES + LOW-GRADE META ROCKS	M519 1-227
6	HUMITE	ORTH.	3.20-3.32	NSI	GROUP NAME, ALSO APPLIED TO SPECIES OF GROUP. IRREGULAR-SHAPED GRAINS. VITR- RESIN. GELAT. SIMILAR TO OLIVINE. CONTACT METAMORPHOSED DOLOMITIC LIMESTONES	M514 1-49
6	CLINOHUMITE	MNCL.	3.21-3.35	NSI	COMPLEX XLS. GELAT. SKARNS, TALC SCHISTS, + SERPENTINITES	M515 1-49

NONMETALLIC BROWN

H	Mineral	System	Sp.Gr.	Class	Description	Ref
6	ZOISITE	ORTH.	3.15-3.37	SSI	DISTINGUISHED FROM CLINOZOISITE BY OPTICAL OR X-RAY ANAL. SAME OCCUR	W446 1-185
6	RAMSAYITE	ORTH.	3.43	ISI	ACICULAR. 1 GOOD CLVG. VEINS IN IGNEOUS ROCKS(E.G.,KOLA PENINSULA)	W401 N
6	DANALITE	ISOM.	3.28-3.44	TSI	HELVITE(SEE) GRP PROPS. FLESH-RED TO GRAY. OCTAH. GRANITES, SKARNS	W351 4-303
6	HELVITE	ISOM.	3.20-3.44	TSI	TETRA,MSV. POOR OCTAH CLVG.GELAT(H2S↑),PYROELECTRIC.BOIL IN HCL + AS2O3,DECANT, WASH,HELVITE IS STAINED CANARY YELLOW (AS2S3). RARE-GNEISS,GRANITE,CNTCT ZONES	W351 4-303
6	PIGEONITE (CLINOPYROXENE)	MNCL.	3.30-3.46	ISI	COMMON PYROXENE OF MANY DOLERITES(INCLUDING DIABASES) + BASALTS	W408 2-143
6	EPIDOTE	MNCL.	3.38-3.49	SSI	GROUP NAME. NO GENERALLY ACCEPTED NOMENCLATURE FOR GROUP MEMBERS. CLOSELY RELATED TO ZOISITE + ALLANITE. PISTACHIO-BLACKISH GREEN-BROWNISH. LONG THIN GROOVED XLS, GRNLR MASSES. VNS, OTHER CAVITIES, REGIONAL META + IG ROCKS	W448 1-193
6	LAVENITE	MNCL.	3.5	NSI	PRISMATIC,TBLR. 1 GOOD CLVG. SOL-HCL(SLOWLY). ALKALIC IGNEOUS RKS	W517 N
6	PIEMONTITE	MNCL.	3.45-3.52	SSI	MN-BEARING EPIDOTE. COLOR GENERALLY DISTINCTIVE(DKR THAN THULITE)	W448 1-194
6	CLINOHYPERSTHENE (CLINOPYROXENE)	MNCL.	3.2 -3.55	ISI	HI-T FORM. ONLY IN METEORITES.	W408 N
6	JOHANNSENITE (CLINOPYROXENE)	MNCL.	3.44-3.55	ISI	MN-ANALOG OF DIOPSIDE-HEDENBERGITE. SKARNS, CNTCT ZONES, VEINS	W411 2-75
6	AEGIRINE-AUGITE (CLINOPYROXENE)	MNCL.	3.40-3.55	ISI	TYPICAL PYROXENE OF ALKALIC ROCKS - ESPECIALLY SYENITIC ONES	W411 2-79
6	HEDENBERGITE (CLINOPYROXENE)	MNCL.	3.50-3.56	ISI	BRWNISH OR DK GREEN TO BLK. SKARNS, META FE-RICH SEDS, ORE DEPS	W411 2-42
6	ACMITE (CLINOPYROXENE)	MNCL.	3.55-3.60	ISI	OCCURRENCES LIKE AEGIRENE + AEGRINE-AUGITE (ALKALIC SYENITES,ETC.)	W411 2-80
6	ARDENNITE	ORTH.	3.6 -3.65	SSI	PRISM XLS. 1 PERF, 1 GOOD-CLVGS. 1 PARTING. RARE-PEGMATITES + VNS	W529 N
6	CLINOFERROSILITE (CLINOPYROXENE)	MNCL.	3.5 -3.7	ISI	HI-T FORM. NOT DEFINITELY IDENTIFIED IN NATURAL MATERIALS	W408 N
6	PSEUDOBROOKITE	ORTH.	4.33-4.39	OXD	SUBMET. TBLR-STRIATED. GRSY. 1 CLVG. VOLCANIC AREAS FUMAROLIC,ETC.	7.5.1.
6	CAPPELENITE	HEXA.	4.4	CSI	GREENISH BRWN. PRISMATIC. RARE-VNS IN SYENITE(LANGESUND, NORWAY)	W509 N
6	COLUMBITE	ORTH.	5.15-5.25	OXD	SUBMET. IN SERIES WITH TANTALITE. RED-BLK STRK. TARN-IRID. TBLR, PRISM. 2 CLEAVAGES. BRITTLE. GRANITIC PEGMATITES	8.3.2.1
6	THOREAULITE	MNCL.	7.6 -7.9	OXD	YEL-GREEN STRK. PRISM. RESIN. 2 CLVGS. RARE-PEG(MONONO, KATANGA)	8.3.7.
6 -6.5	ANORTHOCLASE (PLAGIOCLASE)	TRCL.	2.56-2.62	TSI	DISTINGUISHED FROM ORTHOCLASE + MICROCLINE-OPTICALLY. NA-RICH VOLS	W311 4-7
6 -6.5	MICROCLINE (K-FELDSPAR)	TRCL.	2.56-2.63	TSI	DISTINGUISHED FROM ORTHOCLASE OPTICALLY. PLUTONIC IG RKS, PEGS	W308 4-7
6 -6.5	ORTHOCLASE (K-FELDSPAR)	TRCL.	2.55-2.63	TSI	DISTINGUISHED FROM MICROCLINE OPTICALLY. PLUTONIC IG RKS, PEGS	W303 4-7
6 -6.5	PERTHITE (FELDSPARS)	TRCL.	2.56-2.65	TSI	MEGASCOPICALLY, MICROSCOPICALLY,OR SUBMICROSCOPICALLY INTERDIGITATED MICROCLINE OR ORTHOCLASE + PLAGIOCLASE (TYPICALLY ALBITE)	W299 4-7
6 -6.5	FELDSPAR		2.55-3.39	TSI	INCLUDES PLAGIOCLASE(NA-CA)SERIES, ALKALI(K-NA) AND BA FELDSPARS. MOST EASILY DISTINGUISHED BY NONMEGASCOPIC MEANS. MONOCLINIC ONES HAVE 2 CLVGS AT 90° AND SIMPLE(IF ANY) TWINNING; TRICLINIC ONES HAVE 2 CLVGS UP TO 4° OFF 90° AND TYPICALLY HAVE POLYSYNTHETIC TWINNING (GIVING STRIATED APPEARANCE ON SOME CLVG SURFACES OF PLAGIOCLASE). PEGMATITES; IG, META, AND SOME SED ROCKS	W261 4-1

NONMETALLIC BROWN

HARDNESS	NAME	XL. SYS.	SPECIFIC GRAVITY	CHEM. CLASS	REMARKS	REFERENCES
● 6 -6.5	RUTILE	TETR.	4.21-4.25	OXD	SUBMET. PRISM-STRIATED-ADMN.POOR CLVGS. VNS. ACCESS-META + IG RKS	4.5.1.1
6 -6.5	BRAUNITE	TETR.	4.72-4.83	OXD	SUBMET. OPQ. PYRAMIDAL, GRNLR. PYRAMIDAL CLVG. GELAT. MAG. WITH OTHER MN MINERALS — VEINS, WEATHERING PRODUCT DEPOSITS, ETC.	W60 N
6 -6.5	TAPIOLITE	TETR.	7.85-7.95	OXD	SUBMET. STRK-BRWNISH. PRISM. UNEVN. PEGMATITES, PLACERS	8.3.1.1
6 -6.5	MOSSITE	TETR.	7.85-7.95	OXD	SUBMET. POSSIBLE CB END-MEMBER OF TAPIOLITE SERIES(SEE). RARE-PEGS	8.3.1.2
● 6 -6.5	TANTALITE	ORTH.	7.90-8.00	OXD	SUBMET. IN SERIES WITH COLUMBITE, WHICH SEE. PEGMATITES	8.3.2.2
6 -7	VESUVIANITE	TETR.	3.33-3.43	SSI	PRISM. MSV. 1 POOR CLVG. SUBCONCH. ATTKD-HCL. META-ESP CNTCT LS	W508 1-113
● 6 -7	CASSITERITE	TETR.	6.99	OXD	SUBMET. RADIAL CONCRETIONARY MASSES. ADMN-DULL. HI-T VNS, GREISENS	4.5.1.5
6 -7.5	PYROPE	ISOM.	A 3.58	NSI	MG-AL GARNET. XLS ARE RARE. ALTERS TO KELYPHITE. ULTRABASIC IG RKS	W486 1-97
6 -7.5	GROSSULAR	ISOM.	A 3.59	NSI	CA-AL GARNET. EQUANT XLS. CONTACT + REGIONAL META CALCAREOUS RKS	W489 1-93
6 -7.5	HYDROGROSSULAR	ORTH.	3.13-3.59	NSI	GROSSULAR +H2O. SLOWLY SOL-HCL + HNO3. META CALCAREOUS ROCKS	W493 1-104
● 6 -7.5	ANDRADITE	ISOM.	A 3.86	NSI	CA-FE,TI GARNET. F-MAG GLBL. CONTACT META CALCAREOUS RKS, ALK IGS	W489 1-89
6 -7.5	SPESSARTINE	ISOM.	A 4.19	NSI	MN-AL GARNET. EQUANT CRYSTALS. MN-RICH META RKS + VEINS, PEGS	W486 1-99
6 -7.5	GARNET	ISOM.	3.58-4.32	NSI	GROUP NAME, DOMINANT MOLECULE DETERMINES NAME. DISTINGUISHED BY S.G. + NONMEGA-SCOPIC MEANS. EQUANT XLS.MSV.VITR-DULL.DODEC PARTING. SLIGHTLY- TO IN-SOL IN HF	W483 1-77
6 -7.5	ALMANDINE	ISOM.	A 4.32	NSI	FE-AL GARNET. F-MAG GLBL. META-ARGILLACEOUS RKS, RARE-IG RKS, PEGS	W486 1-85
6.5	AGATE	HEXA.	2.57-2.64	SIL	BANDED OR VARIEGATED-CHALCEDONY(MICROXLINE QTZ). NODULES IN BASALT	D-111 210
● 6.5	CHALCEDONY	HEXA.	2.57-2.64	SIL	MICROXLINE + MICROFIBR QTZ. MAMMILLARY. LO-T VEINS +OTHER CAVITIES	D-111 195
6.5	SARD	HEXA.	2.57-2.64	SIL	BRWNISH ORANGE + RED CHALCEDONY (MICROCRYSTALLINE QUARTZ)	D-111 206
6.5	CARNELIAN	HEXA.	3.15-3.18	SIL	RED TO BRWNISH RED CHALCEDONY (MICROCRYSTALLINE QUARTZ)	D-111 206
6.5	NORBERGITE	ORTH.	3.16-3.26	NSI	TAWNY, CHAMOISLK. DIPYRAMIDAL. GELAT. VITR-RESIN. GELAT. LIKE OTHER HUMITES-SKARNS	W513 1-48
● 6.5	CHONDRODITE	MNCL.	3.16-3.26	NSI	ISOLATED TBLR GRNS. VITR-RESIN. GELAT. LIKE OTHER HUMITES-SKARNS	W513 1-48
6.5	JEREMEJEVITE	HEXA.	3.28	BOR	PRISMS. VITR. CONCH. RARE IN RESIDUUM ABOVE GRANITE(E.SIBERIA)	24.1.7.
6.5	KORNERUPINE	ORTH.	3.27-3.34	NSI	PRISM MASSES. PRISM CLVG. WITH CORDIERITE + SAPPHIRINE IN SCHISTS	W523 N
6.5	GENTHELVITE	ISOM.	3.44-3.70	TSI	HELVITE(SEE) GRP PROPS + OCCURRENCES. STALACTITIC, INCRUSTATIONS	W351 4-303
6.5	KNEBELITE	ORTH.	3.96-4.25	NSI	FAYALITE-TEPHROITE (FE-MN)OLIVINE SERIES HAS NAME KNEBELITE AP- PLIED TO INTERMEDIATE MEMBERS. OCCUR IN META RKS + FE-MN ORE DEPS	W498 1-34
6.5	BADDELEYITE	MNCL.	5.4 -6.02	OXD	FLAT PRISM. VITR. BASAL + 2 POOR CLVGS. DCMP-HOT H2SO4. PLACERS, ASSOCIATED WITH ULTRAMAFIC IGNEOUS ROCKS	5.1.1.
● 6.5-7	JASPER	HEXA.	2.57-2.65	SIL	SUBTRANSLUCENT. SOME IS VARIEGATED OR BANDED. MICROGRNLR. CAVITIES	D-111 224
● 6.5-7	AXINITE	TRCL.	3.26-3.36	CSI	STRIATED WEDGE-SHAPED XLS. 1 GOOD CLVG. SOL-HCL(SLOW). CNTCT META	W505 1-320

162

NONMETALLIC BROWN

Hardness	Mineral	Crystal System	SG	Class	Description	Ref1	Ref2
6.5-7	DIASPORE	ORTH. 3.3 -3.5		OXD	PLTY, FOL-MSV. VITR, PEARLY. WITH CORUNDUM IN EMERY ROCK		7.1.2.1
6.5-7	GADOLINITE	MNCL. 4.0 -4.6		NSI	PRISMATIC. GELAT. COMMONLY METAMICT. WITH FLUORITE IN SOME PEGS		W356 N
6.5-7.5	SILLIMANITE	ORTH. 3.23-3.27		NSI	PRISMATIC XLS, FIBR MASSES. VITR. 1 CLVG. HI-GRADE META ROCKS		W520 1-121
7	QUARTZ	HEXA. 2.65		SIL	DOUBLY TERMINATED HEX PRISMS STRIATED ⊥ LENGTH(LO-T), BIPYRAMIDS (HI-T).MSV.VITR.CONCH.SOL-HF.WDSPRD IN MOST ROCK TYPES + VEINS		D-111 9
7	CITRINE	HEXA. 2.65		SIL	YELLOW. XLINE. RARE--TYPICALLY WITH AMETHYST IN CAVITIES IN BASALT		D-111 181
7	SMOKY QUARTZ	HEXA. 2.65		SIL	YEL-BRWN TO BRWNISH BLK. LOSSES COLOR-CA.225°C. HI-T VNS, PEGS		D-111 182
7	DANBURITE	ORTH. 3.0		TSI	PRISM. 1 POOR CLVG. VITR. GELAT(AFTER IGN). PEGS, WITH FELDSPAR IN DOLOMITE		W258 N
7	DRAVITE	HEXA. 3.03-3.15		CSI	NA- MG-TOURMALINE. FUSES AT 4.TYPICALLY BRWN. META CALCAREOUS RKS		W465 1-300
7	TOURMALINE	HEXA. 3.03-3.25		CSI	GROUP NAME. VARIETIES COMMONLY DISTINGUISHABLE MEGASCOPICALLY ON BASIS OF COLOR(S). LENGTHWISE STRIATED PRISMATIC XLS WITH CROSS-SECTIONS THAT RESEMBLE SPHERICAL TRIANGLES. SOME XLS EXHIBIT BOTH LENGTHWISE + CONCENTRIC COLOR ZONING. VITR-RESIN. BRTL. ELECTRICALLY CHARGED ON HEATING + COOLING. MG-RICH VARIETIES FLUO-YEL (ULTRAVIOLET - SW). PEGS, HI-T VNS, IG AND META ROCKS		W465 1-300
7.5	PHENAKITE	HEXA. 2.98		NSI	RHOMBOHEDRAL(FLAT)-PRISMATIC. VITREOUS. PEGMATITES		W496 N
7.5	STAUROLITE	ORTH. 3.74-3.83		NSI	PRISM-CRUCIFORM TWINS COMMON. FIRES TO MAG POWDER. SCHIST + GNEISS		W522 1-151
7.5	ZIRCON	TETR. 4.6 -4.7		NSI	TERMINATED PRISMS. ADMN. SOME FLUO--YEL-ORANGE. ACCESS-IG, SANDS		W494 1-59
7.5-8	SPINEL	ISOM. 3.55		OXD	GROUP NAME. SOME HAVE DISTINCTIVE COLORS OR OTHER HAND-SPECIMEN PROPS. OCTAH. GRNLR, GLASSY. CONCH. SOME FLUO--RED TO YEL-GREEN. IG,META(ESP CALC RKS),PEGS		7.2.1.1
7.5-8	GAHNITE	ISOM. 4.62		OXD	SPINEL(SEE). OCTAH. SCHISTS, PEGS, CNTCT META-LS		7.2.1.3
10	DIAMOND	ISOM. 3.50-3.53		ELE	OCTAHEDRAL. ADMN-GRSY. BRTL. OCTAH CLVG. ULTRABASIC IGS + PLACERS		1.2.4.1

Table III.

MINERALS
arranged according to
Chemical Compositions

[Minerals on the main part of this table with indicated compositions different from those reported for the "New - minerals considered to be valid species" in Fleischer's compilation (*American Mineralogist*, 1966, v. 51, p. 1247-1357) - printed after the first edition of this table - are marked with a "√" if the alternative composition contains the element of the heading, with a "-" if it does not. The alternative compositions reported by Fleischer are given as an addendum to this table, on pages 208 and 209.

In addition, a sidewise caret ">" is placed in the main table where a mineral, which on the basis of the alternative compositions, should be inserted.]

ALUMINUM

Adularia ..$KAlSi_3O_8$
(K-Feldspar)
Aegirine-augite ..$(Na,Ca)(Fe,Mg,Al,Fe)(Si_2O_6)$
(Clinopyroxene)
Albite..$NaAlSi_3O_8$
(Plagioclase)
Allanite..$(Ca,Ce,La)_2(Al,Fe,$
(=Orthite) $Be,Mg,Mn)_3(OH)Si_3O_{12}$ (?)
Allophane..$Al_4(Si_4O_{10})(OH)_8$ (?)
(Clay)
Almandine ..$Fe_3Al_2(SiO_4)_3$
(Garnet)
Aluminite ..$Al_2(SO_4)(OH)_4 \cdot 7H_2O$
✓ Alumohydrocalcite..$CaAl_2(CO_3)_2(OH) \cdot 2H_2O$ (?)
Alunite ..$KAl_3(SO_4)_2(OH)_6$
Alunogen ..$Al_2(SO_4)_3 \cdot 18H_2O$
Amazonite ..$KAlSi_3O_8$
(K-Feldspar)
Amblygonite ..$(Li,Na)Al(PO_4)(F,OH)$
Amesite..$(Mg_4Al_2)(Si_2Al_2)O_{10}(OH)_8$
(Chlorite)
Aminoffite ..$Ca_8Be_3Al(OH)_3Si_8O_{28} \cdot 4H_2O$ (?)
Ammonia alum..$(NH_4)Al(SO_4)_2 \cdot 12H_2O$
Amphibole..hydrous Fe,Mg,Na,Ca,Mn,K, Al silicates
Analcime ..$Na(AlSi_2O_6) \cdot H_2O$
(Zeolite)
Anauxite..$(Al,H_3)_4(Si_4O_{10})(OH)_8$ (?)
(Clay)
Andalusite..$Al_2O(SiO_4)$
Andesine..An_{30-50} *
(Plagioclase)
Anorthite..$CaAl_2Si_2O_8$
(Plagioclase)
Anorthoclase..$(Na,K)AlSi_3O_8$
(K-Feldspar)
✓ Anthoinite..$Al(WO_4)(OH) \cdot H_2O$
Apjohnite..$MnAl_2(SO_4)_4 \cdot 22H_2O$
Ardennite..$Mn_5Al_5(V,As)(OH)_2Si_5O_{24} \cdot 2H_2O$ (?)
Arfvedsonite.. $.5Na_5Ca(Fe,Mg,Fe,Al)_{10}$
(Amphibole) $Si_{15}AlO_{44}(OH,F)_4$
Armenite..$BaCa_2Si_8Al_6O_{28} \cdot 2H_2O$ (?)
Ashcroftine..$KNa(Ca,Mg,Mn)(Al_4Si_5O_{18}) \cdot 8H_2O$
(Zeolite)
Augelite..$Al_2(PO_4)(OH)_3$
Augite..$(Ca,Mg,Fe,Fe,Ti,Al)_2(Si,Al)_2O_6$
(Clinopyroxene)
Axinite..$(Ca,Mn,Fe)_3Al_2B_2O_3(Si_4O_{12})(OH)$
Banalsite..$BaNa_2(Al_2Si_2O_8)_2$
Barkevikite..$Ca_2(Na,K)(Fe,Mg,Mn)_5$
(Amphibole) $Si_{6.5}Al_{1.5}O_{22}(OH)_2$
Barroisite..Na,K,Ca,Mg,Fe,Ti alumino-silicate
(Amphibole)
Basaltic hornblende..$Ca_2(Na,K)_{0.5-1.0}$
(Amphibole) $(Mg,Fe)_{3-2}(Fe,Al)_{2-3}$
$Si_6Al_2O_{22}(O,OH,F)_2$

Basaluminite..$Al_4(SO_4)(OH)_{10} \cdot 5H_2O$
Batavite..hydrous Mg,Al silicate
(Clay)
Bauxite..hydrous Al oxides
Bavenite..$Ca_4(OH)_4Si_9Al_2BeO_{24}$ (?)
Beaverite..$Pb(Cu,Fe,Al)_3(SO_4)_2(OH)_6$
Beidellite..montmorillonite-like clay
(Clay)
Berlinite..$Al(PO_4)$
Beryl..$Be_3Al_2(Si_6O_{18})$
Biotite..$K(Mg,Fe,Mn)_3(AlSi_3O_{10})(OH,F)_2$
(Mica)
Bityite..$Ca_4(Li,Be)_4Al_8(OH)_{20}$
(Mica) $[(Si,Al)_4O_{10}]_3$ (?)
Boehmite..$AlO(OH)$
Brazilianite..$NaAl_3(PO_4)_2(OH)_4$
Brewsterite..$(Sr,Ba,Ca)(Al_2Si_6O_{16}) \cdot 5H_2O$
(Zeolite)
Brittle Mica ..hydrous Na,Ca,Mg,Al silicates
Brunsvigite..hydrous Fe,Mg,Al silicate
(Chlorite)
Byssolite..hydrous Ca,Mg,Fe alumino-silicate
(Amphibole)
Bytownite...An_{70-90} *
(Plagioclase)
Cadwaladerite..$Al(OH)_2Cl \cdot 4H_2O$
Cancrinite..$(Na,Ca,)_{7-8}(Al_6Si_6O_{24})(CO_3,$
$SO_4,Cl)_{1.5-2} \cdot 1-5H_2O$
Cardenite..hydrous Mg,Fe,Al silicate
(Clay)
Carpholite..$MnAl_2(OH)_4Si_2O_6$
Catoptrite..Mn,Al antimonate-silicate
Celadonite..$(K,Ca,Na)(Al,Fe,Mg)_2$
$(Al_{0.11}Si_{3.89}O_{10})(OH)_2$ (?)
Celesian..$BaAl_2Si_2O_8$
(Feldspar)
Ceruleite..$CuAl_4(AsO_4)_2(OH)_8 \cdot 4H_2O$
Chabazite..$(Ca,Na_2)(Al_2Si_4O_{12}) \cdot 6H_2O$
(Zeolite)
Chalcoalumite..$CuAl_4(SO_4)(OH)_{12} \cdot 3H_2O$
Chalcophyllite..$Cu_{18}Al_2(AsO_4)_3(SO_4)_3(OH)_{27}$
$\cdot 33H_2O$
Chamosite..hydrous Fe,Mg,Al silicate
(Chlorite)
Childrenite..$(Fe,Mn)Al(PO_4)(OH)_2 \cdot H_2O$
Chiolite..$Na_5Al_3F_{14}$
Chloraluminite..$AlCl_3 \cdot 6H_2O$
Chlorite..$(Mg,Al,Fe)_{12}(Si,Al)_8O_{20}(OH)_{16}$
Chloritoid..$(Fe,Mg)_2Al_4(OH)_4Si_2O_{10}$ (?)
Chrysoberyl..$BeAl_2O_4$
Clay..chiefly hydrous silicates of Al or Mg
Clinochlore..hydrous Mg,Fe,Al silicate
(Chlorite)
Clinozoisite..$Ca_2Al_3(OH)Si_3O_{12}$
(Epidote)

---

* On this table An with a subscript refers to anorthite-content.

(ALUMINUM-continued)

Clintonite..$Ca_2(Mg_5Al)Si_2Al_6O_{20}(OH)_4$ (?)
 (Brittle Mica)
Coeruleolactite..$Al_3(PO_4)_2(OH)_3$
Common hornblende..$(Ca,Na,K)_{2-3}(Mg,Fe,Al)_5Si_6(Si,Al)_2O_{22}(OH,F)_2$
 (Amphibole)
Cookeite..Li-bearing chlorite
 (Chlorite)
Cordierite..$Al_3(Mg,Fe)_2(Si_5AlO_{18})$
Corundophilite..hydrous Mg,Fe,Al silicate
 (Chlorite)
Corundum..Al_2O_3
Crandallite..$CaAl_3(PO_4)_2(OH)_5 \cdot H_2O$
Creedite..$Ca_3Al_2F_4(OH,F)_6(SO_4) \cdot 2H_2O$
Crossite..$Na_2(Mg,Fe)_3(Al,Fe)_2Si_8O_{22}(OH,F)_2$
 (Amphibole)
Cryolite..Na_3AlF_6
Cryolithionite..$Na_3Li_3Al_2F_{12}$
Cyanotrichite..$Cu_4Al_2(SO_4)(OH)_{12} \cdot 2H_2O$
Dachiardite..$(Ca,K_2,Na_2)_3Al_4Si_{18}O_{45} \cdot 14H_2O$
 (Zeolite)
Daphnite..hydrous Fe,Al silicate
 (Chlorite)
Davisonite..$Ca_3Al(PO_4)_2(OH)_3 \cdot H_2O$ (?)
Davyne..$(Na,K)_6Ca_2(AlSiO_4)_6(SO_4)_2$ (?)
Dawsonite..$NaAl(CO_3)(OH)_2$
Delessite..hydrous Mg,Fe,Al silicate
 (Chlorite)
✓ Deltaite..$Ca(Al_2Ca)(PO_4)_2(OH)_4 \cdot H_2O$
Diabantite..hydrous Mg,Al silicate
 (Chlorite)
Diallage..$Ca_{14}Fe_2Mg_{13}FeAl_5Si_{29}O_{96}$ (?)
 (Clinopyroxene)
Diaspore..$HAlO_2$
Dickite..$Al_4(Si_4O_{10})(OH)_8$
 (Clay)
Didymolite..$(Ca,Mg,Fe)_2Al_6Si_8O_{27}$ (?)
Dietrichite..$(Zn,Fe,Mn)Al_2(SO_4)_4 \cdot 22H_2O$
Dipyre..Me_{20-50} **
 (Scapolite)
Donbassite..hydrous Na,Ca,Mg,Al silicate
 (Clay)
Dravite..$NaMg_3Al_6(OH)_4(BO_3)_3Si_6O_{18}$
 (Tourmaline)
Dumortierite..$H(Al,Fe)_2Si_3Al_6O_{20}$ (?)
Dundasite..$PbAl_2(CO_3)_2(OH)_4 \cdot 4H_2O$
Durangite..$NaAl(AsO_4)F$
Edenite..$NaCa_2Mg_5AlSi_7O_{22}(OH,F)_2$
 (Amphibole)
Edingtonite..$Ba(Al_2Si_3O_{10}) \cdot 4H_2O$
 (Zeolite)
Ekermannite..$.5Na_5Ca(Mg,Fe,Fe,Al,Li)_{10}Si_{15}AlO_{44}(OH,F)_4$
 (Amphibole)
Ekmannite..$(Fe,Mg,Mn)_3(OH)_2(Si,Al)Si_3O_{10}$ (?)
Elbaite..$Na(Li,Al)_3Al_6(OH,F)_4(BO_3)_3Si_6O_{18}$
 (Tourmaline)
Elpasolite..K_2NaAlF_6

Englishite..$K_2Ca_4Al_8(PO_4)_8(OH)_{10} \cdot 9H_2O$
Enigmatite..$(Ca,Na_2)_2Fe(Al,Fe,Ti)_4(Si_2O_7)_2$ (?)
Ephesite..$(Na,Ca)Al_2[Al(Al,Si)Si_2O_{10}](OH)_2$
 (Brittle Mica)
Epidote..$Ca_2(Al,Fe)Al_2(OH)Si_3O_{12}$
Epistilbite..$Ca(Al_2Si_6O_{16}) \cdot 5H_2O$
 (Zeolite)
Eosphorite..$(Mn,Fe)Al(PO_4)(OH)_2 \cdot H_2O$
Erionite..ca.$(Na_2,K_2,Ca,Mg)_{4.5}Al_9Si_{27}O_{72} \cdot 27H_2O$
 (Zeolite)
Ettringite..$Ca_6Al_2(SO_4)_3(OH)_{12} \cdot 26H_2O$
Euclase..$BeSiAlO_4(OH)$
Eucryptite..$LiSiAlO_4$
Evansite..$Al_3(PO_4)(OH)_6 \cdot 6H_2O$
Fassaite..$Ca(Mg,Fe,Al,Ti)(Si,Al)_2O_6$
 (Clinopyroxene)
Faujasite..ca.$(Na,Ca)_{1.75}Al_{3.5}Si_{8.5}O_{24} \cdot 16H_2O$
 (Zeolite)
Feldspar..K,Na,Ca,Ba alumino-silicates
Felsobanyaite..$Al_4(SO_4)(OH)_{10} \cdot 5H_2O$ (?)
Ferrierite..$(Na,K)_2Mg(Al_3Si_{15}O_{36})(OH) \cdot 9H_2O$
 (Zeolite)
Ferrocarpholite..$FeAl_2(Si_2O_6)(OH)_4$
Ferroedenite..$NaCa_2Fe_5AlSi_7O_{22}(OH,F)_2$
 (Amphibole)
Ferrohastingsite..$NaCa_2Fe_4(Al,Fe)Si_6Al_2O_{22}(OH,F)_2$
 (Amphibole)
Ferrotschermakite..$Ca_2Fe_5Al_2Si_6O_{22}(OH,F)_2$
 (Amphibole)
Florencite..$CeAl_3(PO_4)_2(OH)_6$
Fluellite..$AlF_3 \cdot H_2O$
Gahnite..$ZnAl_2O_4$
 (Spinel)
Galaxite..$MnAl_2O_4$
 (Spinel)
Ganophyllite..$(Na,K)(Mn,Al,Mg,Ca)_3(OH)_4(Si,Al)Si_3O_{10}$ (?)
Garnierite..Ni-bearing chlorite
 (Chlorite)
Gearksutite..$CaAl(OH)F_4 \cdot H_2O$
Gedrite..$(Mg,Fe)_6AlSi_6(Si,Al)_2O_{22}(OH,F)_2$
 (Amphibole)
Gehlenite..$Ca_2(Al_2SiO_7)$
Gibbsite..$Al(OH)_3$
Gismondine..$Ca(Al_2Si_2O_8) \cdot 4H_2O$
 (Zeolite)
Glaucocerinite..$Zn_{13}Al_8Cu_7(SO_4)_2(OH)_{60} \cdot 4H_2O$ (?)
Gmelinite..$(Na_2,Ca)(Al_2Si_4O_{12}) \cdot 6H_2O$
 (Zeolite)
Glauconite..$(K,Ca,Na)(Al,Fe,Mg)_2(Al_{0.35}Si_{3.65}O_{10})(OH)$ (?)
 (Mica)
Glaucophane..$Na_2Mg_3Al_2Si_8O_{22}(OH,F)_2$
 (Amphibole)

** On this table Me with a subscript refers to meionite content.

(ALUMINUM-continued)

Gonnardite..$Na_2Ca[(Al,Si)_5O_{10}]_2 \cdot 6H_2O$
  (Zeolite)
Gorceixite..$BaAl_3(PO_4)_2(OH)_5 \cdot H_2O$
Gordonite..$MgAl_2(PO_4)_2(OH)_2 \cdot 8H_2O$
Goyazite..$SrAl_3(PO_4)_2(OH)_5 \cdot H_2O$
Grandidierite..$H_2Na_2(Mg,Fe)_7(Al,Fe,B)_{15}$
  $Si_7Al_7O_{56}$ (?)
Griphite..$(Na,Ca,Fe,Al)_3Mn_2(PO_4)_{2.5}$
  $(OH,F)_2$ (?)
Grossular..$Ca_3Al_2(SiO_4)_3$
  (Garnet)
Grovesite..$(Mn,Mg,Al)_6(Si,Al)_4O_{10}(OH)_8$
Guembelite..$(K,H_2O)(Al_{1.5}Mg_{0.5})$
  $(AlSi_3O_{10})(OH,H_2O)_2$ (?)
✓ Guildite..$(Cu,Fe)_3(Fe,Al)_4(SO_4)_7(OH)_4 \cdot 15H_2O$
Halloysite..ca.hydrated kaolinite
  (Clay)
Halotrichite..$FeAl_2(SO_4)_4 \cdot 22H_2O$
✓ Harkerite..$Ca(Mg,Al)(Si,BH)O_4 \cdot CaCO_3$
Harmotome..$Ba(Al_2Si_6O_{16})6H_2O$
  (Zeolite)
Harstigite..$(Ca,Mg,Mn)_8Al_2(OH)_4Si_6O_{21}$ (?)
Hastingsite..$NaCa_2Mg_4Al_3Si_6O_{22}(OH)_2$
  (Amphibole)
Hauyne..$(Na,Ca)_{4-8}Al_6Si_6O_{24}(SO_4,S)_{1-2}$
Hellandite..$(Ca,Y,Er,Mn)_3(Al,Fe)Si_2O_4$
  $\cdot H_2O$ (?)
Hematolite..$(Mn,Mg)_4Al(AsO_4)(OH)_8$
Hercynite..$FeAl_2O_4$
  (Spinel)
Heulandite..$(Ca,Na_2)(Al_2Si_7O_{18}) \cdot 6H_2O$
  (Zeolite)
Hinsdalite..$(Pb,Sr)Al_3(PO_4)(SO_4)(OH)_6$
Hoegbomite..$Mg(Al,Fe,Ti)_4O_7$
Hornblende, basaltic..$Ca_2(Na,K)_{0.5-1.0}$
  (Amphibole)
  $(Mg,Fe)_{3-2}(Fe,Al)_{2-3}$
  $Si_6Al_2O_{22}(O,OH,F)_2$
Hornblende, common..$(Ca,Na,K)_{2-3}(Mg,Fe,Al)_5$
  (Amphibole)
  $Si_6(Si,Al)_2O_{22}(OH,F)_2$
Holmquistite..$Li_2(Mg,Fe)_3(Al,Fe)_2Si_8O_{22}$
  (Amphibole)
  $(OH,F)_2$
Hyalophane..$(k,Na,Ba)AlSi_3O_8$
  (Feldspar)
✓ Hydrobasaluminite..$Al_4(SO_4)(OH)_{10} \cdot 36H_2O$ (?)
Hydrobiotite..$(K,H_2O)(Mg,Fe,Mn)_3(H_2O,OH)_2$
  $AlSi_3O_{10}$ (?)
✓ Hydrocalumite..$Ca_4Al_2(OH)_{14} \cdot 6H_2O$
Hydrogrossular..$Ca_3Al_2(SiO_4)_3(OH)_4$
Hydromuscovite..$(K,H_2O)Al_2(H_2O,OH)_2$
  $AlSi_3O_{10}$ (?)
Hydroparagonite..$(Na,H_2O)Al_2(H_2O,OH)_2$
  (=Brammellite)
  $AlSi_3O_{10}$ (?)

Hydrophlogopite..$(K,H_2O)Mg_3(H_2O,OH)_2$
  $AlSi_3O_{10}$ (?)
Hydrotalcite..$Mg_6Al_2(OH)_{16} \cdot CO_3 \cdot 4H_2O$
Illite..$KAl_4Si_7AlO_{20}(OH)_4$
  (Clay)
Indialite..$Mg_2Al_3(AlSi_5O_{18})$
Jadeite..$NaAl(Si_2O_6)$
  (Clinopyroxene)
Jarlite..$NaSr_3Al_3F_{16}$
Jeremejevite..$AlBO_3$
Jezekite..$Na_4CaAl_2(PO_4)_2(OH)_2F_2O$ (?)
Johnstrupite..$(Ca,Y,Na,Ce)_3(Al,Zr,Ti)(F,OH)$
  Si_2O_8
Kaersutite..$Ca_2(Na,K)(Mg,Fe)_4TiSi_6Al_2O_{22}$
  (Amphibole)
  $(OH,F)_2$
Kalinite..$KAl(SO_4)_2 \cdot 11H_2O$
Kalsilite..$K(AlSiO_4)$
Kaliophilite..$K(AlSiO_4)$
Kaolinite..$Al_4(Si_4O_{10})(OH)_8$
  (Clay)
Karinthine..$(Na,K)Ca_{2-3}Mg_8Fe_{1-2}(Al,Fe,Ti)_2$
  (Amphibole)
  $(Al_{3-4}Si_{13-12}O_{44})(OH)_4$
✓ Karpinskiite..$(Na,K,Zn,Mg)_2(OH,H_2O)_{1-2}$
  $(Al,Be)_2Si_4O_{12}$
Katophorite..$Na_2CaFe_4(Fe,Al)Si_7AlO_{22}(OH,F)_2$
  (Amphibole)
Koenenite..$Mg_5Al_2(OH)_{12}Cl_4$
Kolbeckite..Ca,Be,Al,Fe silicate-phosphate
Kornerupine..$Mg_{10}Al_{10}BO_7Si_9O_{36}$ (?)
Kribergite..$Al_4(PO_4)_2(SO_4)_2(OH)_2 \cdot 8H_2O$ (?)
Kyanite..$Al_2O(SiO_4)$
Labradorite..An_{50-70} *
  (Plagioclase)
✓ Labuntsovite..$(K,Ba,Na)Ti(Si,Al)_2(O,OH)_7$
  $\cdot H_2O$ (?)
Lacroixite..Na,Ca,Al fluo-phosphate
✓ Latiumite..$Ca_6(K,Na)_2Al_4(O,CO_3,SO_4)(SiO_4)_6$
Laumontite..$Ca(AlSi_2O_6)_2 \cdot 4H_2O$
  (Zeolite)
Lawsonite..$CaAl_2(Si_2O_7)(OH)_2 \cdot H_2O$
Lazulite..$(Mg,Fe)Al_2(PO_4)_2(OH)_2$
Lazurite..$(Na,Ca)_8(AlSiO_4)_6(SO_4,S,Cl_2)_4$ (?)
Lehiite..$(Na,K)_2Ca_5Al_8(PO_4)_8(OH)_{12} \cdot 6H_2O$ (?)
Leifite..$Na_2FSi_5AlO_{12}$ (?)
Lepidolite..$K_2(Li,Al)_{5-6}(Si_{6-7}Al_{2-1}O_{20})$
  (Mica)
  $(OH,F)_4$
Leucite..$K(AlSi_2O_6)$
Leucophosphite..$K_2(Fe,Al)_7(PO_4)_4(OH)_{11} \cdot 6H_2O$
Levynite..$Ca(Al_2Si_4O_{12}) \cdot 6H_2O$
  (Zeolite)
Liroconite..$Cu_2Al(AsO_4)(OH)_4 \cdot 4H_2O$
Liskeardite..$(Al,Fe)_3(AsO_4)(OH)_6 \cdot 5H_2O$

(ALUMINUM-continued)

Magnesiokatophorite..$Na_2CaMg_4(Fe,Al)$
(Amphibole) $Si_7AlO_{22}(OH,F)_2$
Magnesium Orthite..$CaCeMgAl_2(OH,F)Si_3O_{12}$ (?)
Manandonite..hydrous Li,B,Al silicate
(Chlorite)
Manasseite..$Mg_6Al_2(OH)_{16} \cdot CO_3 \cdot 4H_2O$
Mansfieldite..$Al(AsO_4) \cdot 2H_2O$
Margarite..$CaAl_2(Al_2Si_2O_{10})(OH)_2$
(Brittle Mica)
Marialite..$Na_8(AlSi_3O_8)_6(Cl_2,SO_4,CO_3)$
(Scapolite)
Meionite..$Ca_8(Al_2Si_2O_8)_6(Cl_2,SO_4,CO_3)_2$
(Scapolite)
Melilite..$(Ca,Na)_2(Al,Mg)(Si,Al)_2O_7$
Melinophane..$(Ca,Na)_2(Be,Al)(Si_2O_6F)$
Mellite..$Al_2C_{12}O_{12} \cdot 18H_2O$
Mendozite..$NaAl(SO_4)_2 \cdot 11H_2O$
Mesolite..$Na_2Ca_2(Al_2Si_3O_{10})_3 \cdot 8H_2O$
(Zeolite)
Mesomicrocline..$K(Al,Si)_2Si_2O_8$
(K-Feldspar)
Metavariscite..$Al(PO_4) \cdot 2H_2O$
Metavauxite..$FeAl_2(PO_4)_2(OH)_2 \cdot 8H_2O$
Mica ..complex hydrous alumino-silicates
Microcline..$KAlSi_3O_8$
(K-Feldspar)
Milarite ..$KCa_2Si_{12}Be_2AlO_{30} \cdot 5H_2O$ (?)
Millisite..$(Na,K)CaAl_6(PO_4)_4(OH)_9 \cdot 3H_2O$
— Minnesotaite..$(Fe,Mg,H_2)_3(Si,Al,Fe)_4O_{10}$
$(OH)_2$ (?)
✓ Minyulite..$KAl_2(PO_4)_2(OH) \cdot 3.5H_2O$ (?)
Mizzonite..Me_{50-80} **
(Scapolite)
Montebrasite..$(Li,Na)Al(PO_4)(OH,F)$
Montgomeryite..$Ca_4Al_5(PO_4)_6(OH)_5 \cdot 11H_2O$
Montmorillonite..$Na_{0.7}(Al_{3.3}Mg_{0.7})Si_8O_{20}$
(Clay) $(OH)_4 \cdot nH_2O$
Mordenite..$(Ca,K_2,Na_2)(AlSi_5O_{12})_2 \cdot 7H_2O$
(Zeolite)
Morinite..Na,Ca,Al fluo-phosphate
Mullite..$3Al_2O_3 \cdot 2SiO_2$
Muscovite..$KAl_2(AlSi_3O_{10})(OH,F)_2$
(Mica)
Nacrite..$Al_4(Si_4O_{10})(OH)_8$
(Clay)
Nagatelite..$(Ca,Ce)_2(Al,Fe)_3(OH)(Si,P)_3O_{12}$
Natroalunite..$NaAl_3(SO_4)_2(OH)_6$
Natrolite..$Na_2(Al_2Si_3O_{10}) \cdot 2H_2O$
(Zeolite)
Natromontebrasite..$(Na,Li)Al(PO_4)(F,OH)$
Naujakasite..$Na_4FeAl_4H_4Si_8O_{27}$
Nepheline..$KNa_3(AlSiO_4)_4$
Nontronite..Montmorillonite-like clay
(Clay)
Nosean..$Na_8(AlSiO_4)_6(SO_4)$

Oligoclase..An_{10-30} *
(Plagioclase)
Omphacite..$(Ca,Na)(Mg,Fe,Fe,Ti,Al)(Si_2O_6)$
(Clinopyroxene)
✓ Osumilite..$(K,Na,Ca)(Mg,Fe)_2(Al,Fe)_3$
$(Si,Al)_{12}O_{30} \cdot H_2O$
Orthite..$(Ca,Ce,La)_2(Al,Fe,Be,Mg,Mn)_3(OH)$
Si_3O_{12} (?)
Orthoclase..$KAlSi_3O_8$
(K-Feldspar)
Overite..$Ca_3Al_8(PO_4)_8(OH)_6 \cdot 15H_2O$
Pachnolite..$NCaAlF_6 \cdot H_2O$
✓ Painite..$Al_2O_3 \cdot Ca_2(Si,BH)O_4$
Paracelsian ..$Ba(Al_2Si_2O_8)$
(Feldspar)
Paragonite..$NaAl_2(AlSi_3O_{10})(OH,F)_2$
(Mica)
✓ Paravauxite..$FeAl_2(PO_4)_2(OH)_2 \cdot 8H_2O$
Pargasite..$NaCa_2Mg_4(Al,Fe)Si_6Al_2O_{22}(OH,F)_2$
(Amphibole)
— Parsettensite..$(K,H_2O)(Fe,Mg,Al,Mn)_3$
$Si_4O_{10}(OH)_2$ (?)
Pennantite..low Mg chlorite
(Chlorite)
Penninite..hydrous Mg,Fe,Al silicate
(Chlorite)
Petalite..$Li(AlSi_4O_{10})$
Perthite ..$KAlSi_3O_8 + NaAlSi_3O_8$
(K-Feldspar)
Phillipsite ..$(.5Ca,Na,K)_2Al_3Si_5O_{16} \cdot 6H_2O$
(Zeolite)
Phlogopite ..$KMg_3(AlSi_3O_{10})(OH,F)_2$
(Mica)
Pickeringite ..$MgAl_2(SO_4)_4 \cdot 22H_2O$
Piemontite..$Ca_2(Al,Fe,Mn)_2Al(OH)Si_3O_{12}$
(Epidote)
Pistacite..$Ca_2(Al,Fe)Al_2(OH)Si_3O_{12}$
(Epidote)
Plumbogummite..$PbAl_3(PO_4)_2(OH)_5 \cdot H_2O$
Pollucite..$(Cs,Na)(AlSi_2O_6) \cdot H_2O$ (?)
Potash alum..$KAl(SO_4)_2 \cdot 12H_2O$
Prehnite..$Ca_{2Al_2}(OH)_2Si_3O_{10}$
Prosopite..$CaAl_2(F,OH)_8$
Pseudothuringite..hydrous Fe,Al,Mg silicate
(Chlorite)
✓ Pumpellyite..$Ca_4(Mg,Fe,Mn)(Al,Fe,Ti)_5(OH)_3$
$Si_6O_{23} \cdot 2H_2O$ (?)
Pyrophyllite..$Al_2(OH)_2(Si_4O_{10})$
Pyrope..$Mg_3Al_2(SiO_4)_3$
(Garnet)
Pyroxene..Mg,Fe,Ca,Na,Ti,Al silicates
Ralstonite..ca.$Na(Mg,Al_5)_6F_{12}(OH)_6 \cdot 3H_2O$
Ransomite..$Cu(Fe,Al)_2(SO_4)_4 \cdot 7H_2O$
Redingtonite..$(Fe,Mn,Ni)(Cr,Al)_2(SO_4)_4$
$\cdot 22H_2O$ (?)
Rhipidolite..hydrous Fe,Mg,Al silicate
(Chlorite)
Rhodizite..$NaKLi_4Al_4Be_3B_{10}O_{27}$ (?)

(ALUMINUM-continued)

Richterite..$Na_2Ca(Mg,Fe,Mn,Al,Fe)_5Si_8O_{22}(OH,F)_2$
 (Amphibole)
—Roscherite..$(Ca,Mn,Fe)_2Al(PO_4)_2(OH)\cdot 2H_2O$
Roscoelite..$KV_2(AlSi_3O_{10})(OH,F)_2$ (?)
 (Mica)
Sanidine..$KAlSi_3O_8$
 (K-Feldspar)
Saponite..$(.5Ca,Na)_{0.7}(Al,Mg,Fe)_4(Si,Al)_8O_{20}(OH)_4\cdot nH_2O$
 (Clay)
Sapphirine..$(Mg,Fe)_{15}(Al,Fe)_{34}Si_7O_{80}$
Sarcolite..$(Ca,Na_2)_3Al_2Si_3O_{12}$ (?)
Sauconite..montmorillonite-like clay
 (Clay)
Scapolite..$(Na,Ca,K)_4Al_3(Al,Si)_3Si_6O_{24}(Cl,F,OH,CO_3,SO_4)$
Schorl..$Na(Fe,Mn)_3Al_6(OH,F)_4(BO_3)_3Si_6O_{18}$
 (Tourmaline)
Scolecite..$Ca(Al_2Si_3O_{10})\cdot 3H_2O$
 (Zeolite)
Scorzalite..$(Fe,Mg)Al_2(PO_4)_2(OH)_2$
Serendibite..$Ca_2(Mg,Fe)_4Al_6B_2O_{10}Si_4O_{16}$ (?)
Sericite..$KAl_2(AlSi_3O_{10})(OH,F)_2$
 (Mica)
Sheridanite..hydrous Mg,Fe,Al silicate
 (Chlorite)
Siderophyllite..$KFe_3(AlSi_3O_{10})(OH,F)_2$
 (Mica)
Sillimanite..$Al_2O(SiO_4)$
✓Simpsonite..$Al_2Ta_2O_8$
Soda Alum..$NaAl(SO_4)_2\cdot 12H_2O$
Sodalite..$Na_8(AlSiO_4)_6Cl_2$
Souzalite..$(Mg,Fe)_3(Al,Fe)_4(PO_4)_4(OH)_6\cdot 2H_2O$
Spangolite..$Cu_6Al(SO_4)(OH)_{12}Cl\cdot 3H_2O$
Spessartine..$Mn_3Al_2(SiO_4)_3$
 (Garnet)
Spinel..$MgAl_2O_4$
Spodumene..$LiAl(Si_2O_6)$
Staurolite..$(Fe,Mg)_2(Al,Fe)_9O_6(O,OH)_2(SiO_4)_4$
✓Steigerite..$Al_2(VO_4)_2\cdot 6.5H_2O$
Sterrettite..$Al_6(PO_4)_4(OH)_6\cdot 5H_2O$
Stilbite..$(Ca,Na_2,K_2)(Al_2Si_7O_{18})\cdot 7H_2O$
 (Zeolite)
Stilpnochlorane..$(Ca,K,H_2O)(Al,Fe,Mg)_3(OH,O)_2(Si,P)_4O_{10}$
Stilpnomelane..$(K,Na,Ca)(Fe,Mg,Al,Mn)_3Si_4O_{10}(OH)_2$ (?)
Svanbergite..$SrAl_3(PO_4)(SO_4)(OH)_6$
Tamarugite..$NaAl(SO_4)_2\cdot 6H_2O$
Taramite..$Na_2Ca_2Mg_5Fe_3(Fe,Ti)_2(Al,Fe)_2(OH,F)_4Si_{14}O_{44}$
 (Amphibole)
Taranakite..$K_2Al_6(PO_4)_6(OH)_2\cdot 18H_2O$ (?)
Tavistockite..$Ca_3Al_2(PO_4)_2(OH)_6$

Thomsenolite..$NaCaAlF_6\cdot H_2O$
Thomsonite..$NaCa_2[(Al,Si)_5O_{10}]_2\cdot 6H_2O$
 (Zeolite)
Thorotungstite..$AlFe(Th,Ca,Ce,Zr)WO_3$ (?)
Thulite..$(Ca,Mn)_2Al_3(OH)(SiO_4)_3$
 (var. zoisite)
Thuringite..hydrous Fe,Mg,Al silicate
 (Chlorite)
Titanaugite..$Ca_{14}NaMg_{10}Fe_3Ti_2Fe_2Al_5Si_{27}O_{96}$ (?)
 (Clinopyroxene)
Topaz..$Al_2SiO_4(F,OH)_2$
Torendrikite..hydrous Na,Ca,Mg,Fe alumino-silicate
 (Amphibole)
Tourmaline..$(Na_2,Ca)(Li,Al,Mg,Fe,Mn)_6Al_{12}(OH,F)_8(BO_3)_6Si_{12}O_{36}$
Trudellite..$Al_{10}Cl_{12}(OH)_{12}(SO_4)_3\cdot 30H_2O$
Tscheffkinite..$(Fe,Ca)(Ce,La,Al)_2(Si,Ti)_3O_{10}$ (?)
Tschermakite..$Ca_2Mg_3Al_4Si_6O_{22}(OH,F)_2$
 (Amphibole)
✓Tuhualite..$(Na_2,K_2,Mn)(Al,Fe,Mg,Ti)_2Si_{10}O_{24}$ (?)
Turquois..$CuAl_6(PO_4)_4(OH)_8\cdot 4H_2O$
Uhligite..$Ca_3(Ti,Al,Zr)_9O_{20}$ (?)
Uralite...hydrous Ca,Mg,Fe.alumino-silicate
 (Amphibole)
Uvite..$CaMg_4Al_5B_3Si_6O_{29}\cdot 2H_2O$ (?)
 (Tourmaline)
Variscite..$Al(PO_4)\cdot 2H_2O$
Vashegyite..$Al_4(PO_4)_3(OH)_3\cdot nH_2O$ (?)
✓Vauxite..$FeAl_2(PO_4)_2(OH)_2\cdot 7H_2O$
Vermiculite..$(Mg,Ca)_{0.7}(Mg,Fe,Al)_6$
Vesuvianite..$Ca_{10}Al_4(Mg,Fe)_2Si_9O_{34}(OH,F)_4$
Vishnewite..$(Na,Ca,K)_{6.7}Al_6Si_6O_{24}(SO_4,CO_3,Cl)_{1-5.5}\cdot 1-5H_2O$
Volchonskoite..$(Ca,Mg,Cr,Fe,Al)_2O_3\cdot 3SiO_2\cdot nH_2O$ (?)
 (Clay)
Wairakite..$Ca(AlSi_2O_6)_2\cdot 2H_2O$
Wardite..$Na_4CaAl_{12}(PO_4)_8(OH)_9\cdot 3H_2O$
Wavellite..$Al_3(PO_4)_2(OH)_3\cdot 5H_2O$
Weberite..Na_2MgAlF_7
Woodhouseite..$CaAl_3(PO_4)(SO_4)(OH)_6$
Woodwardite..$Cu_4Al_2(SO_4)(OH)_{12}\cdot 2-4H_2O$ (?)
Xanthophyllite..$Ca_2(Mg_5Al)Si_2Al_6O_{20}(OH)_4$
 (Brittle Mica)
Yugawaralite..hydrous Ca,Al silicate
 (Zeolite)
Zeolite..ca.$(Na_2,K_2,Ca,Ba,Sr)[(Al,Si)O_2]_n\cdot xH_2O$
Zincaluminite..$Zn_3Al_3(SO_4)(OH)_{13}\cdot 2.5H_2O$
Zinnwaldite..$K(Li,Fe,Al)(AlSi_3O_{10})(F,OH)_2$
 (Mica)
✓Zirklerite..Al,Fe basic chloride

(ALUMINUM-continued)

Zoisite..$Ca_2Al_3(OH)(SiO_4)_3$
Zunyite..$Al_{12}AlO_4(OH,F)_{18}ClSi_5O_{16}$

ANTIMONY

Allemontite..AsSb
Andorite..$PbAgSb_3S_6$
Antimony..Sb
Aramayoite..$Ag(Sb,Bi)S_2$
Berthierite..$FeSb_2S_4$
Berthonite..$Pb_2Cu_7Sb_5S_{13}$
Bindheimite..$Pb_2Sb_2O_6(O,OH)$
Boulangerite..$Pb_5Sb_4S_{11}-Pb_2Sb_2S_5$
Bournonite..$PbCuSbS_3$
Breithauptite..NiSb
Catoptrite..Mn,Al antimonate-silicate
Cervantite..Sb_2O_4 (?)
Chalcostibite..$CuSbS_2$
Cylindrite..$Pb_3Sn_4Sb_2S_{14}$
Derbylite..$Fe_6Ti_6Sb_2O_{23}$ (?)
Diaphorite..$Pb_2Ag_3Sb_3S_8$
Dyscrasite..Ag_3Sb
Famatinite..Cu_3SbS_4
Fizelyite..$Pb_5Ag_2Sb_8S_{18}$
Flajolotite..$FeSbO_4 \cdot .75H_2O$ (?)
Franckeite..$Pb_5Sn_3Sb_2S_{14}$
Freieslebenite..$Pb_3Ag_5Sb_5S_{12}$
Fuloppite..$Pb_3Sb_8S_{15}$
Geocronite..$Pb_5(Sb,As)_2S_8$
Goldfieldite..$Cu_{12}Te_3Sb_4S_{16}$ (?)
Gudmundite..FeSbS
Heteromorphite..$Pb_7Sb_8S_{19}$
Horsfordite..Cu_5Sb
Jamesonite..$Pb_4FeSb_6S_{14}$
Kermesite..Sb_2S_2O
Klebelsbergite..Basic Sb sulfate
Kobellite..$Pb_2(Bi,Sb)_2S_5$
Livingstonite..$HgSb_4S_7$
Melanostibian..$(Mn,Fe)_6(SbO_3)_2O_3$
Meneghinite..$Pb_{13}Sb_7S_{23}$
Miargyrite..$AgSbS_2$
Monimolite..$(Pb,Ca)_3Sb_2O_8$ (?)
Nadorite..$PbSbO_2Cl$
Nagyagite..$Pb_5Au(Te,Sb)_4S_{5-8}$ (?)

Owyheeite..$Pb_5Ag_2Sb_6S_{15}$
Plagionite..$Pb_5Sb_8S_{17}$
Polyargyrite..$Ag_{24}Sb_2S_{15}$ (?)
Polybasite..$(Ag,Cu)_{16}Sb_2S_{11}$
Pyrargyrite..Ag_3SbS_3
Pyrostilpnite..Ag_3SbS_3
Ramdohrite..$Pb_3Ag_2Sb_6S_{13}$
Romeite..$Ca_2Sb_2O_6(O,OH,F)$
Samsonite..$Ag_4MnSb_2S_6$
Schafarzikite..$FeSb_2O_4$
✓ Schetelegite..$(Ca,Y,Sb,Mn)_2(Ti,Ta,Cb)_2(O,OH)_7$
Semseyite..$Pb_9Sb_8S_{21}$
Senarmontite..Sb_2O_3
Stephanite..Ag_5SbS_4
Stibiconite..$Sb_3O_6(OH)$
Stibiocolumbite..$SbCbO_4$
✓ Stibiopalladinite..Pd_3Sb
Stibiotantalite..$SbTaO_4$
Stibnite..Sb_2S_3
Stylotypite..$(Ag,Cu,Fe)_3SbS_3$
Swedenborgite..$NaBe_4SbO_7$
Tetrahedrite..$(Cu,Fe)_{12}Sb_4S_{13}$
Tripuhyite..$FeSb_2O_7$ (?)
Ullmannite..NiSbS
Valentinite..Sb_2O_3
Vrbaite..$TlAs_2SbS_5$
Wolfachite..$Ni(As,Sb)S$ (?)
Yeatmanite..$(Mn,Zn)_{16}Sb_2Si_4O_{29}$
Zinkenite..$Pb_6Sb_{14}S_{27}$

ARSENIC

Adamite..$Zn_2(AsO_4)(OH)$
Adelite..$CaMg(AsO_4)(OH,F)$
✓ Akrochordite..$MgMn_4(AsO_4)_2(OH)_8 \cdot 4H_2O$ (?)
Algodonite..Cu_6As
Allactite..$Mn_7(AsO_4)_2(OH)_8$
Allemonite..AsSb
Annabergite..$Ni_3(AsO_4)_2 \cdot 8H_2O$
Ardennite..$Mn_5Al_5(V,As)(OH)_2Si_5O_{24} \cdot 2H_2O$ (?)
Armangite..$Mn_3(AsO_3)_2$

Arsenic..As
Arseniopleite..$(Mn,Ca,Pb,Mg)_9(Mn,Fe)_2(AsO_4)_6(OH)_6$
Arseniosiderite..$Ca_3Fe_4(AsO_4)_4(OH)_4 \cdot 4H_2O$ (?)

(ARSENIC--continued)

✓ Arsenobismite..Bi basic arsenate
Arsenoclasite..$Mn_5(AsO_4)_2(OH)_4$
Arsenolamprite..As
Arsenolite..As_2O_3
Arsenopyrite..FeAsS
Atelestite..$Bi_3(AsO_4)O_2(OH)_2$ (?)
Austinite..$CaZn(AsO_4)(OH)$
Badenite..$(Co,Ni)(As,Bi)_4$ (?)
Baumhauerite..$Pb_4As_6Si_3$
Bayldonite..$(Cu,Pb)_2(AsO_4)(OH)$ (?)
Berzeliite..$(Mg,Mn)_2(Ca,Na)_3(AsO_4)_3$
Beudantite..$PbFe_3(AsO_4)(SO_4)(OH)_6$
Brandtite..$Ca_2Mn(AsO_4)_2 \cdot 2H_2O$
Cahnite..$Ca_2B(OH)_4(AsO_4)$
Carminite..$PbFe_2(AsO_4)_2(OH)_2$
Caryinite..$(Ca,Pb,Na)_5(Mn,Mg)_4(AsO_4)_5$
Ceruleite..$CuAl_4(AsO_4)_2(OH)_8 \cdot 4H_2O$
Chalcophyllite..$Cu_{18}Al_2(AsO_4)_3$
$(SO_4)_3(OH)_{27} \cdot 33H_2O$
Chenevixite..$Cu_2Fe_2(AsO_4)_2(OH)_4 \cdot H_2O$ (?)
Chloanthite..$(Co_2Ni)As_{3-x}$
Chlorophoenicite..$(Zn,Mn)_5(AsO_4)(OH)_7$
Claudetite..As_2O_3
Clinoclase..$Cu_3(AsO_4)(OH)_3$
Cobaltite..CoAsS
Colusite..$Cu_3(Sn,Te,Fe,V,As)S_4$
Conichalcite..$CaCu(AsO_4)(OH)$
Cornwallite..$Cu_5(AsO_4)_2(OH)_4 \cdot H_2O$
Dimorphite..As_4S_3
Dixenite..$Mn_5(OH)_2SiAs_2O_9$
Domeykite..Cu_3As
Dufrenoysite..$Pb_2As_2S_5$
Duftite..$PbCu(AsO_4)(OH)$
Durangite..$NaAl(AsO_4)F$
Dussertite..$BaFe_3(AsO_4)_2(OH)_5 \cdot H_2O$
Ecdemite..$Pb_6As_2O_7Cl_4$ (?)
Enargite..Cu_3AsS_4
Epigenite..$(Cu,Fe)_5AsS_6$ (?)
Erinite..$Cu_5(AsO_4)_2(OH)_4$
Erythrite..$Co_3(AsO_4)_2 \cdot 8H_2O$
Euchroite..$Cu_2(AsO_4)(OH) \cdot 3H_2O$
Fermorite..$(Ca,Sr)_5(P,AsO_4)_3(F,OH)$
Finnemanite..$Pb_5(AsO_3)_3Cl$
Flinkite..$Mn_3(AsO_4)(OH)_4$
Forbesite..$H(Ni,Co)(AsO_4) \cdot 3.5H_2O$ (?)

Freirinite..$Na_3Cu_3(AsO_4)_2(OH)_3 \cdot H_2O$
Geocronite..$Pb_5(Sb,As)_2S_8$
> Georgiadesite..$Pb_3(AsO_4)Cl_3$
Gersdorffite..NiAsS
Glaucodot..$(Co,Fe)AsS$
Gratonite..$Pb_9As_4S_{15}$
Guitermanite..$Pb_{10}As_6S_{19}$
Haidingerite..$CaH(AsO_4) \cdot H_2O$
Hedyphane..$(Ca,Pb)_5(AsO_4)_3Cl$
Heliophyllite..$Pb_6As_2O_7Cl_4$ (?)
Hemafibrite..$Mn_3(AsO_4)(OH)_3 \cdot H_2O$
Hematolite..$(Mn,Mg)_4Al(AsO_4)(OH)_8$
Hoernesite..$Mg_3(AsO_4)_2 \cdot 8H_2O$
Holdenite..$(Mn,Ca)_4(Zn,Mg,Fe)_2(AsO_4)(OH)_5O_2$
Hutchinsonite..$(Pb,Tl)_2(Cu,Ag)As_5S_{10}$
Jordanite..$Pb_{14}As_7S_{24}$
Koettigite..$Zn_3(AsO_4)_2 \cdot 8H_2O$
Lautite..CuAsS
✓ Legrandite..$Zn_{14}(AsO_4)_9(OH) \cdot 12H_2O$
Lengenbachite..$Pb_6(Ag,Cu)_2As_4S_{13}$
Leucochalcite..$Cu_2(AsO_4)(OH) \cdot H_2O$
Lindackerite..$Cu_6Ni_3(AsO_4)_4(SO_4)(OH)_4 \cdot 5H_2O$
Liroconite..$Cu_2Al(AsO_4)(OH)_4 \cdot 4H_2O$
Liskeardite..$(Al,Fe)_3(AsO_4)(OH)_6 \cdot 5H_2O$
Liveingite..$Pb_5As_8S_{17}$
Loellingite..$FeAs_2$
Lorandite..$TlAsS_2$
Manganberzeliite..$(Mn,Mg)_2(Ca,Na)_3(AsO_4)_3$
Mansfieldite..$Al(AsO_4) \cdot 2H_2O$
Maucherite..$Ni_{11}As_8$
Metazeunerite..$Cu(UO_2)_2(AsO_4)_2 \cdot 8H_2O$
Mg-chlorophoenicite..$Mg_5(AsO_4)(OH)_7$
Mimetite..$Pb_5(AsO_4)_3Cl$
Mixite..$Cu_{11}Bi(AsO_4)_5(OH)_{10} \cdot 6H_2O$ (?)
Niccolite..NiAs
Nickel skutterudite..$(Ni,Co)As_3$
Olivenite..$Cu_2(AsO_4)(OH)$
Orpiment..As_2S_3
Pararammelsbergite..$NiAs_2$
Pearceite..$(Ag,Cu)_{16}As_2S_{11}$
Pharmacolite..$CaH(AsO_4) \cdot 2H_2O$
Pharmacosiderite..$Fe_3(AsO_4)_2(OH)_3 \cdot 5H_2O$
Picropharmacolite..$(Ca,Mg)_3(AsO_4)_2 \cdot 6H_2O$ (?)
Pitticite..$Fe_2(AsO_4)(SO_4)(OH) \cdot H_2O$ (?)
Proustite..Ag_3AsS_3
Rammelsbergite..$NiAs_2$
Rathite..$Pb_{13}As_{18}S_{40}$

(ARSENIC—continued)

Realgar..AsS
Retzian..Ca,Rare earths,Mn basic arsenate
Roesslerite..MgH(AsO$_4$)·7H$_2$O
Rooseveltine..Bi(AsO$_4$)
Roselite..Ca$_2$(Co,Mg)(AsO$_4$)$_2$·2H$_2$O
Safflorite..(Co,Fe)As$_2$
Sahlinite..Pb$_{14}$(AsO$_4$)$_2$O$_9$Cl
Sarkinite..Mn$_2$(AsO$_4$)(OH)
Sarmientite..Fe$_2$(AsO$_4$)(SO$_4$)(OH)·5H$_2$O
Sartorite..PbAs$_2$S$_4$
✓ Schallerite..(Mn,Fe)$_8$(OH,Cl)$_{10}$(Si,As)$_6$O$_{15}$ (?)
Schultenite..PbH(AsO$_4$)
Scorodite..Fe(AsO$_4$)·2H$_2$O
Seligmannite..PbCuAsS$_3$
Skutterudite..(Co,Ni)As$_3$
Smaltite..(Co,Ni)As$_{3-x}$
Smithite..AgAsS$_2$
Sperrylite..PtAs$_2$
Svabite..Ca$_5$(AsO$_4$)$_3$(F,OH)
Symplesite..Fe$_3$(AsO$_4$)$_2$·8H$_2$O
Synadelphite..(Mn,Mg,Ca,Pb)$_4$(AsO$_4$)(OH)$_5$
Tennantite..(Cu,Fe)$_{12}$As$_4$S$_{13}$
Tilasite..CaMg(AsO$_4$)F
Trechmannite..AgAsS$_2$
Trichalcite..Cu$_3$(AsO$_4$)$_2$·5H$_2$O (?)
Trigonite..MnPb$_3$H(AsO$_3$)$_3$
Trippkeite..CuAs$_2$O$_4$
Troegerite..(UO$_2$)$_3$(AsO$_4$)$_2$·12H$_2$O
Tyrolite..Cu$_5$Ca(AsO$_4$)$_2$(CO$_3$)(OH)$_4$·6H$_2$O (?)
Uranospinite..Ca(UO$_2$)$_2$(AsO$_4$)$_2$·8H$_2$O
Veszelyite..(Cu,Zn)$_3$(As,PO$_4$)(OH)$_3$·2H$_2$O
Vrbaite..TlAs$_2$SbS$_5$
Walpurgite..Bi$_4$(UO$_2$)(AsO$_4$)O$_4$·3H$_2$O
Wolfachite..Ni(As,Sb)S (?)
Xanthoconite..Ag$_3$AsS$_3$
Zeunerite..Cu(UO$_2$)$_2$(AsO$_4$)$_2$·10-16H$_2$O

BARIUM

Alstonite..CaBa(CO$_3$)$_2$
Armenite..BaCa$_2$Si$_8$Al$_6$O$_{28}$·2H$_2$O (?)
Banalsite..BaNa$_2$(Al$_2$Si$_2$O$_8$)$_2$
Barite..BaSO$_4$
Barylite..BaBeSi$_2$O$_7$
Barytocalcite..CaBa(CO$_3$)$_2$
Benitoite..BaTiSi$_3$O$_9$
Brewsterite..(Sr,Ba,Ca)Al$_2$Si$_6$O$_{16}$·5H$_2$O
 (Zeolite)
Cappelenite..(Ba,Ca,Ce,Na)$_3$(Y,Ce,La)$_6$
  (BO$_3$)$_6$Si$_3$O$_9$ (?)
Celsian..BaAl$_2$Si$_2$O$_8$
 (Feldspar)
Cordylite..Ce$_2$Ba(CO$_3$)$_3$F$_2$
Dussertite..BaFe$_3$(AsO$_4$)$_2$(OH)$_5$·H$_2$O
Edingtonite..Ba(Al$_2$Si$_3$O$_{10}$)·4H$_2$O
 (Zeolite)
Feldspar..K,Na,Ca,Ba alumino-silicates
Garrelsite..(Ba,Ca)$_2$B$_3$SiO$_7$(OH)$_3$
Gillespite..FeBaSi$_4$O$_{10}$
Gorceixite..BaAl$_3$(PO$_4$)$_2$(OH)$_5$·H$_2$O
Harmotome..BaAl$_2$Si$_6$O$_{16}$·6H$_2$O
 (Zeolite)
Hollandite..MnBaMn$_6$O$_{14}$
Hyalophane..(K,Na,Ba)AlSi$_3$O$_8$
 (Feldspar)
Hyalotekite..Ca$_3$Ba$_3$Pb$_3$B$_2$Si$_{12}$O$_{36}$ (?)
Joaquinite..NaBa(Ti,Fe)$_3$Si$_4$O$_{15}$ (?)
✓ Labuntsovite..(K,Ba,Na)Ti(Si,Al)$_2$
          (O,OH)$_7$·H$_2$O (?)
Leucosphenite..(Na$_2$,Ca)$_2$BaTi$_3$BSi$_8$O$_{27}$ (?)
Nitrobarite..Ba(NO$_3$)$_2$
Paracelsian..BaAl$_2$Si$_2$O$_8$
 (Feldspar)
Psilomelane..BaMn$_9$O$_{16}$(OH)$_4$
✓ Sanbornite..Ba$_2$Si$_4$O$_{10}$
Taramellite..BaFe(OH)Si$_2$O$_6$
✓ Todorokite..(Mn,Ba,Ca,Mg)Mn$_3$O$_7$·H$_2$O (?)
Uranocircite..Ba(UO$_2$)$_2$(PO$_4$)$_2$·8H$_2$O
Wad..Hydrous oxide of Mn with Ba,etc.
Witherite..BaCO$_3$
Zeolite..ca.(Na$_2$,K$_2$,Ca,Ba,Sr)
        [(Al,Si)O$_2$]$_n$·xH$_2$O

BERYLLIUM

Allanite..(Ca,Ce,La)$_2$(Al,Fe,
 (=Orthite) Be,Mg,Mn)$_3$(OH)Si$_3$O$_{12}$ (?)
Aminoffite..Ca$_8$Be$_3$Al(OH)$_3$Si$_8$O$_{28}$·4H$_2$O (?)
Barylite..BaBeSi$_2$O$_7$
Bavenite..Ca$_4$(OH)$_4$Si$_9$Al$_2$BeO$_{24}$ (?)
Bertrandite..Be$_4$(OH)$_2$Si$_2$O$_7$
Beryl..Be$_3$Al$_2$(Si$_6$O$_{18}$)
Beryllonite..NaBe(PO$_4$)
Bityite..Ca$_4$(Li,Be)$_4$Al$_8$(OH)$_{20}$
 (Mica)   [(Si,Al)$_4$O$_{10}$]$_3$ (?)
Bromellite..BeO

173

(BERYLLIUM-continued)

Chrysoberyl..$BeAl_2O_4$
Danalite..$Fe_4Be_3Si_3O_{12}S$
Euclase..$BeSiAlO_4(OH)$
Epididymite..$NaBe(OH)Si_3O_7$
Eudidymite..$NaBe(OH)Si_3O_7$
Gadolinite..$FeY_2Be_2(SiO_5)_2$
✓ Genthelvite..$Zn_4(BeSiO_4)_3S$
Hambergite..$Be_2(BO_3)(OH)$
Helvite..$(Mn,Fe,Zn)_4(BeSiO_4)_3S$
Herderite..$CaBe(PO_4)(F,OH)$
Hydroxyl-herderite..$CaBe(PO_4)(OH,F)$
✓ Karpinskiite..$(Na,K,Zn,Mg)_2(OH,H_2O)_{1-2}$
$(Al,Be)_2Si_4O_{12}$
Kolbeckite..Ca,Be,Al,Fe silicate-phosphate
Leucophane..$(Ca,Na)_2BeSi_2(O,OH,F)_7$ (?)
Melinophane..$(Ca,Na)_2(Be,Al)Si_2O_6 \cdot F$
Milarite..$KCa_2Si_{12}Be_2AlO_{30} \cdot 5H_2O$ (?)
Orthite..$(Ca,Ce,La)_2(Al,Fe,Be,Mg,Mn)_3(OH)$
Si_3O_{12} (?)
Phenakite..$BeSiO_4$
> Rhodizite..$NaKLi_4Al_4Be_3B_{10}O_{27}$ (?)
Swedenborgite..$NaBe_4SbO_7$
Trimerite..$(Mn,Ca)BeSiO_4$

BISMUTH

Aikinite..$PbCuBiS_3$
Alaskaite..$Pb(Ag,Cu)_2Bi_4S_8$ (?)
Aramayoite..$Ag(Sb,Bi)S_2$
✓ Arsenobismite..Bi basic arsenate
Atelestite..$Bi_3(AsO_4)O_2(OH)_2$ (?)
Badenite..$(Co,Ni)(As,Bi)_4$ (?)
Beegerite..$Pb_6Bi_2S_9$
✓ Benjaminite...$Pb(Cu,Ag)Bi_2S_4$
✓ Beyerite..$Ca(BiO)_2(CO_3)_2$
Bismite..Bi_2O_3
Bismoclite..$BiOCl$
Bismuth..Bi
Bismuthinite..Bi_2S_3
Bismutite..$(BiO)_2CO_3$
Bismutotantalite..$Bi(Ta,Cb)O_4$
✓ Chiviatite..$Pb_3Bi_8Si_5$
Cosalite..$Pb_2Bi_2S_5$
Daubreeite..$BiO(OH,Cl)$
Djalmaite..$(U,Ca,Pb,Bi,Fe)(Ta,Cb,$
$Ti,Zr)_3O_9 \cdot nH_2O$

Emplectite..$CuBiS_2$
Eulytine..BiS_3O_{12}
Galenobismutite..$PbBi_2S_4$
Gladite..$PbCuBi_5S_9$
✓ Goongarrite..$Pb_4Bi_2S_7$
Gruenlingite..Bi_4TeS_3
Guanajuatite..Bi_2Se_3
Hammarite..$Pb_2Cu_2Bi_4S_9$ (?)
Joseite..Bi_3TeS
✓ Klaprothite..$Cu_6Bi_4S_9$ (?)
Klebelsbergite..Basic Sb sulfate
Kobellite..$Pb_2(Bi,Sb)_2S_5$
Koechlinite..$(BiO)_2(MoO_4)$
✓ Lillianite..$Pb_3Bi_2S_6$
Lindstromite..$PbCuBi_3S_6$
Maldonite..Au_2Bi
Matildite..$AgBiS_2$
Mixite..$Cu_{11}Bi(AsO_4)_5(OH)_{10} \cdot 6H_2O$ (?)
Montanite..$(BiO)_2(TeO_4) \cdot 2H_2O$
> Platynite..$PbBi_2(Se,S)_3$
Pucherite..$BiVO_4$
Rezbanyite..$Pb_3Cu_2Bi_{10}S_{19}$
Rooseveltite..$Bi(AsO_4)$
Russellite..$(Bi_2,W)O_3$
Schirmerite..$PbAg_4Bi_4S_9$
Sillenite..Bi_2O_3
Tellurobismuthite..Bi_2Te_3
Tetradymite..Bi_2Te_2S
Uranosphaerite..$Bi_2O_3 \cdot 2UO_3 \cdot 3H_2O$ (?)
Walpurgite..$Bi_4(UO_2)(AsO_4)_4 \cdot 3H_2O$
Waltherite..Bi basic carbonate
Wehrlite..Bi_3Te_2 (?)
✓ Weilbullite..$PbBi_2(S,Se)_4$
Wittichenite..Cu_3BiS_3 (?)
Wittite..$Pb_5Bi_6(S,Se)_{14}$

BORON

✓ Ammonioborite..$(NH_4)_2B_{10}O_{16} \cdot 5H_2O$ (?)
Avogadrite..$(K,Cs)BF_4$
Axinite..$(Ca,Mn,Fe)_3Al_2BO_3(Si_4O_{12})(OH)$
Bakerite..$Ca_4B_4(BO_4)(SiO_4)_3(OH)_3 \cdot H_2O$
Bandylite..$CuB_2O_4 \cdot CuCl_2 \cdot 4H_2O$
Boracite..$Mg_3B_7O_{13}Cl$
Borax..$Na_2B_4O_7 \cdot 10H_2O$

(BORON-continued)

Cahnite..$Ca_2B(OH)_4AsO_4$
Cappelenite..$(Ba,Ca,Ce,Na)_3(Y,Ce,La)_6(BO_3)_6Si_3O_9$ (?)
Colemanite..$Ca_2B_6O_{11}\cdot 5H_2O$
Danburite..$CaSi_2B_2O_8$
Datolite..$CaB(SiO_4)(OH)$
Dravite..$NaMg_3Al_6(OH)_4(BO_3)_3Si_6O_{18}$
 (Tourmaline)
Dumortierite..$H(Al,Fe)_2Si_3Al_6BO_{20}$ (?)
Elbaite..$Na(Li,Al)_3Al_6(OH,F)_4(BO_3)_3Si_6O_{18}$
 (Tourmaline)
Ferruccite..$NaBF_4$
Fluoborite..$Mg_3BO_3(F,OH)_3$
Garrelsite..$(Ba,Ca)_2B_3SiO_7(OH)_3$
Ginorite..$Ca_2B_{14}O_{23}\cdot 8H_2O$
Grandidierite..$H_2Na_2(Mg,Fe)_7(Al,Fe,B)_{15}Si_7Al_7O_{56}$ (?)
Hambergite..$Be_2BO_3(OH)$
✓ Harkerite..$Ca(Mg,Al)(Si,BH)O_4\cdot CaCO_3$ (?)
Hilgardite..$Ca_8(B_6O_{11})_3Cl_4\cdot 4H_2O$
Homilite..$Ca_2FeB_2(SiO_5)_2$
Horsfordite..$CuSSb$
Howlite..$CaSiB_5O_9(OH)_5$
Hulsite..$(Fe,Ca,Mg)_4(Fe,Sn)_2B_2O_{10}$ (?)
Hyalotekite..$Ca_3Ba_3Pb_3B_2Si_{12}O_{36}$ (?)
Hydroboracite..$CaMgB_6O_{11}\cdot 6H_2O$
Inderborite..$CaMgB_6O_{11}\cdot 11H_2O$
Inderite..$Mg_2B_6O_{11}\cdot 15H_2O$
Inyoite..$Ca_2B_6O_{11}\cdot 13H_2O$
Jeremejevite..$AlBO_3$
Kaliborite..$KMg_2B_{11}O_{19}\cdot 9H_2O$
Kernite..$Na_2B_4O_7\cdot 4H_2O$
Kornerupine..$Mg_{10}Al_{10}BO_7Si_9O_{36}$ (?)
Kotoite..$Mg_3(BO_3)_2$
Kurnakovite..$Mg_2B_6O_{11}\cdot 13H_2O$
Larderellite..$(NH_4)_2B_{10}O_{16}\cdot 5H_2O$ (?)
Leucosphenite..$(Na_2,Ca)_2BaTi_3BSi_8O_{27}$ (?)
Ludwigite..$(Mg,Fe)_2FeBO_5$
Luenebergite..$Mg_3B_2(OH)_6(PO_4)_2\cdot 6H_2O$
Manandonite..hydrous Li,B,Al silicate
 (Chlorite)
Meyerhofferite..$Ca_2B_6O_{11}\cdot 7H_2O$
Nordenskioldine..$CaSn(BO_3)_2$
Paigeite..$(Fe,Mg)_2FeBO_5$
✓ Painite..$Al_2O_3Ca_2(Si,BH)O_4$
Parahilgardite..$Ca_8(B_6O_{11})_3Cl_4\cdot 4H_2O$

Paternoite..$MgB_8O_{13}\cdot 4H_2O$
Pinakiolite..$Mg_3MnMn_2B_2O_{10}$ (?)
Pinnoite..$Mg(BO_2)_2\cdot 3H_2O$
Priceite..$Ca_4B_{10}O_{19}\cdot 7H_2O$ (?)
Probertite..$NaCaB_5O_9\cdot 5H_2O$
Reedmergnerite..$NaBSi_3O_8$
Rhodizite..$NaKLi_4Al_4Be_3B_{10}O_{27}$ (?)
✓ Roweite..$(Mn,Mg,Zn)Ca(BO_2)_2(OH)_2$
Sassolite..$B(OH)_3$
Schorl..$Na(Fe,Mn)_3Al_6(OH)_4(BO_3)_3Si_6O_{18}$
 (Tourmaline)
Seamanite..$Mn_3(PO_4)BO_3\cdot 3H_2O$
Searlesite..$NaBSi_2O_6\cdot H_2O$
Serendibite..$Ca_2(Mg,Fe)_4Al_6B_2O_{10}Si_4O_{16}$ (?)
✓ Stillwellite..$(Ce,...)BSiO_5$
Sulforborite..$Mg_6H_4(BO_3)_4(SO_4)_2\cdot 7H_2O$
Sussexite..$(Mn,Mg)(BO_2)(OH)$
Szaibelyite..$(Mg,Mn)(BO_2)(OH)$
Teepleite..$Na_2B_2O_4\cdot 2NaCl\cdot 4H_2O$
Tincalconite..$Na_2B_4O_7\cdot 5H_2O$
Tourmaline..$(Na_2,Ca)(Li,Al,Mg,Fe,Mn)_6Al_{12}(OH,F)_8(BO_3)_6Si_{12}O_{36}$
Tritomite..$Ca_3(La,Ce)_5Zr_3F_6B_3Si_6O_{27}$ (?)
Ulexite..$NaCaB_5O_9\cdot 8H_2O$
Uvite..$CaMg_4Al_5B_3Si_6O_{29}\cdot 2H_2O$ (?)
 (Tourmaline)
✓ Veatchite..$Sr_3B_{16}O_{27}\cdot 5H_2O$ (?)
Warwickite..$(Mg,Fe)_3TiB_2O_8$

CADMIUM

Cadmium oxide..CdO
Greenockite..CdS
Otavite..$CdCO_3$

CALCIUM

Actinolite..$Ca_2(Mg,Fe)_5Si_8O_{22}(OH,F)_2$
 (Amphibole)
Adelite..$CaMg(AsO_4)(OH,F)$
Aegirine-augite..$(Na,Ca)(Fe,Mg,Fe,Al)(Si_2O_6)$
 (Clinopyroxene)
✓ Afwillite..$(Ca,Pb)_{10}(OH,Cl)_2(Si_2O_7)_3$
Akermanite..$Ca_2Mg(Si_2O_7)$
Allanite..$(Ca,Ce,La)_2(Al,Fe,Be,Mg,Mn)_3(OH)Si_3O_{12}$ (?)
 (=Orthite)

(CALCIUM—continued)

Alstonite..$CaBa(CO_3)_2$
✓ Alumohydrocalcite..$CaAl_2(CO_3)_2(OH)\cdot 2H_2O$ (?)
Aminoffite..$Ca_8Be_3Al(OH)_3Si_8O_{28}\cdot 4H_2O$ (?)
Ampangabeite..$(Y,Er,U,Ca,Th)_2(Cb,Ta,Fe,Ti)_7O_{18}$ (?)
Amphibole..hydrous Fe,Mg,Na,Ca,Mn,K,Al silicates
Anapaite..$Ca_2Fe(PO_4)_2\cdot 4H_2O$
Ancylite..$(Ce,La)_4(Sr,Ca)_3(CO_3)_7(OH)_4\cdot 3H_2O$
Andersonite..$Na_2Ca(UO_2)(CO_3)_3\cdot 6H_2O$
Andesine....An_{30-50} *
  (Plagioclase)
Andradite..$Ca_3(Fe,Ti)_2(SiO_4)_3$
  (Garnet)
Anhydrite..$CaSO_4$
Ankerite..$Ca(Fe,Mg)(CO_3)_2$
Anorthite..$CaAl_2Si_2O_8$
  (Plagioclase)
Apatite..ca.$Ca_{10}(PO_4)_6(F_2,Cl_2,OH_2,CO_3)$ (?)
Apophyllite..$KFCa_4(Si_4O_{10})_2\cdot 8H_2O$
Aragonite..$CaCO_3$

Ardealite..$Ca_2H(PO_4)(SO_4)\cdot 4H_2O$
Arfvedsonite.. .$5Na_5Ca(Fe,Mg,Fe,Al)_{10}Si_{15}AlO_{44}(OH,F)_4$
  (Amphibole)
Armenite..$BaCa_2Si_8Al_6O_{28}\cdot 2H_2O$ (?)
Arseniopleite..$(Mn,Ca,Pb,Mg)_9(Mn,Fe)_2(AsO_4)(OH)_6$
Arseniosiderite..$Ca_3Fe_4(AsO_4)_4(OH)_4\cdot 4H_2O$ (?)
Ashcroftine..$KNa(Ca,Mg,Mn)(Al_4Si_5O_{18})\cdot 8H_2O$
  (Zeolite)
Astrophyllite..$(K_2,Na_2,Ca)(Fe,Mn)_4(Ti,Zr)(OH,F)_2Si_4O_{14}$ (?)
Augite..$(Ca,Mg,Fe,Fe,Ti,Al)_2(Si,Al)_2O_6$

Austinite..$CaZn(AsO_4)(OH)$
Autunite..$Ca(UO_2)_2(PO_4)_2\cdot 10\text{-}12H_2O$
Axinite..$(Ca,Mn,Fe)_3Al_2BO_3(Si_4O_{12})(OH)$
Babingtonite..hydrous Ca,Fe silicate

Bakerite..$Ca_4B_4(BO_4)(SiO_4)_3(OH)_3\cdot H_2O$
Barkevikite..$Ca_2(Na,K)(Fe,Mg,Mn)_5(Si_{6.5}Al_{1.5}O_{22})(OH)_2$
  (Amphibole)
Barroisite..Na,K,Ca,Mg,Fe,Ti alumino-silicate
  (Amphibole)

Barytocalcite..$CaBa(CO_3)_2$
Basaltic hornblende..$Ca_2(Na,K)_{0.5-1.0}(Mg,Fe)_{3-2}(Fe,Al)_{2-3}Si_6Al_2O_{22}(O,OH,F)_2$
  (Amphibole)

Bassanite..$2CaSO_4\cdot H_2O$

Bavenite..$Ca_4(OH)_4Si_9Al_2BeO_{24}$ (?)
> Beckelite..$(Ca,Ce,La,Nd)_4O_3Si_3O_{12}$
Berzeliite..$(Mg,Mn)_2(Ca,Na)_3(AsO_4)_3$
Betafite..$(U,Ca)(Cb,Ta,Ti)_3O_9\cdot nH_2O$
✓ Beyerite..$Ca(BiO)_2(CO_3)_2$
Bityite..$Ca_4(Li,Be)_4Al_8(OH)_{20}[(Si,Al)_4O_{10}]_3$ (?)
  (Mica?)
Borickite..$CaFe_5(PO_4)_2(OH)_{11}\cdot 3H_2O$
Brandtite..$Ca_2Mn(AsO_4)_2\cdot 2H_2O$
✓ Brannerite..$(U,Ca,Fe,Y,Th)_3Ti_5O_{16}$ (?)
Bredigite..Ca_2SiO_4
Brewsterite..$(Sr,Ba,Ca)(Al_2Si_6O_{16})\cdot 5H_2O$
  (Zeolite)
Brittle Mica ..hydrous Na,Ca,Mg,Al silicates
Brushite..$CaH(PO_4)\cdot 2H_2O$
Buetschliite..$K_6Ca_2(CO_3)_5\cdot 6H_2O$
✓ Bultfonteinite..$H_2Ca_2F_2SiO_4$ (?)
Bustamite..$(Mn,Ca,Fe)(SiO_3)$
Byssolite..hydrous Ca,Mg,Fe alumino-silicate
  (Amphibole)
Bytownite....An_{70-90} *
  (Plagioclase)
Cahnite..$Ca_2B(OH)_4(AsO_4)$
Calcioferrite..$Ca_3Fe_3(PO_4)_4(OH)_3\cdot 8H_2O$ (?).
Calciovolborthite..$CuCa(VO_4)(OH)$
Calcite..$CaCO_3$

Calclacite..$CaCl(C_2H_3O_2)\cdot 5H_2O$
Cancrinite..$(Na,Ca)_{7-8}(Al_6Si_6O_{24})(CO_3,SO_4,Cl)_{1.5-2}\cdot 1\text{-}5H_2O$
Cappelenite..$(Ba,Ca,Ce,Na)_3(Y,Ce,La)_6(BO_3)_6Si_3O_9$ (?)

Carbonate-apatite.. $Ca_{10}(PO_4)_6(CO_3)\cdot H_2O$ (?)
Cardenite..hydrous Mg,Ca,Fe,Al silicate
  (Clay)
Caryinite..$(Ca,Pb,Na)_5(Mn,Mg)_4(AsO_4)_5$ (?)
Celadonite..$(K,Ca,Na)(Al,Fe,Mg)_2(Al_{0.11}Si_{3.89}O_{10})(OH)$ (?)
Cerite..$(Ca,Fe)Ce_3Si_3O_{12}\cdot H_2O$ (?)

Chabazite..$(Ca,Na_2)(Al_2Si_4O_{12})\cdot 6H_2O$
  (Zeolite)
Chlorapatite..$Ca_5(PO_4)_3Cl$
Chlorocalcite..$KCaCl_3$
Churchite..$(Ce,Ca)(PO_4)\cdot 2H_2O$
> Clay..chiefly hydrous silicates of Al or Mg
Clinohedrite..$Ca_2Zn_2(OH)_2Si_2O_7\cdot H_2O$
Clinozoisite..$Ca_2Al_3(OH)Si_3O_{12}$
  (Epidote)
Clintonite..$Ca_2(Mg_5Al)Si_2Al_6O_{20}(OH)_4$ (?)
  (Brittle Mica)

176

(CALCIUM—continued)

Colemanite..$Ca_2B_6O_{11} \cdot 5H_2O$

Collinsite..$Ca_2(Mg,Fe)(PO_4)_2 \cdot 2H_2O$

Common hornblende..$(Ca,Na,K)_{2-3}(Mg,Fe,Al)_5Si_6(Si,Al)_2O_{22}(OH,F)_2$
(Amphibole)

Conichalcite..$CaCu(AsO_4)(OH)$

Crandallite..$CaAl_3(PO_4)_2(OH)_5 \cdot H_2O$

Creedite..$Ca_3Al_2F_4(OH,F)_6(SO_4) \cdot 2H_2O$

Cuspidine..$Ca_4(F,OH)_2Si_2O_7$

Dachiardite..$(Ca,K_2,Na_2)_3Al_4Si_{18}O_{45} \cdot 14H_2O$
(Zeolite)

Danburite..$CaSi_2B_2O_8$

Datolite..$CaB(SiO_4)(OH)$

Davisonite..$Ca_3Al(PO_4)_2(OH)_3 \cdot H_2O$ (?)

Davyne..$(Na,K)_6Ca_2(AlSiO_4)_6(SO_4)_2$ (?)

✓ Dehrnite..$(Ca,Na,K)_5(PO_4)_3(OH)$

✓ Deltaite..$Ca(Al,Ca)(PO_4)_2(OH)_4 \cdot H_2O$

Devilline..$Cu_4Ca(SO_4)_2(OH)_6 \cdot 3H_2O$

Diallage..$Ca_{14}Fe_2Mg_{13}FeAl_5Si_{29}O_{96}$ (?)
(Clinopyroxene)

Dickinsonite..$Na_6(Mn,Fe,Ca)_{14}H_2(PO_4)_{12} \cdot H_2O$

Didymolite..$(Ca,Mg,Fe)_2Al_6Si_8O_{27}$ (?)

Dietzeite..$Ca_2(IO_3)_2(CrO_4)$

Diopside..$CaMg(Si_2O_6)$
(Clinopyroxene)

Dipyre....Me_{20-50} ★★
(Scapolite)

Djalmaite..$(U,Ca,Pb,Bi,Fe)(Ta,Cb,Ti,Zr)_3O_9 \cdot nH_2O$

Dolomite..$CaMg(CO_3)_2$

Donbassite..hydrous Na,Ca,Mg,Al silicate
(Clay)

Earlandite..$Ca_3(C_6H_5O_7)_2 \cdot 4H_2O$

Edenite..$NaCa_2Mg_5AlSi_7O_{22}(OH,F)_2$
(Amphibole)

Egueite..$CaFe_{14}(PO_4)_{10}(OH)_{14} \cdot 21H_2O$ (?)

Ekermannite.. $.5Na_5Ca(Mg,Fe,Fe,Al,Li)_{10}Si_{15}AlO_{44}(OH,F)_4$
(Amphibole)

✓ Ellestadite..$Ca_5(Si,S,P)_3(Cl,F,OH)$

Englishite..$K_2Ca_4Al_8(PO_4)_8(OH)_{10} \cdot 9H_2O$

Enigmatite..$(Ca,Na_2)_2Fe(Al,Fe,Ti)_4(Si_2O_7)_2$ (?)

Ephesite..$(Na,Ca)Al_2[Al(Al,Si)Si_2O_{10}](OH)_2$
(Brittle Mica)

Epidote..$Ca_2(Al,Fe)Al_2(OH)Si_3O_{12}$

Epistilbite..$Ca(Al_2Si_6O_{16}) \cdot 5H_2O$
(Zeolite)

Epistolite..$(Na,Ca)(Cb,Ti,Mg,Fe,Mn)SiO_4(OH)$

Erionite..ca.$(Na_2,K_2,Ca,Mg)_{4.5}Al_9Si_{27}O_{72} \cdot 27H_2O$

Eschynite..$(Ce,Ca,Fe,Th)(Ti,Cb)_2O_6$

Ettringite..$Ca_6Al_2(SO_4)_3(OH)_{12} \cdot 26H_2O$

Eudialyte..$(Na,Ca,Fe)_6ZrSi_6O_{18}(OH,Cl)$

Euxenite..$(Y,Ca,Ce,U,Th)(Cb,Ta,Ti)_2O_6$

Fairchildite..$K_2Ca(CO_3)_2$

Fairfieldite..$Ca_2(Mn,Fe)(PO_4)_2 \cdot 2H_2O$

Fassaite..$Ca(Mg,Fe,Al,Ti)(Si,Al)_2O_6$
(Clinopyroxene)

Faujasite..ca.$(Na_2,Ca)_{1.75}Al_{3.5}Si_{8.5}O_{24} \cdot 16H_2O$
(Zeolite)

Feldspar..K,Na,Ca,Ba alumino-silicates

Fermorite..$(Ca,Sr)_5(P,AsO_4)(F,OH)$

> Fernandinite..$CaO \cdot V_2O_4 \cdot 5V_2O_5 \cdot 14H_2O$

Ferroactinolite..$Ca_2Fe_5Si_8O_{22}(OH,F)_2$
(Amphibole)

Ferroedenite..$NaCa_2Fe_5AlSi_7O_{22}(OH,F)_2$
(Amphibole)

Ferrohastingsite..$NaCa_2Fe_4(Al,Fe)Si_6Al_2O_{22}(OH,F)_2$
(Amphibole)

Ferrosalite..$Ca(Fe,Mg)Si_2O_6$
(Clinopyroxene)

Ferrotschermakite..$Ca_2Fe_5Al_2Si_6O_{22}(OH,F)_2$
(Amphibole)

✓ Fersmanite..$Na_4Ca_4Ti_4(O,OH,F)_3Si_3O_{12}$ (?)

Fillowite..$Na_6(Mn,Fe,Ca)_{14}H_2(PO_4)_{12} \cdot H_2O$ (?)

Fluorapatite..$Ca_5(PO_4)_3F$

Fluroite..CaF_2

✓ Formanite..$(U,Zr,Th,Ca)(Ta,Cb,Ti)O_4$

Foshallasite..$Ca_3Si_2O_7 \cdot 3H_2O$

Ganomalite..$(Ca,Pb)_{10}(OH,Cl)_2(Si_2O_7)_3$

Ganophyllite..$(Na,K)(Mn,Al,Mg,Ca)_3(OH)_4(Si,Al)Si_3O_{10}$ (?)

Garrelsite..$(Ba,Ca)_2B_3SiO_7(OH)_3$

Gaylussite..$Na_2Ca(CO_3)_2 \cdot 5H_2O$

Gearksutite..$CaAl(OH)F_4 \cdot H_2O$

Gehlenite..$Ca_2(Al_2SiO_7)$

Ginorite..$Ca_2B_{14}O_{23} \cdot 8H_2O$

Gismondine..$Ca(Al_2Si_2O_8) \cdot 4H_2O$
(Zeolite)

Glauberite..$Na_2Ca(SO_4)_2$

Glaucochroite..$CaMnSiO_4$

Glauconite..$(K,Ca,Na)(Al,Fe,Mg)_2(Al_{0.35}Si_{3.65}O_{10})(OH)$ (?)
(Mica)

Gmelinite..$(Na_2,Ca)(Al_2Si_4O_{12}) \cdot 6H_2O$
(Zeolite)

Gonnardite..$Na_2Ca[(Al,Si)_5O_{10}]_2 \cdot 6H_2O$
(Zeolite)

Graftonite..$(Fe,Mn,Ca)_3(PO_4)_2$

Griphite..$(Na,Ca,Fe,Al)_3Mn_2(PO_4)_{2.5}(OH,F)_2$ (?)

Grossular..$Ca_3Al_2(SiO_4)_3$
(Garnet)

(CALCIUM-continued)

Guarinite..$Ca_2NaZrFSi_2O_8$ (?)
Gypsum..$CaSO_4 \cdot 2H_2O$
Gyrolite..$Ca_4(OH)_2Si_6O_{15} \cdot 3H_2O$
Haidingerite..$CaH(AsO_4) \cdot H_2O$
Hardystonite..$Ca_2ZnSi_2O_7$
✓ Harkerite..$Ca(Mg,Al)(Si,BH)O_4 \cdot CaCO_3$
Harstigite..$(Ca,Mn,Mg)_8Al_2(OH)_4Si_6O_{21}$ (?)
Hastingsite..$NaCa_2Mg_4Al_3Si_6O_{22}(OH)_2$
 (Amphibole)
Hauyne..$(Na,Ca)_{4-8}Al_6Si_6O_{24}(SO_4,S)_{1-2}$
Hedenbergite..$CaFe(Si_2O_6)$
 (Clinopyroxene)
Hedyphane..$(Ca,Pb)_5(AsO_4)_3Cl$
Hellandite..$(Ca,Y,Er,Mn)_3(Al,Fe)Si_2O_4 \cdot H_2O$ (?)
Herderite..$CaBe(PO_4)(F,OH)$
Heulandite..$(Ca,Na_2)(Al_2Si_7O_{18}) \cdot 6H_2O$
 (Zeolite)
Hewettite..$CaV_6O_{16} \cdot 9H_2O$
Hilgardite..$Ca_8(B_6O_{11})_3Cl_4 \cdot 4H_2O$
Hillebrandite..$Ca_2SiO_4 \cdot H_2O$
Holdenite..$(Mn,Ca)_4(Zn,Mg,Fe)(AsO_4)(OH_5)O_2$
Homilite..$Ca_2FeB_2(SiO_5)_2$
Hornblende, basaltic..$Ca_2(Na,K)_{0.5-1.0}$
 (Amphibole) $(Mg,Fe)_{3-2}(Fe,Al)_{2-3}$
 $Si_6Al_2O_{22}(O,OH,F)_2$
Hornblende, common..$(Ca,Na,K)_{2-3}(Mg,Fe,$
 (Amphibole) $Al)_5Si_6(Si,Al)_2O_{22}(OH,F)_2$
Howlite..$Ca_2SiB_5O_9(OH)_5$
✓ Huhnerkobelite..$(Na,Ca)Fe(PO_4)$
Hulsite..$(Fe,Ca,Mg)_4(Fe,Sn)_2B_2O_{10}$ (?)
Hyalotekite..$Ca_3Ba_3Pb_3B_2Si_{12}O_{36}$ (?)
Hydroboracite..$CaMgB_6O_{11} \cdot 6H_2O$
✓ Hydrocalumite..$Ca_4Al_2(OH)_{14} \cdot 6H_2O$
Hydrogrossular..$Ca_3Al_2(SiO_4)_3(OH)_4$
Hydrophilite..$CaCl_2$
Hydroxl-herderite..$CaBe(PO_4)(OH,F)$
Hydroxylapatite..$Ca_5(PO_4)_3(OH)$
Ilvaite..$Ca(Fe,Mn)_2Fe(SiO_4)_2(OH)$
Inderborite..$CaMgB_6O_{11} \cdot 11H_2O$
Inesite..$Ca_2Mn_7(OH)_2Si_{10}O_{28} \cdot 5H_2O$ (?)
Inyoite..$Ca_2B_6O_{11} \cdot 13H_2O$
Isoclasite..$Ca_2(PO_4)(OH) \cdot 2H_2O$
Jezekite..$Na_4CaAl_2(PO_4)_2(OH)_2F_2O$ (?)
Johannsentie..$Ca(Mn,Fe)Si_2O_6$
 (Clinopyroxene)
Johnstrupite..$(Ca,Y,Na,Ce)_3(Al,Zr,Ti)$
 $(F,OH)Si_2O_8$

Kaersutite..$Ca_2(Na,K)(Mg,Fe)_4Ti$
 (Amphibole) $Si_6Al_2O_{22}(OH,F)_2$
Kainosite..$Ca_2(Ce,Y)_2CO_3Si_4O_{12} \cdot 1-2H_2O$
Karinthine..$(Na,K)Ca_{2-3}Mg_8Fe_{1-2}(Al,Fe,Ti)_2$
 (Amphibole) $(Al_{3-4}Si_{13-12}O_{44})(OH)_4$
Katophorite..$Na_2CaFe_4(Fe,Al)Si_7AlO_{22}(OH,F)_2$
 (Amphibole)
✓ Koktaite..$(NH_4)_2Ca(SO_4)_2 \cdot H_2O$
Kolbeckite..Ca,Be,Al,Fe silicate-phosphate
✓ Kupletskite..$(K_2,Na_2,Ca)(Fe,Mn)_4(Ti,Zr)$
 $(OH)_2Si_4O_{14}$
Kutnahorite..$Ca(Mn,Mg)(CO_3)_2$
Labradorite..An_{50-70} *
 (Plagioclase)
Lacroixite..Na,Ca,Al fluo-phosphate
Larnite..Ca_2SiO_4
✓ Latiumite..$Ca_6(K,Na)_2Al_4(O,CO_3,SO_4)(SiO_4)_6$
Laumontite..$Ca(AlSi_2O_6)_2 \cdot 4H_2O$
 (Zeolite)
Lautarite..$Ca(IO_3)_2$
Lavenite..complex fluosilicate
Lawsonite..$CaAl_2(Si_2O_7)(OH)_2 \cdot H_2O$
Lazurite..$(Na,Ca)_8(AlSiO_4)_6(SO_4,S,Cl_2)_4$ (?)
Lehiite..$(Na,K)_2Ca_5Al_8(PO_4)_8(OH)_{12} \cdot 6H_2O$ (?)
Leightonite..$K_2Ca_2Cu(SO_4)_4 \cdot 2H_2O$
Lessingite..$Ca_4(Ce,La,Nd)_7(OH,F)_5Si_6O_{24}$ (?)
Leucophane..$(Ca,Na)_2BeSi_2(O,OH,F)_7$ (?)
Leucosphenite..$(Na_2,Ca)_2BaTi_3BSi_8O_{27}$ (?)
Levynite..$Ca(Al_2Si_4O_{12}) \cdot 6H_2O$
 (Zeolite)
Lewistonite..$(Ca,K,Na)_5(PO_4)_3(OH)$
Liebigite..$Ca_2U(CO_3)_4 \cdot 10H_2O$
Lime..CaO
— Lomonossovite..$(Na,Ca)Ti(O,OH)(S,P)O_4$ (?)
Loranskite..$[Y,Ce,Ca,Zr,(?)][Ta,Zr,(?)]O_4$ (?)
Magnesiokatophorite..$Na_2CaMg_4(Fe,Al)$
 (Amphibole) $Si_7AlO_{22}(OH,F)_2$
Magnesium orthite..$CaCeMgAl_2(OH,F)Si_3O_{12}$
Manganberzeliite..$(Mn,Mg)_2(Ca,Na)_3(AsO_4)_3$
Margarite..$CaAl_2(Al_2Si_2O_{10})(OH)_2$
 (Brittle Mica)
Meionite..$Ca_8(Al_2Si_2O_8)_6(Cl_2,SO_4,CO_3)_2$
 (Scapolite)
✓ Melanovanadite..$Ca_2V_{10}O_{25}$
Melilite..$(Ca,Na)_2(Al,Mg)(Si,Al)_2O_7$
Melinophane..$(Ca,Na)_2(Be,Al)(Si_2O_6F)$
Merwinite..$Ca_3MgSi_2O_8$
Mesolite..$Na_2Ca_2(Al_2Si_3O_{10})_3 \cdot 8H_2O$
 (Zeolite)
Metahewettite..$CaV_6O_{16} \cdot 9H_2O$
Metarossite..$CaV_2O_6 \cdot 2H_2O$

(CALCIUM-continued)

Meyerhofferite..$Ca_2B_6O_{11}·7H_2O$

Mica..complex hydrous alumino-silicates

Microlite..$(Na,Ca)_2Ta_2O_6(O,OH,F)$

Milarite..$KCa_2Si_{12}Be_2AlO_{30}·5H_2O$ (?)

Millisite..$(Na,K)CaAl_6(PO_4)_4(OH)_9·3H_2O$

Mitridatite..$Ca_2Fe(PO_4)_2(OH)_2·H_2O$ (?)

Mizzonite..Me_{50-80} ** (Scapolite)

Monetite..$CaH(PO_4)$

Monimolite..$(Pb,Ca)_3Sb_2O_8$ (?)

Montgomeryite..$Ca_4Al_5(PO_4)_6(OH)_5·11H_2O$

Monticellite..$CaMg(SiO_4)$

Mordenite..$(Ca,K_2,Na_2)(AlSi_5O_{12})_2·7H_2O$ (Zeolite)

Morinite..Na,Ca,Al fluo-phosphate

Mosandrite..$(Ca,Na)_{12}Ce_3(Zr,Ti,Mg)_4F_5Si_{10}O_{40}$ (?)

Nagatelite..$(Ca,Ce)_2(Al,Fe)_3(OH)(Si,P)_3O_{12}$

Nasonite..$(Ca,Pb)_{10}Cl_2(Si_2O_7)_3$

✓ Nekoite..$CaH_2(Si_2O_6)·H_2O$

Nitrocalcite..$Ca(NO_3)_2·4H_2O$

Nocerite..$Ca_3Mg_3F_8O_2$

Nordenskioldine..$CaSn(BO_3)_2$

Okenite..$Ca_2Si_4O_{10}·4H_2O$

Oldhamite..CaS

Oligoclase..An_{10-30} * (Plagioclase)

Omphacite..$(Ca,Na)(Mg,Fe,Fe,Ti,Al)(Si_2O_6)$ (Clinopyroxene)

Orientite..$Ca_4Mn_4(SiO_4)_5·4H_2O$

Orthite..$(Ca,Ce,La)_2(Al,Fe,Be,Mg,Mn)_3(OH)Si_3O_{12}$ (?)

Orthite,Magnesium..$CaCeMgAl_2(OH)Si_3O_{12}$

— Osumilite..$(K,Na,Ca)(Mg,Fe)_2(Al,Fe)_3(SiAl)_{12}O_{30}·H_2O$

Overite..$Ca_3Al_8(PO_4)_8(OH)_6·15H_2O$

Pachnolite..$NaCaAlF_6·H_2O$

✓ Painite..$Al_2O_3·Ca_2(Si,BH)O_4$

Parahilgardite..$Ca_8(B_6O_{11})_3Cl_4·4H_2O$

Parawollastonite..$CaSiO_3$

Pargasite..$NaCa_2Mg_4(Al,Fe)Si_6Al_2O_{22}(OH,F)_2$ (Amphibole)

Parisite..$Ce_2Ca(CO_3)_3F_2$

Pascoite..$Ca_2V_6O_{17}·11H_2O$

Pectolite..$Ca_2NaH(SiO_3)_3$

Pentahydrocalcite..$CaCO_3·5H_2O$

> Perovskite..$CaTiO_3$

Pharmacolite..$CaH(AsO_4)·2H_2O$

Phillipsite..$(.5Ca,Na,K)_3Al_3Si_5O_{16}·6H_2O$ (Zeolite)

Phosphuranylite..Ca uranyl phosphate

Picropharmacolite..$(Ca,Mg)_3(AsO_4)_2·6H_2O$ (?)

Piemontite..$Ca_2(Al,Fe,Mn)_2Al(OH)Si_3O_{12}$ (Epidote)

Pigeonite..$(Mg,Fe,Ca)_2(Si_2O_6)$ (Clinopyroxene)

Pintadoite..$Ca_2V_2O_7·9H_2O$

Pirssonite..$Na_2Ca(CO_3)_2·2H_2O$

Pistacite..$Ca_2(Al,Fe)Al_2(OH)Si_3O_{12}$ (Epidote)

Plombierite..$Ca_5H_2(Si_3O_9)_2·6H_2O$

Polycrase..$(Y,Ca,Ce,U,Th)(Ti,Cb,Ta)_2O_6$

Polyhalite..$K_2Ca_2Mg(SO_4)_4·2H_2O$

Polymignyte..$(Ca,Fe,Y,etc.,Zr,Th)(Cb,Ti,Ta)O_4$

Portlandite..$Ca(OH)_2$

Powellite..$Ca(MoO_4)$

Prehnite..$Ca_2Al_2(OH)_2Si_3O_{10}$

Priceite..$Ca_4B_{10}O_{19}·7H_2O$ (?)

Priorite..$(Y,Er,Ca,Fe,Th)(Ti,Cb)_2O_6$

Probertite..$NaCaB_5O_9·5H_2O$

Prosopite..$CaAl_2(F,OH)_8$

Pseudowollastonite..$CaSiO_3$

✓ Pumpellyite..$Ca_4(Mg,Fe,Mn)(Al,Fe,Ti)_5(OH)_3Si_6O_{23}·2H_2O$ (?)

Pyrochlore..$NaCa(Cb,Ta)_2O_6F$

Pyroxene..Mg,Fe,Ca,Na,Ti,Al silicates

Radiophyllite..$Ca_4(Si_4O_{10})(OH)_4·2H_2O$

Rankinite..$Ca_3Si_2O_7$

Rauvite..$CaU_2V_{12}O_{36}·20H_2O$

Retzian..Ca,Rare earths,Mn basic arsenate

Rhodonite..$(Mn,Fe,Ca)(SiO_3)$

Richellite..$Ca_3Fe_{10}(PO_4)_8(OH)_{12}·H_2O$ (?)

Richterite..$Na_2Ca(Mg,Fe,Mn,Al,Fe)_5Si_8O_{22}(OH,F)_2$ (Amphibole)

Rinkite..$Na(Ca,Ce)_2(Ti,Ce)FSi_2O_8$

✓ Riversideite..$2CaSiO_3·3H_2O$

Roeblingite..$2PbSO_4·H_{10}Ca_7Si_6O_{24}$ (?)

Romeite..$Ca_2Sb_2O_6(O,OH,F)$

✓ Roscherite..$(Ca,Mn,Fe)_2Al(PO_4)_2(OH)·2H_2O$

Roselite..$Ca_2(Co,Mg)(AsO_4)_2·2H_2O$

Rosenbuschite..$(Na,Ca,Mn)_3(Fe,Ti,Zr)FSi_2O_8$

Rossite..$CaV_2O_6·4H_2O$

✓ Roweite..$(Mn,Mg,Zn)Ca(BO_2)_2(OH)_2$

Salite..$Ca(Fe,Mg)Si_2O_6$ (Clinopyroxene)

Samarskite..$(Y,Er,Ce,U,Ca,Fe,Pb,Th)(Cb,Ta,Ti,Sn)_2O_6$

"Samarskite"..$(Ca,Pb,Y,U)(Cb,Ta,Ti,Fe)_2O_6$ (?)

Sampleite..$NaCaCu_5(PO_4)_4Cl·5H_2O$

179

(CALCIUM—continued)

Saponite..$(.5Ca,Na)_{0.7}(Al,Mg,Fe)_4(Si,Al)_8O_{20}(OH)_4 \cdot nH_2O$ (Clay)
Sarcolite..$(Ca,Na_2)_3Al_2Si_3O_{12}$ (?)
Sarcopside..$(Fe,Mn,Ca)_7(PO_4)_4F_2$ (?)
Scapolite..$(Na,Ca,K)_4Al_3(Al,Si)_3Si_6O_{24}(Cl,F,OH,CO_3,SO_4)$
√ Scawtite..$2CaCO_3 \cdot Ca_2Si_3O_8$ (?)
Scheelite..$Ca(WO_4)$
√ Scheteligite..$(Ca,Y,Sb,Mn)_2(Ti,Ta,Cb)_2(O,OH)_7$
Schizolite..$(Ca,Mn)_2NaH(SiO_3)_3$
Schroeckingerite..$NaCa_3(UO_2)(CO_3)_3(SO_4)F \cdot 10H_2O$
> Scolecite..$Ca(Al_2Si_3O_{10}) \cdot 3H_2O$ (Zeolite)
Serendibite..$Ca_2(Mg,Fe)_4Al_6B_2O_{10}Si_4O_{16}$ (?)
Serpierite..$(Zn,Cu,Ca)_5(SO_4)_2(OH)_6 \cdot 3H_2O$ (?)
Shortite..$Na_2Ca_2(CO_3)_3$
√ Sincosite..$Ca(VO)_2(PO_4)_2 \cdot 5H_2O$
Sphene..$CaTi(SiO_4)(OH,F,O)$
Spurrite..$CaCO_3 \cdot 2Ca_2SiO_4$
Stilbite..$(Ca,Na_2,K_2)(Al_2Si_7O_{18}) \cdot 7H_2O$ (Zeolite)
> Stilpnochlorane..$(Ca,K,H_2O)(Al,Fe,Mg)_3(OH,O)_2(Si,P)_4O_{10}$
Stilpnomelane..$(K,Na,Ca)(Fe,Mg,Al,Mn)_3Si_4O_{10}(OH)_2$ (?)
Stokesite..$CaZnSi_3O_9 \cdot 2H_2O$
Svabite..$Ca_5(AsO_4)_3(F,OH)$
Swartzite..$CaMg(UO_2)(CO_3)_3 \cdot 12H_2O$
Synadelphite..$(Mn,Mg,Ca,Pb)_4(AsO_4)(OH)_5$
Synchisite..$CeCa(CO_3)_2F$
Syngenite..$K_2Ca(SO_4)_2 \cdot H_2O$
Tachyhydrite..$CaMg_2Cl_6 \cdot 12H_2O$
Taramite..$Na_2Ca_2Mg_5Fe_3(Fe,Ti)_2(Al,Fe)_2(OH,F)_4Si_{14}O_{44}$ (Amphibole)
Tavistockite..$Ca_3Al_2(PO_4)_2(OH)_6$
Thaumasite..$CaSiO_3 \cdot CaCO_3 \cdot CaSO_4 \cdot 14.5H_2O$
Thomsenolite..$NaCaAlF_6 \cdot H_2O$
Thomsonite..$NaCa_2[(Al,Si)_5O_{10}]_2 \cdot 6H_2O$ (Zeolite)
Thorotungstite..$AlFe(Th,Ca,Ce,Zr)WO_3$ (?)
Thulite..$(Ca,Mn)_2Al_3(OH)(SiO_4)_3$ (var. Zoisite)
Tilasite..$CaMg(AsO_4)F$
√ Tilleyite..$Ca_5(CO_3)_2Si_2O_7$
Titanaugite..$Ca_{14}NaMg_{10}Fe_3Ti_2Fe_2Al_5Si_{27}O_{96}$ (?) (Clinopyroxene)
Titanite..$CaTiSiO_5$ (=Sphene)

Tobermorite..$Ca_5H_2(Si_3O_9)_2 \cdot 4H_2O$
√ Todorokite..$(Mn,Ba,Ca,Mg)Mn_3O_7 \cdot H_2O$ (?)
Torendrikite..hydrous Na,Ca,Mg,Fe alumino-silicate (Amphibole)
Tourmaline..$(Na_2,Ca)(Li,Al,Mg,Fe,Mn)_6Al_{12}(OH,F)_8(BO_3)_6Si_{12}O_{36}$
Tremolite..$Ca_2Mg_5Si_8O_{22}(OH,F)_2$ (Amphibole)
Trihydrocalcite..$CaCO_3 \cdot 3H_2O$
Trimerite..$(Mn,Ca)BeSiO_4$
Tritomite..$Ca_3(La,Ce)_3Zr_3F_6B_3Si_6O_{27}$ (?)
Truscottite..$(Ca,Mg)_4(OH)_2Si_6O_{15} \cdot 3H_2O$ (?)
Tscheffkinite..$(Fe,Ca)(Ce,La,Al)_2(Si,Ti)_3O_{10}$ (?)
Tschermakite..$Ca_2Mg_3Al_4Si_6O_{22}(OH,F)_2$ (Amphibole)
Tyrolite..$Cu_5Ca(AsO_4)_2(CO_3)(OH)_4 \cdot 6H_2O$ (?)
Tyuyamunite..$Ca(UO_2)_2(VO_4)_2 \cdot nH_2O$ (?)
Uhligite..$Ca_3(Ti,Al,Zr)_9O_{20}$ (?)
Ulexite..$NaCaB_5O_9 \cdot 8H_2O$
Uralite..hydrous Ca,Mg,Fe alumino-silicate (Amphibole)
Uranophane..$CaU_2O_3Si_2O_8 \cdot 7H_2O$
Uranospinite..$Ca(UO_2)_2(AsO_4)_2 \cdot 8H_2O$
Uvarovite..$Ca_3Cr_2(SiO_4)_3$ (Garnet)
Uvite..$CaMg_4Al_5B_3Si_6O_{29} \cdot 2H_2O$ (?) (Tourmaline)
√ Varulite..$(Na,Ca,)Mn(PO_4)$
Vaterite..$CaCO_3$
Vermiculite..$(Mg,Ca)_{0.7}(Mg,Fe,Al)_6[(Al,Si)_8O_{20}](OH)_4 \cdot 8H_2O$
Vesuvianite..$Ca_{10}Al_4(Mg,Fe)_2Si_9O_{34}(OH,F)_4$
Vishnewite..$(Na,Ca,K)_{6-7}Al_6Si_6O_{24}(SO_4,CO_3,Cl)_{1-1.5} \cdot 1-5H_2O$
Voglite..$Ca_2CuU(CO_3)_5 \cdot 6H_2O$ (?)
Volchonskoite..$(Ca,Mg,Cr,Fe,Al)_2O_3 \cdot 3SiO_2 \cdot nH_2O$ (?) (Clay)
Wairakite..$Ca(AlSi_2O_6)_2 \cdot 2H_2O$
Wardite..$Na_4CaAl_{12}(PO_4)_8(OH)_9 \cdot 3H_2O$
Wattevilleite..$Na_2Ca(SO_4)_2 \cdot 4H_2O$ (?)
Weddellite..$CaC_2O_4 \cdot 2H_2O$
Whewellite..$CaC_2O_4 \cdot H_2O$
Whitlockite..$Ca_3(PO_4)_2$
Wilkeite..$Ca_5(P,S,Si,CO_4)_3(OH)$
Woehlerite..$NaCa_2(Zr,Cb)FSi_2O_8$
Wollastonite..$Ca(SiO_3)$
Woodhouseite..$CaAl_3(PO_4)(SO_4)(OH)_6$
Xanthophyllite..$Ca_2(Mg_5Al)Si_2Al_6O_{20}(OH)_4$ (Brittle Mica)
Xanthoxenite..$Ca_2Fe(PO_4)_2(OH) \cdot 1.5H_2O$

(CALCIUM-continued)

Xonotlite..$Ca_3Si_3O_9 \cdot H_2O$ (?)
Yttrocrasite..$(Y,Th,U,Ca)_2(Ti,Fe,W)_4O_{11}$ (?)
Yttrotantalite..$(Fe,Y,U,Ca,etc...)(Cb,Ta,Zr,Sn)O_4$
Yugawaralite..hydrous Ca,Al silicate (Zeolite)
Zeolite..ca.$(Na_2,K_2,Ca,Ba,Sr)[(Al,Si)O_2]_n \cdot xH_2O$
Zeophyllite..$Ca_8(OH,F)_{10}Si_6O_{15}$
Zirkelite..$(Ca,Fe,Th,U)_2(Ti,Zr)_2O_5$
✓ Zirklerite..Al,Fe basic chloride
Zoisite..$Ca_2Al_3(OH)(SiO_4)_3$

CESIUM

Avogadrite..$(K,Cs)BF_4$
Pollucite..$(Cs,Na)(AlSi_2O_6) \cdot H_2O$ (?)

CHROMIUM

Barbertonite..$Mg_6Cr_2(OH)_{16} \cdot CO_3 \cdot 4H_2O$
Beresovite..$Pb_6(CrO_4)_3(CO_3)O_2$
Chromite..$FeCr_2O_4$
Clay..chiefly hydrous silicates of Al or Mg
Crocoite..$Pb(CrO_4)$
Daubreelite..Cr_2FeS_4
Dietzeite..$Ca_2(IO_3)_2CrO_4$
Kaemmererite..Cr,Mg chlorite (Chlorite)
Lopezite..$K_2(Cr_2O_7)$
Magnesiochromite..$MgCr_2O_4$
Phoenicochroite..$Pb_3(CrO_4)_2O$ (?)
Redingtonite..$(Fe,Mn,Ni)(Cr,Al)_2(SO_4)_4 \cdot 22H_2O$ (?)
Stichtite..$Mg_6Cr_2(OH)_{16}CO_3 \cdot 4H_2O$
Tarapacaite..K_2CrO_4
Uvarovite..$Ca_3Cr_2(SiO_4)_3$ (Garnet)
Vauquelinite..Pb,Cu chromate-phosphate
Volchonskoite..$(Ca,Mg,Cr,Fe,Al)_2O_3 \cdot 3SiO_2 \cdot nH_2O$ (Clay) (?)

COBALT

> Badenite..$(Co,Ni)(As,Bi)_4$ (?)
Bieberite..$CoSO_4 \cdot 7H_2O$

Carrollite..Co_2CuS_4
Chloanthite..$(Ni,Co)As$ (?)
Cobaltite..$CoAsS$
Cobaltomentite..$CoSeO_3 \cdot 2H_2O$ (?)
Erythrite..$Co_3(AsO_4)_2 \cdot 8H_2O$
Forbesite..$(Ni,Co)H(AsO_4) \cdot 3.5H_2O$ (?)
Glaucodot..$(Co,Fe)AsS$
Julienite..$Na_2Co(SCN)_4 \cdot 8H_2O$
Linnaeite..Co_3S_4
> Nickel skutterudite..$(Ni,Co)As_3$
Roselite..$Ca_2(Co,Mg)(AsO_4)_2 \cdot 2H_2O$
Safflorite..$(Co,Fe)As_2$
Siegenite..$(Co,Ni)_3S_4$
Skutterudite..$(Co,Ni)As_3$
Smaltite..$(Co,Ni)As$ (?)
Spherocobaltite..$CoCO_3$
Stainierite..$CoO(OH)$ (?)

COLUMBIUM

Ampangabeite..$(Y,Er,U,Ca,Th)_2(Cb,Ta,Fe,Ti)_7O_{18}$
Betafite..$(U,Ca)(Cb,Ta,Ti)_3O_9 \cdot nH_2O$
Bismutotantalite..$Bi(Ta,Cb)O_4$
Columbite..$(Fe,Mn)(Cb,Ta)_2O_6$
Djalmaite..$(U,Ca,Pb,Bi,Fe)(Ta,Cb,Ti,Zr)_3O_9 \cdot nH_2O$
Epistolite..$H_3Na_6Cb_3Si_5TiO_{24} \cdot 2H_2O$ (?)
Eschynite..$(Ce,Ca,Fe,Th)(Ti,Cb)_2O_6$
Euxenite..$(Y,Ca,Ce,U,Th)(Cb,Ta,Ti)_2O_6$
Fergusonite..$(Y,Er,Ce,Fe)(Cb,Ta,Ti)O_4$
✓ Formanite..$(U,Zr,Th,Ca)(Ta,Cb,Ti)O_4$
Ishikawaite..$(U,Fe,Y,etc.)(Cb,Ta)O_4$
Mossite..$(Cb,Ta)_2O_6$
Polycrase..$(Y,Ca,Ce,U,Th)(Ti,Cb,Ta)_2O_6$
Polymignyte..$(Ca,Fe,Y,etc.,Zr,Th)(Cb,Ti,Ta)O_4$
Priorite..$(Y,Er,Ca,Fe,Th)(Ti,Cb)_2O_6$
Pyrochlore..$NaCa(Cb,Ta)_2O_6F$
"Samarskite"..$(Ca,Pb,Y,U)(Cb,Ta,Ti,Fe)_2O_6$ (?)
Samarskite..$(Y,Er,Ce,U,Ca,Fe,Pb,Th)(Cb,Ta,Ti,Sn)_2O_6$
✓ Schetelgite..$(Ca,Y,Sb,Mn)_2(Ti,Ta,Cb)_2(O,OH)_7$
> Stibiocolumbite..$SbCbO_4$
Tantalite..$(Fe,Mn)(Ta,Cb)_2O_6$
Woehlerite..$NaCa_2(Zr,Cb)FSi_2O_8$
Yttrotantalite..$(Fe,Y,U,Ca,etc.)(Cb,Ta,Zr,Sn)O_4$

COPPER

Aikinite..$PbCuBiS_3$
Alaskaite..$Pb(Ag,Cu)_2Bi_4S_8$ (?)
Algodonite..Cu_6As
Allopalladium..Pd with traces of Hg,Pt,Ru,Cu,etc.
Andrewsite..$(Cu,Fe)_3Fe_6(PO_4)_4(OH)_{12}$
Antlerite..$Cu_3(SO_4)(OH)_4$
Atacamite..$Cu_2(OH)_3Cl$
Arnimite..$Cu_5(SO_4)_2(OH)_6 \cdot 3H_2O$ (?)
Arsenobismite..Bi basic arsenate
Arzrunite..Pb,Cu sulfate-chloride
Aurichalcite..$(Zn,Cu)_5(CO_3)_2(OH)_6$
Azurite..$Cu_3(CO_3)_2(OH)_2$
Bandylite..$CuB_2O_4 \cdot CuCl_2 \cdot 4H_2O$
Bayldonite..$(Cu,Pb)_2(AsO_4)(OH)$ (?)
Beaverite..$Pb(Cu,Fe,Al)_3(SO_4)_2(OH)_6$
✓ Bellingerite..$Cu(IO_3)_2 \cdot 2/3H_2O$
✓ Benjaminite..$Pb(Cu,Ag)Bi_2S_4$
Berthonite..$Pb_2Cu_7Sb_5S_{13}$
Berzelianite..Cu_2Se
Boleite..$Pb_9Cu_8Ag_3Cl_{21}(OH)_{16} \cdot 2H_2O$ (?)
Boothite..$CuSO_4 \cdot 7H_2O$
Bornite..Cu_5FeS_4
Botallackite..$Cu_2(OH)_3Cl \cdot H_2O$ (?)
Bournonite..$PbCuSbS_3$
Brochantite..$Cu_4(SO_4)(OH)_6$
✓ Buttgenbachite..$Cu_{19}(NO_3)_2(OH)_{32}Cl_4 \cdot 3H_2O$
Calciovolborthite..$CuCa(VO_4)(OH)$
Caledonite..$Cu_2Pb_5(SO_4)_3(CO_3)(OH)_6$
Carrollite..Co_2CuS_4
Ceruleite..$CuAl_4(AsO_4)_2(OH)_8 \cdot 4H_2O$
Chalcanthite..$CuSO_4 \cdot 5H_2O$
Chalcoalumite..$CuAl_4(SO_4)(OH)_{12} \cdot 3H_2O$
Chalcocite..Cu_2S
Chalcocyanite..$CuSO_4$
Chalcomenite..$CuSeO_3 \cdot 2H_2O$
Chalcophyllite..$Cu_{18}Al_2(AsO_4)_3(SO_4)_3(OH)_{27} \cdot 33H_2O$
Chalcopyrite..Cu_2FeS_4
Chalcosiderite..$CuFe_6(PO_4)_4(OH)_8 \cdot 4H_2O$
Chalcostibite..$CuSbS_2$
Chenevixite..$Cu_2Fe_2(AsO_4)_2(OH)_4 \cdot H_2O$ (?)
Chlorothionite..$K_2Cu(SO_4)Cl_2$
✓ Chloroxiphite..$PbCuO_2Cl_2(OH)_2$ (?)

Chrysocolla..$CuSiO_3 \cdot 2H_2O$
Clinoclase..$Cu_3(AsO_4)(OH)_3$
Cocinerite..Cu_4AgS
Colusite..$Cu_3(Sn,Te,Fe,V,As)S_4$
Conichalcite..$CaCu(AsO_4)(OH)$
Connellite..$Cu_{19}(SO_4)Cl_4(OH)_{32} \cdot 3H_2O$ (?)
Copper..Cu
Cornetite..$Cu_3(PO_4)(OH)_3$
Cornwallite..$Cu_5(AsO_4)_2(OH)_4 \cdot H_2O$
Covellite..CuS
Crednerite..$CuMn_2O_4$
Crookesite..$(Cu,Tl,Ag)_2Se$
Cubanite..$CuFe_2S_3$
Cumengite..$Pb_4Cu_4Cl_8(OH)_8 \cdot H_2O$ (?)
Cuprosklodowskite..$CuUO_3Si_2O_8 \cdot 6H_2O$
Cuprotungstite..$Cu_2(WO_4)(OH)_2$
Cuprite..Cu_2O
Cuprocopiapite..$CuFe_4(SO_4)_6(OH)_2 \cdot 20H_2O$
Cyanochroite..$K_2Cu(SO_4)_2 \cdot 6H_2O$
Cyanotrichite..$Cu_4Al_2(SO_4)(OH)_{12} \cdot 2H_2O$
Delafossite..$CuFeO_2$
Devilline..$Cu_4Ca(SO_4)_2(OH)_6 \cdot 3H_2O$
Diaboleite..$Pb_2CuCl_2(OH)_4$
✓ Digenite..$Cu_{2-x}S$
Dioptase..$Cu_6Si_6O_{18} \cdot 6H_2O$
Dolerophanite..$Cu_2(SO_4)O$
Domeykite..Cu_3As
Duftite..$PbCu(AsO_4)(OH)$
Emplectite..$CuBiS_2$
Enargite..Cu_3AsS_4
Epigenite..$(Cu,Fe)_5AsS_6$ (?)
Erinite..$Cu_5(AsO_4)_2(OH)_4$
Eriochalcite..$CuCl_2 \cdot 2H_2O$
Eucairite..CuAgSe
Euchroite..$Cu_2(AsO_4)(OH) \cdot 3H_2O$
Euchlorin..K,Na,Cu basic sulfate
Famatinite..Cu_3SbS_4
Freirinite..$Na_3Cu_3(AsO_4)_2(OH)_3 \cdot H_2O$
Gerhardtite..$Cu_2(NO_3)(OH)_3$
✓ Germanite..Cu_3GeS_4 (?)
Gladite..$PbCuBi_5S_9$
Glaucocerinite..$Zn_{13}Al_8Cu_7(SO_4)_2(OH)_{60} \cdot 4H_2O$ (?)
Goldfieldite..$Cu_{12}Te_3Sb_4S_{16}$ (?)
Guildite..$(Cu,Fe)_3(Fe,Al)_4(SO_4)_7(OH)_4 \cdot 15H_2O$
Hammarite..$Pb_2Cu_2Bi_4S_9$ (?)

(COPPER-continued)

Horsfordite..Cu_5Sb
Hutchinsonite..$(Pb,Tl)_2(Cu,Ag)As_5S_{10}$
Johannite..$Cu(UO_2)_2(SO_4)_2(OH)_2 \cdot 6H_2O$
Kamarezite..$Cu_3(SO_4)(OH)_4 \cdot 6H_2O$ (?)
✓ Klaprothite..$Cu_6Bi_4S_9$ (?)
Klockmannite..$CuSe$
Kroehnkite..$Na_2Cu(SO_4)_2 \cdot 2H_2O$
Langite..$Cu_4(SO_4)(OH)_6 \cdot H_2O$ (?)
Lautite..$CuAsS$
Leightonite..$K_2Ca_2Cu(SO_4)_4 \cdot 2H_2O$
Lengenbachite..$Pb_6(Ag,Cu)_2As_4Si_3$
Leucochalcite..$Cu_2(AsO_4)(OH) \cdot H_2O$
Libethenite..$Cu_2(PO_4)(OH)$
Linarite..$PbCu(SO_4)(OH)_2$
Lindackerite..$Cu_6Ni_3(AsO_4)_4(SO_4)(OH)_4 \cdot 5H_2O$
Lindgrenite..$Cu_3(MoO_4)_2(OH)_2$
Lindstromite..$PbCuBi_3S_6$
Liroconite..$Cu_2Al(AsO_4)(OH)_4 \cdot 4H_2O$
Malachite..$Cu_2(CO_3)(OH)_2$
Marshite..CuI
Metatorbernite..$Cu(UO_2)_2(PO_4)_2 \cdot 8H_2O$
Metazeunerite..$Cu(UO_2)_2(AsO_4)_2 \cdot 8H_2O$
Miersite..$(Ag,Cu)I$
Mitscherlichite..$K_2CuCl_4 \cdot 2H_2O$
Mixite..$Cu_{11}Bi(AsO_4)_5(OH)_{10} \cdot 6H_2O$ (?)
Mottramite..$CuPb(VO_4)(OH)$
Nantokite..$CuCl$
Natrochalcite..$NaCu_2(SO_4)_2(OH) \cdot H_2O$
Olivenite..$Cu_2(AsO_4)(OH)$
Paramelaconite..$Cu_{1-2x}Cu_{2x}O_{1-x}$
Paratacamite..$Cu_2(OH)_3Cl$
Pearceite..$(Ag,Cu)_{16}As_2S_{11}$
✓ Penroseite..$(Ni,Cu,Pb)Se_2$
Percylite..$PbCuCl_2(OH)_2$
Pisanite..$(Fe,Cu)SO_4 \cdot 7H_2O$
Polybasite..$(Ag,Cu)_{16}Sb_2S_{11}$
Pseudoboleite..$Pb_5Cu_4Cl_{10}(OH)_8 \cdot 2H_2O$ (?)
Pseudomalachite..$Cu_5(PO_4)_2(OH)_4 \cdot H_2O$ (?)
Ransomite..$Cu(Fe,Al)_2(SO_4)_4 \cdot 7H_2O$
Rezbanyite..$Pb_3Cu_2Bi_{10}S_{19}$
Rickardite..Cu_4Te_3
Rosasite..$(Cu,Zn)_2(CO_3)(OH)_2$
Salesite..$Cu(IO_3)(OH)$
Sampleite..$NaCaCu_5(PO_4)_4Cl \cdot 5H_2O$
Seligmannite..$PbCuAsS_3$
✓ Sengierite..$Cu(UO_2)(VO_4)(OH) \cdot 4-5H_2O$ (?)

Serpierite..$(Zn,Cu,Ca)_5(SO_4)_2(OH)_6 \cdot 3H_2O$ (?)
Spangolite..$Cu_6Al(SO_4)(OH)_{12}Cl \cdot 3H_2O$
Stannite..Cu_2FeSnS_4
Stromeyerite..$AgCuS$
Stylotypite..$(Ag,Cu,Fe)_3SbS_3$
Sulvanite..Cu_3VS_4
Tagilite..$Cu_2(PO_4)(OH) \cdot H_2O$
✓ Teineite..$Cu_{13}(SO_4)_3(TeO_4)_{10} \cdot 26H_2O$
Tennantite..$(Cu,Fe)_{12}As_4S_{13}$
Tenorite..CuO
Tetrahedrite..$(Cu,Fe)_{12}Sb_4S_{13}$
Torbernite..$Cu(UO_2)_2(PO_4)_2 \cdot 8-12H_2O$
Trichalcite..$Cu_3(AsO_4)_2 \cdot 5H_2O$ (?)
Trippkeite..$CuAs_2O_4$
Tsumebite..$Pb_2Cu(PO_4)(OH)_3 \cdot 3H_2O$
Turanite..$Cu_2(VO_4)(OH)$ (?)
Turquois..$CuAl_6(PO_4)_4(OH)_8 \cdot 4H_2O$
Tyrolite..$Cu_5Ca(AsO_4)_2(CO_3)(OH)_4 \cdot 6H_2O$ (?)
Umangite..Cu_3Se_2
Valleriite..$Cu_2Fe_4S_7$ (?)
Vandenbrandite..$CuO \cdot UO_3 \cdot 2H_2O$
Vauquelinite..Pb,Cu chromate-phosphate
Vernadskite..$Cu_4(SO_4)_3(OH)_2 \cdot 4H_2O$
Veszelyite..$(Cu,Zn)_3(As,PO_4)(OH)_3 \cdot 2H_2O$
Voglite..$Ca_2CuU(CO_3)_5 \cdot 6H_2O$ (?)
Volborthite..$Cu_3(VO_4)_2 \cdot 3H_2O$ (?)
Weissite..Cu_5Te_3
✓ Wherryite..$Pb_4Cu(CO_3)(SO_4)_2(OH,Cl)O$ (?)
Wittichenite..Cu_3BiS_3 (?)
Woodwardite..$Cu_4Al_2(SO_4)(OH)_{12} \cdot 2-4H_2O$ (?)
Zeunerite..$Cu(UO_2)_2(AsO_4)_2 \cdot 10-16H_2O$

GOLD

Aurosmiridium..Ir,Os,Au
Calaverite..$AuTe_2$
Gold..Au
Gold amalgam..Au_2Hg_3 (?)
Krennerite..$AuTe_2$
Maldonite..Au_2Bi
Muthmannite..$(Ag,Au)Te$
Nagyagite..$Pb_5Au(Te,Sb)_4S_{5-8}$ (?)
Petzite..Ag_3AuTe_2
Sylvanite..$(Ag,Au)Te_2$

IRON

Acmite..$NaFe(Si_2O_6)$
 (Clinopyroxene)
Actinolite..$Ca_2(Mg,Fe)_5Si_8O_{22}(OH,F)_2$
 (Amphibole)
Aegirine..$NaFe(Si_2O_6)$
 (Clinopyroxene)
Aegirine-augite..$(Na,Ca)(Fe,Mg,Fe,Al)(Si_2O_6)$
 (Clinopyroxene)
Allanite..$(Ca,Ce,La)_2(Al,Fe,Be,Mg,Mn)_3$
 (=Orthite) $(OH)Si_3O_{12}$ (?)
Alluaudite..$(Na,Fe,Mn)(PO_4)$
Almandine..$Fe_3Al_2(SiO_4)_3$
 (Garnet)
Amarantite..$Fe(SO_4)(OH)\cdot 3H_2O$
Amarillite..$NaFe(SO_4)_2\cdot 6H_2O$
Ammoniojarosite..$(NH_4)Fe_3(SO_4)_2(OH)_6$
Ampangabeite..$(Y,Er,U,Ca,Th)_2(Cb,Ta,Fe,Ti)_7O_{18}$ (?)
Amphibole..hydrous Fe,Mg,Na,Ca,Mn,K,Al silicates
Anapaite..$Ca_2Fe(PO_4)_2\cdot 4H_2O$
Andradite..$Ca_3(Fe,Ti)_2(SiO_4)_3$
 (Garnet)
Andrewsite..$(Cu,Fe)_3Fe_6(PO_4)_4(OH)_{12}$
Ankerite..$Ca(Fe,Mg)(CO_3)_2$
Anthophyllite..$(Mg,Fe)_7Si_8O_{22}(OH,F)_2$
 (Amphibole)
Arfvedsonite..$.5Na_5Ca(Fe,Mg,Fe,Al)_{10}Si_{15}AlO_{44}(OH,F)_4$
 (Amphibole)
Argentojarosite..$AgFe_3(SO_4)_2(OH)_6$
✓ Arizonite..$Fe_2Ti_3O_9$
Arrojadite..$Na_2(Fe,Mn)_5(PO_4)_4$
Arseniopleite..$(Mn,Ca,Pb,Mg)_9(Mn,Fe)_2(AsO_4)(OH)_6$
Arseniosiderite..$Ca_3Fe_4(AsO_4)_4(OH)_4\cdot 4H_2O$ (?)
Arsenopyrite..FeAsS
Astrophyllite..$(K_2,Na_2,Ca)(Fe,Mn)_4(Ti,Zr)(OH,F)_2Si_4O_{14}$ (?)
Attapulgite..hydrous Mg,Fe silicate
 (Clay)
Augite..$(Ca,Mg,Fe,Fe,Ti,Al)_2(Si,Al)_2O_6$
 (Clinopyroxene)
Axinite..$(Ca,Mn,Fe)_3Al_2BO_3(Si_4O_{12})(OH)$
Babingtonite..hydrous Ca,Fe silicate
Badenite..$(Co,Ni,Fe)_3(As,Bi)_4$ (?)
Barkevikite..$Ca_2(Na,K)(Fe,Mg,Mn)_5(Si_{6.5}Al_{1.5}O_{22})(OH)_2$
 (Amphibole)
Barroisite..Na,K,Ca,Mg,Fe,Ti alumino-silicate
 (Amphibole)

Basaltic hornblende..$Ca_2(Na,K)_{0.5-1.0}(Mg,Fe)_{3-2}(Fe,Al)_{2-3}Si_6Al_2O_{22}(O,OH,F)_2$
 (Amphibole)
✓ Bassetite..Ferrous uranyl phosphate
Beaverite..$Pb(Cu,Fe,Al)_3(SO_4)_2(OH)_6$
Beraunite..$Fe_5(PO_4)_3(OH)_5\cdot 3H_2O$ (?)
Bermanite..$(Mn,Mg)_5(Mn,Fe)_8(PO_4)_8(OH)_{10}\cdot 15H_2O$
Berthierite..$FeSb_4S_4$
Beudantite..$PbFe_3(AsO_4)(SO_4)(OH)_6$
> Bilinite..$Fe_3(SO_4)_4\cdot 22H_2O$
Biotite..$K(Mg,Fe,Mn)_3(AlSi_3O_{10})(OH,F)_2$
 (Mica)
Bixbyite..$(Mn,Fe)_2O_3$
Blakeite..Ferric tellurite
Borickite..$CaFe_5(PO_4)_2(OH)_{11}\cdot 3H_2O$
Bornite..Cu_5FeS_4
Botryogen..$MgFe(SO_4)$
Brackebuschite..$Pb_4MnFe(VO_4)_4\cdot 2H_2O$
✓ Brannerite..$(U,Ca,Fe,Y,Th)_3Ti_5O_{16}$ (?)
Bravoit..$(Ni,Fe)S_2$
Bronzite..$(Mg,Fe)(SiO_3)$
 (Orthopyroxene)
Brugnatellite..$Mg_6Fe(OH)_{13}\cdot CO_3\cdot 4H_2O$
Brunsvigite..hydrous Fe,Mg,Al silicate
 (Chlorite)
Bustamite..$(Mn,Ca,Fe)(SiO_3)$
Butlerite..$Fe(SO_4)(OH)\cdot 2H_2O$
Byssolite..hydrous Ca,Mg,Fe alumino-silicate
 (Amphibole)
Cacoxenite..$Fe_4(PO_4)_3(OH)_3\cdot 12H_2O$
Calcioferrite..$Ca_3Fe_3(PO_4)_4(OH)_3\cdot 8H_2O$ (?)
Calderite..$Mn_3Fe_2(SiO_4)_3$
Cardenite..hydrous Mg,Ca,Fe,Al silicate
 (Clay)
Carminite..$PbFe_2(AsO_4)_2(OH)_2$
Carphosiderite..$(H_2O)Fe_3(SO_4)_2[(OH)_5\cdot H_2O]$
Celadonite..$(K,Ca,Na)(Al,Fe,Mg)_2(Al_{0.11}Si_{3.89}O_{10})(OH)_2$ (?)
Cerite..$(Ca,Fe)Ce_3Si_3O_{12}\cdot H_2O$ (?)
Chalcopyrite..$Cu_2Fe_2S_4$
Chalcosiderite..$CuFe_6(PO_4)_4(OH)_8\cdot 4H_2O$
Chamosite..hydrous Fe,Mg,Al silicate
 (Chlorite)
Chenevixite..$Cu_2Fe_2(AsO_4)_2(OH)_4\cdot H_2O$ (?)
Childrenite..$(Fe,Mn)Al(PO_4)(OH)_2\cdot H_2O$
Chlorite..$(Mg,Al,Fe)_{12}(Si,Al)_8O_{20}(OH)_{16}$
Chloritoid..$(Fe,Mg)_2Al_4(OH)_4Si_2O_{10}$ (?)
Chromite..$FeCr_2O_4$
Chrysolite..$(Mg,Fe)_2(SiO_4)$
 (Olivine)
Clay..chiefly hydrous silicates of Al or Mg

(IRON--continued)

Clinochlore..hydrous Fe,Mg,Al silicate
  (Chlorite)
Clinoferrosilite..$FeSiO_3$
  (Clinopyroxene)
Clinohypersthene..$(Mg,Fe)SiO_3$
  (Clinopyroxene)
Clino-ungemachite..$Na_9K_3Fe(SO_4)_6(OH)_3 \cdot 9H_2O$
Cohenite..$(Fe,Ni)_3C$
Collinsite..$Ca_2(Mg,Fe)(PO_4)_2 \cdot 2H_2O$
Columbite..$(Fe,Mn)(Cb,Ta)_2O_6$
Colusite..$Cu_3(Sn,Te,Fe,V,As)S_4$
Common hornblende..$(Ca,Na,K)_{2-3}(Mg,Fe,Al)_5$
  (Amphibole) $Si_6(Si,Al)_2O_{22}(OH,F)_2$
Copiapite..$Fe_5(SO_4)_6(OH)_2 \cdot 20H_2O$
Coquimbite..$Fe_2(SO_4)_3 \cdot 9H_2O$
Cordierite..$Al_3(Mg,Fe)_2(Si_5AlO_{18})$
Corkite..$PbFe_3(PO_4)(SO_4)(OH)_6$
Corundophilite..hydrous Mg,Fe,Al silicate
  (Chlorite)
Crocidolite..$Na_2Fe_5Si_8O_{22}(OH)_2$
  (Amphibole)
Cronstedtite..$Fe_6(Fe_2Si_2O_{10})(OH)_8$
  (Serpentine)
Crossite..$Na(Mg,Fe)_3(Al,Fe)_2Si_8O_{22}(OH,F)_2$
  (Amphibole)
Cubanite..$CuFe_2S_3$
Cummingtonite..$(Fe,Mg)_7Si_8O_{22}(OH)_2$
  (Amphibole)
Cuprocopiapite..$CuFe_4(SO_4)_6(OH)_2 \cdot 20H_2O$
Danalite..$Fe_4Be_3Si_3O_{12}S$
Daphnite..hydrous Fe,Al silicate
  (Chlorite)
Daubreelite..Cr_2FeS_4
Delafossite..$CuFeO_2$
Delessite..hydrous Mg,Fe,Al silicate
  (Chlorite)
Delorenzite..$(Y,U,Fe)(Ti,Sn...)_3O_8$ (?)
Delvauxite..$Fe_2(PO_4)(OH)_3 \cdot xH_2O$ (?)
Derbylite..$Fe_6Ti_6Sb_2O_{23}$ (?)
Diabantite..hydrous Mg,Fe,Al silicate
  (Chlorite)
Diadochite..$Fe(PO_4)(SO_4)(OH) \cdot 5H_2O$
Diallage..$Ca_{14}Fe_2Mg_{13}FeAl_5Si_{29}O_{96}$ (?)
  (Clinopyroxene)
Dickinsonite..$Na_6(Mn,Fe,Ca)_{14}H_2(PO_4)_{12} \cdot H_2O$
Didymolite..$(Ca,Mg,Fe)_2Al_6Si_8O_{27}$ (?)
Dietrichite..$(Zn,Fe,Mn)Al_2(SO_4)_4 \cdot 22H_2O$
Djalmaite..$(U,Ca,Pb,Bi,Fe)(Ta,Cb,Ti,Zr)_3O_9$
                                          $\cdot nH_2O$
Douglasite..$K_2FeCl_4 \cdot 2H_2O$ (?)
Dufrenite..$Fe_5(PO_4)_3(OH)_5 \cdot 2H_2O$ (?)
Dumortierite..$H(Al,Fe)_2Si_3Al_6BO_{20}$ (?)
Dussertite..$BaFe_3(AsO_4)_2(OH)_5 \cdot H_2O$

Egueite..$CaFe_{14}(PO_4)_{10}(OH)_{14} \cdot 21H_2O$ (?)
Ekermannite...$5Na_5Ca(Mg,Fe,Fe,Al,Li)_{10}$
  (Amphibole)
             $Si_{15}AlO_{44}(OH,F)_4$
Ekmannite..$(Fe,Mg,Mn)_3(OH)_2(Si,Al)Si_3O_{10}$ (?)
Emmonsite..$Fe_2(TeO_3)_3 \cdot 2H_2O$ (?)
Enigmatite..$(Ca,Na_2)_2Fe(Al,Fe,Ti)_4$
                                  $(Si_2O_7)_2$ (?)
Eosphorite..$(Mn,Fe)Al(PO_4)(OH)_2 \cdot H_2O$
Epidote..$Ca_2(Al,Fe)Al_2(OH)Si_3O_{12}$
Epigenite..$(Cu,Fe)_5AsS_6$ (?)
Epistolite..$(Na,Ca)(Cb,Ti,Mg,Fe,Mn)SiO_4(OH)$
Erythrosiderite..$K_2FeCl_5 \cdot H_2O$
Eschynite..$(Ce,Ca,Fe,Th)(Ti,Cb)_2O_6$
Eudialyte..$(Na,Ca,Fe)_6ZrSi_6O_{18}(OH,Cl)$ (?)
Eulite..$(Fe,Mg)_2(Si_2O_6)$
  (Orthopyroxene)
Fairfieldite..$Ca(Mn,Fe)(PO_4)_2 \cdot 2H_2O$
Fassaite..$Ca(Mg,Fe,Al,Ti)(Si,Al)_2O_6$
  (Clinopyroxene)
Fayalite..$Fe_2(SiO_4)$
  (Olivine)
Ferberite..$Fe(WO_4)$
Fergusonite..$(Y,Er,Ce,Fe)(Cb,Ta,Ti)O_4$
Ferrimolybdite..$Fe_2(MoO_4)_3 \cdot 8H_2O$ (?)
Ferrinatrite..$Na_3Fe(SO_4)_3 \cdot 3H_2O$
Ferrisepiolite..$(Mg,Fe)_4(H_2O)_3Si_6O_{15}(OH)_2$
  (Clay)
Ferri-sicklerite..$(Li,Fe,Mn)(PO_4)$
✓ Ferritungstite..$Fe_2(WO_4)(OH)_4 \cdot 4H_2O$ (?)
Ferroactinolite..$Ca_2Fe_5Si_8O_{22}(OH,F)_2$
  (Amphibole)
Ferrocarpholite..$FeAl_2(Si_2O_6)(OH)_4$
Ferroedenite..$NaCa_2Fe_5AlSi_7O_{22}(OH,F)_2$
  (Amphibole)
Ferrohastingsite..$NaCa_2Fe_4(Al,Fe)Si_6$
  (Amphibole)
                                $Al_2O_{22}(OH,F)_2$
Ferrohortonolite..$(Fe,Mg)_2(SiO_4)$
  (Olivine)
Ferrohypersthene..$(Fe,Mg)(SiO_3)$
  (Orthopyroxene)
Ferrosalite..$Ca(Fe,Mg)Si_2O_6$
  (Clinopyroxene)
Ferrotschermakite..$Ca_2Fe_5Al_2Si_6O_{22}(OH,F)_2$
  (Amphibole)
Fervanite..$Fe_4V_4O_6 \cdot 5H_2O$
Fibroferrite..$Fe(SO_4)(OH) \cdot 5H_2O$ (?)
Fillowite..$Na_6(Mn,Fe,Ca)_{14}H_2(PO_4)_{12} \cdot H_2O$ (?)
Flajolotite..$FeSbO_4 \cdot .75H_2O$ (?)
Franklinite..$ZnFe_2O_4$
Friedelite..$(Mn,Fe)_{14}(Si_{14}O_{35})(OH,Cl)_{14}$ (?)
Frondelite..$MnFe_4(PO_4)_3(OH)_5$
Gadolinite..$FeY_2Be_2(SiO_5)_2$
Gedrite..$(Mg,Fe)_6AlSi_6(Si,Al)_2O_{22}(OH,F)_2$
> (Amphibole)

185

(IRON-continued)

> Gillespite..$FeBaSi_4O_{10}$
Glaucodot..$(Co,Fe)AsS$
Glauconite..$(K,Ca,Na)(Al,Fe,Mg)_2$
 (Mica) $(Al_{0.35}Si_{3.65}O_{10})(OH)$ (?)
Glockerite..$Fe_4(SO_4)(OH)_{10} \cdot nH_2O$ (?)
Goethite..$Fe_2O_3 \cdot H_2O$
— Gonyerite..low Al chlorite
 (Chlorite)
Graftonite..$(Fe,Mn,Ca)_3(PO_4)_2$
Grandidierite..$H_2Na_2(Mg,Fe)_7(Al,Fe,B)_{15}$
 $Si_7Al_7O_{56}$ (?)
Greenalite..$Fe(Si_4O_{10})(OH)_8$
 (Serpentine)
Griphite..$(Na,Ca,Fe,Al)_3Mn_2(PO_4)_{2.5}$
 $(OH,F)_2$ (?)
Grunerite..$(Mg,Fe)_7Si_8O_{22}(OH)_2$
 (Amphibole)
Gudmundite..$FeSbS$
Guildite..$(Cu,Fe)_3(Fe,Al)_4(SO_4)_7(OH)_4 \cdot 15H_2O$
Halotrichite..$FeAl_2(SO_4)_4 \cdot 22H_2O$
Hedenbergite..$CaFe(Si_2O_6)$
 (Clinopyroxene)
Hellandite..$(Ca,Y,Er,Mn)_3(Al,Fe)Si_2O_4 \cdot H_2O$ (?)
Helvite ..$(Mn,Fe,Zn)_4(BeSiO_4)_3S$
Hematite ..Fe_2O_3
✓ Hematophanite..$Pb(Cl,OH)_2 \cdot 4PbO \cdot 2Fe_2O_3$ (?)
Hercynite..$FeAl_2O_4$
Heterosite..$(Fe,Mn)(PO_4)$
Hisingerite ..$Fe_2O_3 \cdot 2SiO_2 \cdot nH_2O$
Hoegbomite..$Mg(Al,Fe,Ti)_4O_7$
Hohmannite..$Fe_2(SO_4)_2(OH)_2 \cdot 7H_2O$
Holdenite..$(Mn,Ca)_4(Zn,Mg,Fe)(AsO_4)(OH)_5O_2$
Holmquistite..$Li_2(Mg,Fe)_3(Al,Fe)_2Si_8O_{22}(OH,F)_2$
 (Amphibole)
Homilite..$Ca_2FeB_2(SiO_5)_2$
Hortonolite..$(Fe,Mg)_2(SiO_4)$
 (Olivine)
✓ Huhnerkobelite..$(Na,Ca)Fe(PO_4)$
Hulsite ..$(Fe,Ca,Mg)_4(Fe,Sn)_2B_2O_{10}$ (?)
Humboldtine..$FeC_2O_4 \cdot 2H_2O$
Hyalosiderite..$(Mg,Fe)_2(SiO_4)$
 (Olivine)
Hydrobiotite..$(K,H_2O)(Mg,Fe,Mn)_3$
 $(Al,Si_3O_{10})(OH,H_2O)_2$
Hypersthene..$(Fe,Mg)(SiO_3)$
 (Orthopyroxene)
Ilesite..$(Mn,Zn,Fe)SO_4 \cdot 4H_2O$ (?)
Ilmenite..$FeTiO_3$
Ilvaite..$Ca(Fe,Mn)_2Fe(SiO_4)_2(OH)$
Iron..Fe
Ishikawaite..$(U,Fe,Y,etc.)(Cb,TaO_4)$

Jacobsite..$MnFe_2O_4$
Jamesonite..$Pb_4FeSb_6S_{14}$
Jarosite..$KFe_3(SO_4)_2(OH)_6$
Joaquinite..$NaBa(Ti,Fe)_3Si_4O_{15}$ (?)
Johannsenite..$Ca(Mn,Fe)Si_2O_6$
 (Clinopyroxene)
Kaersutite..$Ca_2(Na,K)(Mg,Fe)_4TiSi_6Al_2O_{22}(OH,F)_2$
 (Amphibole)
Kalkowskite..$Fe_2Ti_3O_9$ (?)
Karinthine..$(Na,K)Ca_{2-3}Mg_8Fe_{1-2}(Al,Fe,Ti)_2$
 (Amphibole) $(Al_{3-4}Si_{13-12}O_{44})(OH)_4$
Katophorite..$Na_2CaFe_4(Fe,Al)Si_7AlO_{22}(OH,F)_2$
 (Amphibole)
Kirovite..$(Fe,Mg)SO_4 \cdot 7H_2O$
Klebelsbergite..Basic Sb sulfate
Knebelite..$(Mn,Fe)_2(SiO_4)$
 (Olivine)
Kolbeckite..Ca,Be,Al,Fe silicate-phosphate
Kornelite..$Fe_2(SO_4)_3 \cdot 7H_2O$
Krausite..$KFe(SO_4)_2 \cdot H_2O$
Kremersite..$(NH_4,K)_2FeCl_5 \cdot H_2O$
✓ Kupletskite..$(K_2,Na_2,Ca)(Fe,Mn)_4(Ti,Zr)$
 $(OH)_2Si_4O_{14}$
Landesite..$Fe_6Mn_2O(PO_4)_{16} \cdot 27H_2O$ (?)
Laubmannite..$Fe_9(PO_4)_4(OH)_{12}$
Lausenite..$Fe_2(SO_4)_3 \cdot 6H_2O$
Lawrencite..$FeCl_2$
Lazulite..$(Mg,Fe)Al_2(PO_4)_2(OH)_2$
Lepidocrocite..$FeO(OH)$
Leucophosphite..$K_2(Fe,Al)_7(PO_4)_4(OH)_{11} \cdot 6H_2O$
Limonite..hydrous Fe oxides
Liskeardite ..$(Al,Fe)_3(AsO_4)(OH)_6 \cdot 5H_2O$
Loellingite..$FeAs_2$
Ludlamite..$(Fe,Mg,Mn)_3(PO_4)_2 \cdot 4H_2O$
Ludwigite ..$(Mg,Fe)_2FeBO_5$
✓ Mackayite..$Fe(TeO_3)_3 \cdot xH_2O$ (?)
Maghemite ..Fe_2O_3 (?)
Magnesiocopiapite..$MgFe_4(SO_4)_6(OH)_2 \cdot 20H_2O$
Magnesioferrite..$MgFe_2O_4$
Magnesiokatophorite..$Na_2CaMg_4(Fe,Al)$
 (Amphibole) $Si_7AlO_{22}(OH,F)_2$
Magnetite..Fe_3O_4
✓ Magnetoplumbite..$Pb(Fe,Mn)_6O_{10}$ (?)
Mangan-alluaudite..$(Na,Mn,Fe)(PO_4)$
Marcasite..FeS_2
Melanostibian..$(Mn,Fe)_6(SbO_3)_2O_3$
Melanotekite..$Pb_3Fe_4O_3(SiO_4)_3$
Melanterite..$FeSO_4 \cdot 7H_2O$
Metacinnabar..$(Hg,Fe,Zn)S$

(IRON-continued)

Metahohmannite..$Fe_2(SO_4)_2(OH)_2 \cdot 3H_2O$
Metasideronatrite..$Na_4Fe_2(SO_4)_4(OH)_2 \cdot 3H_2O$
Metastrengite..$Fe(PO_4) \cdot 2H_2O$
Metavauxite..$FeAl_2(PO_4)_2(OH)_2 \cdot 8H_2O$
Metavoltine..$(K,Na,Fe)_5Fe_3(SO_4)_6(OH)_2 \cdot 9H_2O$ (?)
Mica..complex hydrous alumino-silicates
✓ Minnesotaite..$(Fe,Mg,H_2)_3(Si,Al,Fe)_4O_{10}(OH)_2$ (?)
Mitridatite..$CaFe_2(PO_4)_2(OH)_2 \cdot H_2O$ (?)
Molysite..$FeCl_3$
Mossite..$Fe(Cb,Ta)_2O_6$
Nagatelite..$(Ca,Ce)_2(Al,Fe)_3(OH)(Si,P)_3O_{12}$
Narsarsukite..$Na_2(Ti,Fe)(O,OH,F)Si_4O_{10}$
Natrojarosite..$NaFe_3(SO_4)_2(OH)_6$
Naujakasite..$Na_4FeAl_4H_4Si_8O_{27}$
Neptunite..$Na_2FeTi(Si_4O_{12})$
Nickel iron..Ni,Fe
Nontronite..montmorillonite-like clay (Clay)
Olivine..$(Mg,Fe)_2(SiO_4)$
Omphacite..$(Ca,Na)(Mg,Fe,Fe,Ti,Al)(Si_2O_6)$ (Clinopyroxene)
Orthite..$(Ca,Ce,La)_2(Al,Fe,Be,Mg,Mn)_3(OH)Si_3O_{12}$ (?)
Orthoferrosilite..$Fe(SiO_3)$ (Orthopyroxene)
✓ Osumilite..$(K,Na,Ca)(Mg,Fe)_2(Al,Fe)_3(Si,Al)_{12}O_{30} \cdot H_2O$
Paigeite..$(Fe,Mg)_2FeBO_5$
Parabutlerite..$Fe(SO_4)(OH) \cdot 2H_2O$
Paracoquimbite..$Fe_2(SO_4)_3 \cdot 9H_2O$
✓ Paravauxite..$FeAl_2(PO_4)_2(OH)_2 \cdot 8H_2O$
Pargasite..$NaCa_2Mg_4(Al,Fe)Si_6Al_2O_{22}(OH,F)_2$ (Amphibole)
— Parsettensite..$(K,H_2O)(Fe,Mg,Al,Mn)_3Si_4O_{10}(OH)_2$ (?)
Pennantite..low Mg chlorite (Chlorite)
Penninite..hydrous Mg,Fe,Al silicate (Chlorite)
> Pentlandite..$(Fe,Ni)_9S_8$
Pharmacosiderite..$Fe_3(AsO_4)_2(OH)_3 \cdot 5H_2O$
Phosphoferrite..$(Fe,Mn)_3(PO_4)_2 \cdot 3H_2O$
Phosphophyllite..$Zn_2(Fe,Mn)(PO_4)_2 \cdot 4H_2O$
Piemontite..$Ca_2(Al,Fe,Mn)_2Al(OH)Si_3O_{12}$
Pigeonite..$(Mg,Fe,Ca)_2(Si_2O_6)$
Pisanite..$(Fe,Cu)SO_4 \cdot 7H_2O$
Pistacite..$Ca_2(Al,Fe)Al_2(OH)Si_3O_{12}$
Pitticite..$Fe_2(AsO_4)(SO_4)(OH) \cdot H_2O$ (?)

Plumboferrite..$PbFe_4O_7$
Plumbojarosite..$PbFe_6(SO_4)_4(OH)_{12}$
Polymignyte..$(Ca,Fe,Y,etc.,Zr,Th)(Cb,Ta,Ti)O_4$
Priorite..$(Y,Er,Ca,Fe,Th)(Ti,Cb)_2O_6$
Pseudobrookite..Fe_2TiO_5
Pseudothuringite..hydrous Fe,Mg,Al silicate (Chlorite)
✓ Pumpellyite..$Ca_4(Mg,Fe,Mn)(Al,Fe,Ti)_5(OH)_3Si_6O_{23} \cdot 2H_2O$ (?)
Purpurite..$(Mn,Fe)(PO_4)$
Pyrite..FeS_2
Pyroaurite..$Mg_6Fe_2(OH)_{16} \cdot CO_3 \cdot 4H_2O$
Pyrosmalite..$(Mn,Fe)_8(OH,Cl)_{10}Si_6O_{15}$ (?)
Pyroxene..Mg,Fe,Ca,Na,Al,Ti silicates
Pyroxmangite..$(Mn,Fe)(SiO_3)$
Pyrrhotite..$Fe_{1-x}S$
Quenstedtite..$Fe_2(SO_4)_3 \cdot 10H_2O$
Ransomite..$Cu(Fe,Al)_2(SO_4)_4 \cdot 7H_2O$
Reddingite..$(Mn,Fe)_3(PO_4)_2 \cdot 3H_2O$
Redingtonite..$(Fe,Mn,Ni)(Cr,Al)_2(SO_4)_4 \cdot 22H_2O$ (?)
Rhipidolite..hydrous Mg,Fe,Al silicate (Chlorite)
Rhodonite..$(Mn,Fe,Ca)(SiO_3)$
Rhomboclase..$FeH(SO_4)_2 \cdot 4H_2O$
Richellite..$Ca_3Fe_{10}(PO_4)_8(OH)_{12} \cdot H_2O$ (?)
Richterite..$Na_2Ca(Mg,Fe,Mn,Al,Fe)_5Si_8O_{22}(OH,F)_2$ (Amphibole)
Riebeckite..$Na_2Fe_5Si_8O_{22}(OH,F)_2$ (Amphibole)
Rinneite..NaK_3FeCl_6
✓ Rockbridgeite..$Fe_5(PO_4)_3(OH)_5$
Roemerite..$Fe_3(SO_4)_4 \cdot 14H_2O$
✓ Roscherite..$(Ca,Mn,Fe)_2Al(PO_4)_2(OH) \cdot 2H_2O$
Rosenbuschite..$(Na,Ca,Mn)_3(Fe,Ti,Zr)FSi_2O_8$
Safflorite..$(Co,Fe)As_2$
Salite..$(Mg,Fe)Si_2O_6$ (Clinopyroxene)
Salmonsite..$Mn_9Fe_2(PO_4)_8 \cdot 14H_2O$ (?)
Samarskite..$(Y,Er,Ce,U,Ca,Fe,Pb,Th)(Cb,Ta,Ti,Sn)_2O_6$
"Samarskite"..$(Ca,Pb,Y,U)(Cb,Ta,Ti,Fe)_2O_6$ (?)
— Sanmartinite..$(Zn,Fe)(WO_4)$
Saponite..$(.5Ca,Na)_{0.7}(Al,Mg,Fe)_4(Si,Al)_8O_{20}(OH)_4 \cdot nH_2O$ (Clay)
Sapphirine..$(Mg,Fe)_{15}(Al,Fe)_{34}Si_7O_{80}$
Sarcopside..$(Fe,Mn,Ca)_7(PO_4)_4F_2$ (?)
Sarmientite..$Fe_2(AsO_4)(SO_4)(OH) \cdot 5H_2O$
Schafarzikite..$FeSb_2O_4$
✓ Schallerite..$(Mn,Fe)_8(OH,Cl)_{10}(Si,As)O_{15}$ (?)

(IRON-continued)

Schorl..$Na(Fe,Mn)_3Al_6(OH,F)_4(BO_3)_3Si_6O_{18}$
  (Tourmaline)
Schreibersite..$(Fe,Ni)_3P$
Scorodite..$Fe(AsO_4)\cdot 2H_2O$
Scorzalite..$(Fe,Mg)Al_2(PO_4)_2(OH)_2$
Senaite..$(Fe,Mn,Pb)TiO_3$
Serendibite..$Ca_2(Mg,Fe)_4Al_6B_2O_{10}Si_4O_{16}$ (?)
Sheridanite ..hydrous Mg,Fe,Al silicate
  (Chlorite)
Sicklerite ..$(Li,Mn,Fe)(PO_4)$
Siderazot ..Fe_5N_2
Siderite ..$FeCO_3$
Sideronatrite ..$Na_2Fe(SO_4)_2(OH)\cdot 3H_2O$
Siderophyllite ..$KFe_3(AlSi_3O_{10})(OH,F)_2$
  (Mica)
Siderotil ..$FeSO_4\cdot 5H_2O$
Sjogrenite ..$Mg_6Fe_2(OH)_{16}CO_3\cdot 4H_2O$
√ Slavikite ..$(K,Na)_2Fe_{10}(SO_4)_{13}(OH)_6\cdot 63H_2O$ (?)
Souzalite ..$(Mg,Fe)_3(Al,Fe)_4(PO_4)_4(OH)_6\cdot 2H_2O$
Stannite..Cu_2FeSnS_4
Staurolite..$(Fe,Mg)_2(Al,Fe)_9O_6(O,OH)_2(SiO_4)$
Steenstrupine..complex hydrous silicate
Sternbergite..$AgFe_2S_3$
Stevensite..$Mg_{5.76}Mn_{0.04}Fe_{0.04}Si_8O_{20}$
  (Clay) $(OH)_4$...etc. (?)
Stilpnochlorane..$(Ca,K,H_2O)(Al,Fe,Mg)_3$
                                    $(OH,O)_2(Si,P)_4O_{10}$
Stilpnomelane..$(K,H_2O)(Fe,Mg,Mn,Al,Fe)_3$
                                    $(OH)_2Si_4O_{10}$ (?)
Strengite..$Fe(PO_4)\cdot 2H_2O$
Stylotypite ..$(Ag,Cu,Fe)_3SbS_3$
Symplesite ..$Fe_3(AsO_4)_2\cdot 8H_2O$
Szomolnokite..$FeSO_4\cdot H_2O$
Tantalite..$(Fe,Mn)(Ta,Cb)_2O_6$
Tapiolite..$FeTa_2O_6$
Taramellite..$BaFe(OH)Si_2O_6$
Taramite..$Na_2Ca_2Mg_5Fe_3(Fe,Ti)_2(Al,Fe)_2$
  (Amphibole) $(OH,F)_4Si_{14}O_{44}$
Tauriscite..$FeSO_4\cdot 7H_2O$
Tennantite..$(Cu,Fe)_{12}As_4S_{13}$
Tetrahedrite..$(Cu,Fe)_{12}Sb_4S_{13}$
Thorotungsite..$AlFe(Th,Ca,Ce,Zr)WO_3$ (?)
Thuringite..hydrous Fe,Mg,Al silicate
  (Chlorite)
Tinticite..$Fe_3(PO_4)_2(OH)_3\cdot 3.5H_2O$
Titanaugite..$Ca_{14}NaMg_{10}Fe_3Ti_2Fe_2Al_5$
  (Clinopyroxene) $Si_{27}O_{96}$ (?)
Torendrikite..hydrous Na,Ca,Mg,Fe alumino-
  (Amphibole)   silicate

Tourmaline..$(Na_2,Ca)(Li,Al,Mg,Fe,Mn)_6Al_{12}$
                                  $(OH,F)_8(BO_3)_6Si_{12}O_{36}$
Trevorite..$NiFe_2O_4$
Triphylite..$LiFe(PO_4)$
Triplite..$(Mn,Fe)(PO_4)F$
Triploidite..$(Mn,Fe)_2(PO_4)(OH)$
Tripuhyite..$Fe_2Sb_2O_7$ (?)
Tscheffkinite..$(Fe,Ca)(Ce,La,Al)_2$
                                    $(Si,Ti)_3O_{10}$ (?)
√ Tuhualite..$(Na_2,K_2,Mn)(Al,Fe,Mg,Ti)$
                                    $Si_{10}O_{24}$ (?)
√ Ungemachite..$Na_9K_3Fe(SO_4)_6(OH)_3\cdot 9H_2O$
Uralite..hydrous Ca,Mg,Fe alumino-silicate
  (Amphibole)
> Valleriite..$Cu_2Fe_4S_7$ (?)
√ Vauxite..$FeAl_2(PO_4)_2(OH)_2\cdot 7H_2O$
Vermiculite..$(Mg,Ca)_{0.7}(Mg,Fe,Al)_6$
                              $[(Al,Si)_8O_{20}](OH)_4\cdot 8H_2O$
Vesuvianite..$Ca_{10}Al_4(Mg,Fe)_2Si_9O_{34}(OH,F)_4$
Violarite..Ni_2FeS_4
Vivianite..$Fe_3(PO_4)_2\cdot 8H_2O$
Volchonskoite..$(Ca,Mg,Cr,Fe,Al)_2O_3$
  (Clay) $\cdot 3SiO_2\cdot nH_2O$ (?)
Voltaite..$(K,Fe)_3Fe(SO_4)_3\cdot 4H_2O$ (?)
Warwickite..$(Mg,Fe)_3TiB_2O_8$
Wolfeite ..$(Fe,Mn)_2(PO_4)(OH)$
Wolframite..$(Fe,Mn)(WO_4)$
Xanthoxenite..$Ca_2Fe(PO_4)_2(OH)\cdot 1.5H_2O$
Yttrocrasite..$(Y,Th,U,Ca)_2(Ti,Fe,W)_4O_{11}$ (?)
Yttrotantalite..$(Fe,Y,U,Ca,etc...)(Cb,Ta,$
                                            $Zr,Sn)O_4$
Zinnwaldite..$K(Li,Fe,Al)(AlSi_3O_{10})(F,OH)_2$
  (Mica)
Zirkelite..$(Ca,Fe,Th,U)_2(Ti,Zr)_2O_5$
√ Zirklerite..Al,Fe basic chloride

LEAD

— Afwillite..$(Ca,Pb)_{10}(OH,Cl)_2(Si_2O_7)_3$
Aikinite..$PbCuBiS_3$
Alamosite..$PbSiO_3$
Alaskaite..$Pb(Ag,Cu)_2Bi_4S_8$ (?)
Altaite..$PbTe$
Andorite..$PbAgSb_3S_6$
Anglesite..$PbSO_4$
Arseniopleite..$(Mn,Ca,Pb,Mg)_9(Mn,Fe)_2$
                                              $(AsO_4)(OH)_6$

(LEAD-continued)

Arzrunite..Pb,Cu sulfate-chloride
✓ Barysilite..$Pb_3Si_2O_7$
Baumhauerite..$Pb_4As_6S_{13}$
Bayldonite..$(Cu,Pb)_2(AsO_4)(OH)$ (?)
Beaverite..$Pb(Cu,Fe,Al)_3(SO_4)_2(OH)_6$
Beegerite..$Pb_6Bi_2S_9$
✓ Benjaminite..$Pb(Cu,Ag)Bi_2S_4$
Beresovite..$Pb_6(CrO_4)_3(CO_3)O_2$
Berthonite..$Pb_2Cu_7Sb_5S_{13}$
Beudantite..$PbFe_3(AsO_4)(SO_4)(OH)_6$
> Bindheimite..$Pb_2Sb_2O_6(O,OH)$
Boleite..$Pb_9Cu_8Ag_3Cl_{21}(OH)_{16} \cdot 2H_2O$ (?)
Boulangerite..$PbSb_4S_{11} - Pb_2Sb_2S_5$
Bournonite..$PbCuSbS_3$
Brackebuschite..$Pb_4MnFe(VO_4)_4 \cdot 2H_2O$
Caledonite..$Cu_2Pb_5(SO_4)_3(CO_3)(OH)_6$
Caracolite..Na,Pb chloride-sulfate
Carminite..$PbFe_2(AsO_4)_2(OH)_2$
Caryinite..$(Ca,Pb,Na)_5(Mn,Mg)_4(AsO_4)_5$ (?)
Cerussite..$PbCO_3$
Cesarolite..$PbMn_3O_7 \cdot H_2O$
✓ Chiviatite..$Pb_3Bi_8S_{15}$
✓ Chloroxiphite..$PbCuO_2Cl_2(OH)_2$ (?)
> Clausthalite..PbSe
Corkite..$PbFe_3(PO_4)(SO_4)(OH)_6$
Coronadite..$MnPbMn_6O_{14}$
Cosalite..$Pb_2Bi_2S_5$
Cotunnite..$PbCl_2$
Crocoite..$Pb(CrO_4)$
Cumengite..$Pb_4Cu_4Cl_8(OH)_8 \cdot H_2O$ (?)
Curite..$2PbO \cdot 5UO_3 \cdot 4H_2O$ (?)
Cylindrite..$Pb_3Sn_4Sb_2S_{14}$
Daviesite..Pb oxychloride
Descloizite..$ZnPb(VO_4)(OH)$
✓ Dewindtite..$Pb_3(UO_2)_5(PO_4)_4(OH)_4 \cdot 10H_2O$
Diaboleite..$Pb_2CuCl_2(OH)_4$
Diaphorite..$Pb_2Ag_3Sb_3S_8$
Djalmaite..$(U,Ca,Pb,Bi,Fe)(Ta,Cb,Ti,Zr)_3O_9 \cdot nH_2O$
Dufrenoysite..$Pb_2As_2S_5$
Duftite..$PbCu(AsO_4)(OH)$
Dumontite..$Pb_2(UO_2)_3(PO_4)_2(OH)_4 \cdot 3H_2O$
Dundasite..$PbAl_2(CO_3)_2(OH)_4 \cdot 4H_2O$
Ecdemite..$Pb_6As_2O_7Cl_4$ (?)
Fiedlerite..$Pb_3(OH)_2Cl_4$
Finnemanite..$Pb_5(AsO_3)_3Cl$

Fizelyite..$Pb_5Ag_2Sb_8S_{18}$
✓ Fourmarierite..$PbO \cdot 4UO_3 \cdot 5H_2O$ (?)
Franckeite..$Pb_5Sn_3Sb_2S_{14}$
Freieslebenite..$Pb_3Ag_5Sb_5S_{12}$
Fuloppite..$Pb_3Sb_8S_{15}$
Galena..PbS
Galenobismutite..$PbBi_2S_4$
Ganomalite..$(Ca,Pb)_{10}(OH,Cl_2)(Si_2O_7)_3$
Geocronite..$Pb_5(Sb,As)_2S_8$
Geogiadesite..$Pb_3(AsO_4)Cl_3$
Gladite..$PbCuBi_5S_9$
Goongarrite..$Pb_4Bi_2S_7$
Gratonite..$Pb_9As_4S_{15}$
Guitermanite..$Pb_{10}As_6S_{19}$
Hammarite..$Pb_2Cu_2Bi_4S_9$ (?)
Hedyphane..$(Ca,Pb)_5(AsO_4)_3Cl$
Heliophyllite..$Pb_6As_2O_7Cl_4$ (?)
✓ Hematophanite..$Pb(Cl,OH)_2 \cdot 4PbO \cdot 2Fe_2O_3$ (?)
Heteromorphite..$Pb_7Sb_8S_{19}$
Hinsdalite..$(Pb,Sr)Al_3(PO_4)(SO_4)(OH)_6$
Hutchinsonite..$(Pb,Tl)_2(Cu,Ag)As_5S_{10}$
Hyalotekite..$Ca_3Ba_3Pb_3B_2Si_{12}O_{36}$ (?)
Hydrocerussite..$Pb_3(CO_3)_2(OH)_2$
Jamesonite..$Pb_4FeSb_6S_{14}$
Jordanite..$Pb_{14}As_7S_{24}$
✓ Kasolite..$Pb_2U_2O_4Si_2O_8 \cdot H_2O$ (?)
Kentrolite..$Pb_3Mn_4O_3(SiO_4)_3$
Kerstenite..$PbSeO_3 \cdot 2H_2O$ (?)
Kobellite..$Pb_2(Bi,Sb)_2S_5$
Lanarkite..$Pb_2(SO_4)O$
Larsenite..$PbZnSiO_4$
Laurionite..$Pb(OH)Cl$
Lead..Pb
Leadhillite..$Pb_4(CO_3)_2(OH)_2(SO_4)$
Lengenbachite..$Pb_6(Ag,Cu)_2As_4S_{13}$
Lillianite..$Pb_3Bi_2S_6$
Linarite..$PbCu(SO_4)(OH)_2$
Lindstromite..$PbCuBi_3S_6$
Litharge..PbO
Liveingite..$Pb_5As_8S_{17}$
Lorettoite..$Pb_7O_6Cl_2$
✓ Magnetoplumbite..$Pb(Fe,Mn)_6O_{10}$ (?)
Massicot..PbO
Matlockite..PbFCl
Melanotekite..$Pb_3Fe_4O_3(SiO_4)_3$
Mendipite..$Pb_3O_2Cl_2$
Meneghinite..$Pb_{13}Sb_7S_{23}$

(LEAD-continued)

Mimetite..$Pb_5(AsO_4)_3Cl$
Minium..Pb_3O_4
Molybdophyllite..$Pb_2Mg_2(OH)_2Si_2O_7$
Monimolite..$(Pb,Ca)_3Sb_2O_8$ (?)
Mottramite..$CuPb(VO_4)(OH)$
Nadorite..$PbSbO_2Cl$
Nagyatite..$Pb_5Au(Te,Sb)_4S_{5-8}$ (?)
Nasonite..$(Ca,Pb)_{10}Cl_2(Si_2O_7)_3$
Owyheeite..$Pb_5Ag_2Sb_6S_{15}$
Palmierite..$(K,Na)_2Pb(SO_4)_2$
Paralaurionite...$Pb(OH)Cl$
> Parsonsite..$Pb_2(UO_2)(PO_4)_2 \cdot 2H_2O$
Penfieldite..$Pb_2(OH)Cl_3$
Penroseite..$(Ni,Cu,Pb)Se_2$
Percylite..$PbCuCl_2(OH)_2$
Phoenicochroite..$Pb_3(CrO_4)_2O$ (?)
Phosgenite..$Pb_2(CO_3)Cl_2$
Pilbarite..$2PbO \cdot 3ThO_2 \cdot 4UO_2 \cdot 8SiO_2 \cdot 21H_2O$ (?)
Plagionite..$Pb_5Sb_8S_{17}$
Plattnerite..PbO_2
Platynite..$PbBi_2(Se,S)_3$
Plumboferrite..$PbFe_4O_7$
Plumbogummite..$PbAl_3(PO_4)_2(OH)_5 \cdot H_2O$
Plumbojarosite..$PbFe_6(SO_4)_4(OH)_{12}$
Pseudoboleite..$Pb_5Cu_4Cl_{10}(OH)_8 \cdot 2H_2O$ (?)
Pseudocotunnite..K_2PbCl_4
Pyrobelonite..$MnPb(VO_4)(OH)$
Pyromorphite..$Pb_5(PO_4)_3Cl$
Quenselite..$PbMnO_2(OH)$
Ramdohrite..$Pb_3Ag_2Sb_6S_{13}$
Raspite..$Pb(WO_4)$
Rathite..$Pb_{13}As_{18}S_{40}$
Renardite..$Pb(UO_2)_4(PO_4)_2(OH)_4 \cdot 7H_2O$
Rezbanyite..$Pb_3Cu_2Bi_{10}S_{19}$
Roeblingite..$2PbSO_4 \cdot H_{10}Ca_7Si_6O_{24}$ (?)
Sahlinite..$Pb_{14}(AsO_4)_2O_9Cl$
Samarskite..$(Y,Er,Ce,U,Ca,Fe,Pb,Th)(Cb,Ta,Ti,Sn)_2O_6$
"Samarskite"..$(Ca,Pb,Y,U)(Cb,Ta,Ti,Fe)_2O_6$
Sartorite..$PbAs_2S_4$
Schirmerite..$PbAg_4Bi_4S_9$
Schultenite..$PbH(AsO_4)$
Schwartzembergite..$Pb_5(IO_3)Cl_3O_3$
Seligmannite..$PbCuAsS_3$
Semseyite..$Pb_9Sb_8S_{21}$
Senaite..$(Fe,Mn,Pb)TiO_3$

Stolzite..$Pb(WO_4)$
Susannite..$Pb_4(CO_3)_2(OH)_2(SO_4)$
Synadelphite..$(Mn,Mg,Ca,Pb)_4(AsO_4)(OH)_5$
Teallite..$PbSnS_2$
Trigonite..$MnPb_3H(AsO_3)_3$
Tsumebite..$Pb_2Cu(PO_4)(OH)_3 \cdot 3H_2O$
Vanadinite..$Pb_5(VO_4)_3Cl$
Vauquelinite..Pb,Cu chromate-phosphate
✓ Weibullite..$PbBi_2(S,Se)_4$
✓ Wherryite..$Pb_4Cu(CO_3)(SO_4)_2(OH,Cl)O$ (?)
Wittite..$Pb_5Bi_6(S,Se)_{14}$
Wulfenite..$Pb(MoO_4)$
Zinkenite..$Pb_6Sb_{14}S_{27}$

LITHIUM

Amblygonite..$(Li,Na)Al(PO_4)(F,OH)$
Bityite..$Ca_4(Li,Be)_4Al_8(OH)_{20}[(Si,Al)_4O_{10}]_3$ (?)
 (Mica)
Clay..chiefly hydrous silicates of Al or Mg
Cookeite ..Li-bearing chlorite
 (Chlorite)
Cryolithionite ..$Na_3Li_3Al_2F_{12}$
Elbaite..$Na(Li,Al)_3Al_6(OH,F)_4(BO_3)_3Si_6O_{18}$
Ekermannite.. $.5Na_5Ca(Mg,Fe,Fe,Al,Li)_{10}Si_{15}AlO_{44}(OH,F)_4$
 (Amphibole)
Eucryptite..$LiSiAlO_4$
Ferri-sicklerite..$(Li,Fe,Mn)PO_4$
✓ Hectorite..montmorillonite-like clay
 (Clay)
Holmquistite..$Li_2(Mg,Fe)_3(Al,Fe)_2Si_8O_{22}(OH,F)_2$
 (Amphibole)
Lepidolite..$K_2(Li,Al)_{5-6}(Si_{6-7}Al_{2-1}O_{20})(OH,F)_4$
 (Mica)
Lithiophilite..$LiMnPO_4$
Manandonite ..hydrous Li,B,Al silicate
 (Chlorite)
Montebrasite ..$(Li,Na)Al(PO_4)(OH,F)$
Natromontebrasite ..$(Na,Li)Al(PO_4)(F,OH)$
Petalite ..$Li(AlSi_4O_{10})$
Rhodizite ..$NaKLi_4Al_4Be_3B_{10}O_{27}$ (?)
Sicklerite ..$(Li,Mn,Fe)PO_4$
Spodumene..$LiAl(Si_2O_6)$
Tourmaline..$(Na_2,Ca)(Li,Al,Mg,Fe,Mn)_6Al_{12}(OH,F)_8(BO_3)_6Si_{12}O_{36}$
Triphylite..$LiFePO_4$
Zinnwaldite..$K(Li,Fe,Al)(AlSi_3O_{10})(F,OH)_2$
 (Mica)

MAGNESIUM

Actinolite..$Ca_2(Mg,Fe)_5Si_8O_{22}(OH,F)_2$
 (Amphibole)
Adelite..$CaMg(AsO_4)(OH,F)$
Aegirine-augite..$(Na,Ca)(Fe,Mg,Al)(Si_2O_6)$
 (Clinopyroxene)
Akermanite..$Ca_2Mg(Si_2O_7)$
✓ Akrochordite..$MgMn_4(AsO_4)_2(OH)_4 \cdot 4H_2O$ (?)
Allanite..$(Ca,Ce,La)_2(Al,Fe,Be,Mg,Mn)_3$
 (=Orthite) $(OH)Si_3O_{12}$
Amesite..$(Mg_4Al_2)(Si_2Al_2)O_{10}(OH)_8$
 (Chlorite)
Amphibole..hydrous Fe,Mg,Na,K,Ca,Mn,
  Al silicates
Ankerite..$Ca(Fe,Mg)(CO_3)_2$
Anthophyllite..$(Mg,Fe)_7Si_8O_{22}(OH,F)_2$
 (Amphibole)
Antigorite..$Mg_3(Si_2O_5)(OH)_4$
 (Serpentine)
Arfvedsonite.. $.5Na_5Ca(Fe,Mg,Fe,Al)_{10}$
 (Amphibole) $Si_{15}AlO_{44}(OH,F)_4$
Arseniopleite..$(Mn,Ca,Pb,Mg)_9(Mn,Fe)_2$
 $(AsO_4)(OH)_6$
Artinite..$Mg_2(CO_3)(OH)_2 \cdot 3H_2O$
Ashcroftine..$KNa(Ca,Mg,Mn)(Al_4Si_5O_{18}) \cdot 8H_2O$
 (Zeolite)
Attapulgite ..hydrous Mg,Fe silicate
 (Clay)
Augite ..$(Ca,Mg,Fe,Fe,Ti,Al)_2(Si,Al)_2O_6$
Barbertonite..$Mg_6Cr_2(OH)_{16} \cdot CO_3 \cdot 4H_2O$
Barkevikite..$Ca_2(Na,K)(Fe,Mg,Mn)_5$
 (Amphibole) $(Si_{6.5}Al_{1.5}O_{22})(OH)_2$
Barroisite..Na,K,Ca,Mg,Fe,Ti alumino-silicate
 (Amphibole)
Basaltic hornblende..$Ca_2(Na,K)_{0.5-1.0}$
 (Amphibole) $(Mg,Fe)_{3-2}(Fe,Al)_{2-3}$
 $Si_6Al_2O_{22}(O,OH,F)_2$
Batavite..hydrous Mg,Al alumino-silicate
 (Clay)
✓ Bayleyite..$Mg(UO_2)(CO_3)_3 \cdot 18H_2O$
Bermanite..$(Mn,Mg)_5(Mn,Fe)_8(PO_4)_8$
 $(OH)_{10} \cdot 15H_2O$ (?)
Berzeliite..$(Mg,Mn)_2(Ca,Na)_3(AsO_4)_3$
Biotite ..$K(Mg,Fe,Mn)_3(AlSi_3O_{10})(OH,F)_2$
 (Mica)
Bischofite ..$MgCl_2 \cdot 6H_2O$
Bloedite ..$Na_2Mg(SO_4)_2 \cdot 4H_2O$
Bobierrite ..$Mg_3(PO_4)_2 \cdot 8H_2O$
Boracite ..$Mg_3B_7O_{13}Cl$
Botryogen ..$MgFe(SO_4)_2(OH) \cdot 7H_2O$
Boussingaultite ..$(NH_4)_2Mg(SO_4)_2 \cdot 6H_2O$
Bradleyite ..$Na_3Mg(CO_3)(PO_4)$

Brittle mica ..hydrous Na,Ca,Al,Mg silicates
Bronzite ..$(Mg,Fe)(SiO_3)$
 (Orthopyroxene)
Brucite ..$Mg(OH)_2$
Brugnatellite ..$Mg_6Fe(OH)_{13} \cdot CO_3 \cdot 4H_2O$
Brunsvigite ..hydrous Fe,Mg,Al silicate
 (Chlorite)
Byssolite..hydrous Ca,Mg,Fe alumino-silicate
 (Amphibole)
Cardenite..hydrous Ca,Mg,Fe,Al alumino-
 (Clay) silicate
Carnallite..$KMgCl_3 \cdot 6H_2O$
Caryinite..$(Ca,Pb,Na)_5(Mn,Mg)_4(AsO_4)_5$ (?)
Celadonite..$(K,Ca,Na)(Al,Fe,Mg)_2$
 $(Al_{0.11}Si_{3.89}O_{10})(OH)_2$ (?)
Chamosite..hydrous Fe,Mg,Al silicate
 (Chlorite)
Chlorite ..$(Mg,Al,Fe)_{12}(Si,Al)_8O_{20}(OH)_{16}$
Chloritoid ..$(Fe,Mg)_2Al_4(OH)_4Si_2O_{10}$ (?)
Chloromagnesite ..$MgCl_2$
Chondrodite ..$Mg(OH,F)_2 \cdot 2Mg_2(SiO_4)$
Chrysolite ..$(Mg,Fe)_2(SiO_4)$
 (Olivine)
Chrysotile ..ca.$Mg_3(Si_2O_5)(OH)_4$
 (Serpentine)
Clay ..chiefly hydrous silicates of Al or Mg
Clinochlore ..hydrous Mg,Fe,Al silicate
 (Chlorite)
Clinoenstatite ..$MgSiO_3$
 (Clinopyroxene)
Clinohumite ..$Mg(OH,F)_2 \cdot 4Mg_2(SiO_4)$
Clinohypersthene ..$(Mg,Fe)SiO_3$
 (Clinopyroxene)
Clino-sklodowskite ..$Mg(H_3O)_2UO_2Si_2O_8 \cdot 3H_2O$
Clintonite ..$Ca_2(Mg_5Al)Si_2Al_6O_{20}(OH)_4$ (?)
 (Brittle Mica)
Collinsite..$Ca_2(Mg,Fe)(PO_4)_2 \cdot 2H_2O$
Common hornblende..$(Ca,Na,K)_{2-3}(Mg,Fe,$
 (Amphibole) $Al)_5Si_6(Si,Al)_2O_{22}(OH,F)_2$
Cordierite..$Al_3(Mg,Fe)_2(Si_5AlO_{18})$
Corundophilite ..hydrous Mg,Al silicate
 (Chlorite)
Crossite ..$Na_2(Mg,Fe)_3(Al,Fe)_2Si_8O_{22}(OH,F)_2$
 (Amphibole)
Cummingtonite..$(Fe,Mg)_7Si_8O_{22}(OH)_2$
 (Amphibole)
Delessite..hydrous Mg,Fe,Al silicate
 (Chlorite)
Diabantite..hydrous Mg,Fe,Al silicate
 (Chlorite)
Diallage..$Ca_{14}Fe_2Mg_{13}FeAl_5Si_{29}O_{96}$ (?)
 (Clinopyroxene)
Didymolite..$(Ca,Mg,Fe)_2Al_6Si_8O_{27}$ (?)
Diopside..$CaMg(Si_2O_6)$
 (Clinopyroxene)
Dolomite..$CaMg(CO_3)_2$
Donbassite..hydrous Na,Ca,Mg,Al silicate
 (Clay)
Dravite..$NaMg_3Al_6(OH)_4(BO_3)_3Si_6O_{18}$
 (Tourmaline)

(MAGNESIUM-continued)

Edenite..$NaCa_2Mg_5AlSi_7O_{22}(OH,F)_2$
  (Amphibole)
Ekermannite.. $.5Na_5Ca(Mg,Fe,Fe,Al,Li)_{10}$
  (Amphibole) $Si_{15}AlO_{44}(OH,F)_4$
Ekmannite..$(Fe,Mg,Mn)_3(OH)_2(Si,Al)Si_3O_{10}$ (?)
Enstatite..$Mg(SiO_3)$
  (Orthopyroxene)
Epistolite..$(Na,Ca)(Cb,Ti,Mg,Fe,Mn)SiO_4(OH)$
Epsomite..$MgSO_4 \cdot 7H_2O$
Erionite..ca.$(Na_2,K_2,Ca,Mg)_{4.5}Al_9Si_{27}O_{72} \cdot 27H_2O$
  (Zeolite)
Eulite..$(Fe,Mg)_2(Si_2O_6)$
  (Orthopyroxene)
Fassaite..$Ca(Mg,Fe,Al,Ti)(Si,Al)_2O_6$
  (Clinopyroxene)
Ferrierite..$(Na,K)_2Mg(Al_3Si_{15}O_{36})(OH) \cdot 9H_2O$
  (Zeolite)
Ferrisepiolite..$(Mg,Fe)_4Si_6O_{15}(OH)_2 \cdot 3H_2O$
  (Clay)
Ferrohortonolite..$(Fe,Mg)_2(SiO_4)$
  (Olivine)
Ferrohypersthene..$(Fe,Mg)(SiO_3)$
  (Orthopyroxene)
Ferrosalite..$Ca(Fe,Mg)Si_2O_6$
  (Clinopyroxene)
Fluoborite..$Mg_3(BO_3)(F,OH)_3$
Forsterite..$Mg_2(SiO_4)$
  (Olivine)
Ganophyllite..$(Na,K)(Mg,Al,Mg,Ca)_3(OH)_4$
  $(Si,Al)Si_3O_{10}$ (?)
Garnierite..Ni-bearing chlorite
  (Chlorite)
Gedrite..$(Mg,Fe)_6AlSi_6(Si,Al)_2O_{22}(OH,F)_2$
  (Amphibole)
Geikielite..$MgTiO_3$
Glauconite..$(K,Ca,Na)(Al,Fe,Mg)_2$
  (Mica) $(Al_{0.35}Si_{3.65}O_{10})(OH)$ (?)
Glaucophane..$Na_2Mg_3Al_2Si_8O_{22}(OH,F)_2$
  (Amphibole)
✓ Gonyerite..low Al chlorite
  (Chlorite)
Gordonite..$MgAl_2(PO_4)_2(OH)_2 \cdot 8H_2O$
Grandidierite..$H_2Na_2(Mg,Fe)_7(Al,Fe,B)_{15}$
  $Si_7Al_7O_{56}$ (?)
Grovesite..$(Mn,Mg,Al)_6(Si,Al)_4O_{10}(OH)_8$
  (Serpentine)
Grunerite..$(Mg,Fe)_7Si_8O_{22}(OH)_2$
  (Amphibole)
Guembelite..$(K,H_2O)(Al_{1.5}Mg_{0.5})$
  $(AlSi_3O_{10})(OH,H_2O)_2$ (?)
Hannayite..$Mg_3(NH_4)_2H_4(PO_4)_4 \cdot 8H_2O$
✓ Harkerite..$Ca(Mg,Al)(Si,BH)O_4 \cdot CaCO_3$ (?)
Harstigite..$(Ca,Mn,Mg)_8Al_2(OH)_4Si_6O_{21}$ (?)
Hastingsite..$NaCa_2Mg_4Al_3Si_6O_{22}(OH)_2$
  (Amphibole)
✓ Hectorite.. montmorillonite-like clay
  (Clay)
Hematolite..$(Mn,Mg)_4Al(AsO_4)(OH)_8$

Hexahydrite..$MgSO_4 \cdot 6H_2O$
Hoegbomite..$Mg(Al,Fe,Ti)_4O_7$
Hoernesite..$Mg_3(AsO_4)_2 \cdot 8H_2O$
Holdenite..$(Mn,Ca)_4(Zn,Mg,Fe)(AsO_4)(OH)_5O_2$
Holmquistite..$Li_2(Mg,Fe)_3(Al,Fe)_2Si_8O_{22}(OH,F)_2$
  (Amphibole)
Hortonolite..$(Fe,Mg)_2(SiO_4)$
  (Olivine)
Hulsite..$(Fe,Ca,Mg)_4(Fe,Sn)_2B_2O_{10}$ (?)
Humite..$Mg(OH,F)_2 \cdot 3Mg_2(SiO_4)$
Hyalosiderite..$(Mg,Fe)_2(SiO_4)$
  (Olivine)
Hydrobiotite..$(K,H_2O)(Mg,Fe,Mn)_3$
  $(AlSi_3O_{10})(OH,H_2O)_2$
Hydroboracite..$CaMgB_6O_{11} \cdot 6H_2O$
Hydromagnesite ..$Mg_4(CO_3)_3(OH)_2 \cdot 3H_2O$
Hydrophlogopite ..$(K,H_2O_2)Mg_3(AlSi_3O_{10})(OH,H_2O)$
Hydrotalcite ..$Mg_6Al_2(OH)_{16} \cdot CO_3 \cdot 4H_2O$
Hypersthene ..$(Fe,Mg)(SiO_3)$
  (Orthopyroxene)
Inderborite ..$CaMgB_6O_{11} \cdot 11H_2O$
Inderite ..$Mg_2B_6O_{11} \cdot 15H_2O$
Indialite..$Mg_2Al_3(AlSi_5O_{18})$
Kaemmererite ..Cr,Mg chlorite
  (Chlorite)
Kaersutite..$Ca_2(Na,K)(Mg,Fe)_4Ti$
  (Amphibole) $Si_6Al_2O_{22}(OH,F)_2$
Kaliborite ..$KMg_2B_{11}O_{19} \cdot 9H_2O$
Kainite..$KMg(SO_4)Cl \cdot 3H_2O$
Karinthine..$(Na,K)Ca_{2-3}Mg_8Fe_{1-2}(Al,Fe,$
  (Amphibole) $Ti)_2(OH)_4(Al_{3-4}Si_{13-12}O_{44})$
✓ Karpinskiite..$(Na,K,Zn,Mg)_2(OH,H_2O)_{1-2}$
  $(Al,Be)_2Si_4O_{12}$
Kieserite..$MgSO_4 \cdot H_2O$
Kirovite ..$(Fe,Mg)SO_4 \cdot 7H_2O$
Klebelsbergite ..Basic Sb sulfate
Koenenite ..$Mg_5Al_2(OH)_{12}Cl_4$
Kornerupine ..$Mg_{10}Al_{10}BO_7Si_9O_{36}$ (?)
Kotoite ..$Mg_3(BO_3)_2$
Kurnakovite ..$Mg_2B_6O_{11} \cdot 13H_2O$
Kutnahorite ..$Ca(Mn,Mg)(CO_3)_2$
Langbeinite ..$K_2Mg_2(SO_4)_3$
Lansfordite ..$MgCO_3 \cdot 5H_2O$
Lazulite ..$(Mg,Fe)Al_2(PO_4)_2(OH)_2$
Leonite ..$K_2Mg(SO_4)_2 \cdot 4H_2O$
Lizardite ..ca.$Mg_3(Si_2O_5)(OH)_4$
  (Serpentine)
Loeweite ..$Na_2Mg(SO_4)_2 \cdot 2.5H_2O$
Ludlamite ..$(Fe,Mg,Mn)_3(PO_4)_2 \cdot 4H_2O$
Ludwigite ..$(Mg,Fe)_2FeBO_5$
Luenebergite ..$Mg_3B_2(OH)_6(PO_4)_2 \cdot 6H_2O$

(MAGNESIUM-continued)

Magnesiochromite ..$MgCr_2O_4$
Magnesiocopiapite ..$MgFe_4(SO_4)_6(OH)_2 \cdot 20H_2O$
Magnesioferrite..$MgFe_2O_4$
Magnesiokatophorite..$Na_2CaMg_4(Fe,Al)$
 (Amphibole) $Si_7AlO_{22}(OH,F)_2$
Magnesite..$MgCO_3$
Magnesium orthite..$(Ca,Ce)_2(Al,Fe)_3(OH)$
 $(Si,P)_3O_{12}$
Manasseite..$Mg_6Al_2(OH)_{16} \cdot CO_3 \cdot 4H_2O$
Manganberzeliite ..$(Mn,Mg)_2(Ca,Na)_3(AsO_4)_3$
Melilite ..$(Ca,Na)_2(Al,Mg)(Si,Al)_2O_7$
Merwinite ..$Ca_3MgSi_2O_8$
Mg-chlorophoenicite ..$Mg_5(AsO_4)(OH)_7$
Mica..complex hydrous alumino-silicates
✓ Minnesotaite..$(Fe,Mg,H_2)_3(Si,Al,Fe)_4$
 $O_{10}(OH)_2$ (?)
Molybdophyllite..$Pb_2Mg_2(OH)_2Si_2O_7$
Monticellite..$CaMg(SiO_4)$
 (Olivine)
Montmorillonite..$Na_{0.7}(Al_{3.3}Mg_{0.7})$
 (Clay) $Si_8O_{20}(OH)_4 \cdot nH_2O$
Mooreite..$(Mg,Mn,Zn)_8(SO_4)(OH)_{14} \cdot 4H_2O$
Mosandrite..$(Ca,Na)_{12}Ce_3(Zr,Ti,Mg)_4F_5$
 $Si_{10}O_{40}$ (?)
Nesquehonite..$MgCO_3 \cdot 3H_2O$
Newberyite ..$MgH(PO_4) \cdot 3H_2O$
Nitromagnesite ..$Mg(NO_3)_2 \cdot 6H_2O$
Nocerite ..$Ca_3Mg_3F_8O_2$
Norbergite ..$Mg(OH,F)_2 \cdot Mg_2(SiO_4)$
Northupite ..$Na_3Mg(CO_3)_2Cl$
Olivine..$(Mg,Fe)_2(SiO_4)$
Omphacite..$(Ca,Na)(Mg,Fe,Fe,Ti,Al)(Si_2O_6)$
 (Clinopyroxene)
Orthite..$(Ca,Ce,La)_2(Al,Fe,Be,Mg,Mn)_3(OH)$
 Si_3O_{12} (?)
Osumilite..$(K,Na,Ca)(Mg,Fe)_2(Al,Fe)_3$
 $(Si,Al)_{12}O_{30} \cdot H_2O$
Paigeite..$(Fe,Mg)_2FeBO_5$
Palygorskite..hydrous Mg silicate
 (Clay)
Pargasite..$NaCa_2Mg_4(Al,Fe)Si_6Al_2O_{22}(OH,F)_2$
 (Amphibole)
— Parsettensite..$(K,H_2O)(Fe,Mg,Al,Mn)_3$
 $Si_4O_{10}(OH)_2$ (?)
Paternoite..$MgB_8O_{13} \cdot 4H_2O$
Penninite ..hydrous Mg,Fe,Al silicate
 (Chlorite)
Pentahydrite ..$MgSO_4 \cdot 5H_2O$
Periclase ..MgO

Phlogopite ..$KMg_3(AlSi_3O_{10})(OH,F)_2$
 (Mica)
Phosphorroesslerite ..$MgH(PO_4) \cdot 7H_2O$
Pickeringite ..$MgAl_2(SO_4)_4 \cdot 22H_2O$
Picromerite ..$K_2Mg(SO_4)_2 \cdot 6H_2O$
Picropharmacolite ..$(Ca,Mg)_3(AsO_4)_2 \cdot 6H_2O$ (?)
Pigeonite ..$(Mg,Fe,Ca)_2(Si_2O_6)$
 (Clinopyroxene)
Pinakiolite ..$Mg_3MnMn_2B_2O_{10}$ (?)
Pinnoite ..$Mg(BO_2)_2 \cdot 3H_2O$
Polyhalite ..$K_2Ca_2Mg(SO_4)_4 \cdot 2H_2O$
Pseudothuringite..hydrous Mg,Fe,Al silicate
 (Chlorite)
✓ Pumpellyite..$Ca_4(Mg,Fe,Mn)(Al,Fe,Ti)_5(OH)_3$
 $Si_6O_{23} \cdot 2H_2O$ (?)
Pyroaurite..$Mg_6Fe_2(OH)_{16} \cdot CO_3 \cdot 4H_2O$
Pyrope ..$Mg_3Al_2(SiO_4)_3$
 (Garnet)
Pyroxene ..Mg,Fe,Ca,Na,Ti,Al silicates
Ralstonite..ca.$Na(Mg,Al_5)_6F_{12}(OH)_6 \cdot 3H_2O$
Rhipidolite ..hydrous Mg,Fe,Al silicate
 (Chlorite)
Richterite ..$Na_2Ca(Mg,Fe,Mn,Al)_5Si_8O_{22}(OH,F)_2$
 (Amphibole)
Roesslerite ..$MgH(AsO_4) \cdot 7H_2O$
Roselite ..$Ca_2(Co,Mg)(AsO_4)_2 \cdot 2H_2O$
— Roweite..$(Mn,Mg,Zn)Ca(BO_2)_2(OH)_2$
✓ Saleeite ..$Mg(UO_2)_2(PO_4)_2 \cdot 10H_2O$
Salite..$(Mg,Fe)Si_2O_6$
 (Clinopyroxene)
Saponite..$(1/2Ca,Na)_{0.7}(Al,Mg,Fe)_4$
 (Clay) $(Si,Al)_8O_{20}(OH)_4 \cdot nH_2O$
Sapphirine..$(Mg,Fe)_{15}(Al,Fe)_{34}Si_7O_{80}$ (?)
Sauconite..montmorillonite-like clay
 (Clay)
Scorzalite..$(Fe,Mg)Al_2(PO_4)_2(OH)_2$
Sellaite ..MgF_2
Sepiolite ..$Mg_3Si_4O_{11} \cdot nH_2O$ (?)
 (Clay)
Serendibite ..$Ca_2(Mg,Fe)_4Al_6B_2O_{10}Si_4O_{16}$ (?)
Serpentine ..ca.$Mg_3(Si_2O_5)(OH)_4$
Sheridanite ..hydrous Mg,Fe,Al silicate
 (Chlorite)
Sjogrenite ..$Mg_6Fe_2(OH)_{16} \cdot CO_3 \cdot 4H_2O$
✓ Sklodowskite ..$MgU_2O_3Si_2O_8 \cdot 6H_2O$
> Souzalite ..$(Mg,Fe)_3(Al,Fe)_4(PO_4)_4(OH)_6 \cdot 2H_2O$
Spinel ..$MgAl_2O_4$
Staurolite..$(Fe,Mg)_2(Al,Fe)_9O_6(SiO_4)_4(O,OH)_2$
Stevensite..$Mg_{5.76}Mn_{0.04}Fe_{0.04}Si_8O_{20}$
 (Clay) $(OH)_4 \ldots$etc. (?)
Stichtite..$Mg_6Cr_2(OH)_{16} \cdot CO_3 \cdot 4H_2O$
Stilpnochlorane..$(Ca,K,H_2O)(Al,Fe,Mg)_3$
 $(OH,O)_2(Si,P)_4O_{10}$

193

(MAGNESIUM—continued)

Stilpnomelane..$(K,Na,Ca)(Fe,Mg,Al,Mn)_3 Si_4O_{10}(OH)_2$ (?)
Struvite ..$(NH_4) Mg(PO_4) \cdot 6H_2O$
Sulfoborite ..$Mg_6H_4(BO_3)_4(SO_4)_2 \cdot 7H_2O$
Sussexite ..$(Mn,Mg)(BO_2)(OH)$
Swartzite ..$CaMg(UO_2)(CO_3)_3 \cdot 12H_2O$
Synadelphite ..$(Mn,Mg,Ca,Pb)_4(AsO_4)(OH)_5$
Szaibelyite ..$(Mg,Mn)(BO_2)(OH)$
Tachyhydrite ..$CaMg_2Cl_6 \cdot 12H_2O$
Talc..$Mg_3(Si_4O_{10})(OH)_2$
Taramite..$Na_2Ca_2Mg_5Fe_3(Fe,Ti)_2(Al,Fe)_2 Si_{14}O_{44}(OH,F)_4$
 (Amphibole)
Thuringite..hydrous Fe,Mg,Al silicate
 (Chlorite)
Tilasite..$CaMg(AsO_4)F$
Titanaugite..$Ca_{14}NaMg_{10}Fe_3Ti_2Fe_2Al_5 Si_{27}O_{96}$ (?)
 (Clinopyroxene)
✓ Todorokite..$(Mn,Ba,Ca,Mg)Mn_3O_7 \cdot H_2O$ (?)
Torendrikite..hydrous Na,Ca,Fe,Mg alumino-
 (Amphibole)   silicate
Torreyite..$(Mg,Mn,Zn)_7(SO_4)(OH)_{12} \cdot 4H_2O$
Tourmaline..$(Na_2,Ca)(Li,Al,Mg,Fe,Mn)_6 Al_{12}(OH,F)_8(BO_3)_6Si_{12}O_{36}$
Tremolite..$Ca_2Mg_5Si_8O_{22}(OH,F)_2$
 (Amphibole)
Truscottite..$(Ca,Mg)_4(OH)_2Si_6O_{15} \cdot 3H_2O$ (?)
Tschermakite..$Ca_2Mg_3Al_4Si_6O_{22}(OH,F)_2$
 (Amphibole)
Tuhualite..$(Na_2,K_2,Mn)(Al,Fe,Mg,Ti)_2 Si_{10}O_{24}$ (?)
Tychite..$Na_6Mg_2(CO_3)_4(SO_4)$
Uralite..hydrous Ca,Mg,Fe alumino-silicate
 (Amphibole)
Uvite....$CaMg_4Al_5B_3Si_6O_{29} \cdot 2H_2O$ (?)
 (Tourmaline)
Vanthoffite..$Na_6Mg(SO_4)_4$
Vermiculite..$(Mg,Ca)_{0.7}(Mg,Fe,Al)_6 [(Al,Si)_8O_{20}](OH)_4 \cdot 8H_2O$
Vesuvianite..$Ca_{10}Al_4(Mg,Fe)_2Si_9O_{34}(OH,F)_4$
Volchonskoite..$(Ca,Mg,Cr,Fe,Al)_2O_3 \cdot 3SiO_2 \cdot nH_2O$ (?)
 (Clay)
Wagnerite..$Mg_2(PO_4)F$
Warwickite ..$(Mg,Fe)_3TiB_2O_8$
Weberite ..Na_2MgAlF_7
Xanthophyllite ..$Ca_2(Mg_5Al)Si_2Al_6O_{20}(OH)_4$
 (Brittle Mica)
✓ Zirklerite ..Al,Fe basic chloride

194

MANGANESE

✓ Akrochordite..$MgMn_4(AsO_4)_2(OH)_4 \cdot 4H_2O$ (?)
Alabandite..MnS
Allactite..$Mn_7(AsO_4)_2(OH)_8$
Allanite..$(Ca,Ce,La)_2(Al,Fe,Be,Mg,Mn)_3 (OH)Si_3O_{12}$ (?)
 (=Orthite)
Alleghanyite..$Mn(OH,F)_2 \cdot 2Mn_2(SiO_4)$
Alluaudite..$(Na,Fe,Mn)(PO_4)$
Amphibole..Fe,Mg,Na,Ca,Mn,K hydrous alumino-
                                      silicates
Apjohnite..$MnAl_2(SO_4)_4 \cdot 22H_2O$
Ardennite..$Mn_5Al_5(V,As)(OH)_2Si_5O_{24} \cdot 2H_2O$ (?)
Armangite..$Mn_3(AsO_3)_2$
Arrojadite..$Na_2(Fe,Mn)_5(PO_4)_4$
Arseniopleite..$(Mn,Ca,Pb,Mg)_9(Mn,Fe)_2 (AsO_4)(OH)_6$
Arsenoclasite..$Mn_5(AsO_4)_2(OH)_4$
Ashcroftine..$KNa(Ca,Mg,Mn)(Al_4Si_5O_{18}) \cdot 8H_2O$
 (Zeolite)
Astrophyllite..$(K_2,Na_2,Ca)(Fe,Mn)_4 (Ti,Zr)(OH,F)_2Si_4O_{14}$ (?)
Axinite..$(Ca,Mn,Fe)_3Al_2BO_3(Si_4O_{12})(OH)$
Barkevikite..$Ca_2(Na,K)(Fe,Mg,Mn)_5 Si_{16.5}Al_{1.5}O_{22}(OH)_2$
> (Amphibole)
Bermanite..$(Mn,Mg)_5(Mn,Fe)_8(PO_4)_8(OH)_{10} \cdot 15H_2O$
Berzeliite ..$(Mg,Mn)_2(Ca,Na)_3(AsO_4)_3$
Biotite ..$K(Mg,Fe,Mn)_3(AlSi_3O_{10})(OH,F)_2$
 (Mica)
Bixbyite ..$(Mn,Fe)_2O_3$
Brackebuschite ..$Pb_4MnFe(VO_4)_4 \cdot 2H_2O$
Brandtite ..$Ca_2Mn(AsO_4)_2 \cdot 2H_2O$
Braunite ..$(Mn,Si)_2O_3$
Bustamite ..$(Mn,Ca,Fe)(SiO_3)$
Calderite ..$Mn_3Fe_2(SiO_4)_3$
Carpholite ..$MnAl_2(OH)_4Si_2O_6$
Caryinite ..$(Ca,Pb,Na)_5(Mn,Mg)_4(AsO_4)_5$ (?)
Catoptrite ..Mn,Al antimonate-silicate
Cesarolite ..$PbMn_3O_7 \cdot H_2O$
Chalcophanite ..$ZnMn_2O_5 \cdot 2H_2O$ (?)
Childrenite ..$(Fe,Mn)Al(PO_4)(OH)_2 \cdot H_2O$
Chlormanganokalite ..K_4MnCl_6
Chlorophoenicite ..$(Zn,Mn)_5(AsO_4)(OH)_7$
Columbite ..$(Fe,Mn)(Cb,Ta)_2O_6$
Coronadite ..$MnPbMn_6O_{14}$
Crednerite ..$CuMn_2O_4$
Dickinsonite ..$Na_6(Mn,Fe,Ca)_{14}H_2(PO_4)_{12} \cdot H_2O$

(MANGANESE-continued)

Dietrichite ..$(Zn,Fe,Mn)Al_2(SO_4)_4 \cdot 22H_2O$
Dixenite ..$Mn_5(OH)_2SiAs_2O_9$
Ekmannite ..$(Fe,Mg,Mn)_3(OH)_2(Si,Al)Si_3O_{10}$ (?)
Eosphorite ..$(Mn,Fe)Al(PO_4)(OH)_2 \cdot H_2O$
Epistolite ..$(Na,Ca)(Cb,Ti,Mg,Fe,Mn)SiO_4(OH)$
Fairfieldite ..$Ca_2(Mn,Fe)(PO_4)_2 \cdot 2H_2O$
Ferri-sicklerite ..$(Li,Fe Mn)(PO_4)$
Fillowite ..$Na_6(Mn,Fe,Ca)_{14}H_2(PO_4)_{12} \cdot H_2O$ (?)
Flinkite ..$Mn_3(AsO_4)(OH)_4$
Friedelite ..$(Mn,Fe)_{14}(Si_{14}O_{35})(OH,Cl)_{14}$
Frondelite..$MnFe_4(PO_4)_3(OH)_5$

Galaxite..$MnAl_2O_4$
Ganophyllite..$(Na,K)(Mn,Al,Mg,Ca)_3(OH)_4$
$(Si,Al)Si_3O_{10}$ (?)
> Glaucochroite..$CaMnSiO_4$
(Olivine)
✓ Gonyerite..low Al chlorite
(Chlorite)
Graftonite..$(Fe,Mn,Ca)_3(PO_4)_2$
Griphite..$(Na,Ca,Fe,Al)_3Mn_2(PO_4)_{2.5}$
$(OH,F)_2$ (?)
Grovesite..$(Mn,Mg,Al)_6(Si,Al)_4O_{10}(OH)_8$
(Serpentine)
Harstigite ..$(Ca,Mn,Mg)_8Al_2(OH)_4Si_6O_{21}$ (?)
Hauerite ..MnS_2

Hausmannite..$MnMn_4O_4$
Hellandite..$(Ca,Y,Er,Mn)_3(Al,Fe)$
$Si_2O_4 \cdot H_2O$ (?)
Helvite..$(Mn,Fe,Zn)_4(BeSiO_4)_3S$
Hemafibrite ..$Mn_3(AsO_4)(OH)_3 \cdot H_2O$
Hematolite ..$(Mn,Mg)_4Al(AsO_4)(OH)_8$
Hetaerolite ..$ZnMn_2O_4$
Heterosite ..$(Fe,Mn)(PO_4)$
Hodgkinsonite ..$MnZn_2(OH)_2SiO_4$
Holdenite ..$(Mn,Ca)_4(Zn,Mg,Fe)_2(AsO_4)(OH)_5O_2$
Hollandite ..$MnBaMn_6O_{14}$
> Huebnerite ..$Mn(WO)_4$
Hureaulite..$Mn_5H_2(PO_4)_4 \cdot 4H_2O$
Hydrobiotite..$(K,H_2O)(Mg,Fe,Mn)_3$
$(AlSi_3O_{10})(OH,H_2O)_2$
Hydrohetaerolite..$Zn_2Mn_4O_8 \cdot H_2O$

Ilesite ..$(Mn,Zn,Fe)SO_4 \cdot 4H_2O$ (?)
Ilvaite ..$Ca(Fe,Mn)_2Fe(SiO_4)_2OH$
Inesite ..$Ca_2Mn_7(OH)_2Si_{10}O_{28} \cdot 5H_2O$ (?)
Jacobsite ..$MnFe_2O_4$
Johannsenite ..$Ca(Mn,Fe)Si_2O_6$
(Clinopyroxene)

Kempite ..$Mn_2(OH)_3Cl$
Kentrolite ..$Pb_3Mn_4O_3(SiO_4)_3$
Knebelite..$(Mn,Fe)_2(SiO_4)$
(Olivine)
✓ Kupletskite..$(K_2,Na_2,Ca)(Fe,Mn)_4(Ti,Zr)$
$(OH)_2Si_4O_{14}$
Kutnahorite..$Ca(Mn,Mg)(CO_3)_2$
Landesite ..$Fe_6Mn_2O(PO_4)_{16} \cdot 27H_2O$ (?)
Langbanite ..$Mn_4Mn_3O_8 \cdot SiO_4$ (?)
Lavenite ..complex fluosilicate
Leucophenicite ..$Mn_7(OH)_2Si_3O_{12}$
Lithiophilite ..$LiMn(PO_4)$
Loseyite ..$(Mn,Zn)_7(CO_3)_2(OH)_{10}$
Ludlamite ..$(Fe,Mg,Mn)_3(PO_4)_2 \cdot 4H_2O$
✓ Magnetoplumbite ..$Pb(Fe,Mn)_6O_{10}$ (?)
Mallardite ..$MnSO_4 \cdot 7H_2O$
Mangan-alluaudite ..$(Na,Mn,Fe)(PO_4)$
Manganberzeliite ..$(Mn,Mg)_2(Ca Na)_3(AsO_4)_3$
Manganite ..$MnO(OH)$
Manganolangbeinite ..$K_2Mn_2(SO_4)_3$
Manganosite ..MnO
Melanostibian..$(Mn,Fe)_6(SbO_3)_2O_3$
Mica ..complex hydrous alumino-silicates
Mooreite ..$(Mg,Mn,Zn)_8(SO_4)(OH)_{14} \cdot 4H_2O$
Natrophilite ..$NaMn(PO_4)$
Orientite ..$Ca_4Mn_4(SiO_4)_5 \cdot 4H_2O$
Orthite..$(Ca,Ce,La)_2(Al,Fe,Be,Mg,Mn)_3$
$(OH)Si_3O_{12}$ (?)
✓ Parsettensite..$(K,H_2O)(Fe,Mg,Al,Mn)_3$
$Si_4O_{10}(OH)_2$ (?)
Pennantite..low-Mg chlorite
(Chlorite)
Phosphoferrite ..$(Fe,Mn)_3(PO_4)_2 \cdot 3H_2O$
Phosphophyllite ..$Zn_2(Fe,Mn)(PO_4)_2 \cdot 4H_2O$
Piemontite ..$Ca_2(Al,Fe,Mn)_2Al(OH)Si_3O_{12}$
(Epidote)
Pinakiolite ..$Mg_3MnMn_2B_2O_{10}$ (?)
Psilomelane..$BaMn_9O_{16}(OH)_4$
✓ Pumpellyite..$Ca_4(Mg,Fe,Mn)(Al,Fe,Ti)_5$
$(OH)_3Si_6O_{23} \cdot 2H_2O$ (?)
Purpurite..$(Mn,Fe)(PO_4)$
Pyrobelonite..$MnPb(VO_4)(OH)$
Pyrochroite..$Mn(OH)_2$
Pyrolusite..MnO_2
Pyrophanite..$MnTiO_3$
Pyrosmalite..$(Mn,Fe)_8(OH,Cl)_{10}Si_6O_{15}$ (?)
Pyroxmangite..$(Mn,Fe)(SiO_3)$
Quenselite..$PbMnO_2(OH)$
Reddingite..$(Mn,Fe)_3(PO_4)_2 \cdot 3H_2O$

(MANGANESE-continued)

Redingtonite..$(Fe,Mn,Ni)(Cr,Al)_2(SO_4)_4 \cdot 22H_2O$ (?)
Retzian..Ca,Rare earths,Mn basic arsenate
Rhodochrosite..$MnCO_3$
Rhodonite..$(Mn,Fe,Ca)(SiO_3)$
Richterite..$Na_2Ca(Mg,Fe,Mn,Al)_5Si_8O_{22}(OH,F)_2$
 (Amphibole)
✓ Roscherite..$(Ca,Mn,Fe)_2Al(PO_4)_2(OH) \cdot 2H_2O$
Rosenbuschite..$(Na,Ca,Mn)_3(Fe,Ti,Zr)FSi_2O_8$
✓ Roweite..$(Mn,Mg,Zn)Ca(BO_2)_2(OH)_2$
Salmonsite..$Mn_9Fe_2(PO_4)_8 \cdot 14H_2O$ (?)
Samsonite..Ag_4MnSbS_6
Sarcopside..$(Fe,Mn,Ca)_7(PO_4)_4F_2$ (?)
Sarkinite..$Mn_2(AsO_4)(OH)$
Scacchite..$MnCl_2$
✓ Schallerite..$(Mn,Fe)_8(OH,Cl)_{10}(Si,As)O_{15}$ (?)
✓ Scheteligite..$(Ca,Y,Sb,Mn)_2(Ti,Ta,Cb)_2(O,OH)_7$
Schizolite..$(Ca,Mn)_2NaH(SiO_3)_3$
Schorl..$Na(Fe,Mn)_3Al_6(OH,F)_4(BO_3)_3Si_6O_{18}$
 (Tourmaline)
Seamanite..$Mn_3(PO_4)(BO_3) \cdot 3H_2O$
Senaite..$(Fe,Mn,Pb)TiO_3$
✓ Serandite..$Mn_2NaH(SiO_3)_3$
Sicklerite..$(Li,Mn,Fe)(PO_4)$
Spessartine..$Mn_3Al_2(SiO_4)_3$
 (Garnet)
Steenstrupine..complex hydrous silicate
Stevensite..$Mn_{5.76}Mn_{0.04}Fe_{0.04}Si_8O_{20}(OH)_4$...etc. (?)
 (Clay)
Stewartite..Mn hydrous phosphate
Stilpnomelane..$(K,Na,Ca)(Fe,Mg,Al,Mn)_3Si_4O_{10}(OH)_2$ (?)
Sussexite..$(Mn,Mg)(BO_2)(OH)$
Synadelphite..$(Mn,Mg,Ca,Pb)_4(AsO_4)(OH)_5$
Szaibelyite..$(Mg,Mn)(BO_2)(OH)$
Szmikite..$MnSO_4 \cdot H_2O$
Tantalite..$(Fe,Mn)(Ta,Cb)_2O_6$
Tephroite..$Mn_2(SiO_4)$
 (Olivine)
Thulite..$(Ca,Mn)_2Al_3(OH)(SiO_4)_3$
✓ Todorokite..$(Mn,Ba,Ca,Mg)Mn_3O_7 \cdot H_2O$ (?)
Torreyite..$(Mg,Mn,Zn)_7(SO_4)(OH)_{12} \cdot 4H_2O$
Tourmaline..$(Na_2,Ca)(Li,Al,Mg,Fe,Mn)_6Al_{12}(OH,F)_8(BO_3)_6Si_{12}O_{36}$
Trigonite..$MnPb_3H(AsO_3)_3$
Trimerite..$(Mn,Ca)BeSiO_4$
Triplite..$(Mn,Fe)(PO_4)F$
Triploidite..$(Mn,Fe)_2PO_4(OH)$

Tuhualite..$(Na_2,K_2,Mn)(Al,Fe,Mg,Ti)_2Si_{10}O_{24}$ (?)
✓ Varulite..$(Na,Ca)Mn(PO_4)$
Wad..Hydrous oxides of Mn,with Ba,etc.
Wolfeite..$(Fe,Mn)_2(PO_4)(OH)$
Wolframite..$(Fe,Mn)(WO_4)$
Yeatmanite..$(Mn,Zn)_{16}Sb_2Si_4O_{29}$

MERCURY

Allopalladium..Pd with traces of Hg, Pt,Ru,Cu,etc.
Calomel..$HgCl$
Cinnabar..HgS
Coccinite..HgI_2
Coloradoite..$HgTe$
Eglestonite..Hg_4OCl_2
Gold amalgam..Au_2Hg_3 (?)
Kleinite..Hg,NH_4,Cl,SO_4
Livingstonite..$HgSb_4S_7$
Mercury..Hg
Metacinnabar..$(Hg,Fe,Zn)S$
Montroydite..HgO
Moschellandsbergite..Ag_2Hg_3
Mosesite..Hg,NH_4,Cl,SO_4
✓ Potarite..Pd_3Hg_2 (?)
Terlinguaite..Hg_2OCl
Tiemannite..HgSe

MOLYBDENUM

Ferrimolybdite..$Fe_2(MoO_4)_3 \cdot 8H_2O$ (?)
Ilsemannite..$Mo_3O_8 \cdot nH_2O$ (?)
Koechlinite..$(BiO)_2(MoO_4)$
Lindgrenite..$Cu_3(MoO_4)_2(OH)_2$
Molybdenite..MoS_2
Powellite..$Ca(MoO_4)$
Wulfenite..$Pb(MoO_4)$

NICKEL

✓ Ahlfeldite..$Ni(SeO_4) \cdot 6H_2O$ (?)
Annabergite..$Ni_3(AsO_4)_2 \cdot 8H_2O$
Badenite..$(Co,Ni)(As,Bi)_4$ (?)
Braggite..$(Pt,Pd,Ni)S$

(NICKEL-continued)

Bravoite..$(Ni,Fe)S_2$
Breithauptite..$NiSb$
Bunsenite..NiO
Chloanthite..$(Co_2Ni)As_{3-x}$
Cohenite..$(Fe,Ni)_3C$
Forbesite..$(Ni,Co)H(AsO_4)\cdot 3.5H_2O$ (?)
Garnierite..Ni-bearing chlorite
  (Chlorite)
Gersdorffite..$NiAsS$
Kolovratite..Ni vanadate (?)
Lindackerite..$Cu_6Ni_3(AsO_4)_4(SO_4)(OH)_4\cdot 5H_2O$
Maucherite..$Ni_{11}As_8$
Melonite..$NiTe_2$
Millerite..NiS
Morenosite..$NiSO_4\cdot 7H_2O$
Niccolite..$NiAs$
Nickel iron..Ni,Fe
Nickel skutterudite..$(Ni,Co)As_3$
Pararammelsbergite..$NiAs_2$
√ Parkerite..NiS_2 (?)
√ Penroseite..$(Ni,Cu,Pb)Se_2$
Pentlandite..$(Fe,Ni)_9S_8$
Polydymite..Ni_3S_4
Rammelsbergite..$NiAs_2$
Redingtonite..$(Fe,Mn,Ni)(Cr,Al)_2(SO_4)_4$
  $\cdot 22H_2O$ (?)
Retgersite..$NiSO_4\cdot 6H_2O$
Schreibersite..$(Fe,Ni)_3P$
Siegenite..$(Co,Ni)_3S_4$
Skutterudite..$(Co,Ni)As_3$
Smaltite..$(Co,Ni)As_{3-x}$
Trevorite..$NiFe_2O_4$
Ullmannite..$NiSbS$
Violarite..Ni_2FeS_4
Wolfachite..$Ni(As,Sb)S$ (?)
Zaratite..$Ni_3CO_3(OH)_4\cdot 4H_2O$

PLATINUM

Allopalladium..Pd with traces of Hg,Pt,
  Ru,Cu,etc.
Braggite..$(Pt,Pd,Ni)S$
√ Cooperite..PtS
√ Niggliite..$PtTe_3$ (?)
Platiniridium..Pt,Ir
Platinum..Pt
Sperrylite..$PtAs_2$

POTASSIUM

Adularia..$KAlSi_3O_8$
  (K-Feldspar)
Alunite..$KAl_3(SO_4)_2(OH)_6$
Amazonite..$KAlSi_3O_8$
  (K-Feldspar)
Amphibole..complex hydrous alumino-silicates
Anorthoclase..$(Na,K)AlSi_3O_8$
  (K-Feldspar)
Aphthitalite..$(K,Na)_3Na(SO_4)_2$
Apophyllite..$KFCa_4(Si_4O_{10})_2\cdot 8H_2O$
Arcanite..K_2SO_4
Ashcroftine..$KNa(Ca,Mg,Mn)(Al_4Si_5O_{18})\cdot 8H_2O$
  (Zeolite)
Astrophyllite..$(K_2,Na_2,Ca)(Fe,Mn)_4$
  $(Ti,Zr)(OH,F)_2Si_4O_{14}$ (?)
Avogadrite..$(K,Cs)BF_4$
Barkevikite..$Ca_2(Na,K)(Fe,Mg,Fe,Mn)_5$
  (Amphibole)
  $Si_{6.5}Al_{1.5}O_{22}(OH)_2$
Barroisite..Na,K,Ca,Mg,Fe,Ti alumino-silicate
  (Amphibole)
Basaltic hornblende..$Ca_2(Na,K)_{0.5-1.0}$
  (Amphibole)
  $(Mg,Fe)_{3-2}(Fe,Al)_{2-3}$
  $Si_6Al_2O_{22}(O,OH,F)_2$
Biotite..$K(Mg,Fe,Mn)_3(AlSi_3O_{10})(OH,F)_2$
  (Mica)
Buetschliite..$K_6Ca_2(CO_3)_5\cdot 6H_2O$
Carnallite..$KMgCl_3\cdot 6H_2O$
Carnotite..$K_2(UO_2)_2(VO_4)_2\cdot 3H_2O$
Celadonite..$(K,Ca,Na)(Al,Fe,Mg)_2$
  $(Al_{0.11}Si_{3.89}O_{10})(OH)_2$ (?)
Chlormanganokalite..K_4MnCl_6
Chlorocalcite..$KCaCl_3$
Chlorothionite..$K_2Cu(SO_4)Cl_2$
Clay..chiefly hydrous silicates of Al or Mg
Clino-ungemachite..$Na_9K_3Fe(SO_4)_6(OH)_3\cdot 9H_2O$
Common hornblende..$(Ca,Na,K)_{2-3}(Mg,Fe,$
  (Amphibole)
  $Al)_5Si_6(Si,Al)_2O_{22}(OH,F)_2$
Cyanochroite..$K_2Cu(SO_4)_2\cdot 6H_2O$
Dachiardite..$(Ca,K_2,Na_2)_3Al_4Si_{18}O_{45}\cdot 14H_2O$
  (Zeolite)
Dalyite..$K_2Zr(Si_6O_{15})$
Davyne..$(Na,K)_6Ca_2(AlSiO_4)_6(SO_4)_2$ (?)
√ Dehrnite..$(Ca,Na,K)_5(PO_4)_3(OH)$
Douglasite..$K_2FeCl_4\cdot 2H_2O$ (?)
Elpasolite..K_2NaAlF_6
Englishite..$K_2Ca_4Al_8(PO_4)_8(OH)_{10}\cdot 9H_2O$
Erionite..ca. $(Na_2,K_2,Ca,Mg)_{4.5}Al_9Si_{27}O_{72}\cdot 27H_2O$
  (Zeolite)
Erythrosiderite..$K_2FeCl_5\cdot H_2O$
Euchlorin..K,Na,Cu basic sulfate
Fairchildite..$K_2Ca(CO_3)_2$

(POTASSIUM-continued)

Feldspar ..K,Na,Ca,Ba alumino-silicates
Ferrierite..$(Na,K)_2Mg(Al_3Si_{15}O_{36})(OH) \cdot 9H_2O$
  (Zeolite)
Ganophyllite..$(Na,K)(Mn,Al,Mg,Ca)_3(OH)_4$
                            $(Si,Al)Si_3O_{10}$ (?)
Glauconite..$(K,Ca,Na)(Al,Fe,Mg)_2$
  (Mica)
                 $(Al_{0.35}Si_{3.65}O_{10})(OH)$ (?)
Guembelite..$(K,H_2O)(Al_{1.5}Mg_{0.5})(AlSi_3O_{10})$
                            $(OH,H_2O)_2$ (?)
Hanksite..$Na_{22}K(SO_4)_9(CO_3)_2Cl$
Hieratite..K_2SiF_6
Hornblende, basaltic..$Ca_2(Na,K)_{0.5-1.0}$
  (Amphibole)
                    $(Mg,Fe)_{3-2}(Fe,Al)_{2-3}$
                    $Si_6Al_2O_{22}(O,OH,F)_2$
Hornblende, common..$(Ca,Na,K)_{2-3}(Mg,Fe,$
  (Amphibole)
                    $Al)_5Si_6(Si,Al)_2O_{22}(OH,F)_2$
Hyalophane..$(K,Na,Ba)AlSi_3O_8$
  (Feldspar)
Hydrobiotite..$(K,H_2O)(Mg,Fe,Mn)_3$
                    $(AlSi_3O_{10})(OH,H_2O)_2$
Hydromuscovite..$(K,H_2O)Al_2(H_2O,OH)_2$
                    $AlSi_3O_{10}$ (?)
Hydrophlogopite..$(K,H_2O)Mg_3(AlSi_3O_{10})$
                    $(OH,H_2O)_2$
Illite..$KAl_4Si_7AlO_{20}(OH)_4$
  (Clay)
Jarosite..$KFe_3(SO_4)_2(OH)_6$
Kaersutite..$Ca_2(Na,K)(Mg,Fe)_4TiSi_6$
  (Amphibole)
                    $Al_2O_{22}(OH,F)_2$
Kainite..$KMg(SO_4)Cl \cdot 3H_2O$
Kaliborite..$KMg_2B_{11}O_{19} \cdot 9H_2O$
Kalicinite..$KHCO_3$
Kaliophilite..$K(AlSiO_4)$
Kalinite..$KAl(SO_4)_2 \cdot 11H_2O$
Kalsilite..$K(AlSiO_4)$
Karinthine..$(Na,K)Ca_{2-3}Mg_8Fe_{1-2}(Al,Fe,$
  (Amphibole)
                    $Ti)_2(OH)_4(Al_{3-4}S_{13-12}O_{44})$
—Karpinskiite..$(Na,K,Zn,Mg)_2(OH,H_2O)_{1-2}$
                    $(Al,Be)_2Si_4O_{12}$
Klebelsbergite..Basic Sb sulfate
Krausite ..$KFe(SO_4)_2 \cdot H_2O$
Kremersite..$(NH_4,K)_2FeCl_5 \cdot H_2O$
Kupletskite..$(K_2,Na_2,Ca)(Fe,Mn)_4(Ti,Zr)$
                    $(OH)_2Si_4O_{14}$
√ Labuntsovite..$(K,Ba,Na)Ti(Si,Al)_2$
                    $(O,OH)_7 \cdot H_2O$ (?)
Langbeinite..$K_2Mg_2(SO_4)_3$

√ Latiumite ..$Ca_6(K,Na)_2Al_4(O,CO_3,SO_4)(SiO_4)_6$
Lecontite ..$Na(NH_4,K)(SO_4) \cdot 2H_2O$
Leightonite ..$K_2Ca_2Cu(SO_4)_4 \cdot 2H_2O$
Lehiite ..$(Na,K)_2Ca_5Al_8(PO_4)_8(OH)_{12} \cdot 6H_2O$ (?)
Leonite..$K_2Mg(SO_4)_2 \cdot 4H_2O$ (?)
Lepidolite..$K_2(Li,Al)_{5-6}(Si_{6-7}Al_{2-1}O_{20})$
  (Mica)
                            $(OH,F)_4$
Leucite..$K(AlSi_2O_6)$
Leucophosphite ..$K_2(Fe,Al)_7(PO_4)_4(OH)_{11} \cdot 6H_2O$
Lewistonite ..$(Ca,K,Na)_5(PO_4)_3(OH)$
Lithidionite ..$(Na,K)_2(Si_3O_7)$ (?)
Lopezite ..$K_2(Cr_2O_7)$
Manganolangbeinite..$K_2Mn_2(SO_4)_3$
Mercallite..$KHSO_4$
Mesomicrocline..$K(Al,Si)_2Si_2O_8$
  (K-Feldspar)
Metavoltine..$(K,Na,Fe)_5Fe_3(SO_4)_6(OH)_2$
                            $\cdot 9H_2O$ (?)
Mica..complex hydrous alumino-silicates
Microcline ..$KAlSi_3O_8$
  (K-Feldspar)
Milarite ..$KCa_2Si_{12}Be_2AlO_{30} \cdot \cdot 5H_2O$ (?)
Millisite ..$(Na,K)_2Ca_5Al_8(PO_4)_8(OH)_{12} \cdot 6H_2O$
√ Minyulite ..$KAl_2(PO_4)_2(OH) \cdot 3.5(H_2O)$ (?)
Misenite..$K_8H_6(SO_4)_7$
Mitscherlichite ..$K_2CuCl_4 \cdot 2H_2O$
Mordenite ..$(Ca,K_2,Na_2)(AlSi_5O_{12})_2 \cdot 7H_2O$
  (Zeolite)
Muscovite ..$KAl_2(AlSi_3O_{10})(OH,F)_2$
  (Mica)
Nepheline ..$KNa_3(AlSiO_4)_4$
Niter ..$K(NO_3)$
Orthoclase..$KAlSi_3O_8$
  (K-Feldspar)
√ Osumilite..$(K,Na,Ca)(Mg,Fe)_2(Al,Fe)_3$
                            $(Si,Al)_{12}O_{30} \cdot H_2O$
Palmierite..$(K,Na)_2Pb(SO_4)_2$
—Parsettensite..$(K,H_2O)(Fe,Mg,Al,Mn)_3$
                            $Si_4O_{10}(OH)_2$
Perthite..$KAlSi_3O_8 + NaAlSi_3O_8$
  (K-Feldspar)
Phillipsite..$(.5Ca,Na,K)_2Al_3Si_5O_{16} \cdot 6H_2O$
  (Zeolite)
Phlogopite ..$KMg_3(AlSi_3O_{10})(OH,F)_2$
  (Mica)
Picromerite ..$K_2Mg(SO_4)_2 \cdot 6H_2O$
Polyhalite ..$K_2Ca_2Mg(SO_4)_4 \cdot 2H_2O$
Potash alum ..$KAl(SO_4)_2 \cdot 12H_2O$
Pseudocotunnite ..K_2PbCl_4
Rhodizite..$NaKLi_4Al_4Be_3B_{10}O_{27}$ (?)
Rinneite ..NaK_3FeCl_6
Roscoelite ..$KV_2(AlSi_3O_{10})(OH,F)_2$ (?)
  (Mica)

(POTASSIUM-continued)

Sanidine..$KAlSi_3O_8$
(K-Feldspar)
Scapolite..$(Na,Ca,K)_4Al_3(Al,Si)_3Si_6O_{24}$
(Cl,F,OH,CO_3,SO_4)
Sericite..$KAl_2(AlSi_3O_{10})(OH,F)_2$
(Mica)
Siderophyllite..$KFe_3(AlSi_3O_{10})(OH,F)_2$
(Mica)
—Slavikite..$(K,Na)_2Fe_{10}(SO_4)_{13}(OH)_6 \cdot 63H_2O$ (?)
Stilbite..$(Ca,Na_2,K_2)(Al_2Si_7O_{18}) \cdot 6H_2O$
(Zeolite)
Stilpnochlorane..$(Ca,K,H_2O)(Al,Fe,Mg)_3$
$(OH,O)_2(Si,P)_4O_{10}$
Stilpnomelane..$(K,Na,Ca)(Fe,Mg,Al,Mn)_3$
$Si_4O_{10}(OH)_2$ (?)
Sylvite..KCl
Syngenite..$K_2Ca(SO_4)_2 \cdot H_2O$
Taranakite..$K_2Al_6(PO_4)_6(OH)_2 \cdot 18H_2O$ (?)
Tarapacaite..$K_2(CrO_4)$
Taylorite..$(K,NH_4)_2SO_4$ (?)
✓ Tuhualite..$(Na_2,K_2,Mn)(Al,Fe,Mg,Ti)_2$
$Si_{10}O_{24}$ (?)
✓ Ungemachite..$Na_9K_3Fe(SO_4)_6(OH)_3 \cdot 9H_2O$
Vishnewite..$(Na,Ca,K)_{6-7}Al_6Si_6O_{24}$
$(SO_4,CO_3,Cl)_{1.0-1.5} \cdot 1-5H_2O$
Voltaite..$(K,Fe)_3Fe(SO_4)_3 \cdot 4H_2O$ (?)
✓ Wadeite..$K_2ZrSi_3O_9$ (?)
Zeolite..ca.$(Na_2,K_2,Ca,Ba,Sr)[(Al,Si)O_2]_n \cdot xH_2O$
Zinnwaldite..$K(Li,Fe,Al)(AlSi_3O_{10})(F,OH)_2$
(Mica)

RARE EARTHS

Allanite..$(Ca,Ce,La)_2(Al,Fe,Be,Mg,Mn)_3$
(=Orthite)
$(OH)Si_3O_{12}$ (?)
Ampangabeite..$(Y,Er,U,Ca,Th)_2(Cb,Ta,$
$Fe,Ti)_7O_{18}$ (?)
Ancylite..$(Ce,La)_4(Sr,Ca)_3(CO_3)_7(OH)_4 \cdot 3H_2O$
Bastnasite..$Ce(CO_3)F$
Beckelite..$(Ca,Ce,La,Nd)_4O_3Si_3O_{12}$
Cappelenite..$(Ba,Ca,Ce,Na)_3(Y,Ce,La)_6$
$(BO_3)_6Si_3O_9$ (?)
Cerite..$(Ca,Fe)Ce_3Si_3O_{12} \cdot H_2O$ (?)
Churchite..$(Ce,Ca)(PO_4) \cdot 2H_2O$
Clinotscheffkinite..$(Ce,La)_2Ti_2Si_2O_{11}$ (?)
Cordylite..$Ce_2Ba(CO_3)_3F_2$

Eschynite..$(Ce,Ca,Fe,Th)(Ti,Cb)_2O_6$
Euxenite..$(Y,Ca,Ce,U,Th)(Cb,Ta,Ti)_2O_6$
Fergusonite..$(Y,Er,Ce,Fe)(Cb,Ta,Ti)O_4$
Florencite..$CeAl_3(PO_4)_2(OH)_6$
Fluocerite..$(Ce,La,Nd)F_3$
Hellandite..$(Ca,Y,Er,Mn)_3(Al,Fe)Si_2O_4$
$\cdot H_2O$ (?)
Johnstrupite..$(Ca,Y,Na,Ce)_3(Al,Zr,Ti)$
$(F,OH)Si_2O_8$
Kainosite..$Ca_2(Ce,Y)_2CO_3Si_4O_{12} \cdot 1-2H_2O$
Lanthanite..$(La,Ce)_2(CO_3)_3 \cdot 8H_2O$
Lessingite..$Ca_4(Ce,La,Nd)_7(OH,F)_5Si_6O_{24}$ (?)
Loranskite..$(Y,Ce,Ca,Zr,?)(Ta,Zr,?)O_4$ (?)
Magnesium orthite..$CaCeMgAl_2(OH,F)Si_3O_{12}$ (?)
Monazite..$(Ce,La,Y,Th)(PO_4)$
Mosandrite..$(Ca,Na)_{12}Ce_3(Zr,Ti,Mg)_4F_5$
Nagatelite..$(Ca,Ce)_2(Al,Fe)_3(OH)(Si,P)_3O_{12}$
Orthite..$(Ca,Ce,La)_2(Al,Fe,Be,Mg,Mn)_3$
$(OH)Si_3O_{12}$ (?)
Parisite..$Ce_2Ca(CO_3)_3F_2$
✓ Perrierite..$Ce_2Ti_2Si_2O_{11}$ (?)
Polycrase..$(Y,Ca,Ce,U,Th)(Ti,Nb,Ta)_2O_6$
Polymignyte..$(Ca,Fe,Y,etc.Zr,Th)(Cb,Ti,Ta)O_4$
Priorite..$(Y,Er,Ca,Fe,Th)(Ti,Nb)_2O_6$
Retzian..Ca,Rare earths,Mn basic arsenate
Rhabdophane..$(Ce,Y)(PO_4) \cdot H_2O$
Rinkite..$Na(Ca,Ce)_2(Ti,Ce)FSi_2O_8$
Samarskite..(Y,Er,Ce,U,Ca,Fe,Pb,Th)
$(Cb,Ta,Ti,Sn)_2O_6$
Steenstrupine..complex hydrous silicate
✓ Stillwellite..$(Ce,...)BSiO_5$
Synchisite..$CeCa(CO_3)_2F$
Thorotungstite..$AlFe(Th,Ca,Ce,Zr)WO_3$ (?)
Tritomite..$Ca_3(La,Ce)_3Zr_3F_6B_3Si_6O_{27}$ (?)
Tscheffkinite..$(Fe,Ca)(Ce,La,Al)_2$
$(Si,Ti)_3O_{10}$ (?)
Weinschenkite..$(Y,Er)(PO_4) \cdot 2H_2O$

SILVER

Acanthite..Ag_2S
Aguilarite..$Ag_2(Se,S)$
Alaskaite..$Pb(Ag,Cu)_2Bi_4S_8$ (?)
Andorite..$PbAgSb_3S_6$
Aramayoite..$Ag(Sb,Bi)S_2$
Argentite..Ag_2S

(SILVER-continued)

Argentojarosite..$AgFe_3(SO_4)_2(OH)_6$
Argyrodite..Ag_8GeS_6
✓ Benjaminite..$Pb(Cu,Ag)Bi_2S_4$
Boleite..$Pb_9Cu_8Ag_3Cl_{21}(OH)_{16} \cdot 2H_2O$ (?)
Bromargyrite..$AgBr$
Canfieldite..Ag_8SnS_6
Chlorargyrite..$AgCl$
Cocinerite..Cu_4AgS
Crookesite..$(Cu,Tl,Ag)_2Se$
Diaphorite..$Pb_2Ag_3Sb_3S_8$
Dyscrasite..Ag_3Sb
Empressite..$AgTe$
Eucairite..$CuAgSe$
Fizelyite..$Pb_5Ag_2Sb_8S_{18}$
Freieslebenite..$Pb_3Ag_5Sb_5S_{12}$
Hessite..Ag_2Te
Hutchinsonite..$(Pb,Tl)_2(Cu,Ag)As_5S_{10}$
Iodargyrite..AgI
Lengenbachite..$Pb_6(Ag,Cu)_2As_4S_{13}$
Matildite..$AgBiS_2$
Miargyrite..$AgSbS_2$
Miersite..$(Ag,Cu)I$
Moschellandsbergite..Ag_2Hg_3
Muthmannite..$(Ag,Au)Te$
Naumannite..Ag_2Se
Owyheeite..$Pb_5Ag_2Sb_6S_{15}$
Pearceite..$(Ag,Cu)_{16}As_2S_{11}$
Petzite..Ag_3AuTe_2
Polyargyrite..$Ag_{24}Sb_2S_{15}$
Polybasite..$(Ag,Cu)_{16}Sb_2S_{11}$
Proustite..Ag_3AsS_3
Pyrargyrite..Ag_3SbS_3
Pyrostilpnite..Ag_3SbS_3
Ramdohrite..$Pb_3Ag_2Sb_6S_{13}$
Samsonite..$Ag_4MnSb_2S_6$
Schirmerite..$PbAg_4Bi_4S_9$
Silver..Ag
Smithite..$AgAsS_2$
Stephanite..Ag_5SbS_4
Sternbergite..$AgFe_2S_3$
Stromeyerite..$AgCuS$
Stylotypite..$(Ag,Cu,Fe)_3SbS_3$
Sylvanite..$(Ag,Au)Te_2$
Trechmannite..$AgAsS_2$
Xanthoconite..Ag_3AsS_3

200

SODIUM

Acmite..$NaFe(Si_2O_6)$
 (Clinopyroxene)
Aegirine..$NaFe(Si_2O_6)$
 (Clinopyroxene)
Aegirine-augite..$(Na,Ca)(Fe,Fe,Mg,Al)(Si_2O_6)$
 (Clinopyroxene)
Albite..$NaAlSi_3O_8$
 (Plagioclase)
Alluaudite..$(Na,Fe,Mn)(PO_4)$
Amarillite..$NaFe(SO_4)_2 \cdot 6H_2O$
Amblygonite..$(Li,Na)Al(PO_4)(F,OH)$
Amphibole..hydrous Fe,Mg,Na,Ca,Mn,K
 alumino-silicates
Analcime..$Na(AlSi_2O_6) \cdot H_2O$
 (Zeolite)
Andersonite..$Na_2Ca(UO_2)(CO_3)_3 \cdot 6H_2O$
Andesine..An_{30-50} *
 (Plagioclase)
Anorthoclase..$(Na,K)AlSi_3O_8$
 (K-Feldspar)
Aphthitalite..$(K,Na)_3Na(SO_4)_2$
Arfvedsonite.. $.5Na_5Ca(Fe,Mg,Fe,Al)_{10}$
 (Amphibole) $Si_{15}AlO_{44}(OH,F)_4$
Arrojadite..$Na_2(Fe,Mn)_5(PO_4)_4$
Ashcoftine..$KNa(Ca,Mg,Mn)(Al_4Si_5O_{18}) \cdot 8H_2O$
 (Zeolite)
Astrophyllite..$(K_2,Na_2,Ca)(Fe,Mn)_4$
 $(Ti,Zr)(OH,F)_2Si_4O_{14}$ (?)
Banalsite..$BaNa_2(Al_2Si_2O_8)_2$
Basaltic hornblende..$Ca_2(Na,K)_{0.5-1.0}$
 (Amphibole) $(Mg,Fe)_{3-2}(Fe,Al)_{2-3}$
 $Si_6Al_2O_{22}(O,OH,F)_2$
Barkevikite..$Ca_2(Na,K)(Fe,Mg,Fe,Mn)_5$
 (Amphibole) $(Si_{6.5}Al_{1.5}O_{22})(OH)_2$
Barroisite..Na,K,Ca,Mg,Fe,Ti alumino-silicate
 (Amphibole)
Beidellite..montmorillonite-like clay
 (Clay)
Beryllonite..$NaBe(PO_4)$
Berzeliite..$(Mg,Mn)_2(Ca,Na)_3(AsO_4)_3$
Bloedite..$Na_2Mg(SO_4)_2 \cdot 4H_2O$
Borax..$Na_2B_4O_7 \cdot 10H_2O$
Bradleyite..$Na_3Mg(CO_3)(PO_4)$
Brazilianite..$NaAl_3(PO_4)_2(OH)_4$
Brittle Mica..hydrous Na,Ca,Mg,Al silicate
Burkeite..$Na_6(SO_4)_2(CO_3)$
Bytownite..An_{70-90} *
 (Plagioclase)
Cancrinite..$(Na,Ca)_{7-8}(Al_6Si_6O_{24})$
 $(CO_3,SO_4,Cl)_{1.5-2} \cdot 1-5H_2O$
Cappelenite..$(Ba,Ca,Ce,Na)_3(Y,Ce,La)_6$
 $(BO_3)_6Si_3O_9$ (?)

(SODIUM-continued)

Caracolite..Na,Pb chloride sulfate
Caryinite ..$(Ca,Pb,Na)_5(Mn,Mg)_4(AsO_4)_5$ (?)
Catapleite..$Na_2ZrSi_3O_9 \cdot 2H_2O$ (?)
Celadonite..$(K,Ca,Na)(Al,Fe,Mg)_2$
$(Al_{0.11}Si_{3.89}O_{10})(OH)_2$ (?)
Chabazite..$(Ca,Na_2)(Al_2Si_4O_{12}) \cdot 6H_2O$
(Zeolite)
> Chiolite ..$Na_5Al_3F_{14}$
Clay ..chiefly hydrous silicates of Al or Mg
Clino-ungemachite..$Na_9K_3Fe(SO_4)_6(OH)_3 \cdot 9H_2O$
Common hornblende..$(Ca,Na,K)_{2-3}(Mg,Fe,$
(Amphibole) $Al)_5Si_6(Si,Al)_2O_{22}(OH,F)_2$
Crocidolite..$Na_2Fe_5Si_8O_{22}(OH)_2$
(Amphibole)
Crossite ..$Na_2(Mg,Fe)_3(Al,Fe)_2Si_8O_{22}(OH,F)_2$
(Amphibole)
Cryolite ..Na_3AlF_6
Cryolithionite ..$Na_3Li_3Al_2F_{12}$
Dachiardite ..$(Ca,K_2,Na_2)_3Al_4Si_{18}O_{45} \cdot 14H_2O$
(Zeolite)
Darapskite ..$Na_3(NO_3)(SO_4) \cdot H_2O$
Davyne ..$(Na,K)_6Ca_2(AlSiO_4)_6(SO_4)_2$ (?)
Dawsonite ..$NaAl(CO_3)(OH)_2$
✓ Dehrnite ..$(Ca,Na,K)_5(PO_4)_3(OH)$
Dickinsonite..$Na_6(Mn,Fe,Ca)_{14}H_2(PO_4)_{12} \cdot H_2O$
Dipyre ..Me_{20-50} **
(Scapolite)
Donbassite ..hydrous Na,Ca,Mg,Al silicate
(Clay)
Dravite ..$NaMg_3Al_6(OH)_4(BO_3)_3Si_6O_{18}$
(Tourmaline)
Durangite ..$NaAl(AsO_4)F$
Edenite..$NaCa_2Mg_5AlSi_7O_{22}(OH,F)_2$
(Amphibole)
Ekermannite.. .$5Na_5Ca(Mg,Fe,Fe,Al,Li)_{10}$
(Amphibole) $Si_{15}AlO_{44}(OH,F)_4$
Elbaite..$Na(Li,Al)_3Al_6(OH,F)_4(BO_3)_3Si_6O_{18}$
(Tourmaline)
Elpasolite..K_2NaAlF_6
Elpidite..$NaZrSi_6O_{15} \cdot 3H_2O$
Enigmatite..$(Ca,Na_2)_2Fe(Al,Fe,Ti)_4$
$(Si_2O_7)_2$ (?)
Ephesite ..$(Na,Ca)Al_2[Al(Al,Si)Si_2O_{10}](OH)_2$
(Brittle Mica)
Epididymite ..$NaBe(OH)Si_3O_7$
Epistolite..$(Na,Ca)(Cb,Ti,Mg,Fe,Mn)SiO_4(OH)$
Erionite..ca.$(Na_2,K_2,Ca,Mg)_{4.5}Al_9Si_{27}O_{72} \cdot 27H_2O$
(Zeolite)
Euchlorin..K,Na,Cu basic sulfate
Eudialyte ..$(Na,Ca,Fe)_6ZrSi_6O_{18}(OH,Cl)$
Eudidymite ..$NaBe(OH)Si_3O_7$
Faujasite ..ca.$(Na_2,Ca)_{1.75}Al_{3.5}Si_{8.5}O_{24} \cdot 16H_2O$
(Zeolite)
Feldspar ..K,Na,Ca,Ba alumino-silicates

Ferrierite ..$(Na,K)_2Mg(Al_3Si_{15}O_{36})(OH) \cdot 9H_2O$
(Zeolite)
Ferrinatrite ..$Na_3Fe(SO_4)_3 \cdot 3H_2O$
Ferroedenite..$NaCa_2Fe_5AlSi_7O_{22}(OH,F)_2$
((Amphibole)
Ferrohastingsite..$NaCa_2Fe_4(Al,Fe)$
(Amphibole) $Si_6Al_2O_{22}(OH,F)_2$
Ferruccite..$NaBF_4$
✓ Fersmantite ..$Na_4Ca_4Ti_4(O,OH,F)_3Si_3O_{12}$ (?)
Fillowite ..$Na_6(Mn,Fe,Ca)_{14}H_2(PO_4)_{12} \cdot H_2O$ (?)
Frierinite..$Na_3Cu_3(AsO_4)_2(OH)_3 \cdot H_2O$
Ganophyllite..$(Na,K)(Mn,Al,Mg,Ca)_3(OH)_4$
$(Si,Al)Si_3O_{10}$ (?)
Gaylussite..$Na_2Ca(CO_3)_2 \cdot 5H_2O$
Glauberite..$Na_2Ca(SO_4)_2$
Glauconite..$(K,Ca,Na)(Al,Fe,Mg)_2$
(Mica) $(Al_{0.35}Si_{3.65}O_{10})(OH)$ (?)
Glaucophane..$Na_2Mg_3Al_2Si_8O_{22}(OH,F)_2$
(Amphibole)
Gmelinite ..$(Na_2,Ca)(Al_2Si_4O_{12}) \cdot 6H_2O$
(Zeolite)
Gonnardite ..$Na_2Ca[(Al,Si)_5O_{10}]_2 \cdot 6H_2O$
(Zeolite)
Grandidierite..$H_2Na_2(Mg,Fe)_7(Al,Fe,B)_{15}$
$Si_7Al_7O_{56}$ (?)
Griphite..$(Na,Ca,Fe,Al)_3Mn_2(PO_4)_{2.5}$
$(OH,F)_2$ (?)
Guarinite..$Ca_2NaZrFSi_2O_8$ (?)
Halite ..NaCl
Hanksite ..$Na_{22}K(SO_4)_9(CO_3)_2Cl$
Hastingsite ..$NaCa_2Mg_4Al_3Si_6O_{22}(OH)_2$
(Amphibole)
Hauyne ..$(Na,Ca)_{4-8}Al_6Si_6O_{24}(SO_4,S)_{1-2}$
✓ Hectorite ..montmorillonite-like clay
(Clay)
Heulandite ..$(Ca,Na_2)(Al_2Si_7O_{18}) \cdot 6H_2O$
(Zeolite)
Hornblende, basaltic..$Ca_2(Na,K)_{0.5-1.0}$
(Amphibole) $(Mg,Fe)_{3-2}(Fe,Al)_{2-3}$
$Si_6Al_2O_{22}(O,OH,F)_2$
Hornblende, common..$(Ca,Na,K)_{2-3}(Mg,Fe,$
(Amphibole) $Al)_5Si_6(Si,Al)_2O_{22}(OH,F)_2$
✓ Huhnerkobelite..$(Na,CA)Fe(PO_4)$
Hyalophane..$(K,Na,Ba)AlSi_3O_8$
(Feldspar)
Hydroparagonite..$(Na,H_2O)Al_2(H_2O,OH)_2$
$AlSi_3O_{10}$ (?)
Jadeite ..$NaAl(Si_2O_6)$
(Clinopyroxene)
Jarlite ..$NaSr_3Al_3F_{16}$
Jezekite..$Na_4CaAl_2(PO_4)_2(OH)_2F_2O$ (?)
Joaquinite..$NaBa(Ti,Fe)_3Si_4O_{15}$ (?)
Johnstrupite..$(Ca,Y,Na,Ce)_3(Al,Zr,Ti)$
$(F,OH)Si_2O_8$

(SODIUM–continued)

Julienite..$Na_2CO(SCN)_4 \cdot 8H_2O$
Kaersutite..$Ca_2(Na,K)(Mg,Fe)_4TiSi_6Al_2O_{22}(OH,F)_2$
 (Amphibole)
Karinthine..$(Na,K)Ca_{2-3}Mg_8Fe_{1-2}(Al,Fe,Ti)_2(OH)_4(Al_{3-4}Si_{13-12}O_{44})$
 (Amphibole)
✓ Karpinskiite..$(Na,K,Zn,Mg)_2(OH,H_2O)_{1-2}(Al,Be)_2Si_4O_{12}$
Katophorite..$Na_2CaFe_4(Fe,Al)Si_7AlO_{22}(OH,F)_2$
 (Amphibole)
Kernite..$Na_2B_4O_7 \cdot 4H_2O$
Klebelsbergite..Basic Sb Sulfate
Kroehnkite..$Na_2Cu(SO_4)_2 \cdot 2H_2O$
✓ Kupletskite..$(K_2,Na_2,Ca)(Fe,Mn)_4(Ti,Zr)(OH)_2Si_4O_{14}$
Labradorite..An_{50-70} *
 (Plagioclase)
✓ Labuntsovite..$(K,Ba,Na)Ti(Si,Al)_2(O,OH)_7 \cdot H_2O$ (?)
Lacroixite..Na,Ca,Al fluo-phosphate
Lamprophyllite..$Na_3SrTi_3O_2(OH)Si_3O_{12}$ (?)
✓ Latiumite..$Ca_6(K,Na)_2Al_4(O,CO_3,SO_4)(SiO_4)_6$ (?)
Lavenite..complex fluosilicate
Lazurite..$(Na,Ca)_8(AlSiO_4)_6(SO_4,S,Cl_2)_4$ (?)
Lecontite..$Na(NH_4,K)(SO_4) \cdot 2H_2O$
Lehiite..$(Na,K)_2Ca_5Al_8(PO_4)_8(OH)_{12} \cdot 6H_2O$ (?)
Leifite..$Na_2FSi_5AlO_{12}$ (?)
Leucophane..$(Ca,Na)_2BeSi_2(O,OH,F)_7$ (?)
Leucosphenite..$(Na_2,Ca)_2BaTi_3BSi_8O_{27}$ (?)
Lewistonite..$(Ca,K,Na)_5(PO_4)_3(OH)$
Lithidionite..$(Na,K)_2(Si_3O_7)$ (?)
Loeweite..$Na_2Mg(SO_4)_2 \cdot 2.5H_2O$
✓ Lomonossovite..$(Na,Ca)Ti(O,OH)(S,P)O_4$ (?)
Magnesiokatophorite..$Na_2CaMg_4(Fe,Al)Si_7AlO_{22}(OH,F)_2$
 (Amphibole)
Malladrite..Na_2SiF_6
Mangan-alluaudite..$(Na,Mn,Fe)(PO_4)$
Manganberzeliite..$(Mn,Mg)_2(Ca,Na)_3(AsO_4)_3$
Marialite..$Na_8(AlSi_3O_8)_6(Cl_2,SO_4,CO_3)$
 (Scapolite)
Meionite..$Ca_8(Al_2Si_2O_8)_6(Cl_2,SO_4,CO_3)_2$
 (Scapolite)
Melilite..$(Ca,Na)_2(Al,Mg)(Si,Al)_2O_7$
Melinophane..$(Ca,Na)_2(Be,Al)(Si_2O_6F)$
Mendozite..$NaAl(SO_4)_2 \cdot 11H_2O$
Mesolite..$Na_2Ca_2(Al_2Si_3O_{10})_3 \cdot 8H_2O$
 (Zeolite)
Metasideronatrite..$Na_4Fe_2(SO_4)_4(OH)_2 \cdot 3H_2O$

Metavoltine..$(K,Na,Fe)_5Fe_3(SO_4)_6(OH)_2 \cdot 9H_2O$ (?)
Mica..hydrous Na,K,Ca,Mg,Fe,V,Mn,Al silicates
Mica..hydrous Na,Ca,Mg,Al silicates
 (Brittle)
Microlite..$(Na,Ca)_2Ta_2O_6(O,OH,F)$
Millisite..$(Na,K)CaAl_6(PO_4)_4(OH)_9 \cdot 3H_2O$
Mirabilite..$Na_2SO_4 \cdot 10H_2O$
Mizzonite..Me_{50-80} **
 (Scapolite)
Montebrasite..$(Li,Na)Al(PO_4)(OH,F)$
Montmorillonite..$Na_{0.7}(Al_{3.3}Mg_{0.7})Si_8O_{20}(OH)_4 \cdot nH_2O$
 (Clay)
Mordenite..$(Ca,K_2,Na_2)(AlSi_5O_{12})_2 \cdot 7H_2O$
 (Zeolite)
Morinite..Na,Ca,Al fluo-phosphate
Mosandrite..$(Ca,Na)_{12}Ce_3(Zr,Ti,Mg)_4F_5Si_{10}O_{40}$ (?)
Murmanite..$Na_2Ti_2(OH)_4Si_2O_7$ (?)
Nahcolite..$NaHCO_3$
Narsarsukite..$Na_2(Ti,Fe)(O,OH,F)Si_4O_{10}$
Natroalunite..$NaAl_3(SO_4)_2(OH)_6$
Natrochalcite..$NaCu_2(SO_4)_2(OH) \cdot H_2O$
Natrojarosite..$NaFe_3(SO_4)_2(OH)_6$
Natrolite..$Na_2(Al_2Si_3O_{10}) \cdot 2H_2O$
 (Zeolite)
Natromontebrasite..$(Na,Li)Al(PO_4)(F,OH)$
Natron..$Na_2CO_3 \cdot 10H_2O$
Natrophilite..$NaMn(PO_4)$
Naujakasite..$Na_4FeAl_4H_4Si_8O_{27}$
Nepheline..$KNa_3(AlSiO_4)_4$
Neptunite..$Na_2FeTi(Si_4O_{12})$
Nontronite..montmorillonite-like clay
 (Clay)
Northupite..$Na_3Mg(CO_3)_2Cl$
Nosean..$Na_8(AlSiO_4)_6(SO_4)$
Oligoclase..An_{10-30} *
 (Plagioclase)
Omphacite..$(Ca,Na)(Mg,Fe,Fe,Ti,Al)(Si_2O_6)$
 (Clinopyroxene)
✓ Osumilite..$(K,Na,Ca)(Mg,Fe)_2(Al,Fe)_3(Si,Al)_{12}O_{30} \cdot H_2O$
Pachnolite..$NaCaAlF_6 \cdot H_2O$
Palmierite..$(K,Na)_2Pb(SO_4)_2$
Paragonite..$NaAl_2(AlSi_3O_{10})(OH,F)_2$
 (Mica)
Pargasite..$NaCa_2Mg_4(Al,Fe)Si_6Al_2O_{22}(OH,F)_2$
 (Amphibole)
Pectolite..$Ca_2NaH(SiO_3)_3$
Perthite..$KAlSi_3O_8 + NaAlSi_3O_8$
 (K-Feldspar)
Phillipsite..$(.5Ca,K,Na)_3Al_3Si_5O_{16} \cdot 6H_2O$
 (Zeolite)

(SODIUM-continued)

Pirssonite..$Na_2Ca(CO_3)_2 \cdot 2H_2O$
Pollucite..$(Cs,Na)(AlSi_2O_6) \cdot H_2O$ (?)
Probertite..$NaCaB_5O_9 \cdot 5H_2O$
Pyrochlore..$NaCa(Cb,Ta)_2O_6F$
Pyroxene..Mg,Fe,Ca,Na,Ti,Al silicates
Ralstonite..ca.$Na(Mg,Al_5)_6F_{12}(OH)_6 \cdot 3H_2O$
Ramsayite..$Na_2TiSi_2TiO_9$ (?)
Reedmergnerite..$Na(BSi_3O_8)$
Rhodizite..$NaKLi_4Al_4Be_3B_{10}O_{27}$ (?)
Richterite..$Na_2Ca(Mg,Fe,Mn,Al)_5Si_8O_{22}(OH,F)_2$
 (Amphibole)
Riebeckite..$Na_2Fe_5Si_8O_{22}(OH,F)_2$
 (Amphibole)
Rinkite..$Na(Ca,Ce)_2(Ti,Ce)FSi_2O_8$
Rinneite..NaK_3FeCl_6
Rosenbuschite..$(Na,Ca,Mn)_3(Fe,Ti,Zr)FSi_2O_8$
Sampleite..$NaCaCu_5(PO_4)_4Cl \cdot 5H_2O$
Sanidine..$KAlSi_3O_8$
Saponite..$(.5Ca,Na)_{0.7}(Al,Mg,Fe)_4$
 (Clay) $(Si,Al)_8O_{20}(OH)_4 \cdot nH_2O$

Sarcolite..$(Ca,Na_2)_3Al_2Si_3O_{12}$ (?)
Sauconite..montmorillonite-like clay
 (Clay)
Scapolite..$(Na,Ca,K)_4Al_3(Al,Si)_3$
 $Si_6O_{24}(Cl,F,OH,CO_3,SO_4)$
Schairerite..$Na_3(SO_4)(F,Cl)$
Schizolite..$(Ca,Mn)_2NaH(SiO_3)_3$
Schroeckingerite..$NaCa_3(UO_2)(CO_3)_3$
 $(SO_4)F \cdot 10H_2O$
Schorl..$Na(Fe,Mn)_3Al_6(OH,F)_4(BO_3)_3Si_6O_{18}$
 (Tourmaline)
Searlesite..$NaBSi_2O_6 \cdot H_2O$
✓ Serandite..$Mn_2NaH(SiO_3)_3$
Shortite..$Na_2Ca_2(CO_3)_3$
Sideronatrite..$Na_2Fe(SO_4)_2(OH) \cdot 3H_2O$
— Slavikite..$(K,Na)_2Fe_{10}(SO_4)_{13}(OH)_6 \cdot 63H_2O$ (?)
Soda-alum..$NaAl(SO_4)_2 \cdot 12H_2O$
Soda-niter..$Na(NO_3)$
Sodalite..$Na_8(AlSiO_4)_6Cl_2$
Steenstrupine..complex hydrous silicate
Stercorite..$Na(NH_4)H(PO_4) \cdot 4H_2O$
Stilbite..$(Ca,Na_2,K_2)(Al_2Si_7O_{18}) \cdot 7H_2O$
 (Zeolite)
Stilpnomelane..$(K,Na,Ca)(Fe,Mg,Al,Mn)_3$
 $Si_4O_{10}(OH)_2$ (?)
Sulfohalite..$Na_6ClF(SO_4)_2$
Swedenborgite..$NaBe_4SbO_7$

Tamarugite..$NaAl(SO_4)_2 \cdot 6H_2O$
Taramite..$Na_2Ca_2Mg_5Fe_3(Fe,Ti)_2(Al,Fe)_2$
 (Amphibole) $Si_{14}O_{44}(OH,F)_4$
Teepleite..$Na_2B_2O_4 \cdot 2NaCl \cdot 4H_2O$
Thenardite..Na_2SO_4
Thermonatrite..$Na_2CO_3 \cdot H_2O$
Thomsenolite..$NaCaAlF_6 \cdot H_2O$
Thomsonite..$NaCa_2[(Al,Si)_5O_{10}]_2 \cdot 6H_2O$
 (Zeolite)
Tincalconite..$Na_2B_4O_7 \cdot 5H_2O$
Titanaugite..$Ca_{14}NaMg_{10}Fe_3Ti_2Fe_2Al_5$
 (Clinopyroxene) $Si_{27}O_{96}$ (?)
Torendrikite..hydrous Na,Ca,Mg,Fe alumino-
 (Amphibole) silicate
Tourmaline..$(Na_2,Ca)(Li,Al,Mg,Fe,Mn)$
 $Al_{12}(OH,F)_8(BO_3)_6Si_{12}O_{36}$
Trona..$Na_3H(CO_3)_2 \cdot 2H_2O$
✓ Tuhualite..$(Na_2,K_2,Mn)(Al,Fe,Mg,Ti)_2$
 $Si_{10}O_{24}$ (?)
Tychite..$Na_6Mg_2(CO_3)_4(SO_4)$
Ulexite..$NaCaB_5O_9 \cdot 8H_2O$
✓ Ungemachite..$Na_9K_3Fe(SO_4)_6(OH)_3 \cdot 9H_2O$
Uralite..hydrous Ca,Mg,Fe alumino-silicate
 (Amphibole)
Vanthoffite..$Na_6Mg(SO_4)_4$
✓ Varulite..$(Na,Ca)Mn(PO_4)$
Villiaumite..NaF
Vishnewite..$(Na,Ca,K)_{6-7}Al_6Si_6O_{24}$
 $(SO_4,CO_3,Cl)_{1-1.5} \cdot 1-5H_2O$
Wardite..$Na_4CaAl_{12}(PO_4)_8(OH)_9 \cdot 3H_2O$
Wattevilleite..$Na_2Ca(SO_4)_2 \cdot 4H_2O$ (?)
Weberite..Na_2MgAlF_7
Woehlerite..$NaCa_2(Zr,Cb)FSi_2O_8$
Zeolite..ca.(Na_2,K_2,Ca,Ba,Sr)
 $[(Al,Si)O_2]_n \cdot xH_2O$

STRONTIUM

Ancylite..$(Ce,La)_4(Sr,Ca)_3(CO_3)_7(OH)_4 \cdot 3H_2O$
Brewsterite..$(Sr,Ba,Ca)(Al_2Si_6O_{16}) \cdot 5H_2O$
 (Zeolite)
Celestite..$SrSO_4$
Fermorite..$(Ca,Sr)_5(P,AsO_4)_3(F,OH)$
Goyazite..$SrAl_3(PO_4)_2(OH)_5 \cdot H_2O$
Hinsdalite..$(Pb,Sr)Al_3(PO_4)(SO_4)(OH)_6$
Jarlite..$NaSr_3Al_3F_{16}$
Lamprophyllite..$Na_3SrTi_3O_2(OH)Si_3O_{12}$ (?)
Strontianite..$SrCO_3$

203

(STRONTIUM-continued)

Svanbergite..$SrAl_3(PO_4)(SO_4)(OH)_6$
✓ Veatchite..$Sr_3Bi_6O_{27} \cdot 5H_2O$ (?)
Zeolite..ca.$(Na_2,K_2,Ca,Ba,Sr)[(Al,Si)O_2]_n \cdot xH_2O$

THALLIUM

Crookesite..$(Cu,Tl,Ag)_2Se$
Hutchinsonite..$(Pb,Tl)_2(Cu,Ag)As_5S_{10}$
Lorandite..$TlAsS_2$
Vrbaite..$TlAs_2SbS_5$

THORIUM

Ampangabeite..$(Y,Er,U,Ca,Th)_2(Cb,Ta,Fe,Ti)_7O_{18}$ (?)
✓ Brannerite..$(U,Ca,Fe,Y,Th)_3Ti_5O_{16}$ (?)
Eschynite..$(Ce,Ca,Fe,Th)(Ti,Cb)_2O_6$
Euxenite..$(Y,Ca,Ce,U,Th)(Cb,Ta,Ti)_2O_6$
Formanite..$(U,Zr,Th,Ca)(Ta,Cb,Tl)O_4$
Huttonite..$Th(SiO_4)$
> Monazite..$(Ce,La,Y,Th)(PO_4)$
Pilbarite..$2PbO \cdot 3ThO_2 \cdot 4UO_2 \cdot 8SiO_2 \cdot 21H_2O$ (?)
Polycrase..$(Y,Ca,Ce,U,Th)(Ti,Cb,Ta)_2O_6$
Polymignyte..$(Ca,Fe,Y,etc.,Zr,Th)(Cb,Tl,Ta)O_4$
Priorite..$(Y,Er,Ca,Fe,Th)(Ti,Cb)_2O_6$
Samarskite..$(Y,Er,Ce,U,Ca,Fe,Pb,Th)(Cb,Ta,Ti,Sn)_2O_6$
Thorianite..ThO_2
Thorite..$ThSiO_4$
Thorogummite..$(Th,U)(Si,H_4)O_4$
Thorotungstite..$AlFe(Th,Ca,Ce,Zr)WO_3$ (?)
Uranothorite..$(Th,U)(SiO_4)$
Yttrailite..$(Y,Th)_2(Si_2O_7)$
Yttrocrasite..$(Y,Th,U,Ca)_2(Ti,Fe,W)_4O_{11}$ (?)
Zirkelite..$(Ca,Fe,Th,U)_2(Ti,Zr)_2O_5$

TIN

Canfieldite..Ag_8SnS_6
Cassiterite..SnO_2
Colusite..$Cu_3(Sn,Te,Fe,V,As)S_4$

Cylindrite..$Pb_3Sn_4Sb_2S_{14}$
Delorenzite..$(Y,U,Fe)(Ti,Sn...)_3O_8$ (?)
Franckeite..$Pb_5Sn_3Sb_2S_{14}$
Herzenbergite..SnS
Hulsite..$(Fe,Ca,Mg)_4(Fe,Sn)_2B_2O_{10}$ (?)
> Nordenskioldine..$CaSn(BO_3)_2$
Samarskite..$(Y,Er,Ce,U,Ca,Fe,Pb,Th)(Cb,Ta,Ti,Sn)_2O_6$
Stannite..Cu_2FeSnS_4
Teallite..$PbSnS_2$
Thoreaulite..$SnTa_2O_7$
Tin..Sn
Yttrotantalite..$(Fe,Y,U,Ca,etc.)(Cb,Ta,Zr,Sn)O_4$

TITANIUM

Ampangabeite..$(Y,Er,U,Ca,Th)_2(Cb,Ta,Fe,Ti)_7O_{18}$
Anatase..TiO_2
Andradite..$Ca_3(Fe,Ti)_2(SiO_4)_3$
Arizonite..$Fe_2Ti_3O_9$
Astrophyllite..$(K_2,Na_2,Ca)(Fe,Mn)_4(Ti,Zr)(OH,F)Si_4O_{14}$ (?)
Augite..$(Ca,Mg,Fe,Ti,Al)_2(Si,Al)_2O_6$ (Clinopyroxene)
Barroisite..Na,K,Ca,Mg,Fe,Ti alumino-silicate (Amphibole)
Benitoite..$BaTiSi_3O_9$
Betafite..$(U,Ca)(Cb,Ta,Ti)_3O_9 \cdot nH_2O$
✓ Brannerite..$(U,Ca,Fe,Y,Th)_3Ti_5O_{16}$ (?)
Brookite..TiO_2
Clinotscheffkinite..$(Ce,La)_2Ti_2Si_2O_{11}$ (?)
Delorenzite..$(Y,U,Fe)(Ti,Sn...)_3O_8$ (?)
Derbylite..$Fe_6Ti_6Sb_2O_{23}$ (?)
Djalmaite..$(U,Ca,Pb,Bi,Fe)(Ta,Cb,Ti,Zr)_3O_9 \cdot nH_2O$
Enigmatite..$(Ca,Na_2)_2Fe(Al,Fe,Ti)_4(Si_2O_7)_2$ (?)
Epistolite..$(Na,Ca)(Cb,Ti,Mg,Fe,Mn)SiO_4(OH)$
Eschynite..$(Ce,Ca,Fe,Th)(Ti,Cb)_2O_6$
Euxenite..$(Y,Ca,Ce,U,Th)(Cb,Ta,Ti)_2O_6$
Fassaite..$Ca(Mg,Fe,Al,Ti)(Si,Al)_2O_6$ (Clinopyroxene)
Fergusonite..$(Y,Er,Ce,Fe)(Cb,Ta,Ti)O_4$
✓ Fersmanite..$Na_4Ca_4Ti_4(O,OH,F)_3Si_3O_{12}$ (?)
✓ Formanite..$(U,Zr,Th,Ca)(Ta,Cb,Ti)O_4$
Geikielite..$MgTiO_3$
Hoegbomite..$Mg(Al,Fe,Ti)_4O_7$

204

(TITANIUM-continued)

Ilmenite ..$FeTiO_3$
Joaquinite..$NaBa(Ti,Fe)_3Si_4O_{15}$ (?)
Johnstrupite..$(Ca,Y,Na,Ce)_3(Al,Zr,Ti)$
$(F,OH)Si_2O_8$
Kaersutite..$Ca_2(Na,K)(Mg,Fe)_4Ti$
(Amphibole) $Si_6Al_2O_{22}(OH,F)_2$
Kalkowskite..$Fe_2Ti_3O_9$ (?)
Karinthine..$(Na,K)Ca_{2-3}Mg_8Fe_{1-2}(Al,Fe,$
(Amphibole) $Ti)_2(OH)_4(Al_{3-4}Si_{13-12}O_{44})$
Kupletskite..$(K_2,Na_2,Ca)(Fe,Mn)_4(Ti,Zr)$
$(OH)_2Si_4O_{14}$
✓ Labuntsovite..$(K,Ba,Na)Ti(Si,Al)_2$
$(O,OH)_7 \cdot H_2O$ (?)
Lamprophyllite..$Na_3SrTi_3O_2(OH)Si_3O_{12}$ (?)
Leucosphenite ..$(Na_2,Ca)_2BaTi_3BSi_8O_{27}$ (?)
✓ Lomonossovite ..$(Na,Ca)Ti(O,OH)(S,P)O_4$ (?)
Mosandrite..$(Ca,Na)_{12}Ce_3(Zr,Ti,Mg)_4F_5$
$Si_{10}O_{40}$ (?)
Murmanite..$Na_2Ti_2(OH)_4Si_2O_7$ (?)
Narsarsukite..$Na_2(Ti,Fe)(O,OH,F)Si_4O_{10}$
Neptunite..$Na_2FeTi(Si_4O_{12})$
Oliveiraite..$Zr_3Ti_2O_{10} \cdot 2H_2O$
Omphacite..$(Ca,Na)(Mg,Fe,Fe,Ti,Al)(Si_2O_6)$
(Clinopyroxene)
Osbornite..TiN
Perovskite..$CaTiO_3$
✓ Perrierite..$Ce_2Ti_2Si_2O_{11}$ (?)
Polycrase..$(Y,Ca,Ce,U,Th)(Ti,Cb,Ta)_2O_6$
Polymignyte..$(Ca,Fe,Y,etc.Zr,Th)(Cb,Ti,Ta)O_4$
Priorite..$(Y,Er,Ca,Fe,Th)(Ti,Cb)_2O_6$
Pseudobrookite..Fe_2TiO_5
— Pumpellyite..$Ca_4(Mg,Fe,Mn)(Al,Fe,Ti)_5(OH)_3$
$Si_6O_{23} \cdot 2H_2O$ (?)
Pyrophanite..$MnTiO_3$
Pyroxene..Mg,Fe,Ca,Na,Al,Ti silicates
Ramsayite..$Na_2TiSi_2TiO_9$ (?)
Rinkite..$Na(Ca,Ce)_2(Ti,Ce)FSi_2O_8$
Rosenbuschite..$(Na,Ca,Mn)_3(Fe,Ti,Zr)FSi_2O_8$
Rutile..TiO_2
Samarskite..(Y,Er,Ce,U,Ca,Fe,Pb,Th)
$(Cb,Ta,Ti,Sn)_2O_6$
"Samarskite"..$(Ca,Pb,Y,U)(Cb,Ta,Ti,Fe)_2O_6$ (?)
✓ Scheteligite..$(Ca,Y,Sb,Mn)_2(Ti,Ta,Cb)_2(O,OH)_7$
Senaite..$(Fe,Mn,Pb)TiO_3$
Sphene..$CaTi(SiO_4)(OH,F,O)$

Taramite..$Na_2Ca_2Mg_5Fe_3(Fe,Ti)_2(Al,Fe)_2$
(Amphibole) $Si_{14}O_{44}(OH,F)_4$
Titanaugite....$Ca_{14}NaMg_{10}Fe_3Ti_2Fe_2Al_5$
(Clinopyroxene) $Si_{27}O_{96}$ (?)
Titanite..$CaTi(SiO_4)(OH,F,O)$
(=Sphene)
Tscheffkinite..$(Fe,Ca)(Ce,La,Al)_2$
$(Si,Ti)_3O_{10}$ (?)
Tuhualite..$(Na_2,K_2,Mn)(Al,Fe,Mg,Ti)_2$
$Si_{10}O_{24}$ (?)
Uhligite..$Ca_3(Ti,Al,Zr)_9O_{20}$ (?)
Warwickite..$(Mg,Fe)_3TiB_2O_8$
Yttrocrasite..$(Y,Th,U,Ca)_2(Ti,Fe,W)_4O_{11}$ (?)
Zirkelite..$(Ca,Fe,Th,U)_2(Ti,Zr)_2O_5$

TUNGSTEN

✓ Anthoinite..$Al(WO_4)(OH) \cdot H_2O$
Cuprotungstite..$Cu_2(WO_4)(OH)_2$
Ferberite..$Fe(WO_4)$
✓ Ferritungstite..$Fe_2(WO_4)(OH)_4 \cdot 4H_2O$ (?)
Huebnerite..$Mn(WO_4)$
Powellite.. $Ca(Mo,WO_4)$
Raspite..$Pb(WO_4)$
Russellite..$(Bi_2W)O_3$
✓ Sanmartinite..$(Zn,Fe)(WO_4)$
> Scheelite..$Ca(WO_4)$
Stolzite..$Pb(WO_4)$
Thorotungstite..$AlFe(Th,Ca,Ce,Zr)WO_3$ (?)
Tungstenite..WS_2
Tungstite..$WO_3 \cdot H_2O$ (?)
Wolframite..$(Fe,Mn)(WO_4)$
Yttrocrasite..$(Y,Th,U,Ca)_2(Ti,Fe,W)_4O_{11}$ (?)

URANIUM

Ampangabeite..$(Y,Er,U,Ca,Th)_2(Cb,Ta,$
$Fe,Ti)_7O_{18}$ (?)
Andersonite..$Na_2CaUO_2(CO_3)_3 \cdot 6H_2O$
Autunite..$Ca(UO_2)_2(PO_4)_2 \cdot 10-12H_2O$
✓ Bassetite..Ferrous uranyl phosphate
Bayleyite..$Mg_2(UO_2)(CO_3)_3 \cdot 18H_2O$
✓ **Becquerelite**..$2UO_3 \cdot 3H_2O$ (?)
Betafite..$(U,Ca)(Cb,Ta,Ti)_3O_9 \cdot nH_2O$
✓ Brannerite..$(U,Ca,Fe,Y,Th)_3Ti_5O_{16}$ (?)

(URANIUM-continued)

Carnotite..$K_2(UO_2)_2(VO_4)_2·3H_2O$
√ Clarkeite..$UO_3·nH_2O$ (?)
Clino-sklodowskite..$Mg(H_3O)_2UO_2Si_2O_8·3H_2O$
Coffinite..$U(Si,H_4)O_4$
Cuprosklodowskite..$CuU_2O_3Si_2O_8·6H_2O$
Curite..$2PbO·5UO_3·4H_2O$ (?)
Delorenzite..$(Y,U,Fe)(Ti,Sn,...)_3O_8$ (?)
√ Dewindtite..$Pb_3(UO_2)_5(PO_4)_4(OH)_4·10H_2O$
Djalmaite..$(U,Ca,Pb,Bi,Fe)(Ta,Cb,Ti,Zr)_3O_9·nH_2O$
Dumontite..$Pb_2(UO_2)_3(PO_4)_2(OH)_4·3H_2O$
Euxenite..$(Y,Ca,Ce,U,Th)(Cb,Ta,Ti)_2O_6$
Ferghanite..$U_3(VO_4)_2·6H_2O$
√ Formanite..$(U,Zr,Th,Ca)(Ta,Cb,Ti)O_4$
√ Fourmarierite..$PbO·4UO_3·5H_2O$ (?)
Gummite..$UO_3·nH_2O$ (?)
√ Ianthinite..$2UO_2·7H_2O$
Ishikawaite..$(U,Fe,Y,etc.)(Cb,Ta)O_4$
Johannite..$Cu(UO_2)_2(SO_4)_2(OH)$
√ Kasolite..$Pb_2U_2O_4Si_2O_8·H_2O$ (?)
Liebigite..$Ca_2U(CO_3)_4·10H_2O$
Metatorbernite..$Cu(UO_2)_2(PO_4)_2·8H_2O$
Meta-uranopilite..$(UO_2)_6SO_4(OH)_{10}·5H_2O$
Metazeunerite..$Cu(UO_2)_2(AsO_4)_2·8H_2O$
Parsonsite..$Pb_2(UO_2)(PO_4)_2·2H_2O$
Phosphuranylite..Ca uranyl phosphate
Pilbarite..$2PbO·3ThO_2·4UO_2·8SiO_2·21H_2O$ (?)
Polycrase..$(Y,Ca,Ce,U,Th)(Ti,Cb,Ta)_2O_6$
Rauvite..$CaU_2V_{12}O_{36}·20H_2O$
Renardite..$Pb(UO_2)_4(PO_4)_2(OH)_4·7H_2O$
Rutherfordine..$(UO_2)(CO_3)$ (?)
√ Saleeite..$Mg(UO_2)_2(PO_4)_2·10H_2O$
Samarskite..$(Y,Er,Ce,U,Ca,Fe,Pb,Th)(Cb,Ta,Ti,Sn)_2O_6$
"Samarskite"..$(Ca,Pb,Y,U)(Cb,Ta,Ti,Fe)_2O_6$ (?)
Schoepite..$4UO_3·9H_2O$ (?)
Schroeckingerite..$NaCa_3(UO_2)(CO_3)_3(SO_4)F·10H_2O$
√ Sengierite..$Cu(UO_2)(VO_4)(OH)·4-5H_2O$ (?)
Sharpite..$(UO_2)_6(CO_3)_5(OH)_2·6H_2O$ (?)
√ Sklodowskite..$MgU_2O_3Si_2O_8·6H_2O$
Soddyite..$U_5Si_2O_{19}·6H_2O$ (?)
Swartzite..$Ca,Mg(UO_2)(CO_3)_3·12H_2O$
Thorogummite..$(Th,U)(Si,H_4)O_4$
Torbernite..$Cu(UO_2)_2(PO_4)_2·8-12H_2O$
Troegerite..$(UO_2)_3(AsO_4)_2·12H_2O$

Tyuyamunite..$Ca(UO_2)_2(VO_4)_2·nH_2O$ (?)
Uraninite..UO_2
Uranocircite..$Ba(UO_2)_2(PO_4)_2·8H_2O$
Uranophane..$CaU_2O_3Si_2O_8·7H_2O$
Uranopilite..$(UO_2)_6(SO_4)(OH)_{10}·12H_2O$
Uranosphaerite..$Bi_2O_3·2UO_3·3H_2O$ (?)
Uranospinite..$Ca(UO_2)_2(AsO_4)_2·8H_2O$
Uranothorite..$(Th,U)(SiO_4)$
Uvanite..$U_2V_6O_{21}·15H_2O$
Vanderbrandite..$CuO·UO_3·2H_2O$
Voglite..$Ca_2CuU(CO_3)_5·6H_2O$ (?)
Walpurgite..$Bi_4(UO_2)(AsO_4)O_4·3H_2O$
Waltherite..Bi basic carbonate
Yttrocrasite..$(Y,Th,U,Ca)_2(Ti,Fe,W)_4O_{11}$ (?)
Yttrotantalite..$(Fe,Y,U,Ca,etc.)(Cb,Ta,Zr,Sn)O_4$
Zeunerite..$Cu(UO_2)_2(AsO_4)_2·10-16H_2O$
Zippeite..$(UO_2)_2SO_4(OH)_2·4H_2O$
Zirkelite..$(Ca,Fe,Th,U)_2(Ti,Zr)_2O_5$

VANADIUM

Ardennite..$Mn_5Al_5(V,As)(OH)_2Si_5O_{24}·2H_2O$ (?)
Brackebuschite..$Pb_4MnFe(VO_4)_4·2H_2O$
Calciovolborthite..$CuCa(VO_4)(OH)$
Carnotite..$K_2(UO_2)_2(VO_4)_2·3H_2O$
Colusite..$Cu_3(Sn,Te,Fe,V,As)S_4$
Corvusite..$V_2V_{12}O_{34}·nH_2O$ (?)
Descloizite..$ZnPb(VO_4)(OH)$
Ferghanite..$U_3(VO_4)_2·6H_2O$
Fernandinite..$CaO·V_2O_4·5V_2O_5·14H_2O$
Fervanite..$Fe_4V_4O_6·5H_2O$
Hewettite..$CaV_6O_{16}·9H_2O$
Kolovratite..Na vanadate?
Melanovanadite..$Ca_2V_{10}O_{25}$
Metahewettite..$CaV_6O_{16}·9H_2O$
Metarossite..$CaV_2O_6·2H_2O$
Mica..complex hydrous alumino-silicates
Minasragrite..$(VO)_2H_2(SO_4)_3·15H_2O$
Mottramite..$CuPb(VO_4)(OH)$
Pascoite..$Ca_2V_6O_{17}·11H_2O$
Pintadoite..$Ca_2V_2O_7·9H_2O$
Pyrobelonite..$MnPb(VO_4)(OH)$
Pucherite..$Bi(VO_4)$
Rauvite..$CaU_2V_{12}O_{36}·20H_2O$

(VANADIUM—continued)

Roscoelite..$KV_2(AlSi_3O_{10})(OH,F)_2$ (Mica)
Rossite..$CaV_2O_6 \cdot 4H_2O$
✓ Sengierite..$Cu(UO_2)(VO_4)(OH) \cdot 4-5H_2O$ (?)
✓ Sincosite..$Ca(VO)_2(PO_4)_2 \cdot 5H_2O$
✓ Steigerite..$Al_2(VO_4)_2 \cdot 6.5H_2O$
Sulvanite..Cu_3VS_4
Turanite..$Cu_2(VO_4)(OH)$ (?)
Tyuyamunite..$Ca(UO_2)_2(VO_4)_2 \cdot nH_2O$ (?)
Uvanite..$U_2V_6O_{21} \cdot 15H_2O$
Vanadinite..$Pb_5(VO_4)_3Cl$
Vanoxite..$V_4V_2O_{13} \cdot 8H_2O$ (?)
Volborthite..$Cu_3(VO_4)_2 \cdot 3H_2O$ (?)

YTTRIUM

Ampangabeite..$(Y,Er,U,Ca,Th)_2(Cb,Ta,Fe,Ti)_7O_{18}$ (?)
✓ Brannerite..$(U,Ca,Fe,Y,Th)_3Ti_5O_{16}$ (?)
Cappelenite..$(Ba,Ca,Ce,Na)_3(Y,Ce,La)_6(BO_3)_6Si_3O_9$ (?)
Delorenzite..$(Y,U,Fe)(Ti,Sn...)_3O_8$ (?)
Euxenite..$(Y,Ca,Ce,U,Th)(Cb,Ta,Ti)_2O_6$
Fergusonite..$(Y,Er,Ce,Fe)(Cb,Ta,Ti)O_4$
> Gadolinite..$FeY_2Be_2(SiO_5)_2$
Hellandite..$(Ca,Y,Er,Mn)_3(Al,Fe)Si_2O_4 \cdot H_2O$ (?)
Ishikawaite..$(U,Fe,Y,etc.)(Cb,Ta)O_4$
Johnstrupite..$(Ca,Y,Na,Ce)_3(Al,Zr,Ti)(F,OH)Si_2O_8$
Kainosite..$Ca_2(Ce,Y)_2CO_3Si_4O_{12} \cdot 1-2H_2O$
Loranskite..$(Y,Ce,Ca,Zr,?)(Ta,Zr,?)O_4$ (?)
Monazite..$(Ce,La,Y,Th)(PO_4)$
Polycrase..$(Y,Ca,Ce,U,Th)(Ti,Cb,Ta)_2O_6$
Polymignyte..$(Ca,Fe,Y,etc.,Zr,Th)(Cb,Ti,Ta)O_4$
Priorite..$(Y,Er,Ca,Fe,Th)(Ti,Cb)_2O_6$
Rhabdophane..$(Ce,Y)(PO_4) \cdot H_2O$
Samarskite..$(Y,Er,Ce,U,Ca,Fe,Pb,Th)(Cb,Ta,Ti,Sn)_2O_6$
"Samarskite"..$(Ca,Pb,Y,U)(Cb,Ta,Ti,Fe)_2O_6$ (?)
✓ Scheteligite..$(Ca,Y,Sb,Mn)_2(Ti,Ta,Cb)_2(O,OH)_7$
Thalenite..$Y_2Si_2O_7$
Thortveitite..$(Sc,Y)_2Si_2O_7$
Weinschenkite..$(Y,Er)(PO_4) \cdot 2H_2O$
Xenotime..$Y(PO_4)$
Yttrialite..$(Y,Th)_2Si_2O_7$
Yttrocrasite..$(Y,Th,U,Ca)_2(Ti,Fe,W)_4O_{11}$ (?)
Yttrotantalite..$(Fe,Y,U,Ca,etc.)(Cb,Ta,Zr,Sn)O_4$

ZINC

Adamite..$Zn_2(AsO_4)(OH)$
Aurichalcite..$(Zn,Cu)_5(CO_3)_2(OH)_6$
Austinite..$CaZn(AsO_4)(OH)$
✓ Bianchite..$ZnSO_4 \cdot 6H_2O$
Calamine..$Zn_4(OH)_2Si_2O_7 \cdot H_2O$ (=Hemimorphite)
Chalcophanite..$ZnMnO_5 \cdot 2H_2O$ (?)
Chlorophoenicite..$(Zn,Mn)_5(AsO_4)(OH)_7$
Clinohedrite..$Ca_2Zn_2(OH)_2Si_2O_7 \cdot H_2O$
Descloizite..$ZnPb(VO_4)(OH)$
Dietrichite..$(Zn,Fe,Mn)Al_2(SO_4)_4 \cdot 22H_2O$
Franklinite..$ZnFe_2O_4$
Gahnite..$ZnAl_2O_4$
✓ Genthelvite..$Zn_4(BeSiO_4)_3S$
Glaucocerinite..$Zn_{13}Al_8Cu_7(SO_4)_2(OH)_{60} \cdot 4H_2O$ (?)
Goslarite..$ZnSO_4 \cdot 7H_2O$
Hardystonite..$Ca_2ZnSi_2O_7$
Helvite..$(Mn,Fe,Zn)_4(BeSiO_4)_3S$
Hemimorphite..$Zn_4(OH)_2Si_2O_7 \cdot H_2O$
Hetaerolite..$ZnMn_2O_4$
Hodgkinsonite..$MnZn_2(OH)_2SiO_4$
Holdenite..$(Mn,Ca)_4(Zn,Mg,Fe)_2(AsO_4)(OH)_5O_2$
Hopeite..$Zn_3(PO_4)_2 \cdot 4H_2O$
Hydrohetaerolite..$Zn_2Mn_4O_8 \cdot H_2O$
Hydrozincite..$Zn_5(CO_3)_2(OH)_6$
Ilesite..$(Mn,Zn,Fe)SO_4 \cdot 4H_2O$ (?)
✓ Karpinskiite..$(Na,K,Zn,Mg)_2(OH,H_2O)_{1-2}(Al,Be)_2Si_4O_{12}$
Koettigite..$Zn_3(AsO_4)_2 \cdot 8H_2O$
Larsenite..$PbZnSiO_4$
✓ Legrandite..$Zn_{14}(AsO_4)_9(OH) \cdot 12H_2O$
Lorseyite..$(Mn,Zn)_7(CO_3)_2(OH)_{10}$
Metacinnabar..$(Hg,Fe,Zn)S$
Mooreite..$(Mg,Mn,Zn)_8(SO_4)(OH)_{14} \cdot 4H_2O$
Parahopeite..$Zn_3(PO_4)_2 \cdot 4H_2O$
Phosphophyllite..$Zn_2(Fe,Mn)(PO_4)_2 \cdot 4H_2O$
Rosasite..$(Cu,Zn)_2(CO_3)(OH)_2$
— Roweite..$(Mn,Mg,Zn)Ca(BO_2)_2(OH)_2$
✓ Sanmartinite..$(Zn,Fe)(WO_4)$

(ZINC-continued)

Serpierite ..$(Zn,Cu,Ca)_5(SO_4)_2(OH)_6 \cdot 3H_2O$ (?)
Smithsonite ..$ZnCO_3$
Spencerite ..$Zn_4(PO_4)_2(OH)_2 \cdot 3H_2O$
Sphalerite ..ZnS
Stokesite..$CaZnSi_3O_9 \cdot 2H_2O$
> Tarbuttite ..$Zn(PO_4)(OH)$
Torreyite ..$(Mg,Mn,Zn)_7(SO_4)(OH)_{12} \cdot 4H_2O$
Veszelyite..$(Cu,Zn)_3(As,PO_4)(OH)_3 \cdot 2H_2O$
Voltzite ..Zn_5S_4O
Willemite ..Zn_2SiO_4
Wurtzite ..ZnS
Yeatmanite ..$(Mn,Zn)_{16}Sb_2Si_4O_{29}$
Zinc ..Zn
Zincaluminite ..$Zn_3Al_3(SO_4)(OH)_{13} \cdot 2.5H_2O$
Zincite ..ZnO

ZIRCONIUM

Astrophyllite..$(K_2,Na_2,Ca)(Fe,Mn)_4(Ti,Zr)(OH,F)_2Si_4O_{14}$ (?)
Baddeleyite..ZrO_2
Catapleite..$Na_2ZrSi_3O_9 \cdot 2H_2O$ (?)
Dalyite..$K_2Zr(Si_6O_{15})$
Djalmaite..$(U,Ca,Pb,Bi,Fe)(Ta,Cb,Ti,Zr)_3O_9 \cdot nH_2O$
Elpidite..$Na_2ZrSi_6O_{15} \cdot 3H_2O$
Eudialyte..$(Na,Ca,Fe)_6ZrSi_6O_{18}(OH,Cl)$
√ Formanite..$(U,Zr,Th,Ca)(Ta,Cb,Ti)O_4$
Guarinite..$Ca_2NaZrFSi_2O_8$ (?)
Johnstrupite..$(Ca,Y,Na,Ce)_3(Al,Zr,Ti)(F,OH)Si_2O_8$
Kupletskite..$(K_2,Na_2,Ca)(Fe,Mn)_4(Ti,Zr)(OH)_2Si_4O_{14}$
Lavenite..complex fluosilicate
Loranskite..$[Y,Ce,Ca,Zr,(?)][Ta,Zr,(?)]O_4$ (?)
Mosandrite..$(Ca,Na)_{12}Ce_3(Zr,Ti,Mg)_4F_5Si_{10}O_{40}$ (?)
Oliveiraite..$Zr_3Ti_2O_{10} \cdot 2H_2O$
Polymignyte..$(Ca,Fe,Y,etc.,Zr,Th)(Cb,Ti,Ta)O_4$
Rosenbuschite..$(Na,Ca,Mn)_3(Fe,Ti,Zr)FSi_2O_8$
Thorotungstite..$AlFe(Th,Ca,Ce,Zr)WO_3$ (?)
Tritomite..$Ca_3(La,Ce)_3Zr_3F_6B_3Si_6O_{27}$ (?)
Uhligite..$Ca_3(Ti,Al,Zr)_9O_{20}$ (?)
√ Wadeite..$K_2ZrSi_3O_9$ (?)
Woehlerite..$NaCa_2(Zr,Cb)FSi_2O_8$
Yttrotantalite..$(Fe,Y,U,Ca,etc.)(Cb,Ta,Zr,Sn)O_4$
Zircon..$ZrSiO_4$
Zirkelite..$(Ca,Fe,Th,U)_2(Ti,Zr)_2O_5$

ADDENDUM

Afwillite..$Ca_3Si_2O_7 \cdot 3H_2O$
Ahlfeldite..$(Ni,Co)SeO_3 \cdot 2H_2O$
Akrochordite..$(Mn,Mg)_5(AsO_4)_2(OH)_4 \cdot 4H_2O$
Alumohydrocalcite..$CaAl_2(CO_3)_2(OH)_4 \cdot 3H_2O$ (?)
Ammonioborite..$(NH_4)_2B_{10}O_{16} \cdot 5\ 1/3 H_2O$
Anthoinite..$Al_2W_2O_9 \cdot 3H_2O$
Arsenobismite..$Bi_2(AsO_4)(OH)_3$

Barysilite..$Pb_4MnSi_3O_{11}$
Bassetite..$Fe(UO_2)_2(PO_4)_2 \cdot 8H_2O$
Bayleyite..$Mg_2(UO_2)(CO_3)_3 \cdot 18H_2O$
Becquerelite..$CaU_6O_{19} \cdot 11H_2O$
Beegerite..A mixture
Bellingerite..$Cu(IO_3)_2$
Benjaminite..$(Cu,Ag)_2Pb_2Bi_4S_9$
Beyerite..$(Ca,Pb)Bi_2(CO_3)_2O_2$
Bianchite..$(Zn,Fe)SO_4 \cdot 6H_2O$
Brannerite..$(U,Ca,Th,Y)(Ti,Fe)_2O_6$
Bultfonteinite..$Ca_2SiO_2(OH,F)_4$
Buttgenbachite..$Cu_{19}(NO_3)_2(OH)_{32}Cl_4 \cdot 2H_2O$

Chiviatite..A mixture (?)
Chloroxiphite..$K_2Cu(SO_4)Cl_2$
Clarkeite..$(Na,Ca,Pb)_2U_2(O,OH)_7$
Cooperite..$(Pt,Pd)S$

Dehrnite..$(Ca,Na,K)_5(PO_4,CO_3)_3(OH)$
Deltaite..A mixture (?)
Dewindite..$Pb(UO_2)_2(PO_4)_2 \cdot 3H_2O$
Digenite..Cu_9S_5

Ellestadite..$Ca_5[(Si,S,P)O_4]_3(O,OH,Cl,F)$

Ferritungstite..$Ca_2Fe_2Fe_2(WO_4)_7 \cdot 9H_2O$
Fersmanite..$Na_4Ca_4Ti_4(O,F)_2Si_3O_{18}$
Formanite..$(U,Zr,Y,Th,Ca)(Ta,Cb,Ti)O_4$
Fourmarierite..$PbU_4O_{13} \cdot 4H_2O$

Genthelvite..$(Zn,Fe,Mn)_4Be_3(SiO_4)_3S$
Germanite..$Cu_3(Ge,Ga,Fe)(S,As)_4$
Gonyerite..$(Mn,Mg)_6Si_4O_{10}(OH)_8$
Goongarite..A mixture

Harkerite..$Ca_7(Mg,Al)_5(B,Si)_6(O,OH,Cl)_{24} \cdot 5CaCO_3$
Hectorite..$(Mg,Li)_3Si_4O_{10}(OH,F)_2 \cdot Na_{.33}$
Hematophanite..$Pb_5Fe_4O_{10}(OH,Cl)_2$
Huhnerkobelite..$(Na,Ca)(Fe,Mn)_2(PO_4)_2$
Hydrobasaluminite..$Al_4(SO_4)(OH)_{10} \cdot nH_2O$
Hydrocalumite..$Ca_4Al_2O_9 \cdot 12H_2O$

Ianthinite..$UO_2 \cdot 5UO_3 \cdot 10H_2O$

Karpinskiite..$Na_2(Be,Zn,Mg)Al_2Si_6O_{16}(OH)_2$
Kasolite..$Pb(UO_2)SiO_4 \cdot H_2O$
Klaprothite..A mixture
Koktaite..$(NH_4)_2Ca(SO_4)_2 \cdot 2H_2O$
Kupletskite..$(K_2,Na_2,Ca)(Fe,Mn)_4(Ti,Zr,Nb)(OH)_2Si_4O_{14}$

Labuntsovite..$(K,Ba,Na)(Ti,Nb)(Si,Al)_2(O,OH)_7 \cdot H_2O$ (?)
Latiumite..$(Ca,K)_8(Al,Mg,Fe)(Si,Al)_{10}O_{25}(SO_4)$
Legrandite..$Zn_2(AsO_4)(OH) \cdot H_2O$
Lomonosovite..$Na_2Ti_2Si_2O_9 \cdot Na_3PO_4$

Mackayite..$Fe_2(TeO_3)_3 \cdot xH_2O$ (?)

209

Magnetoplumbite..$(Pb,Mn)_2Fe_6O_{11}$
Melanovanadinite..$Ca_2V_{10}O_{25} \cdot nH_2O$
Minnesotaite..$(Fe,Mg)_3Si_4O_{10}(OH)_2$
Minyulite..$KAl_2(PO_4)_2(OH,F) \cdot 4H_2O$

Nekoite..$Ca_3Si_{10}O_{15} \cdot 8H_2O$
Niggliite..PtSn or PtTe (?)

Osumilite..$(K,Na)(Mg,Fe)_2(Al,Fe)_3$
$(Si,Al)_{12}O_{30} \cdot H_2O$

Painite..$Ca_4BAl_{20}SiO_{38}$ (?)
Paravauxite..$FeAl_2(PO_4)(OH)_2 \cdot 10H_2O$
Parkerite..$Ni_3(Bi,Pb)_2S_2$
Parsettensite..$Mn_3Si_6O_{13}(OH)_8$ (?)
Penroseite..$(Ni,Cu,Co)Se_2$
Perrierite..$(Ce,Ca,Th)_2(Ti,Fe)_2Si_2O_{11}$
Potarite..PdHg
Pumpellyite..$Ca_4Al_4(Al,Fe,Fe,Mg,Mn)_2$
$Si_6O_{23}(OH)_3 \cdot 2H_2O$

Riversideite..$Ca_5Si_6O_{16}(OH)_2 \cdot 2H_2O$
Rockbridgeite..$FeFe_6(PO_4)_4(OH)_8$
Roscherite..$(Ca,Mn,Fe)_3Be_3(PO_4)_3$
$(OH)_3 \cdot 2H_2O$

Roweite..$H_2CaMn(BO_3)_2$

Saleeite..$Mg(UO_2)_2(PO_4)_2 \cdot 8H_2O$
Sanbornite..$BaSi_2O_5$
Sanmartinite..$ZnWO_4$
Scawtite..$Ca_6Si_6O_{18} \cdot CaCO_3 \cdot 2H_2O$
Schallerite..$(Mn,Fe)_8AsS_6(O,OH,Cl)_{26}$

Scheteligite..$(Ca,Y,Sb,Mn)_2$
$(Ti,Ta,W,Cb)_2(O,OH)_7$
Sengierite..$Cu(UO_2)_2(VO_4)_2 \cdot 8\text{-}10H_2O$
Serandite..$Na(Mn,Ca)_2Si_3O_8(OH)$
Sincosite..$CaV_2(PO_4)_2(OH)_4 \cdot 3H_2O$
Simpsonite..$Al_4(Ta,Cb)_3(O,OH,F)_{14}$
Sklodowskite..$MgU_2O_3Si_2O_8 \cdot 7H_2O$
Slavikite..$MgFe_3(SO_4)_4(OH)_3 \cdot 18H_2O$
Steigerite..$AlVO_4 \cdot 3H_2O$
Stibiopalladinite..Pd_2Sb
Stillwellite..$(Ce,La,Ca)BSiO_5$

Teineite..$CuTeO_2 \cdot 2H_2O$
Todorokite..(Mn,Ba,Ca,Mg,Zn)
$Mn_3O_7 \cdot H_2O$ (?)
Tuhualite..$(Na,K)_2(Fe,Fe,Al)_3$
$Si_7O_{18}(OH)_2$

Ungemachite..$Na_9K_3Fe(SO_4)_6(OH)_3 \cdot$
$10H_2O$

Varulite..$(Na_2,Ca)(Mn,Fe)_2(PO_4)_2$
Vauxite..$FeAl_2(PO_4)_2(OH)_2$
Veatchite..$SrB_6O_{10} \cdot 2H_2O$

Wadeite..$K_2CaZrSi_4O_{12}$
Weibullite..A mixture
Wherryite..$Pb_4Cu(CO_3)(SO_4)_2$
$(OH,Cl)_2O$ (?)

Zirklerite..$(Fe,Mg,Ca)_9Al_4Cl_{18}$
$(OH)_{12} \cdot 14H_2O$ (?)

APPENDIX I.   ABBREVIATIONS

Several abbreviations are used in the "Remarks" section of Tables I and II. Those not in general use are listed in this appendix even though they were fashioned so that their meanings would be apparent. Periods are not used after abbreviations, although they do appear coincidently after those which occur as the final word of a categorical entry. Plurals, indicated by the addition of "s" to an abbreviation have been used sporadically.

Chemical symbols are also used as a kind of abbreviation (even as modifiers, *e.g.*, Sn-wht = tin-white). They are used in their widely accepted forms modified only in that both letters of two letter designations are capitalized. The lack of lower-case letters along with the impossibility of printing subscripts as computer output with the utilized IBM-7040 has led to entries like HCL for hydrochloric acid and H2SO4 for sulfuric acid.

The employed abbreviations and the word(s) for which each stands are:

ACCESS - accessory mineral
ADMN - adamantine
AGGS - aggregates
ALK - Alkalic (*e.g.*, rock) *or* alkaline (*e.g.*, taste)
ALTER - alteration product
AN - anorthite-content (*e.g.*, AN 30 indicates plagioclase with the composition Anorthite - 30 per cent and Albite - 70 per cent)
ANAL - analysis
AQ-REG - aqua regia
ARTIF - artificial
ASSOC - associated (with)
ATTKD - attacked (by)
BLDD - bladed
BLDS - blades
BLK - black
BLKISH - blackish
BLU - blue
BRTL - brittle
BRWN - brown
CALC - calcareous
CHALCOPY - chalcopyrite
CHEM - chemical(ly)
CHOC - chocolate
CLVG - cleavage
CNTCT - contact (as applied to rocks metamorphosed near contacts with igneous material rocks)
CONC - concentrated
CONCH - conchoidal
CONFIRM - confirmation
CVTS - cavities (in)
DCMP - decomposes in
DEP - deposit

DISSEM - disseminated
DK - dark
DKR - darker (than)
DLQSNT - deliquescent
DODEC - dodecahedral
DOLO - dolomite (the rock)
DUCT - ductile
EFF - effervescent
EFFLOR - efflorescent
ELAST - elastic
ESP - especially
EXPOS - exposure or exposed
F - fuses
FA - fayalite-content (*e.g.*, FA30 indicates olivine with the composition Fayalite - 30 per cent and Forsterite - 70 per cent)
FERROMAG - ferromagnesian mineral
FIBR - fibrous
FL - flame (color)
FLEX - flexible (*e.g.*, to flexible sheets)
FLKS - flakes
FLUO - fluoresces (in ultra-violet light unless otherwise indicated) *or* fluorescent
FM - form *or* formation
.... FM - form (as a suffix - *e.g.*, collofm)
FMED - formed
FOL - foliated
FRACT - fracture
FRANKLN - Franklin (New Jersey)
FS - ferrosilite-content (*e.g.*, FS30 indicates pyroxene of enstatite-ferrosilite series with a composition of Ferrosilite - 30 per cent and Enstatite - 70 per cent)

211

GELAT - gelatinizes (in HCl unless otherwise noted)
GLBL - globular
GLDN - golden
GRN - grain
GRND - grained
GRNLR - granular
GRP - group
GRSY - greasy
GYP - gypsum
H - hardness
HBLDE - hornblende
HEMIMORPH - hemimorphic
HEX - hexagonal
HI-T - high temperature (*e.g.*, as referred to metamorphic processes)
IG - igneous (rocks)
IGN - ignition
IMPERF - imperfect
INCRUST - incrustation
INELAST - inelastic
IRID - iridescent
LIQ - liquid
..... LK - like (as a suffix - *e.g.*, lenslk)
LO-T - low temperature (*e.g.*, as referred to metamorphic processes)
LS - limestone
LW - long wave (as referred to ultra-violet radiation)
MAG - magnetic
MAL - malleable
MESO-T - mesothermal
MET - metallic
META - metamorphism *or* metamorphic
METEOR - meteoritic
MICRO - microcrystals
MIN - mineral
MOD-T - moderate temperature (*e.g.*, as referred to metamorphic processes)
MSS - masses
MSV - massive
NEPH - nepheline
OCCUR - occurrence
OCTAH - octahedral *or* octahedra
OPQ - opaque
OR - orange
PEG - pegmatite
PERF - perfect
PHOSPHO - phosphorescent
PHYS - physical
PLAG - plagioclase feldspar
PLTS - plates
PLTY - platy
PRISM - prismatic

PROD - product
PROPS - properties
PROV - province
PYROELECT - pyroelectric
PYRRH - pyrrhotite
QTZ - quartz
RDACTIV - radioactive
RESID - residue or residuum
RESIN - resinous
RHOMB - rhombohedral
RIEB - riebeckite
RK - rock
RSMBLS - resembles
S-VNS - sulfide (nor sulfur) veins
SACCH - saccharoidal
SCNDRY - secondary
SECT - sectile
SED - sediment, sedimentary, *or* sedimentary rock
SER - series
SERP - serpentine
S.G. - specific gravity
SHPD - shaped
SIL - siliceous
SOL - soluble (in)
SOLUT - solution
SPHAL - sphalerite
SQ - square
SS - sandstone
STALAC - stalactitic
STL - steel
STRK - streak
SUFS - sulfides
SURF - surface
SW - short wave (ultra-violet radiation)
SYS - system
TARN - tarnish or tarnished
TBLR - tabular
TERRES - terrestrial
TETRA - tetrahedral *or* tetrahedra
THERMOLUM - thermoluminescent
TRIBOLUM - triboluminscent
UNEVN - uneven
VAR - variety
VITR - vitreous
VN - vein
VNLETS - veinlets
VOL - volcanic or volcanic rock
WDSPRD - widespread
WHT - white
WM - warm (*e.g.*, warm aqua regia)
WXED - weathered
WXING - weathering
XL - crystal
YEL - yellow
ZEOL-ASSOC - typical zeolite association
ZNS - zones

APPENDIX II.   GLOSSARY

(The following are meanings implied by their usage in the introduction and in Tables I and II of this book; they should not be considered all-inclusive or hard-and-fast definitions. If further information is desired, the reader is referred to textbooks in the diverse fields of the Geological Sciences and to the Glossary of Geology and Related Sciences with Supplement, published by the American Geological Institute, Washington, D. C., 1960.)

Absorbent, having tendency or capacity to take in or suck up, $e.g.$, as a blotter does ink.
Accessory mineral, any mineral constituent present in small amounts and disregarded in naming and classifying the rock of which it is a part.
Acicular, needle-like (said of crystals).
Acidic igneous rock, igneous rock that contains more than 65 per cent silica.
Acrid, sharp and harsh or bitterly pungent in taste or odor.
Adamantine, with a brilliant luster like that of a diamond.
Alkalic rock, igneous rock in which relatively large proportional quantities of the alkali constituents, chiefly potassium and sodium, have caused the formation of distinctive minerals such as nepheline, leucite, and sodalite and/or soda-rich amphiboles.
Alkaline solution, solution exhibiting basic properties, $e.g.$, a pH > 7.
Alteration, change in mineralogical composition as a result of processes like weathering or metasomatism.
Amorphous, having no apparent crystalline structure.
Amphibolite, metamorphic rock, typically foliated, composed chiefly of an actinolitic amphibole and a plagioclase feldspar, commonly andesine.
Amygdule, a filled vesicle, typically in basalt.
Anhedral, having a form which has been determined by the shapes of the surrounding grains, $i.e.$, with none of its own crystal faces developed (cf. euhedral and subhedral).
Aphanitic, having a cryptocrystalline or microcrystalline texture which is so fine that individual grains cannot be distinguished even with the aid of a hand-lens.
Argillaceous, containing abundant clay.
Asbestiform, fibrous like asbestos.
Asteriated, exhibiting a star-shaped figure when observed in certain directions in reflected or transmitted light.
Astringent, having a taste that puckers the mouth.
Authigenic, pertaining to minerals that have originated within sediments during or after deposition; the term is used for minerals formed at low temperature ($e.g.$, by diagenesis) but not for those formed by metamorphism.
Axis (crystallographic), reference line used to describe the properties of a crystal, $e.g.$, in general, the unique (tetragonal and hexagonal systems) or longest (orthorhombic, monoclinic, and triclinic systems) axis is designated the c axis; each of the reference lines, axes, is commonly parallel to intersections between crystal faces.

Basal cleavage, cleavage parallel to the basal pinacoid, $i.e.$, parallel to the face that cuts only the c axis (q.v.)
Basalt, aphanitic igneous rock composed chiefly of calcic plagioclase, typically labradorite, and augite or pigeonite plus or minus olivine and minor glass; the coarse-grained, chemical equivalent is gabbro.

Basic rock, igneous rock that contains less than approximately 55 per cent silica and is relatively rich in iron, magnesium, and calcium.
Bentonite, montmorillonitic clay formed by decomposition of volcanic ash; it commonly has ability to absorb relatively large quantities of water and thus to swell.
Bipyramid, having the form of two pyramids arranged base to base.
Botryoidal, having the shape of a bunch of grapes.

Calcareous, containing a noteworthy part of calcium carbonate.
Carbonaceous, containing a noteworthy part of amorphous carbon or some solid organic hydrocarbon(s).
Clay, hydrous aluminum or magnesium silicates that typically occur as particles of very fine size (< 2 microns).
Cleavage, tendency of crystalline substances to break or split along structurally controlled plane surfaces.
Colloform, having rounded or hemispherical prominences (includes botryoidal, globular, mammillary, and reniform).
Compound, a distinct chemical substance composed of two or more chemical elements.
Conchoidal, capable of fracturing along smoothly curving fracture surfaces.
Concretion, ellipsoidal or irregular-shaped mass, typically developed in sedimentary rocks by localized deposition from solution around some nucleus.
Contact metamorphism, changes related to igneous activity and typically occurring at or near contacts between the igneous rock formed as the result of the igneous activity and the juxtaposed rocks.
Cruciform, having a shape like that of a Latin cross.
Cryptocrystalline, submicroscopically crystalline (*not* glassy).
Crystal, a solid, bounded by plane surfaces, which is the outward manifestation of a regular, periodic arrangement of constituent atoms, ions, or molecules.
Crystalline, describes materials, such as granular rocks, composed wholly of substances in which there is a regular three-dimensional arrangement of constituent atoms.
Crystal habit, crystal form that may be considered typical for a given mineral.
Cubic, synonym for isometric (q.v.).

Dacite, aphanitic igneous rock with essential plagioclase, typically oligoclase-andesine, and quartz; the coarse-grained chemical equivalent is tonalite (quartz diorite).
Deep-sea red clay, extremely fine-grained, deep-sea sediment which typically consists of quartz, mica, and several clay minerals stained by iron oxide; it is generally considered to be of terrigenous derivation.
Dehydrated, rendered free from water.
Deliquescent, capable of absorbing water from the air and thereby becoming a liquid.
Detrital, describes fragments of preexisting rocks that constitute some sediments and sedimentary rocks.
Deuteric, describes alterations in an igneous rock that occur in response to activities of the magma itself during the late stages of its cooling and consolidation.
Diabase, a fine-grained igneous rock that consists chiefly of randomly arranged, bladelike grains of calcic plagioclase surrounded by augite, glass, *etc*.
Diagenesis, changes that take place in sediments after their deposition but before, and commonly contributing to, their conversion to solid rocks.
Diaphaneity, state or quality of transmitting light; degrees of diaphaneity are described by terms like transparent, translucent, and opaque.

Diorite, coarse-grained igneous rock composed of intermediate plagioclase, such as andesine, and one quarter to one half ferromagnesian minerals, typically hornblende with or without biotite.

Dodecahedral, describes either of two crystal forms bounded by twelve equal four- or five-sided plane faces, respectively, or by cleavages parallel thereto.

Dolerite, fine-grained gabbro (q.v.); some consider dolerite and diabase to be synonyms.

Dolomite, sedimentary rock composed chiefly of the mineral dolomite.

Druse, encrustation consisting of minute crystals.

Ductile, capable of being drawn into a fine wire without breaking.

Eclogite, granular metamorphic rock composed chiefly of pyrope-rich garnet and jadeitic pyroxene; the rock is commonly believed to indicate formation under conditions of high pressure.

Effervesce, to foam or bubble as gas escapes, *e.g.*, when HCl is put on calcite.

Efflorescence, development or formation of an incrustation or other deposit of mineral grains or powder as the result of a gain or loss of water of crystallization.

Elastic, capable of resuming the original shape when a deforming strain is removed.

Element, a substance of fixed atomic structure that cannot be separated into substances different from itself by ordinary chemical means.

Emery, granular rock composed essentially of corundum and a spinel-type mineral.

Equant, describes grains that are more or less equidimensional.

Euhedral, bounded by its own crystal faces (cf. anhedral and subhedral).

Evaporite, rock formed by evaporation of water that contains dissolved solids (*e.g.*, sea water).

Exfoliation, process of peeling or breaking off in layers, typically curved like those of an onion; the term is generally applied to rocks rather than to minerals.

Ferromagnesian mineral, mineral that contains iron and magnesium, (*e.g.*, an amphibole, biotite, olivine, or pyroxene).

Ferruginous, containing iron.

Fetid, having the foul odor of rotten eggs, *etc.*

Flame color, color imparted to a flame when powder of a material is ignited.

Flexible, describes material that bends without breaking and does not tend to resume its original shape.

Fluorescence, emission of visible energy by a material as a result of exposing it to ultra-violet light, *x*-ray beams, *etc.*

Foliation, laminated appearance of a metamorphic or igneous rock that manifests the segregation and/or preferred orientation of platy or rod-like minerals.

Fracture, breaking along rough or irregular surfaces that are generally believed to have no particular relation to the arrangement of the constituent atoms.

Friable, describes granular rocks that are easily crumbled into constituent grains.

Fumarole, natural orifice in a volcanic region from which hot gases issue (cf. solfatara).

Fuse, to make fluid by heating.

Gabbro, coarse-grained igneous rock composed chiefly of calcic plagioclase, typically labradorite, and augite or pigeonite plus or minus minor olivine and/or hypersthene (cf. norite).

Gangue, the worthless minerals with which the useful minerals of an ore deposit are admixed.

215

Gelatinize, to dissolve with the formation of a jelly-like substance.
Gliding, slippage along certain planes of atoms in crystalline materials.
Globular, having irregular spheroidal shape.
Gneiss, relatively coarse-grained metamorphic rock with a rough foliation or banding.
Gossan, iron-oxide-rich capping formed above a sulfide-rich vein as a result of weathering.
Gouge, fine-grained material formed when rocks are ground up within a fault zone.
Grade (metamorphic), degree or intensity of metamorphism; the "higher" the grade, the higher the temperature and/or pressure conditions under which the minerals of the metamorphic rock are believed to have been formed or to have attained equilibrium.
Granitic, term used to refer in the broad sense to any relatively light-colored, quartz-bearing, coarse-grained igneous rock.
Granular, consisting of more or less equidimensional, megascopically discernible grains.
Granulite, nonfoliated granular metamorphic rock; some petrologists also require formation at relatively high temperatures and pressures (cf. gneiss).
Greasy, having a luster like that of pitch.
Greensand, sediment made up in a noteworthy proportion of the mineral glauconite.
Greenschist, foliated metamorphic rock composed in a noteworthy proportion of chlorite; it is generally considered to reflect metamorphism under conditions of relatively low temperature and low pressure.
Greisen, an altered rock generally believed to have resulted from the passage of fluids rich in fluorine, lithium, *etc.*; it commonly contains notable mica, topaz, fluorite, and accessory cassiterite, rutile, *etc.*

Hardness, resistance to abrasion.
Hemimorphic, describes crystal forms developed at only one end or differently on opposite ends of a symmetry axis.
Hexagonal system, crystal system which can be related to three equal axes that lie in one plane and intersect at 120° and a fourth axis which is not equal to the other three and is perpendicular to them.
High-temperature metamorphism, metamorphism at temperatures just below those needed for partial melting of the involved rock composition; the resultant rocks are commonly characterized by minerals such as perthite, pyroxene, and sillimanite.
High-temperature vein (or ore deposit), vein or other deposit generally believed to have been formed between 300°C and 500°C (=hypothermal).
Hydrothermal, describes hot water solutions carrying dissolved mineral matter and the deposits and other products produced by such solutions.
Hydrous, containing chemically combined water.
Hygroscopic, capable of absorbing moisture from the air.

Igneous rock, rock formed by the cooling and consolidation of magma (q.v.).
Iridescent, exhibiting multicolored reflections.
Ironstone, stratified clay- or mud-stone rich in iron oxide or siderite.
Isometric system, crystal system which can be related to three equal, mutually perpendicular axes.
Isomorphous, describes compounds of the same structure that form solid solutions, *i.e.*, the atoms, ions, or molecules of one may replace those of the other without changing the basic structural arrangement.

Lacustrine, pertaining to lakes.
Lagoon, shallow part of a sea which is nearly surrounded by land.

Laterite, residual product rich in aluminum and/or iron hydroxides, generally believed to have been formed in well-drained areas as the result of subtropical and tropical, chemical weathering.
Limestone, sedimentary rock that consists chiefly of calcite or aragonite.
Long-wave ultra-violet radiation, radiation within the wave-length range of approximately 3100-4000 Angstrom units; with the equipment usually employed, there is a predominance of radiation with wave length of 3650 Angstrom units (1 Angstrom unit = 10^{-8} cm.).
Low-temperature metamorphism, metamorphism at temperatures only slightly above those at which no new minerals would be expected to form in abundance in rocks of the given composition; characteristic new minerals are chlorite, albite, and muscovite (sericite).
Low-temperature vein (or ore deposit), vein or other deposit generally believed to have been formed between 100°C and 200°C (=epithermal).
Luminescence, emission of light not caused by incandescence.
Luster, appearance in reflected light.

Magma, molten or partially molten rock material.
Magnetic, capable of being attracted to a magnet.
Malleable, capable of being extended or shaped by hammering.
Mammillary, describes rounded masses resembling portions of spheres (see colloform).
Marble, metamorphic rock consisting chiefly of recrystallized calcite or dolomite.
Massive, compact, essentially homogeneous aggregate without distinctive form.
Matrix, the finer grained material of certain rocks within which relatively large grains or fragments are embedded.
Megascopic, distinguishable by the unaided eye (as opposed to microscopic).
Mesothermal, describes ore deposition at intermediate temperatures, *i.e.*, 200-300°C; the deposits are commonly characterized by sericite, carbonates, and certain sulfides.
Metallic luster, having the appearance of a metal in reflected light.
Metamict, describes any mineral in which radioactive decay of one or more of its constituents has destroyed all or part of the mineral's originally ordered molecular arrangement.
Metamorphic rock, rock formed by the transformation of a preëxisting rock in response to changes in temperature and/or pressure, sometimes accompanied by changes in chemical environment.
Metamorphism, processes responsible for transformation of a rock into a metamorphic rock; included are thermally and dynamically activated processes and metasomatism.
Metasomatism, the essentially simultaneous, chemical introduction and removal of certain rock constituents; these cause changes in the mineral and chemical compositions of the affected rocks.
Micaceous, relatively easily separated into thin, more or less elastic folia or sheets.
Microcrystalline, having individual grains which can be distinguished only with the aid of a hand-lens or microscope.
Microscopic, distinguishable only with the aid of a microscope (cf. megascopic).
Mineralized, describes rocks or zones into which new minerals have been introduced by such processes as hydrothermal activity or metasomatism.
Moderate-temperature metamorphism, metamorphism at temperatures intermediate to those associated with high- and low-temperature metamorphism (q.v.); these rocks are commonly characterized by minerals such as biotite, amphiboles, epidote, feldspar, and staurolite.
Moderate-temperature vein (or ore deposit), vein or other deposit generally believed to have formed between 200°C and 300°C (=mesothermal).

Monoclinic system, crystal system which can be related to three unequal axes one of which is perpendicular to the other two which intersect obliquely.
Moonstone, feldspar - generally orthoclase, oligoclase, or labradorite - that exhibits opalescence.

Native, adjective referred to minerals (*e.g.*, sulfur, copper, and gold) when they occur as chemical elements in the pure state, uncombined with other elements.
Nodular, having the appearance of small, more or less rounded lumps.
Nonmetallic luster, having an appearance unlike that of a metal in reflected light.
Norite, coarse-grained igneous rock composed of essential calcic plagioclase, typically labradorite, and in which the principal pyroxene is hypersthene, rather than augite or pigeonite as in ordinary gabbro.
Nugget, lump of gold, silver, platinum, *etc.* that has been shaped in part by battering during stream transport.

Octahedral, describes a crystal or cleavage form, in the isometric system, which is bounded by eight triangular faces of equal size.
Oolite, rock consisting of small nearly spherical masses, which may resemble fish roe and are typically composed of calcium carbonate.
Opalescent, exhibiting internal milky or multicolored reflections.
Opaline, resembling opal.
Opaque, will not transmit light.
Optical methods, in mineralogy, the methods that consist chiefly of determining such things as index(es) of refraction and optical class by utilizing the polarizing microscope and its accessories.
Ore, aggregate of useful and worthless minerals from which the former can be separated at a profit.
Organic association, describes minerals, (*e.g.*, whewellite and mellite) occurring with decayed organic material.
Orthorhombic system, crystal system which can be related to three unequal axes at right angles to each other.
Oxidation product, material formed, commonly during weathering, as the result of the combination of some substance with oxygen.
Oxidized, describes a mineral, rock, zone, deposit, *etc.* in which some of the original constituents have been altered to oxides, hydroxides, *etc.*

Parting, tendency of certain minerals to break along composition planes of twinned crystals (sometimes it is described as breakage along planes that are not cleavage planes *but*, as is apparent, according to the aforementioned definition, parting could be parallel to cleavage).
Pegmatite, exceptionally coarse-grained igneous rock or multimineralic hydrothermal vein that typically consists of quartz, alkali feldspars, and mica with or without minerals rich in elements such as boron, lithium, uranium, and the rare earths.
Percussion figure, figure, characterized by radiating lines, that may be formed in such minerals as mica by striking them with a sharp instrument (cf. pressure figure).
Perthite, intimate mixture of microcline or orthoclase and albite or oligoclase.
Phenocryst, name given the relatively large grains surrounded by the groundmass (matrix) in a porphyry (q.v.).
Phonolite, aphanitic igneous rock that consists essentially of sanidine or Na-rich orthoclase plus nepheline with or without sodalite, cancrinite, and soda-amphiboles and/or soda-pyroxenes; it is the aphanitic equivalent of nepheline syenite.

Phosphate rock, sedimentary rock consisting in a noteworthy proportion of calcium phosphate.

Phosphorescent, having a luminescence (caused by exposure to ultra-violet or other types of radiation) that continues after the source of excitation has been removed.

Piezoelectric, capable of developing surface charges of electricity as a result of the application of pressure.

Placer, mineral deposit in which the useful mineral (*e.g.*, gold, platinum, or monazite) is a minor constituent of unconsolidated sedimentary material such as stream gravel or beach sand.

Plastic, 1. capable of being molded into and retaining a form (said of some clays); 2. capable of being permanently deformed without fracture (see ductile, malleable, and sectile).

Playa, a shallow, ephemeral lake in a desert region; the short-life is a result of rapid evaporation.

Plutonic, formed relatively deep (but not necessarily more than two-three miles) within the Earth's crust.

Porphyry, igneous rock in which relatively large and conspicuous grains, phenocrysts, are surrounded by a finer grained matrix, generally called the groundmass.

Pressure figure, figure, characterized by radiating lines, that may be produced in certain minerals, such as mica, by compressing them between blunt points; these figures are similar in appearance but not necessarily in orientation to percussion figures (q.v.).

Prismatic, describes minerals in which the predominant crystal faces are parallel to the $\overline{c}$ crystallographic axis.

Pseudo..., prefix meaning false, *e.g.*, pseudo-hexagonal could be used to describe a material that appears to be but is not hexagonal.

Pyramidal, describes minerals in which the predominant crystal faces cut the $\overline{c}$ axis and one or two of the other crystallographic axes.

Pyroelectric, capable of developing electricity or electrical polarity as a result of temperature changes.

Radiating, having parts extending in many or all directions from a point (said of crystal aggregates).

Radioactive, capable of spontaneous disintegration with emission of alpha or beta particles and/or gamma radiation.

Regional metamorphism, metamorphism that extends over large areas and is not necessarily associated spatially with nearby igneous activity.

Resinous, having a nonmetallic luster that resembles the appearance of freshly broken resin.

Rhombohedral, characterized by rhombohedron faces, *i.e.*, the diamond-shaped faces in the hexagonal system that completely bound six-sided polyhedra.

Rock, a natural, solid aggregate composed of mineral grains or natural glass or a combination of these.

Rosette, a flowerlike mineral growth, *e.g.*, of barite.

"Ruby Silver", term applied to the reddish-colored silver minerals proustite and pyrargyrite.

Saccharoidal, with a granular texture similar to that of a sugar cube.

Saline, 1. describes evaporite deposits; 2. describes water with a salinity of more than 3,000 parts per million of total dissolved solids.

Salt, 1. any compound that could have been produced as the result of wholly or partially replacing the acid hydrogen of an acid by a metal (*e.g.*, $FeSO_4$ may be considered to be an iron salt of H_2SO_4, sulfuric acid); 2. halite (NaCl), *i.e.* "common salt."

Schist, metamorphic rock with well developed, relatively closely spaced foliation; micaceous minerals constitute a large, if not predominant, proportion of these rocks.
Secondary, describes any second generation of minerals or rocks that have been formed by changes of or substitutions for the original material(s).
Sectile, capable of being severed smoothly, *i.e.* without fracturing, by a knife.
Sedimentary rock, any rock formed by the consolidation of fragments accumulated on the earth's surface or of chemically or biochemically precipitated minerals.
Series, term applied to individual groups of isomorphous minerals, *e.g.*, the plagioclase feldspar series.
Serpentinite, rock consisting chiefly or wholly of serpentine minerals.
Shale, a consolidated clay- or silt-sized sediment that exhibits fissility, *i.e.* it tends to split along parallel or subparallel planes because of a depositional, preferred orientation of the constituent lamellar mineral grains.
Short-wave ultra-violet radiation, radiation within the general wavelength range of 2300-3100 Angstrom units; with the equipment usually employed, there is a predominance of radiation with wave length of 2537 Angstrom units (1 Angstrom unit = 10^{-8} cm.).
Siliceous, consisting in a noteworthy part of silica (SiO_2).
Solfatara, fissure or other volcanic orifice from which chiefly sulfurous vapors are emitted (cf. fumarole).
Soluble, describes substances that may be dissolved.
Solution, liquid combination formed by dissolving a solid or fluid (the solute) in a liquid (the solvent).
Specific Gravity, ratio of the weight of an object to the weight of an equal volume of water.
Spherulite, a small, spheroidal aggregation of one or more minerals; typically there is both a radiating and a concentric arrangement of the constituents.
Stalactitic, formed or shaped like a stalactite (cave "icicle").
Streak, color of the powder of a mineral.
Striated, characterized by fine grooves (*e.g.*, those commonly parallel to the lengths of tourmaline crystals).
Subhedral, having a form determined in part by its own crystal faces and in part by the shapes of the adjacent minerals (intermediate between anhedral and euhedral, q.v.).
Sublimate, anything deposited directly from a gas or vapor.
Submetallic luster, luster that may be termed "imperfect metallic", *e.g.*, the luster of some sphalerite.
Subtranslucent, a rather poor term, sometimes used to describe substances that are opaque in all but extremely thin pieces.
Sunstone, an oligoclase which is typically salmon-colored and contains numerous inclusions which by reflection yield a "play of colors."
Supergene, describes minerals formed or deposited by groundwater in near-surface mantle above unaltered bedrock.
"Swelling clay," any clay that swells greatly upon wetting; it commonly contains montmorillonite.
Syenite, coarse-grained igneous rock that consists chiefly of an alkali feldspar (microcline, orthoclase, or perthite) and minor amounts of ferromagnesian minerals such as biotite and hornblende; quartz is sparse or absent.

Tarnish, thin coating produced by chemical reactions between the coated substance and air.
Tetragonal system, crystal system which can be related to three mutually perpendicular axes, two of which are equal and the third of which is either longer or shorter than the other two.

Tetrahedron, a solid bounded by four equilateral triangles of equal size.
Thermoluminescent, having a luminescence caused by heating.
Triboluminescent, having a luminescence caused by friction (crushing, rubbing, scratching, *etc.*).
Triclinic system, crystal system which can be related to three unequal, mutually oblique axes.
Trigonal, term applied to the subdivision of the hexagonal system that has one direction of threefold symmetry, *i.e.* it has the same appearance in three positions during one rotation around the $\bar{c}$ crystallographic axis (q.$\underline{v}$.); the rhombohedron is the simplest representative form.
Twinning, the distinctive characteristic of any crystal that appears to consist of two or more parts arranged so that the unit looks like two or more symmetrically united or intergrown crystals.

Ultrabasic rock, igneous rock containing less than 45 per cent silica, *e.g.*, dunite.
Unctuous, having a soapy feel when rubbed or touched, *e.g.*, talc.
Uneven, describes irregular, non-uniform fractures.

Variety, a distinctive type, *e.g.*, amethyst is a variety of quartz.
Vesicle, small typically spheroidal-shaped cavity which has been formed by the expansion of gas bubbles or steam during solidification of magma to an aphanitic or glassy igneous rock (cf. amygdule).
Vitreous, having the luster of broken glass.

Weathering, describes the processes and products of those chemical and mechanical changes that are caused by exposure to the elements.

X-ray methods, in mineralogy, the methods of x-ray diffraction used to determine the internal arrangement of constituent atoms, ions, and molecules and of x-ray fluorescence used to determine chemical composition.

"Zeolite association," two or more zeolites together with commonly associated minerals, such as prehnite; these commonly occur in cavities such as vesicles and cracks in basaltic rocks.
Zone, a belt or elongate region more or less distinct in some respect(s) from those adjacent to it.

INDEX

Acanthite, 29,199-C*
Acmite, 48,161,184-C
Actinolite, 83,175-C
Adamite, 43,62,78,98,106,121,153,171-C
Adelite, 66,82,93,109,135,171-C
Adularia, 111,126,166-C
Aegirine, 86,146,184-C
Aegirine-augite, 69,86,145,161,166-C
Afwillite, 107,175-C
Agate, 49,127,138,146,162
Aguilarite, 30,199-C
Ahlfeldite, 181-C
Aikinite, 12,14,16,23,29,174-C
Akermanite, 83,110,135,158,175-C
Akrochordite, 42,61,153,171-C
Alabandite, 31,34,141,154,194-C
Alamosite, 108,123,188-C
Alaskaite, 22,34,174-C
Albite, 95,111,126,137,166-C
Alexandrite (dichroic chrysoberyl - see)
Algodonite, 20,26,171-C
Allactite, 45,98,155,171-C
Allanite (=Orthite), 144,159,166-C
Alleghanyite, 47,136,194-C
Allemontite, 12,20,25,171-C
Allopalladium, 182-C
Allophane, 73,89,117,149,166-C
Alluaudite, 66,82,143,157,184-C
Almandine, 49,146,162,166-C
Alstonite, 44,64,107,123,133,173-C
Altaite, 14,19,188-C
Aluminite, 114,166-C
Aluminum-bearing minerals, 166
Alumohydrocalcite, 58,90,97,118,130, 166-C
Alunite, 43,62,92,122,132,166-C
Alunogen, 38,55,100,115,166-C
Amarantite, 40,51,150,184-C
Amarillite, 59,76,184-C
Amazonite, 86,166-C
Amblygonite, 48,68,84,94,111,126,136, 159,166-C
Amesite, 74,90,166-C
Amethyst, 50,95,99
Aminoffite, 166-C
Ammonia Alum, 100,115,166-C
Ammonioborite, 174-C
Ammoniojarosite, 60,184-C
Ampangabeite, 63,142,155,176-C

Amphibole, 47,83,94,99,125,136,143, 158,166-C
Analcime, 47,125,136,166-C
Anapaite, 78,121,176-C
Anatase, 13,15,16,17,18,27,32,35,48, 68,85,95,99,111,137,145,160,204-C
Anauxite, 101,166-C
Ancylite, 52,64,133,155,176-C
Andalusite, 50,99,127,138,166-C
Andersonite, 176-C
Andesine, 86,111,126,137,166-C
Andorite, 14,25,171-C
Andradite, 49,69,146,162,176-C
Andrewsite, 80,92,182-C
Anglesite, 59,76,91,103,119,131,188-C
Anhydrite, 42,92,97,106,132,153,176-C
Ankerite, 43,62,122,153,176-C
Annabergite, 72,115,129,171-C
Anorthite, 112,127,137,166-C
Anorthoclase, 111,126,137,161,166-C
Anthionite, 114,166-C
Anthophyllite, 68,85,126,136,160,184-C
Antigorite, 60,76,91,119,131,191-C
Antimony, 19,171-C
Antimony-bearing minerals, 171
Antlerite, 78,182-C
Apatite, 46,66,82,93,98,109,124,135, 156,176-C
Aphthitalite, 41,77,91,104,120,131, 197-C
Apjohnite, 38,54,72,100,115,166-C
Apophyllite, 45,65,81,108,123,176-C
Aquamarine (blue to greenish-blue beryl - see)
Aragonite, 43,62,79,92,98,122,133,176-C
Aramayoite, 29,171-C
Arcanite, 104,119,197-C
Ardealite, 176-C
Ardennite, 69,161,166-C
Arfvedsonite, 83,135,143,166-C
Argentite, 23,29,199-C
Argentojarosite, 60,151,184-C
Argyrodite, 12,17,18,24,30,200-C
Arizonite, 27,136,184-C
Armangite, 141,171-C
Armenite, 166-C
Arnimite, 182-C
Arrojadite, 82,184-C
Arsenic, 20,171-C
Arsenic-bearing minerals, 171

---
*-C *after a page number indicates that the reference is to a page upon which the chemical composition is given. Only one reference within the Chemical Composition listing is given for each mineral.*

Arseniopleite, 42,152,171-C
Arseniosiderite, 38,55,140,148,171-C
Arsenobismite, 172-C
Arsenoclasite, 47,172-C
Arsenolamprite, 22,172-C
Arsenolite, 38,54,89,115,172-C
Arsenopyrite, 21,27,172-C
Artinite, 118,191-C
Arzrunite, 182-C
Ashcroftine, 42,166-C
Astrophyllite, 14,34,60,151,176-C
Atacamite, 77,141,182-C
Atelestite, 65,81,172-C
Attapulgite, 184-C
Augelite, 45,65,108,123,166-C
Augite, 84,99,136,144,159,166-C
Aurichalcite, 72,89,182-C
Aurosmiridium, 21,28,183-C
Austinite, 64,107,123,172-C
Autunite, 56,73,176-C
Avogadrite, 174-C
Axinite, 70,87,99,138,162,166-C
Azurite, 92,182-C

Babingtonite, 84,145,159,176-C
Baddeleyite, 49,70,87,112,146,162, 208-C
Badenite, 172-C
Bakerite, 123,174-C
Banalsite, 126,166-C
Bandylite, 75,90,174-C
Bararite, 118
Barbertonite, 38,97,181-C
Barite, 41,61,77,92,105,120,132,173-C
Barium-bearing minerals, 173
Barkevikite, 166-C
Barroisite, 166-C
Barylite, 113,128,173-C
Barysilite, 105,189-C
Barytocalcite, 63,80,107,122,133,173-C
Basaltic hornblende, 143,158,166-C
Basaluminite, 166-C
Bassanite, 176-C
Bassetite, 56,184-C
Bastnasite, 44,64,155,199-C
Batavite, 166-C
Baumhauerite, 25,172-C
Bauxite, 40,57,74,117,130,150,166-C
Bavenite, 125,166-C
Bayldonite, 65,81,172-C
Bayleyite, 191-C
Beaverite, 166-C
Beckelite, 66,157,176-C
Becquerelite, 57,149,176-C
Beegerite, 174-C
Beidellite, 54,72,114,148,166-C
Bellingerite, 80,182-C
Benitoite, 95,99,112,173-C
Benjaminite, 12,15,26,174-C
Beraunite, 43,79,153,184-C

Beresovite, 181-C
Berlinite, 49,112,138,166-C
Bermanite, 42,153,184-C
Berthierite, 23,171-C
Berthonite, 26,31,171-C
Bertrandite, 68,111,173-C
Beryl, 50,71,88,96,113,128,166-C
Beryllium-bearing minerals, 173
Beryllonite, 68,126,173-C
Berzelianite, 19,22,182-C
Berzeliite, 52,65,172-C
Beta Sulfur, 56,102,149
Betafite, 15,16,31,34,64,80,142,155, 176-C
Beudantite, 79,141,154,172-C
Beyerite, 57,74,117,130,174-C
Bianchite, 56,117,184-C
Bieberite, 38,181-C
Bilinite, 55,115,184-C
Bindheimite, 44,64,80,123,133,155, 171-C
Biotite, 40,76,140,151,166-C
Bischofite, 100,114,191-C
Bismite, 65,81,134,174-C
Bismoclite, 57,90,117,130,174-C
Bismuth, 12,19,23,174-C
Bismuth-bearing minerals, 174
Bismuthinite, 14,19,22,174-C
Bismutite, 60,76,91,119,131,140,151, 174-C
Bismutotantalite, 13,15,32,46,66,143, 174-C
Bityite, 67,125,166-C
Bixbyite, 33,146,184-C
Blakeite, 40,150,184-C
Bloedite, 40,76,91,103,191-C
Bloodstone (=Heliotrope), 49,87,138
Blue Quartz, 95,99,138
Bobierrite, 101,116,191-C
Boehmite, 120,151,166-C
Boleite, 92,141,182-C
Boothite, 89,182-C
Boracite, 71,87,96,113,128,139,174-C
Borax, 73,89,101,116,129,174-C
Borickite, 42,152,176-C
Bornite, 12,18,34,182-C
Boron-bearing minerals, 174
Botallackite, 182-C
Botryogen, 39,51,184-C
Boulangerite, 14,17,24,171-C
Bournonite, 24,30,171-C
Boussingaultite, 38,55,101,191-C
Brackebuschite, 184-C
Bradleyite, 191-C
Braggite, 196-C
Brammallite (see hydro paragonite)
Brandtite, 106,121,172-C
Brannerite, 65,142,156,176-C
Braunite, 28,33,36,138,146,162,194-C
Bravoite, 27,184-C
Brazilianite, 67,84,166-C

Bredigite, 176-C
Breithauptite, 13,18,171-C
Brewsterite, 62,122,153,166-C
Brittle mica, 43,63,79,107,132,154,
  166-C
Brochantite, 79,141,182-C
Bromargyrite, 58,75,103,130,200-C
Bromellite, 128,173-C
Bronzite, 16,35,83,158,184-C
Brookite, 48,68,145,160,204-C
Brucite, 75,90,181,130,191-C
Brugnatellite, 39,55,116,149,184-C
Brunsvigite, 74,166-C
Brushite, 58,102,150,176-C
Buetschliite, 176-C
Bultfonteinite, 45,176-C
Bunsenite, 84,197-C
Burkeite, 121,132,152,200-C
Bustamite, 48,160,176-C
Butlerite, 51,184-C
Buttgenbachite, 91,182-C
Byssolite, 83,166-C
Bytownite, 112,127,137,166-C

Cacoxenite, 42,61,77,152,184-C
Cadmium oxide, 141,175-C
Cadmium-bearing minerals, 175
Cadwaladerite, 166-C
Cahnite, 104,120,172-C
Calamine (=Hemimorphite), 82,93,109,
  124,135,156,207-C
Calaverite, 14,19,25,183-C
Calcioferrite, 58,75,118,176-C
Calciovolborthite, 61,78,176-C
Calcite, 41,52,60,77,91,97,104,119,
  131,141,151,176-C
Calcium-bearing minerals, 175
Calclacite, 176-C
Calderite, 184-C
Caledonite, 76,91,182-C
Calomel, 54,100,115,129,148,196-C
Cancrinite, 47,67,93,110,125,135,
  166-C
Canfieldite, 12,17,18,24,30,200-C
Cappelenite, 86,161,173-C
Caracolite, 81,108,134,189-C
Carbonate-apatite, 46,82,93,98,109,
  124,134,156,176-C
Cardenite, 166-C
Carminite, 43,153,172-C
Carnallite, 40,57,90,102,118,191-C
Carnelian, 49,162
Carnotite, 57,74,197-C
Carpholite, 66,166-C
Carphosiderite, 64,184-C
Carrollite, 13,27,181-C
Caryinite, 63,154,172-C
Cassiterite, 13,16,18,21,28,33,36,49,
  69,112,127,138,146,162,204-C
Catapleite, 68,95,160,201-C

Catoptrite, 32,166-C
Celadonite, 72,114,166-C
Celestite, 41,77,92,105,120,152,203-C
Celsian, 69,112,127,166-C
Cerargyrite (see Chlorargyrite)
Cerite, 47,136,159,176-C
Ceruleite, 166-C
Cerussite, 77,92,105,120,132,141,189-C
Cervantite, 64,123,171-C
Cesarolite, 26,134,189-C
Cesium-bearing minerals, 181
Chabazite, 45,64,80,107,123,166-C
Chalcanthite, 75,90,182-C
Chalcedony, 49,70,86,95,138,162
Chalcoalumite, 75,90,130,166-C
Chalcocite, 24,30,182-C
Chalcocyanite, 62,78,92,106,153,182-C
Chalcomenite, 89,182-C
Chalcophanite, 17,29,194-C
Chalcophyllite, 73,89,166-C
Chalcopyrite, 15,182-C
Chalcosiderite, 80,182-C
Chalcostibite, 16,17,25,171-C
Chamosite, 56,74,166-C
Chenevixite, 63,79,172-C
Chiastolite (andalusite - see - with
  regularly arranged inclusions)
Childrenite, 66,156,166-C
Chiolite, 106,122,166-C
Chiviatite, 23,174-C
Chloanthite, 21,27,172-C
Chloraluminite, 166-C
Chlorapatite, 46,82,93,98,109,124,134,
  156,176-C
Chlorargyrite, 58,97,103,130,150,200-C
Chlorite, 39,51,57,74,90,102,117,130,
  140,149,166-C
Chloritoid, 86,166-C
Chlormanganokalite, 58,194-C
Chlorocalcite, 97,119,176-C
Chloromagnesite, 191-C
Chlorophoenicite, 42,78,98,132,172-C
Chlorothionite, 90,182-C
Chloroxiphite, 75,182-C
Chondrodite, 49,70,162,191-C
Chromite, 32,181-C
Chromium-bearing minerals, 181
Chrysoberyl, 50,71,88,166-C
Chrysocolla, 72,89,140,148,182-C
Chrysolite, 70,87,184-C
Chrysoprase, 70,86,138
Chrysotile, 58,75,118,130,191-C
Churchite, 41,131,176-C
Cinnabar, 12,23,34,39,130,149,196-C
Citrine, 70,163
Clarkeite, 44,155,206-C
Claudetite, 103,118,172-C
Clausthalite, 17,24,189-C
Clay, 38,51,54,72,89,100,114,129,148,
  166-C

225

Clino-sklodowskite, 191-C
Clino-ungemachite, 58,102,185-C
Clinochlore, 39,74,102,117,166-C
Clinoclase, 76,91,140,172-C
Clinoenstatite, 68,85,111,145,160,
  191-C
Clinoferrosilite, 86,161,185-C
Clinohedrite, 99,110,125,176-C
Clinohumite, 68,160,191-C
Clinohypersthene, 85,161,185-C
Clinotscheffkinite, 199-C
Clinozoisite, 70,86,112,138,166-C
Clintonite, 44,63,79,107,154,167-C
Cobaltite, 13,18,21,27,172-C
Cobaltomenite, 181-C
Cobalt-bearing minerals, 181
Coccinite, 196-C
Cocinerite, 24,29,182-C
Coeruleolactite, 93,124,167-C
Coesite, 113,128
Coffinite, 206-C
Cohenite, 15,21,185-C
Colemanite, 64,108,123,134,155,175-C
Collinsite, 153,177-C
Coloradoite, 24,30,196-C
Columbite, 13,33,36,48,146,161,181-C
Columbium-bearing minerals, 181
Colusite, 15,34,172-C
Common hornblende, 83,143,167-C
Conichalcite, 65,81,172-C
Connellite, 91,182-C
Cookeite, 39,56,73,117,167-C
Cooperite, 26,197-C
Copiapite, 51,59,76,185-C
Copper, 12,34,182-C
Copper-bearing minerals, 182
Coquimbite, 57,75,97,185-C
Cordierite, 96,138,167-C
Cordylite, 65,108,173-C
Corkite, 63,79,185-C
Cornetite, 80,93,182-C
Cornwallite, 81,142,172-C
Coronadite, 26,31,134,142,189-C
Corundophilite, 74,167-C
Corundum, 50,71,88,96,99,113,167-C
Corvusite, 91,140,151,206-C
Cosalite, 24,174-C
Cotunnite, 58,75,103,119,189-C
Covellite, 12,14,17,38,55,89,182-C
Crandallite, 66,124,134,167-C
Crednerite, 31,182-C
Creedite, 98,106,122,167-C
Crestmorite (see Riversideite)
Cristobalite, 127
Crocidolite, 94,135,143,185-C
Crocoite, 41,52,59,181-C
Cronstedtite, 78,141,153,185-C
Crookesite, 24,182-C
Crossite, 94,99,135,143,167-C
Cryolite, 40,103,118,140,150,167-C

Cryolithionite, 103,119,167-C
Cryptohalite, 102,118,130
Cubanite, 15,182-C
Cumengite, 91,182-C
Cummingtonite, 84,158,185-C
Cuprite, 13,31,43,141,182-C
Cuprocopiapite, 52,59,76,182-C
Cuprosklodowskite, 182-C
Cuprotungstite, 80,182-C
Curite, 45,52,189-C
Cuspidine, 47,110,177-C
Cyanochroite, 182-C
Cyanotrichite, 167-C
Cylindrite, 23,29,171-C

Dachiardite, 122,167-C
Dalyite, 197-C
Danalite, 48,69,137,161,174-C
Danburite, 70,112,163,175-C
Daphnite, 74,90,167-C
Darapskite, 102,201-C
Datolite, 46,66,82,109,125,175-C
Daubreeite, 57,90,117,130,174-C
Daubreelite, 29,181-C
Daviesite, 189-C
Davisonite, 123,167-C
Davyne, 109,167-C
Dawsonite, 104,119,167-C
Dehrnite, 81,108,124,134,177-C
Delafossite, 32,182-C
Delessite, 39,74,140,167-C
Delorenzite, 145,185-C
Deltaite, 66,93,98,134,167-C
Delvauxite, 41,151,185-C
Derbylite, 142,171-C
Descloizite, 42,52,105,141,152,189-C
Devilline, 75,90,177-C
Dewindtite, 189-C
Diabantite, 74,167-C
Diaboleite, 91,182-C
Diadochite, 42,61,77,120,152,185-C
Diallage, 85,145,167-C
Daimond, 50,53,71,88,96,99,113,128,
  139,147,163
Diaphorite, 24,171-C
Diaspore, 50,70,87,95,99,112,127,138,
  163,167-C
Dickinsonite, 62,79,153,177-C
Dickite, 56,116,167-C
Didymolite, 134,167-C
Dietrichite, 55,115,148,167-C
Dietzeite, 62,177-C
Digenite, 17,30,182-C
Dimorphite, 14,172-C
Diopside, 85,111,126,177-C
Dioptase, 82,182-C
Dipyre, 47,53,67,82,94,98,110,125,
  135,157,167-C
Dixenite, 141,172-C
Djalmaite, 68,84,144,159,174-C

226

Dolerophanite, 141,152,182-C
Dolomite, 43,62,79,106,122,132,153, 177-C
Domeykite, 19,25,172-C
Donbassite, 118,167-C
Douglasite, 185-C
Dravite, 147,163,167-C
Dufrenite, 43,79,141,154,185-C
Dufrenoysite, 25,172-C
Duftite, 77,131,172-C
Dumontite, 189-C
Dumortierite, 50,87,96,99,167-C
Dundasite, 116,167-C
Durangite, 46,52,82,167-C
Dussertite, 78,172-C
Dyscrasite, 15,20,26,171-C

Earlandite, 177-C
Ecdemite, 59,76,172-C
Edenite, 83,143,167-C
Edingtonite, 46,109,124,157,167-C
Eglestonite, 51,58,140,150,196-C
Egueite, 177-C
Ekermannite, 83,94,167-C
Ekmannite, 136,167-C
Elbaite, 50,70,87,96,113,167-C
Ellestadite, 45,177-C
Elpasolite, 103,167-C
Elpidite, 50,70,127,201-C
Emerald (green beryl - see)
Emmonsite, 65,81,185-C
Emplectite, 19,22,174-C
Empressite, 15,34,200-C
Enargite, 25,30,172-C
Englishite, 104,167-C
Enigmatite, 144,167-C
Enstatite, 67,83,110,135,158,192-C
Eosphorite, 46,167-C
Ephesite, 47,167-C
Epididymite, 110,174-C
Epidote, 48,69,85,111,137,145,161, 167-C
Epigenite, 17,26,31,172-C
Epistilbite, 44,63,107,122,133,154, 167-C
Epistolite, 54,114,129,177-C
Epsomite, 39,73,101,116,192-C
Erinite, 81,167-C
Eriochalcite, 58,75,90,182-C
Erionite, 177-C
Erythrite, 38,51,172-C
Erythrosiderite, 185-C
Eschynite, 15,32,35,67,144,158,177-C
Ettringite, 101,116,167-C
Eucairite, 19,24,182-C
Euchlorin, 182-C
Euchroite, 79,172-C
Euclase, 87,96,113,167-C
Eucryptite, 167-C
Eudialyte, 47,177-C

Eudidymite, 111,174-C
Eulite, 84,158,185-C
Eulytine, 65,134,156,174-C
Euxenite, 16,33,35,85,145,160,177-C
Evansite, 42,61,77,92,120,152,167-C

Fairchildite, 177-C
Fairfieldite, 52,61,78,121,177-C
Famatinite, 12,26,171-C
Fassaite, 85,167-C
Faujasite, 61,105,167-C
Fayalite, 70,87,185-C
Feldspar, 49,53,69,86,95,112,127,138, 161,167-C
Felsobanyaite, 54,115,167-C
Ferberite, 31,142,185-C
Ferghanite, 57,206-C
Fergusonite, 15,27,33,35,68,137,145, 160,181-C
Fermorite, 46,124,172-C
Fernandinite, 177-C
Ferri-sicklerite, 63,154,185-C
Ferrierite, 122,167-C
Ferrimolybdite, 185-C
Ferrinatrite, 75,90,97,102,118,130, 185-C
Ferrisepiolite, 185-C
Ferritungstite, 177-C
Ferroactinolite, 83,143,177-C
Ferrocarpholite, 167-C
Ferroedenite, 83,143,167-C
Ferrohastingsite, 83,143,167-C
Ferrohortonolite, 70,87,185-C
Ferrohypersthene, 84,158,185-C
Ferrosalite, 85,126,145,160,177-C
Ferrotschermakite, 83,143,167-C
Ferruccite, 104,119,175-C
Fersmanite, 159,177-C
Fervanite, 185-C
Fibroferrite, 56,73,117,130,185-C
Fiedlerite, 105,121,189-C
Fillowite, 45,64,108,155,177-C
Finnemanite, 131,140,172-C
Fizelyite, 22,171-C
Flajolotite, 171-C
Flinkite, 81,155,172-C
Florencite, 67,167-C
Fluellite, 60,104,119,167-C
Fluoborite, 106,121,175-C
Fluocerite, 64,123,155,199-C
Fluorapatite, 46,82,93,98,109,124,134, 156,177-C
Fluorite, 44,63,79,92,98,107,122,133, 142,154,177-C
Forbesite, 118,130,172-C
Formanite, 15,27,33,35,68,137,145,160, 177-C
Forsterite, 70,87,192-C
Foshallasite, 103,177-C
Fourmarierite, 42,61,152,189-C

Franckeite, 24,30,171-C
Franklinite, 17,33,35,95,145,160, 185-C
Freieslebenite, 19,23,171-C
Freirinite, 172-C
Friedelite, 44,107,185-C
Frondelite, 45,80,142,155,185-C
Fuloppite, 14,17,23,34,171-C

Gadolinite, 87,146,163,174-C
Gahnite, 71,88,96,163,167-C
Galaxite, 147,167-C
Galena, 24,189-C
Galenobismutite, 14,19,25,174-C
Gamma Sulfur, 54,100
Ganomalite, 104,131,177-C
Ganophyllite, 154,167-C
Garnet, 49,53,69,86,127,138,146,162
Garnierite, 54,72,167-C
Garrelsite, 173-C
Gaylussite, 59,103,119,131,177-C
Gearksutite, 116,167-C
Gedrite, 68,85,126,137,160,167-C
Gehlenite, 83,110,135,158,167-C
Geikielite, 32,35,144,158,192-C
Genthelvite, 49,99,162,174-C
Geocronite, 17,24,171-C
Georgiadesite, 62,122,153,172-C
Gerhardtite, 73,182-C
Germanite, 13,26,172-C
Gersdorffite, 21,27,32,172-C
Gibbsite, 41,60,76,119,131,167-C
Gillespite, 41,173-C
Ginorite, 121,175-C
Gismondine, 42,61,105,120,152,167-C
Gladite, 23,174-C
Glauberite, 40,59,103,131,177-C
Glaucocerinite, 72,89,114,129,148, 167-C
Glaucochroite, 85,95,177-C
Glaucodot, 13,20,27,31,172-C
Glauconite, 55,73,89,101,167-C
Glaucophane, 95,99,137,167-C
Glockerite, 186-C
Gmelinite, 45,64,80,107,123,167-C
Goethite, 13,15,32,35,46,67,143,157, 186-C
Gold, 14,19,183-C
Gold amalgam, 15,20,183-C
Goldfieldite, 25,171-C
Gold-bearing minerals, 183
Gonnardite, 45,108,124,156,168-C
Gonyerite, 149,186-C
Goongarrite, 25,174-C
Gorceixite, 160,168-C
Gordonite, 42,78,107,121,132,168-C
Goslarite, 73,89,101,116,149,207-C
Goyazite, 45,65,108,168-C
Graftonite, 46,157,177-C

Grandidierite, 87,96,168-C
Graphite, 17,22,177-C
Gratonite, 24,172-C
Greenalite, 57,74,149,186-C
Greenockite, 52,61,175-C
Griphite, 144,159,168-C
Grossular, 69,86,127,162,168-C
Grovesite, 168-C
Gruenlingite, 22,174-C
Grunerite, 84,158,186-C
Guanajuatite, 17,25,174-C
Guarinite, 67,94,178-C
Gudmundite, 21,28,171-C
Guembelite, 114,168-C
Guildite, 150,168-C
Guitermanite, 17,25,172-C
Gummite, 41,52,60,140,151,206-C
Gypsum, 55,101,116,129,149,178-C
Gyrolite, 121,178-C

Haidingerite, 101,117,172-C
Halite, 39,55,89,97,101,116,129,201-C
Halloysite, 116,168-C
Halotrichite, 54,72,100,115,168-C
Hambergite, 71,113,128,139,174-C
Hammarite, 12,25,174-C
Hanksite, 60,105,131,198-C
Hannayite, 54,192-C
Hardystonite, 83,110,135,158,178-C
Harkerite, 168-C
Harmotome, 45,64,108,123,134,168-C
Harstigite, 110,168-C
Hastingsite, 83,143,168-C
Hauerite, 13,31,34,195-C
Hausmannite, 32,35,144,159,195-C
Hauyne, 84,94,125,136,168-C
Hectorite, 114,190-C
Hedenbergite, 86,145,161,178-C
Hedyphane, 65,93,123,156,172-C
Heliophyllite, 55,73,172-C
Heliotrope (=Bloodstone), 87,138
Hellandite, 47,159,168-C
Helvite, 48,69,85,161,173-C
Hemafibrite, 41,141,151,174-C
Hematite, 13,27,47,136,186-C
Hematolite, 12,31,34,42,141,153,168-C
Hematophanite, 12,34,40,150,186-C
Hemimorphite, 82,93,109,124,135,156, 207-C
Hercynite, 147,168-C
Herderite, 66,82,109,124,174-C
Herzenbergite, 204-C
Hessite, 23,200-C
Hetaerolite, 33,146,195-C
Heteromorphite, 30,171-C
Heterosite, 44,98,186-C
Heulandite, 43,62,106,122,132,153, 168-C
Hewettite, 178-C

Hexahydrite, 192-C
Hieratite, 102,118,130,198-C
Hilgardite, 109,175-C
Hillebrandite, 84,125,178-C
Hinsdalite, 81,108,168-C
Hisingerite, 41,60,141,186-C
Hodgkinsonite, 44,155,195-C
Hoegbomite, 33,168-C
Hoernesite, 114,172-C
Hohmannite, 41,52,151,186-C
Holdenite, 44,63,172-C
Hollandite, 27,32,173-C
Holmquistite, 94,99,143,168-C
Homilite, 142,156,175-C
Hopeite, 60,105,120,131,207-C
Hornblende, basaltic, 143,158,168-C
Hornblende, common, 83,143,168-C
Horsfordite, 20,26,171-C
Hortonolite, 70,87,186-C
Howlite, 121,175-C
Huebnerite, 13,15,34,44,63,155,195-C
Huhnerkobelite, 66,82,157,178-C
Hulsite, 30,140,175-C
Humboldtine, 55,148,186-C
Humite, 48,53,68,160,192-C
Hureaulite, 42,52,62,97,106,132,153,
 195-C
Hutchinsonite, 12,172-C
Huttonite, 204-C
Hyalophane, 69,112,127,168-C
Hyalosiderite, 70,87,186-C
Hyalotekite, 109,135,173-C
Hydrobasaluminite, 168-C
Hydrobiotite, 72,114,148,168-C
Hydroboracite, 102,117,175-C
Hydrocalumite, 76,104,168-C
Hydrocerussite, 78,106,121,132,189-C
Hydrogrossular, 86,127,138,162,168-C
Hydrohetaerolite, 32,35,144,158,195-C
Hydromagnesite, 106,121,192-C
Hydromuscovite, 114,168-C
Hydroparagonite, 100,168-C
Hydrophilite, 178-C
Hydrophlogopite, 114,148,168-C
Hydrotalcite, 115,149,168-C
Hydroxyl-herderite, 66,82,109,125,174-C
Hydroxylapatite, 46,81,93,98,109,124,
 134,156,178-C
Hydrozincite, 39,56,97,117,129,149,
 207-C
Hypersthene, 84,158,186-C

Ianthinite, 18,29,97,140,206-C
Ice (see water)
Idocrase (see vesuvianite)
Ilesite, 186-C
Illite, 114,168-C
Ilmenite, 27,32,136,144,186-C
Ilsemannite, 196-C
Ilvaite, 137,145,178-C

Inderborite, 105,121,175-C
Inderite, 41,104,119,175-C
Indialite, 168-C
Inesite, 48,178-C
Inyoite, 101,115,175-C
Iodargyrite, 54,72,100,129,148,200-C
Iridosmine, 21,28
Iron, 26,31,186-C
Iron-bearing minerals, 184
Ishikawaite, 144,181-C
Isoclasite, 100,115,178-C

Jacobsite, 32,35,145,160,186-C
Jadeite, 85,95,111,126,168-C
Jamesonite, 23,29,171-C
Jarlite, 107,123,133,168-C
Jarosite, 60,151,186-C
Jasper, 50,70,87,95,146,162
Jeremejevite, 70,112,162,168-C
Jezekite, 44,107,123,168-C
Joaquinite, 67,159,173-C
Johannite, 73,183-C
Johannsenite, 85,137,145,161,178-C
Johnstrupite, 79,154,168-C
Jordanite, 25,172-C
Joseite, 22,29,174-C
Julienite, 181-C

Kaemmererite, 39,97,181-C
Kaersutite, 143,158,168-C
Kainite, 40,59,91,103,131,192-C
Kainosite, 46,66,157,178-C
Kaliborite, 44,107,122,155,175-C
Kalicinite, 198-C
Kalinite, 116,168-C
Kaliophilite, 111,168-C
Kalkowskite, 31,34,141,153,186-C
Kalsilite, 111,126,137,168-C
Kamarezite, 77,183-C
Kaolinite, 39,89,116,149,168-C
Karinthine, 168-C
Karpinskiite, 168-C
Kasolite, 64,155,189-C
Katophorite, 46,93,142,157,168-C
Katoptrite, see Catoptrite
Kempite, 78,195-C
Kentrolite, 46,157,189-C
Kermesite, 12,38,171-C
Kernite, 102,118,175-C
Kerstenite, 120,189-C
Kieserite, 61,106,121,132,192-C
Kirovite, 55,73,115,186-C
Klaprothite, 14,23,174-C
Klebelsbergite, 171-C
Kleinite, 43,52,63,196-C
Klockmannite, 17,25,30,183-C
Knebelite, 138,146,162,186-C
Kobellite, 24,30,171-C
Koechlinite, 174-C
Koenenite, 38,54,100,168-C

229

Koettigite, 41,172-C
Koktaite, 178-C
Kolbeckite, 92,132,168-C
Kolovratite, 197-C
Kornelite, 186-C
Kornerupine, 49,70,86,127,162,168-C
Kotoite, 112,175-C
Krausite, 58,186-C
Kremersite, 186-C
Krennerite, 14,19,23,183-C
Kribergite, 168-C
Kroehnkite, 76,91,183-C
Kunzite (pink spodumene - see)
Kupletskite, 178-C
Kurnakovite, 119,175-C
Kutnahorite, 43,122,178-C
Kyanite, 85,95,126,137,168-C

Labradorite, 95,111,126,137,168-C
Labuntsovite, 168-C
Lacroixite, 64,80,123,168-C
Lamprophyllite, 57,130,149,202-C
Lanarkite, 56,73,117,129,189-C
Landesite, 60,152,186-C
Langbanite, 146,195-C
Langbeinite, 43,62,79,98,106,132, 192-C
Langite, 76,91,183-C
Lansfordite, 102,118,192-C
Lanthanite, 40,59,103,119,199-C
Lapis-lazuli (chiefly hauyne - see)
Larderellite, 175-C
Larnite, 178-C
Larsenite, 104,120,189-C
Latiumite, 168-C
Laubmannite, 79,133,153,186-C
Laumontite, 41,60,105,120,152,168-C
Laurionite, 105,121,189-C
Laurite, 28,33
Lausenite, 186-C
Lautarite, 62,107,178-C
Lautite, 12,25,30,41,132,141,172-C
Lavenite, 69,111,161,178-C
Lawrencite, 72,114,148,186-C
Lawsonite, 95,111,126,168-C
Lazulite, 84,94,126,168-C
Lazurite, 93,168-C
Lead, 19,22,189-C
Leadhillite, 59,76,91,104,119,131, 189-C
Lead-bearing minerals, 188
Lecontite, 101,198-C
Legrandite, 65,108,172-C
Lehiite, 125,136,168-C
Leifite, 111,168-C
Leightonite, 77,91,178-C
Lengenbachite, 22,172-C
Leonite, 59,103,192-C
Lepidocrocite, 13,35,46,157,186-C

Lepidolite, 41,97,104,168-C
Lessingite, 45,65,81,178-C
Leucite, 125,136,168-C
Leucochalcite, 172-C
Leucophane, 63,79,122,174-C
Leucophenicite, 48,99,195-C
Leucophosphite, 168-C
Leucosphenite, 127,173-C
Levynite, 45,64,80,108,123,168-C
Lewistonite, 81,108,124,178-C
Libethenite, 80,142,183-C
Liebigite, 59,76,178-C
Lillianite, 23,174-C
Lime, 121,132,178-C
Limonite, 45,52,64,142,155,186-C
Linarite, 91,183-C
Lindackerite, 73,172-C
Lindgrenite, 65,81,183-C
Lindstromite, 25,174-C
Linnaeite, 13,26,181-C
Liroconite, 73,89,168-C
Liskeardite, 168-C
Litharge, 39,189-C
Lithidionite, 198-C
Lithiophilite, 44,52,64,155,190-C
Lithium-bearing minerals, 190
Liveingite, 12,172-C
Livingstonite, 22,29,171-C
Lizardite, 75,118,192-C
Loellingite, 20,27,172-C
Loeweite, 40,59,103,192-C
Lomonossovite, 202-C
Lopezite, 40,51,181-C
Lorandite, 12,14,18,23,172-C
Loranskite, 31,142,178-C
Lorettoite, 41,59,189-C
Loseyite, 91,120,151,195-C
Ludlamite, 78,186-C
Ludwigite, 82,175-C
Luenebergite, 115,148,175-C

Mackayite, 81,186-C
Maghemite, 157,186-C
Magnesiochromite, 15,16,32,35,181-C
Magnesiocopiapite, 52,59,76,186-C
Magnesioferrite, 32,35,145,160,186-C
Magnesiokatophorite, 46,157,169-C
Magnesite, 63,107,122,133,154,193-C
Magnesium orthite, 144,159,169-C
Magnesium-bearing minerals, 191
Magnetite, 32,35,145,160,186-C
Magnetoplumbite, 28,33,137,146,186-C
Malachite, 79,141,183-C
Maldonite, 12,19,22,29,174-C
Malladrite, 202-C
Mallardite, 38,195-C
Manandonite, 102,169-C
Manasseite, 89,115,129,149,169-C
Mangan-alluaudite, 66,82,143,157,186-C

Manganberzeliite, 45,52,172-C
Manganese-bearing minerals, 194
Manganite, 26,31,133,142,195-C
Manganolangbeinite, 195-C
Manganosite, 84,144,195-C
Mansfieldite, 122,133,169-C
Marcasite, 16,21,186-C
Margarite, 44,63,79,133,169-C
Marialite, 47,53,67,82,93,98,110,125,
    135,157,169-C
Marshite, 40,58,103,183-C
Mascagnite, 56,101,129
Massicot, 39,56,189-C
Matildite, 24,30,174-C
Matlockite, 59,76,104,151,189-C
Maucherite, 13,27,172-C
Meerschaum (see sepiolite)
Meionite, 47,53,67,83,94,99,110,125,
    135,158,169-C
Melanostibian, 142,171-C
Melanotekite, 146,186-C
Melanovanadite, 140,178-C
Melanterite, 55,73,89,115,186-C
Melilite, 67,83,158,169-C
Melinophane, 66,169-C
Mellite, 39,56,116,149,169-C
Melonite, 12,19,34
Mendipite, 40,58,91,103,119,130,189-C
Mendozite, 104,119,169-C
Meneghinite, 24,30,171-C
Mercallite, 198-C
Mercury, 19,22,196-C
Mercury-bearing minerals, 196
Merwinite, 85,111,178-C
Mesolite, 45,66,108,124,134,169-C
Mesomicrocline, 169-C
Meta-uranopilite, 206-C
Metacinnabar, 25,30,186-C
Metahewettite, 178-C
Metahohmannite, 187-C
Metarossite, 178-C
Metasideronatrite, 58,187-C
Metastrengite, 43,79,98,106,187-C
Metatorbernite, 75,183-C
Metavariscite, 77,121,169-C
Metavauxite, 76,104,119,169-C
Metavoltine, 51,58,75,150,187-C
Metazeunerite, 73,172-C
Meyerhofferite, 101,116,175-C
Mg-chlorophoenicite, 172-C
Miargyrite, 23,29,171-C
Mica, 40,57,74,90,97,102,117,130,140,
    150,169-C
Microcline, 48,53,69,86,111,126,137,
    161,169-C
Microlite, 46,67,82,157,179-C
Miersite, 58,183-C
Milarite, 84,110,169-C
Millerite, 15,197-C

Millisite, 125,136,169-C
Mimetite, 52,63,107,122,154,172-C
Minasragrite, 206-C
Minium, 40,59,150,190-C
Minnesotaite, 103,118,187-C
Minyulite, 106,121,169-C
Mirabilite, 100,115,202-C
Misenite, 198-C
Mitridatite, 57,74,140,150,179-C
Mitscherlichite, 75,90,183-C
Mixite, 77,92,121,172-C
Mizzonite, 47,53,67,83,94,98,110,125,
    135,158,169-C
Moissanite, 16,17,33,88,96,147
Molybdenite, 22,196-C
Molybdenum-bearing minerals, 196
Molybdophyllite, 78,105,190-C
Molysite, 38,54,72,97,148,187-C
Monazite, 46,67,82,125,157,199-C
Monetite, 61,121,179-C
Monimolite, 65,81,134,156,171-C
Montanite, 174-C
Montebrasite, 48,68,84,94,110,126,136,
    159,169-C
Montgomeryite, 79,107,169-C
Monticellite, 110,136,179-C
Montmorillonite, 38,51,54,72,114,169-C
Montroydite, 40,150,196-C
Mooreite, 104,193-C
Mordenite, 42,61,105,120,152,169-C
Morenosite, 73,116,197-C
Morinite, 43,169-C
Mosandrite, 44,154,179-C
Moschellandsbergite, 20,26,196-C
Mosesite, 61,78,196-C
Mossite, 33,36,146,162,181-C
Mottramite, 77,183-C
Mullite, 49,112,127,138,169-C
Murmanite, 18,202-C
Muscovite, 41,76,104,151,169-C
Muthmannite, 14,19,23,183-C

Nacrite, 116,169-C
Nadorite, 62,133,154,171-C
Nagatelite, 144,159,169-C
Nagyagite, 22,29,171-C
Nahcolite, 118,130,150,202-C
Nantokite, 75,103,118,130,183-C
Narsarsukite, 68,95,111,187-C
Nasonite, 122,179-C
Natroalunite, 43,62,92,122,133,169-C
Natrochalcite, 81,183-C
Natrojarosite, 60,151,187-C
Natrolite, 45,66,108,124,134,169-C
Natromontebrasite, 48,68,84,94,111,
    126,136,159,169-C
Natron, 54,100,114,129,202-C
Natrophilite, 65,195-C
Naujakasite, 169-C

231

Naumannite, 30,200-C
Nekoite, 179-C
Nepheline, 110,125,136,169-C
Neptunite, 47,143,187-C
Nesquehonite, 102,118,193-C
Newberyite, 105,193-C
Niccolite, 13,27,32,172-C
Nickel iron, 20,27,187-C
Nickel skutterudite, 21,27,172-C
Nickel-bearing minerals, 196
Niggliite, 19,25,197-C
Niter, 101,116,129,198-C
Nitrobarite, 173-C
Nitrocalcite, 179-C
Nitromagnesite, 193-C
Nocerite, 179-C
Nontronite, 51,54,72,148,169-C
Norbergite, 162,193-C
Nordenskioldine, 68,111,175-C
Northupite, 62,106,132,153,193-C
Nosean, 84,94,159,169-C

Octahedrite (see anatase)
Okenite, 66,93,124,156,179-C
Oldhamite, 34,179-C
Oligoclase, 49,53,111,126,137,169-C
Oliveiraite, 205-C
Olivenite, 60,77,120,131,152,172-C
Olivine, 67,84,94,110,136,144,159,
    187-C
Omphacite, 83,169-C
Onyx (black and white agate - see)
Opal, 48,53,68,85,95,111,126,137,145,
    160
Orientite, 156,179-C
Orpiment, 14,34,172-C
Orthite, 144,159,169-C
Orthoclase, 48,53,69,86,111,126,137,
    161,169-C
Orthoferrosilite, 84,158,187-C
Osbornite, 205-C
Osumilite, 169-C
Otavite, 175-C
Overite, 79,106,169-C
Owyheeite, 14,19,23,171-C
Oxammite, 57,117

Pachnolite, 104,120,169-C
Paigeite, 82,142,175-C
Painite, 169-C
Palladium, 20,26
Palmierite, 190-C
Palygorskite, 193-C
Parabutlerite, 51,150,187-C
Paracelsian, 69,112,169-C
Paracoquimbite, 97,187-C
Paragonite, 58,102,118,169-C
Parahilgardite, 109,175-C
Parahopeite, 107,207-C
Paralaurionite, 72,97,100,114,190-C

Paramelaconite, 18,31,183-C
Pararammelsbergite, 20,172-C
Paratacamite, 77,141,183-C
Paravauxite, 77,104,119,169-C
Parawollastonite, 65,124,134,179-C
Pargasite, 158,169-C
Parisite, 65,134,156,179-C
Parkerite, 19,34,174-C
Parsettensite, 61,195-C
Parsonsite, 59,190-C
Pascoite, 39,51,56,179-C
Paternoite, 175-C
Pearceite, 30,175-C
Pectolite, 108,124,179-C
Penfieldite, 190-C
Pennantite, 39,51,169-C
Penninite, 39,51,74,97,117,149,169-C
Penroseite, 24,181-C
Pentahydrite, 193-C
Pentahydrocalcite, 179-C
Pentlandite, 15,187-C
Percylite, 90,183-C
Periclase, 110,125,136,193-C
Peridot (see olivine)
Perovskite, 15,27,32,35,67,136,144,
    159,179-C
Perrierite, 144,159,199-C
Perthite, 53,69,86,126,137,161,169-C
Petalite, 49,86,127,138,169-C
Petzite, 24,30,183-C
Pharmacolite, 101,116,129,172-C
Pharmacosiderite, 40,58,75,150,172-C
Phenakite, 50,71,113,163,174-C
Phillipsite, 44,63,107,123,133,169-C
Phlogopite, 39,56,73,101,149,169-C
Phoenicochroite, 42,61,181-C
Phosgenite, 40,57,74,102,117,130,150,
    190-C
Phosphoferrite, 77,152,187-C
Phosphophyllite, 77,92,105,187-C
Phosphorroesslerite, 57,102,193-C
Phosphuranylite, 56,179-C
Pickeringite, 38,54,100,115,169-C
Picromerite, 40,57,102,118,130,193-C
Picropharmacolite, 172-C
Piemontite, 48,145,161,169-C
Pigeonite, 85,145,161,179-C
Pilbarite, 60,190-C
Pinakiolite, 33,175-C
Pinnoite, 61,78,175-C
Pintadoite, 179-C
Pirssonite, 105,120,131,179-C
Pisanite, 55,89,115,183-C
Pistacite, 69,85,137,169-C
Pitticite, 39,56,117,130,140,149,172-C
Plagionite, 23,29,171-C
Plasma, 87
Platiniridium, 16,21,28,197-C
Platinum, 20,26,31,197-C
Platinum-bearing minerals, 197

Plattnerite, 32,35,190-C
Platynite, 23,29,174-C
Plombierite, 179-C
Plumboferrite, 135,143,187-C
Plumbogummite, 45,65,81,93,124,134,
    156,169-C
Plumbojarosite, 148,187-C
Pollucite, 112,169-C
Polyargyrite, 24,30,171-C
Polybasite, 29,171-C
Polycrase, 16,33,35,85,145,160,179-C
Polydymite, 13,27,197-C
Polyhalite, 42,106,121,132,179-C
Polymignyte, 33,146,179-C
Portlandite, 101,179-C
Potarite, 20,26,196-C
Potash alum, 101,116,169-C
Potassium-bearing minerals, 197
Powellite, 62,79,92,122,133,141,154,
    179-C
Prase, 87
Prehnite, 69,86,127,137,169-C
Priceite, 120,175-C
Priorite, 15,32,35,67,144,158,179-C
Probertite, 105,175-C
Prosopite, 80,123,134,169-C
Proustite, 39,172-C
Pseudoboleite, 91,183-C
Pseudobrookite, 13,33,35,48,146,161,
    187-C
Pseudocotunnite, 190-C
Pseudomalachite, 81,93,142,183-C
Pseudothuringite, 187-C
Pseudowollastonite, 109,179-C
Psilomelane, 27,32,136,144,173-C
Pucherite, 44,63,155,174-C
Pumpellyite, 85,95,160,169-C
Purpurite, 44,98,187-C
Pyrargyrite, 40,171-C
Pyrite, 16,187-C
Pyroaurite, 58,75,102,118,150,187-C
Pyrobelonite, 43,190-C
Pyrochlore, 46,67,143,157,179-C
Pyrochroite, 75,90,103,140,150,195-C
Pyrolusite, 17,23,29,195-C
Pyromorphite, 43,62,63,79,107,133,154,
    190-C
Pyrope, 49,146,162,169-C
Pyrophanite, 13,47,195-C
Pyrophyllite, 54,72,89,114,148,169-C
Pyrosmalite, 80,155,187-C
Pyrostilpnite, 39,171-C
Pyroxene, 47,67,84,94,99,110,125,136,
    144,159,169-C
Pyroxmangite, 48,145,160,187-C
Pyrrhotite, 15,34,187-C

Quartz, 50,53,70,87,95,99,112,127,138,
    147,163
Quenselite, 30,140,190-C

Quenstedtite, 40,97,187-C

Radiophyllite, 117,179-C
Ralstonite, 64,108,123,169-C
Ramdohrite, 22,29,171-C
Rammelsbergite, 13,21,172-C
Ramsayite, 145,161,203-C
Rankinite, 179-C
Ransomite, 90,169-C
Rare Earths-bearing minerals, 199
Raspite, 60,131,151,190-C
Rathite, 25,172-C
Rauvite, 179-C
Realgar, 38,51,55,173-C
Reddingite, 41,61,105,120,187-C
Redingtonite, 169-C
Reedmergnerite, 175-C
Renardite, 63,190-C
Retgersite, 74,90,197-C
Retzian, 154,173-C
Rezbanyite, 24,174-C
Rhabdophane, 43,62,121,153,199-C
Rhipidolite, 74,130,169-C
Rhodizite, 71,113,128,139,169-C
Rhodochrosite, 43,133,153,196-C
Rhodonite, 48,160,187-C
Rhomboclase, 55,73,89,101,116,129,
    187-C
Richellite, 39,56,149,179-C
Richterite, 47,67,83,158,170-C
Rickardite, 12,18,183-C
Riebeckite, 93,142,187-C
Rinkite, 66,157,179-C
Rinneite, 41,60,97,104,151,187-C
Riversideite, 119,179-C
Rockbridgeite, 45,80,142,155,187-C
Roeblingite, 104,179-C
Roemerite, 60,97,152,187-C
Roesslerite, 102,117,173-C
Romeite, 48,68,160,171-C
Rooseveltite, 133,173-C
Rosasite, 80,93,183-C
Roscherite, 80,155,174-C
Roscoelite, 75,150,170-C
Rose quartz, 50
Roselite, 42,173-C
Rosenbuschite, 53,135,179-C
Rossite, 56,179-C
Roweite, 156,175-C
Ruby (red corundum - see)
"Ruby silver" (see proustite and
    pyrargyrite)
Russellite, 62,78,174-C
Rutherfordine, 68,206-C
Rutile, 13,15,16,17,18,33,36,49,69,
    86,95,99,146,162,205-C

Safflorite, 20,26,173-C
Sahlinite, 57,173-C
Salammoniac, 55,100,115,129,148

233

Saleeite, 57,193-C
Salesite, 77,91,183-C
Salite, 85,126,145,160,179-C
Salmonsite, 63,154,187-C
Samarskite, 15,32,35,67,144,159,
    179-C
"Samarskite", 15,32,35,67,144,159,
    179-C
Sampleite, 79,92,179-C
Samsonite, 23,29,171-C
Sanbornite, 109,173-C
Sanidine, 111,126,137,160,170-C
Sanmartinite, 205-C
Saponite, 89,114,170-C
Sapphire (yellow,green,blue,and
    white corundum - see)
Sapphirine, 87,96,139,170-C
Sarcolite, 48,170-C
Sarcopside, 44,80,92,98,154,180-C
Sard, 49,53,162
Sarkinite, 44,64,173-C
Sarmientite, 173-C
Sartorite, 25,173-C
Sassolite, 54,114,129,148,175-C
Sauconite, 38,148,170-C
Scacchite, 38,100,148,196-C
Scapolite, 47,53,67,83,94,98,110,
    125,135,158,170-C
Scawtite, 108,180-C
Schafarzikite, 43,153,171-C
Schairerite, 106,203-C
Schallerite, 156,173-C
Scheelite, 45,52,65,81,108,124,134,
    156,180-C
Scheteligite, 144,171-C
Schirmerite, 22,29,174-C
Schizolite, 46,157,180-C
Schoepite, 57,206-C
Schorl, 147,170-C
Schreibersite, 16,21,28,36,188-C
Schroeckingerite, 58,75,180-C
Schultenite, 103,173-C
Schwartzembergite, 39,56,89,190-C
Scolecite, 66,108,124,134,170-C
Scorodite, 62,79,92,98,106,133,153,
    173-C
Scorzalite, 85,94,126,170-C
Seamanite, 63,175-C
Searlesite, 121,175-C
Selen-tellurium, 23,29
Selenium, 22
Seligmannite, 25,30,173-C
Sellaite, 109,124,193-C
Semseyite, 24,29,171-C
Senaite, 33,146,188-C
Senarmontite, 101,117,129,171-C
Sengierite, 58,75,183-C
Sepiolite, 39,56,73,116,193-C
Serandite, 180-C
Serendibite, 95,170-C

Sericite, 76,103,151,170-C
Serpentine, 60,76,91,119,131,140,151,
    193-C
Serpierite, 180-C
Sharpite, 57,74,206-C
Sheridanite, 56,73,102,117,170-C
Shortite, 60,104,180-C
Sicklerite, 63,154,188-C
Siderazot, 188-C
Siderite, 43,63,79,122,133,154,188-C
Sideronatrite, 51,55,148,188-C
Siderophyllite, 41,76,140,151,170-C
Siderotil, 188-C
Siegenite, 13,26,181-C
Sillenite, 54,72,129,174-C
Sillimanite, 70,112,127,163,170-C
Silver, 19,25,30,200-C
Silver-bearing minerals, 199
Simpsonite, 180-C
Sincosite, 180-C
Siserskite, 28
Sjogrenite, 58,118,150,188-C
Sklodowskite, 193-C
Skutterudite, 21,27,173-C
Slavikite, 188-C
Smaltite, 21,27,173-C
Smithite, 38,51,173-C
Smithsonite, 64,80,93,107,123,133,
    155,208-C
Smoky Quartz, 147,163
Soda alum, 104,170-C
Soda-niter, 38,55,100,115,129,148,
    203-C
Sodalite, 48,68,84,94,136,170-C
Soddyite, 61,206-C
Sodium-bearing minerals, 200
Souzalite, 84,170-C
Spangolite, 74,90,170-C
Spencerite, 120,208-C
Sperrylite, 21,173-C
Spessartine, 49,53,146,162,170-C
Sphalerite, 12,15,16,18,20,31,34,
    208-C
Sphene, 66,82,109,157,180-C
Spherocobaltite, 44,133,142,154,
    181-C
Spinel, 50,71,88,96,113,147,163,
    170-C
Spodumene, 70,87,99,112,127,138,170-C
Spurrite, 109,180-C
Stainierite, 26,31,133,142,181-C
Stannite, 17,26,31,183-C
Staurolite, 50,71,163,170-C
Steenstrupine, 142,154,188-C
Steigerite, 170-C
Stephanite, 29,171-C
Stercorite, 55,115,149,203-C
Sternbergite, 17,18,34,188-C
Sterrettite, 66,108,170-C
Stevensite, 114,129,148,188-C

Stewartite, 196-C
Stibiconite, 45,64,123,171-C
Stibiocolumbite, 47,68,84,159,171-C
Stibiopalladinite, 13,20,26,171-C
Stibiotantalite, 47,68,84,159,171-C
Stibnite, 22,171-C
Stichtite, 38,97,181-C
Stilbite, 43,62,106,122,132,153,170-C
Stillwellite, 180-C
Stilpnochlorane, 42,61,152,170-C
Stilpnomelane, 42,61,77,141,152,170-C
Stokesite, 111,180-C
Stolzite, 41,60,76,131,151,190-C
Strengite, 43,98,106,188-C
Stromeyerite, 17,24,183-C
Strontianite, 43,62,78,106,121,132, 153,203-C
Strontium-bearing minerals, 203
Struvite, 55,101,115,149,194-C
Stylotypite, 30,171-C
Sulfoborite, 44,107,175-C
Sulfohalite, 61,78,106,132,203-C
Sulfur, 38,55,72,148
Sulfur (beta),56,102,149
Sulfur (gamma),54,100
Sulvanite, 15,34,183-C
Susannite, 190-C
Sussexite, 61,120,152,175-C
Svabite, 64,80,107,123,133,173-C
Svanbergite, 46,66,109,156,170-C
Swartzite, 180-C
Swedenborgite, 71,113,171-C
Sylvanite, 14,19,22,183-C
Sylvite, 38,55,89,101,115,129,199-C
Symplesite, 74,90,140,173-C
Synadelphite, 45,108,142,155,173-C
Synchisite, 65,134,156,180-C
Syngenite, 58,102,118,180-C
Szaibelyite, 60,120,152,175-C
Szmikite, 38,115,196-C
Szomolnokite, 40,58,90,103,150,188-C

Tachyhydrite, 55,100,180-C
Tagilite, 77,183-C
Talc, 72,100,114,148,194-C
Tamarugite, 104,170-C
Tantalite, 13,33,36,49,146,162,181-C
Tantalum, 16,28
Tapiolite, 33,36,146,162,188-C
Taramellite, 47,159,173-C
Taramite, 170-C
Taranakite, 170-C
Tarapacaite, 181-C
Tarbuttite, 43,62,107,154,208-C
Tauriscite, 188-C
Tavistockite, 170-C
Taylorite, 55,115,199-C
Teallite, 22,29,190-C
Teepleite, 105,120,152,175-C

Teineite, 90,130,183-C
Tellurite, 56,116
Tellurium, 117,129
Tellurobismuthite, 22,174-C
Tennantite, 12,25,31,173-C
Tenorite, 26,31,183-C
Tephroite, 95,137,196-C
Terlinguaite, 59,75,150,196-C
Teschemacherite, 54,100,115
Tetradymite, 22,174-C
Tetrahedrite, 12,25,31,171-C
Thalenite, 49,207-C
Thallium-bearing minerals, 204
Thaumasite, 105,180-C
Thenardite, 40,59,103,119,131,151, 203-C
Thermonatrite, 54,100,114,129,203-C
Thomsenolite, 39,101,116,149,170-C
Thomsonite, 46,109,124,157,170-C
Thoreaulite, 161,204-C
Thorianite, 28,33,138,146,204-C
Thorite, 52,65,142,156,204-C
Thorium-bearing minerals, 204
Thorogummite, 64,93,204-C
Thorotungstite, 170-C
Thortveitite, 86,138,207-C
Thulite, 48,170-C
Thuringite, 74,149,170-C
Tiemannite, 24,30,196-C
Tilasite, 82,98,135,173-C
Tilleyite, 180-C
Tin, 19,22,204-C
Tincalconite, 175-C
Tinticite, 56,74,117,188-C
Tin-bearing minerals, 204
Titanaugite, 48,85,99,145,170-C
Titanite (=Sphene), 66,82,109,157, 180-C
Titanium-bearing minerals, 204
Tobermorite, 180-C
Todorokite, 29,173-C
Topaz, 71,113,128,139,170-C
Torbernite, 73,183-C
Torendrikite, 94,99,143,170-C
Torreyite, 91,119,194-C
Tourmaline, 50,70,87,96,113,128,147, 163,170-C
Trechmannite, 38,173-C
Tremolite, 110,135,180-C
Trevorite, 31,35,143,157,188-C
Trichalcite, 75,91,173-C
Tridymite, 112,127
Trigonite, 57,150,173-C
Trihydrocalcite, 180-C
Trimerite, 49,174-C
Triphylite, 80,93,133,188-C
Triplite, 46,143,157,188-C
Triploidite, 45,65,156,188-C
Trippkeite, 173-C

Tripuhyite, 65,81,156,171-C
Tritomite, 68,159,175-C
Troegerite, 57,173-C
Trona, 59,103,131,203-C
Trudellite, 57,170-C
Truscottite, 120,180-C
Tscheffkinite, 144,159,170-C
Tschermakite, 83,143,170-C
Tsumebite, 78,183-C
Tuhualite, 170-C
Tungstenite, 24,205-C
Tungsten-bearing minerals, 205
Tungstite, 58,75,205-C
Turanite, 81,183-C
Turquois, 83,94,135,170-C
Tychite, 122,194-C
Tyrolite, 72,89,173-C
Tyuyamunite, 55,73,180-C

Uhligite, 32,170-C
Ulexite, 102,118,175-C
Ullmannite, 20,27,171-C
Umangite, 12,17,18,183-C
Ungemachite, 58,102,188-C
Uralite, 83,135,143,170-C
Uraninite, 32,35,144,159,206-C
Uranium-bearing minerals, 205
Uranocircite, 56,73,173-C
Uranophane, 57,74,180-C
Uranopilite, 206-C
Uranosphaerite, 40,51,57,174-C
Uranospinite, 57,74,173-C
Uranothorite, 204-C
Uvanite, 206-C
Uvarovite, 86,180-C
Uvite, 127,170-C

Valentinite, 41,59,103,119,131,151,
    171-C
Valleriite, 183-C
Vanadinite, 41,52,59,151,190-C
Vanadium-bearing minerals, 206
Vandenbrandite, 80,142,183-C
Vanoxite, 207-C
Vanthoffite, 106,194-C
Variscite, 80,92,107,170-C
Varulite, 66,82,157,180-C
Vashegyite, 56,73,117,149,170-C
Vaterite, 180-C
Vauquelinite, 76,140,151,181-C
Vauxite, 78,92,170-C
Veatchite, 101,116,175-C
Vermiculite, 54,72,100,148,170-C
Vernadskite, 78,183-C
Vesuvianite, 49,69,86,95,162,170-C
Veszelyite, 79,92,173-C
Villiaumite, 39,101,116,203-C
Violarite, 13,18,27,188-C
Vishnevite, 46,67,93,109,125,135,
    170-C

Vivianite, 72,89,100,140,188-C
Voglite, 180-C
Volborthite, 62,78,207-C
Volchonskoite, 75,90,170-C
Voltaite, 77,141,188-C
Voltzite, 44,64,92,208-C
Vrbaite, 17,26,92,132,171-C

Wad, 90,140,150,173-C
Wadeite, 199-C
Wagnerite, 46,66,82,135,194-C
Wairakite, 110,170-C
Walpurgite, 62,173-C
Waltherite, 80,155,174-C
Wardite, 81,93,109,170-C
Warwickite, 12,31,34,43,141,153,175-C
Water, 72,89,100,115
Wattevilleite, 180-C
Wavellite, 61,77,92,105,120,141,152,
    170-C
Weberite, 132,170-C
Weddellite, 63,107,122,154,180-C
Wehrlite, 19,22,174-C
Weibullite, 23,174-C
Weinschenkite, 105,121,199-C
Weissite, 17,30,183-C
Wernerite, 47,53,83,94,98,110,125,135,
    158
Wherryite, 183-C
Whewellite, 59,103,150,180-C
Whitlockite, 66,109,124,134,180-C
Wilkeite, 45,65,180-C
Willemite, 47,68,84,125,208-C
Witherite, 61,77,105,120,131,152,173-C
Wittichenite, 19,23,174-C
Wittite, 23,174-C
Woehlerite, 68,159,180-C
Wolfachite, 20,26,31,171-C
Wolfeite, 45,81,156,188-C
Wolframite, 26,31,34,133,142,155,196-C
Wollastonite, 81,108,124,134,180-C
Woodhouseite, 45,93,108,123,170-C
Woodwardite, 170-C
Wulfenite, 41,52,59,76,131,151,190-C
Wurtzite, 141,154,208-C

Xanthoconite, 40,51,57,149,173-C
Xanthophyllite, 44,63,79,107,154,170-C
Xanthoxenite, 56,149,180-C
Xenotime, 44,64,80,123,134,155,207-C
Xonotlite, 49,112,127,181-C

Yeatmanite, 92,171-C
Yttrialite, 85,145,160,204-C
Yttrium-bearing minerals, 207
Yttrocrasite, 32,35,181-C
Yttrotantalite, 27,135,181-C
Yugawaralite, 170-C

Zaratite, 78,197-C
Zeolite, 42,61,78,105,121,132,152, 170-C
Zeophyllite, 77,104,181-C
Zeunerite, 73,173-C
Zinc, 19,22,208-C
Zincaluminite, 91,119,170-C
Zincite, 44,52,63,208-C
Zinc-bearing minerals, 207
Zinkenite, 25,171-C
Zinnwaldite, 60,97,131,151,170-C
Zippeite, 52,60,206-C
Zircon, 50,53,71,87,113,139,163,208-C
Zirconium-bearing minerals, 208
Zirkelite, 144,181-C
Zirklerite, 170-C
Zoisite, 48,85,137,161,171-C
Zunyite, 112,171-C

NOTES

NOTES

NOTES